CIRCUIT ORIENTED ELECTROMAGNETIC MODELING USING THE PEEC TECHNIQUES

CIRCUIT ORIENTED ELECTROMAGNETIC MODELING USING THE PEEC TECHNIQUES

ALBERT E. RUEHLI
Missouri University of Science and Technology, Rolla, MO

GIULIO ANTONINI
Università degli Studi dell'Aquila, Italy

LIJUN JIANG
The University of Hong Kong, Pokfulam, Hong Kong

IEEE PRESS

WILEY

Published by John Wiley & Sons, Inc., Hoboken, New Jersey.

Published simultaneously in Canada.

Library of Congress Cataloging-in-Publication Data:

Names: Ruehli, A. E. (Albert E.), 1937- author. | Antonini, Giulio, 1969- author. | Jiang, Lijun 1970- author.
Title: The partial element equivalent circuit method for electro-magnetic and circuit problems : a paradigm for EM modeling / Albert E. Ruehli, Giulio Antonini, Lijun Jiang.
Description: Hoboken, New Jersey : John Wiley & Sons, 2016. | Includes bibliographical references and index.
Identifiers: LCCN 2016026830 (print) | LCCN 2016049198 (ebook) | ISBN 9781118436646 (cloth) | ISBN 9781119078395 (pdf) | ISBN 9781119078401 (epub)
Subjects: LCSH: Electric circuits–Mathematical models. | Electromagnetism–Mathematical models.
Classification: LCC TK3001 .R68 2016 (print) | LCC TK3001 (ebook) | DDC 621.301/51–dc23
LC record available at https://lccn.loc.gov/2016026830

Cover Design: Wiley
Cover Image: © Vectorig/Gettyimages

MIX
Paper from responsible sources
FSC® C013604
FSC www.fsc.org

CONTENTS

The book is dedicated to our families without whose patience and support it would have been impossible to write.

Albert: to my wife, Kristina, and all members of my extended family. This includes not only the children but also the grandchildren who are our future.

Giulio: to my wife Francesca, "the one, the beloved one, the most beautiful" (Ancient Egyptian Poem) and to our loved son Andrea.

Lijun: to my dear wife, Tao, and to my blessed daughters.

PREFACE

GENERAL ASPECTS

Electromagnetic (EM) modeling has been of interest to the authors of this book for a large portion of their careers. Giulio Antonini has been involved with partial element equivalent circuit (PEEC) for over 15 years at the Università degli Studi dell'Aquila, Italy, where he is now a professor. Both Albert Ruehli and Lijun Jiang worked as Research Staff members at the IBM Research Laboratory in Yorktown Heights, New York, on electrical interconnect and package modeling and electromagnetic compatibility (EMC) issues. Lijun Jiang is now a professor at the University of Hong Kong, Hong Kong, and Albert Ruehli is now an adjunct professor at the University of Science and Technology, Rolla, Missouri. We all continue to work today on different aspects of the PEEC method.

We welcome the opportunity to share the product of our experience with our readers. Fortunately, electromagnetic modeling (EMM) is a field of increasing importance. Electronic systems have been and will continue to increase in complexity over the years leading to an ever increasing set of new problems in the EM and circuit modeling areas. The number of electronic systems and applications expands every day. This leads to an ever-increasing need for electrical modeling of such systems.

EMM has been a key area of interest to the authors for quite a while. About 40 years ago, the general field of EMM was very specialized and more theoretical. The number of tools in this area and consequent applications were much more limited. Research is driven by the desire to discover new ways and potential applications as well as the need for solutions of real life problems.

Waveguides that mostly were interesting mechanically complex structures were physically large due to the lower frequencies involved. Some of the main topics of interest were antennas and waveguides as well as transmission lines. EM textbooks usually demanded an already high level of education in the theory and they were sometimes removed from realistic problems.

Transmission lines were the most accessible devices from both a theoretical and a practical point of view. Very few tools were available for practical computations especially before computers were widely available. Computers were mostly used for specialized applications. Problems were solved with a combination of theoretical analysis and measurements as well as insight that was a result of years of experience.

In contrast, today electromagnetic solver tools are available for the solution of a multitude of problems. Hence, the theoretical and intuitively ascertained solutions have been replaced with numerical method-based results. However, this does not eliminate the need for a thorough understanding of the EM fundamentals and the methods used in EM tools. The advanced capabilities available in the tools require a deeper understanding of the formulations on which the tools are based. We are well aware that the interaction of tools and theory leads to advances.

Textbooks such as Ramo and Whinnery [1] have evolved over many years. Meanwhile, many new excellent introductory textbooks have been written that treat different special subjects such as EMC [2]. Our book is oriented toward a diverse group of students at the senior to graduate level as well as professionals working in this general area. In our text, we clearly want to emphasize the utility of the concepts for real-life applications, and we tried to include as many relevant references as possible.

FUNDAMENTALS OF EMM SOLUTION METHODS

We have to distinguish between two fundamentally different types of circuit models for electromagnetics. Some of them are based on a differential equation (DE) formulation of Maxwell's equations, while others are based on integral equation (IE) form.

The DE forms are commensurate with the system of equations that results from the formulation of a problem in terms of DEs. This results in circuit models that have neighbor-to-neighbor coupling only. The most well-known form is the finite difference time domain (FDTD) method, which is a direct numerical solution of Maxwell's equations. The advantages of DE methods is that very sparse systems of equation result. At the same time, these systems are larger than the ones obtained from IE-based methods.

On the other hand, the IE-based methods will result in systems that have element-to-element couplings. Hence, this results in smaller, denser systems of equations. The finite element (FE) method is a somewhat hybrid technique since it involves local integrations while the overall coupling is local as in the DE methods. This also results in a large and sparse system of equations. Among the formulations used today, there are two circuit-oriented ones: (a) the DE-based transmission line modeling (TLM) method; and (b) the PEEC method. In this text, we mainly consider the IE-based PEEC method.

The PEEC method has evolved over the years from its start in the early 1970s [3–5]. Interestingly, this is about the same time when the other circuit-oriented EM approach – the TLM method – was first published [6]. Some early circuit-oriented work for DE solutions of Maxwell's equations was done by Kron in the 1940s [7]. However, the solution of the large resultant systems was impossible to solve without a computer. Hence, the work was of little practical importance. Recently, matrix stamps for FDTD models have been presented [8].

Around the same time, numerical DE methods made important progress. The FDTD method was conceived in 1966 [9]. Also, the finite integration technique (FIT) technique was published in 1977 [10]. All these methods have made substantial progress since the early work was published.

MORE ABOUT THE PEEC METHOD

The PEEC method evolved in a time span of more than 40 years. From the start, the approach has been tailored for EMM of electronic packages or Electronic Interconnect and Packaging

called signal integrity (SI). Power integrity (PI) and noise integrity (NI) as well as EMC problems. In the beginning, only high-performance computer system modeling needed accurate models for the electrical performance of the interconnects and power distribution in the package and chips. In main frame computes the speed of the circuits was much faster than that of conventional computer circuits such as the early personal computers.

Quasistatic solutions were adequate then even for the highest performance systems. Problems such as the transient voltage drops due to large switching currents were discovered very early. This prompted and extended the work on partial inductance calculations for problems of an ever-increasing size. In the 1990s, the modeling of higher performance chips and packages became an issue with the race for higher clock rates in computer chips. This led to the need for full-wave solutions. As a consequence, stability and passivity issues became important. Today, aspects such as skin-effect loss and dielectric loss models are required for realistic models.

Numerous problems can be solved besides package and interconnect and microwave problems. Approximate physics-based PEEC equivalent circuit models can be constructed, which are very helpful for a multitude of purposes. Further, PEEC is one of the methods used in some of the EMM tools. Fortunately, PEEC models can easily be augmented with a multitude of additional circuit models. This leads to other real advantages. Further, techniques have been found to improve the efficiency of these methods. As we show, PEEC is ideally suited for small simple models. Also, the wealth of circuit solution techniques that are available today can be employed. One example of this is the use of the modified nodal analysis (MNA) approach, which helps PEEC for low-frequency and a *dc* solution that other techniques may not provide.

TEACHING ASPECTS

We hope that this text can be used as an effective tool to introduce EM to new students. We think that a key advantage of the PEEC method is its suitability for an introductory course in EM.

The teaching of the PEEC method can be approached from several different points of view. It may be used as a way to introduce EMM, since most engineering students are more familiar with circuit theory rather than EM theory. This is also the case since circuit courses are taught at a lower level than EM courses. Alternatively, one may want to start with the introduction of the quasistatic PEEC models in a first EM course.

We prefer to use concepts that can be understood in lieu of the introduction of more advanced topics and mathematical notation. As a second course, general PEEC methods could be covered. This could be done, perhaps, in conjunction with introduction of concepts such as interconnect modeling and other chip and package design concepts.

Albert E. Ruehli,
Windham, New Hampshire,
USA
Giulio Antonini,
L'Aquila,
Italy
Lijun Jiang,
Hong Kong
January, 2017

REFERENCES

1. S. Ramo, J. R. Whinnery, and T. Van Duzer. *Fields and Waves in Communication Electronics*. John Wiley and Sons, Inc., New York, 1994.
2. C. R. Paul. *Introduction to Electromagnetic Compatibility*. John Wiley and Sons, Inc., New York, 1992.
3. A. E. Ruehli. Inductance calculations in a complex integrated circuit environment. *IBM Journal of Research and Development*, 16(5):470–481, September 1972.
4. A. E. Ruehli and P. A. Brennan. Efficient capacitance calculations for three-dimensional multiconductor systems. *IEEE Transactions on Microwave Theory and Techniques*, 21(2):76–82, February 1973.
5. A. E. Ruehli. Equivalent circuit models for three dimensional multiconductor systems. *IEEE Transactions on Microwave Theory and Techniques*, MTT-22(3):216–221, March 1974.
6. P. B. Johns and R. L. Beurle. Numerical solution of 2-dimensional scattering problems using a transmission-line matrix. *Proceedings of the IEEE*, 59(9):1203–1208, September 1971.
7. G. Kron. Equivalent circuit for the field equations of Maxwell. *Proceedings of the IRE*, 32(5):289–299, May 1944.
8. A. Ramachandran, A. Ramachandran, and A. C. Cangellaris. SPICE-compatible stamps for semi-discrete approximations of Maxwell's equations. In *International Journal of Numerical Modelling: Electronic Networks, Devices and Fields*, Volume 21, pp. 265–277, October 2008.
9. K. S. Yee. Numerical solution of initial boundary value problems involving Maxwell's equations in isotropic media. *IEEE Transactions on Antennas and Propagation*, 14(5):302–307, May 1966.
10. T. Weiland. Eine Methode zur Loesung der Maxwellschen Gleichungen fuer sechskompoentige Felder auf diskreter Basis. *Archiv der Elektrischen Ubertragung*, 31:116–120, 1977.

ACKNOWLEDGMENTS

The authors of this book are aware that progress in any area, including science and technology, is driven by the requirement for innovation. As part of this contribution, we also learn that it is not pursued in isolation. Our own thinking is greatly impacted in this process in our various areas of research.

We acknowledge contributions by many people who helped to debug the book to various degrees. To remove the errors was a daunting challenge. In particular, Giulio Antonini wants to acknowledge Professor Antonio Orlandi for having initially introducing him to the PEEC method. Giulio Antonini is then deeply grateful to Albert E. Ruehli for his guidance in the PEEC modeling, his long lasting friendship and for the many evenings spent together, making integrals and solving equations, having a lot of fun.

Kristina Ruehli read and corrected each of the chapters multiple times and modified the text written by Swiss-German, Italian, and Chinese thinkers to conform with English grammatical forms of speech, syntax, and punctuation.

We decided to find an alternate approach for acknowledgment of contributions by other researchers, since it is clearly impossible to add all the appropriate references to the book chapters. We added a list of M.S. and Ph.D. works that have contributed to the partial element equivalent circuit (PEEC) method and thereby acknowledge the contributions by their advisors.

Albert E. Ruehli	University of Vermont	PhD 1972
Chien-Nan Kuo	National Taiwan University (NTU)	MS 1990
Jean-Luc Schanen	Université Grenoble Alpes – CNRS	PhD 1994
Carla Giachino	Politecnico di Torino	MSc 1996
Andrea Giuliano	Politecnico di Torino	MSc 1996
Edith Clavel	Université Grenoble Alpes – CNRS	PhD 1996
Jan E. Garrett	The University of Kentucky	PhD 1997
William P. Pinello	The University of Arizona	PhD 1997
Hao Shi	Missouri University of Science and Technology	PhD 1997
Joel Reuben Phillips	Massachusetts Institute of Technology	PhD 1997

Giulio Antonini	Università degli Studi dell'Aquila	PhD 1998
Youssef Moez	Università Grenoble Alpes – CNRS	PhD 1998
Mattan Kamon	Massachusetts Institute of Technology	PhD 1998
Nicola Caporale	Università degli Studi dell'Aquila	MS 1999
Nadége Piette	Université Grenoble Alpes – CNRS	PhD 1999
Gabriele Di Fazio	Università degli Studi dell'Aquila	MS 1999
Junfeng Wang	Massachusetts Institute of Technology	PhD 1999
Yehia Mahmoud Massoud	Massachusetts Institute of Technology	PhD 1999
Jun Fan	Missouri University Science and Technology	PhD 2000
Karen M. Coperich	University of Illinois at Urbana – Champaign	PhD 2001
Jean-Michel Guichon	Université Grenoble Alpes – CNRS	PhD 2001
Maxime Besacier	Université Grenoble Alpes – CNRS	PhD 2001
Andre Görisch	Otto von Guericke University Magdeburg	PhD 2002
Ting-Yi Huang	National Taiwan University (NTU)	MS 2002
Valerio Di Fulvio	Università degli Studi dell'Aquila	MS 2003
Chun-Te Wu	National Taiwan University (NTU)	PhD 2002
Jonas Ekman	Lulea University of Technology	PhD 2003
Luca Daniel	University of California, Berkeley	PhD 2003
Gernot Steinmair	Johannes Kepler University, Linz (A)	PhD 2003
Enrico Vialardi	Politecnico di Torino	PhD 2003
Zhenhai Zhu	Massachusetts Institute of Technology	PhD 2004
Carlo Pariset	Università degli Studi dell'Aquila	MS 2004
Andrea Viaggi	Università degli Studi dell'Aquila	MS 2004
Luca Goffi	Università degli Studi dell'Aquila	MS 2004
Corrado Pistilli	Università degli Studi dell'Aquila	MS 2004
Dipanjan Gope	University of Washington	PhD 2005
Lina Colangelo	Università degli Studi dell'Aquila	MS 2005
Jean-Paul Gonnet	Université Grenoble Alpes – CNRS	PhD 2005
Christian Martin	Université Grenoble Alpes – CNRS	PhD 2005
Thomas Jonas Klemas	Massachusetts Institute of Technology	PhD 2005
Frederik Schmid	Lulea University of Technology	MSc 2005
Shaofeng Yuan	Missouri University of Science and Technology	PhD 2005
Mattia Di Prinzio	Università degli Studi dell'Aquila	MS 2005
Chuanyi Yang	University of Washington	PhD 2005
Amilcare Tiberti	Università degli Studi dell'Aquila	MS 2005
Swagato Chakraborty	University of Washington	PhD 2005
Marco Sabatini	Università degli Studi dell'Aquila	MS 2005
Chuanyi Yang	University of Washington	PhD 2005
Xin Hu	Massachusetts Institute of Technology	PhD 2006
Magnus Olovsson	Lulea University of Technology	MSc 2006
Liang Li	Missouri University Science and Technology	MS 2006

Francesco Ferranti	Università degli Studi dell'Aquila	PhD 2007
Michael Cracraft	Missouri University of Science and Technology	PhD 2007
Martin Ludwig Zitzmann	Friedrich-Alexander-Universität Erlangen-Nürnberg	PhD 2007
Alfredo Centinaro	Università degli Studi dell'Aquila	MS 2008
Xu Gao	Missouri University of Science and Technology	MS 2008
Giuseppe Miscione	Università degli Studi dell'Aquila	MS 2008
Carla Monterisi	Università degli Studi dell'Aquila	MS 2008
Sergey Kochetov	Otto von Guericke University Magdeburg	Habil. 2008
Luca De Camillis	Università degli Studi dell'Aquila	MS 2008
Tore Lindgren	Lulea University of Technology	PhD 2009
Jeremy Aimé	Université Grenoble Alpes – CNRS	PhD 2009
Volker Vahrenholt	Technische Universität Hamburg-Harburg	PhD 2009
Miao Zhang	Tsinghua University	MS 2009
Fan Zhou	Missouri University of Science and Technology	MS 2010
Vincent Ardon	Université Grenoble Alpes – CNRS	PhD 2010
Peter Scholz	Technische Universität Darmstadt	PhD 2010
Eleonora Palma	Università degli Studi dell'Aquila	MS 2011
Tarek A. El Moselhy	Massachusetts Institut of Technology	PhD 2010
Ruey-Bo Sun	National Taiwan University (NTU)	PhD 2011
Hsiang-Yuan Cheng	National Taiwan University (NTU)	PhD 2011
Tung Le Duc	Université Grenoble Alpes – CNRS	PhD 2011
Pasquale Caravaggio	Università degli Studi dell'Aquila	MS 2012
Danesh Daroui	Lulea University of Technology	PhD 2012
Ali Jazzar	Université Grenoble Alpes – CNRS	PhD 2012
Andrea De Luca	Università degli Studi dell'Aquila	MS 2012
Hanfeng Wang	Missouri University of Science and Technology	PhD 2012
Michela Miliacca	Università degli Studi dell'Aquila	MS 2012
Trung Son Nguyen	Université Grenoble Alpes – CNRS	PhD 2012
Maria De Lauretis	Università degli Studi dell'Aquila	MS 2012
Anca Goleanu	Université Grenoble Alpes – CNRS	PhD 2012
Andreas M. Muessing	ETH Zürich	PhD 2012
Thomas de Oliveira	Université Grenoble Alpes – CNRS	PhD 2012
Ivana F. Kovaćević	ETH Zürich	PhD 2012
Nenghong Xia	Hong Kong Polytechnic University	MS 2013
Tao Wang	Missouri University of Science and Technology	MS 2013
Xu Gao	Missouri University of Science and Technology	PhD 2014
Thanh Trung Nguyen	Université Grenoble Alpes – CNRS	PhD 2014
Sohrab Safavi	Lulea University of Technology	PhD 2014
Leihao Wei	Rose-Hulman Institute of Technology	MS 2015

Natalia Bondarenko	Missouri University of Science and Technology	PhD 2015
Tamar Makharashvili	Missouri University of Science and Technology	MS 2015
Daniele Romano	Università degli Studi dell'Aquila	PhD 2017
Yves Hackl	Technische Universität Darmstadt	PhD 2016
Luigi Lombardi	Università degli Studi dell'Aquila	PhD 2018
Carmine Gianfagna	Università degli Studi dell'Aquila	PhD 2019

Albert, Giulio, Lijun

ACRONYMS

ABC	absorbing boundary condition
BE	backward Euler method, BD1
BD2	backward differentiation method, Gear 2
CAD	computer-aided design
Ckt	circuit
EFIE	electric field integral equation
EM	electromagnetic
EMM	electromagnetic modeling
FDTD	finite difference time domain
FE	forward Euler method
FEM	finite element method
FFT	fast Fourier transform
FIR	digital filter nonrecursive
FIT	finite integration technique
IIR	digital filter with feedback
KCL	Kirchhoff's current law
KVL	Kirchhoff's voltage law
MFM	multifunction method
MFIE	magnetic field integral equation
MNA	modified nodal analysis
MOR	model order reduction
NI	noise integrity
PCB	printed circuit board
PEEC	partial element equivalent circuit
PI	power integrity
PDE	partial differential equation
PEC	perfect electric conductor
PMC	perfect magnetic conductor
PML	perfect matched layer
PPP	parallel plane PEEC model
PWTD	plane wave time domain

RCS	radar cross section
ROM	reduced order model
SI	signal integrity
SPICE	Simulation Program with Integrated Circuit Emphasis
TEM	transverse electromagnetic
Theta,Θ	theta integration method
TL	transmission line
TLM	transmission line modeling method
TR	trapezoidal method
VFI	volume filament
WRM	weighted residual method

1

INTRODUCTION

The history of the fundamental general techniques that are applied in electromagnetic (EM)-solvers today is interesting. Many techniques were devised in the decades between 1960 and 1980. The first work in the EM field for the finite element (FE) technique was presented in 1965 [1]. A paper on the foundation of the numerical differential equation (DE)-based finite difference time domain (FDTD) method [2] was published in 1966.

Interestingly, a second set of techniques used in commercial solvers today had been devised in the decade from 1970 to 1980. The circuit-oriented DE-based transmission line matrix (TLM) method originated in 1971 [3]. The circuit-oriented partial element equivalent circuit (PEEC) method, which is based on an integral equation (IE) formulation, originated in 1972 [4]. Finally, the DE-based finite integration (FIT) approach was devised in 1977 [5]. Since then, many different submethods have been developed based on all the cited techniques.

We note that most of the numerical solutions are solved using the weighted residuals method (WRM) [6]. The notation is not always the same since sometimes the notation MWR is used. Importantly, the WRM applies to the majority of fundamental solution techniques that include FE, DE, or IE based, for example, Refs [6, 7].

In 1968, a key paper on the numerical implementation of integral equation-based approach was presented in Ref. [8] and it was called the method of moments (MoM). The name is used in an inconsistent way since the MoM is a subset of the WRM methods that applies to a subclass of all formulations [9]. The MoM name originated in 1932 [10] as one of the WRM methods where the two approximation functions are not the same unlike for the Galerkin method. Further, there is no relation between MoM and the moment matching method used for macromodeling [11, 12]. To avoid confusions, we do not use MoM as a name for impedance-type IE solution as is done by some researchers.

Circuit Oriented Electromagnetic Modeling Using the PEEC Techniques, First Edition.
Albert E. Ruehli, Giulio Antonini, and Lijun Jiang.

In this book, we consistently use the WRM notation. We also use it for the PEEC method since the finite, circuit-based solution of the PEEC method can be viewed as being solved using a WRM technique.

Several book chapters have been written on PEEC techniques. An introduction to the capacitance circuit elements and partial inductances is given in Ref. [13] and a book dedicated to partial inductances is [14], while sections of other books are dedicated to PEEC methods in Refs [15–17], and [18].

Making the teaching of PEEC for electromagnetic solution techniques easier and more concrete is one of the aims of the authors. Many entry-level students are discouraged with the complexity of the EM subject. Quite a number of excellent textbooks on the fundamentals of electromagnetic exist today, such as Balanis [19], Chew [20], Collin [21], Kong [22], Paul [23], Plonsey and Collin [24] and Ramo et al. [25]. Hence, the underlying physics and the derivation of Maxwell's equations are well covered in these books. Most of them do an excellent job in presenting the subject in a comprehensive way.

The aim of this book is different. The fact that the PEEC approach is circuits based should help many students and readers understand the electromagnetic concepts. Most electrical engineering students will learn basic circuit analysis before taking a course in electromagnetics. Our aim is to present the concepts in an application-oriented way. Hence, the understanding of the concepts presented in this book should be of practical as well as of theoretical interest.

The PEEC method has evolved from the initial work, for example, Ref. [26] to a multitude of works by many researchers. The technique has expanded from its original focus in the interconnect modeling area to a wider range of applications. The modeling of combined electromagnetic and circuit (EM/Ckt) problems is one of the strengths of this technique. For this reason, we also included an introduction to the necessary circuit concepts in this book.

Many new systems involve circuit aspects as well as EM parts. Therefore, it is very desirable to be able to solve combined EM/Ckt problems. Circuit-based EM approaches are usually differential equation based. This includes the transmission line method (TLM) method. Other equivalent circuit methods are derived from DEs. The IE-based PEEC method is treated in this book.

In this text, we are interested in all types of electronic systems. Unlike some other approaches, PEEC also provides a stable *dc* solution, which is important for many realistic EM/Ckt problems. Today, many applications for low-frequency problems are in the power engineering area. Quasistatic PEEC solutions lead to conventional SPICE-type circuits [27]. For higher frequencies above the quasistatic frequency range, the approach results in full-wave solutions. Unfortunately, conventional SPICE solvers cannot be employed for full-wave solutions since the resultant PEEC circuits include delays. This leads to circuit solvers with delays.

Full-wave solutions are becoming more relevant in many cases. Miniaturization and other aspects are important for the growth in electrical systems. Due to the improvements and miniaturization of the semiconductor devices, the maximum frequencies are reaching into the 1000 GHz range. At these higher frequencies and component densities, coupling among the components has become a key issue. As an extreme case, power engineering systems may include integrated circuits with currents in the microampere range in the vicinity of bars conducting hundreds of amperes. The sizes of the coupled subsystems range from micrometers to meters.

Full-wave solutions may also be of importance due to the increasing spectrum in the noise signal frequencies and due to physical largeness of the systems. Electromagnetic compatibility (EMC) is another area of growing interest where the ever-increasing frequencies represent new challenges. The use of semiconductor devices in power electronic systems leads to higher frequency noise. From an EM-modeling point of view, these challenges represent many new and interesting problems to be solved.

The fundamental technical idea of the PEEC approach is to convert an IE-based solution of Maxwell's equations into appropriate equivalent circuits, which can then be used in conjunction with other different linear or nonlinear circuits in a circuit solver mode. We should not assume that this will compromise the solution from an electromagnetic point of view. In many situations, the opposite is true. Solutions can very often be found in the circuit domain, which are much more difficult to obtain without circuits. Besides, we can borrow from the large number of techniques that are available from circuit theory as well as from the implementation of today's SPICE-type circuit solvers.

The recent rapid increase in performance of today's PC computers provides many new possibilities for the EM-modeling area. In addition, the processor and increased memory size have made EM-modeling affordable for everyone. The availability of a large memory is key for the solution of larger problems since the complexity of the solution increases rapidly with the problem size. The speedup of PEEC solutions for large models is one aspect that we do not cover in this text. Parallel processing is a way to enhance the compute power. Fortunately, processors also have become widely available at a low cost. All EM-modeling techniques need to be tailored to computer systems to take full advantage of these changes. However, these issues for PEEC are not included in this book.

In this book, we aim to introduce electromagnetic PEEC models in a practical useful way. The book is written such that the concepts are ready for real-life applications. We hope that this helps the understanding of the fundamental concepts. Also, it should make the book useful to industry and will help to emphasize the importance of the subject to new students in the EM field. We are also attempting to make the mathematical formulations as transparent and readable as possible. This should extend the overall readability of the book for self-study for everyone.

Besides the necessary introduction of the basic concepts in the circuit and electromagnetic theories, the techniques are presented first starting with the circuit concepts. All aspects of building PEEC models are presented in a logical way. We find it useful to implement PEEC in a SPICE circuit solver-like implementation such that both the time and frequency domain solutions can be provided without the need for the Fourier transform of the frequency domain solution. Linear equivalent circuits can lead to models that can be used equally well in both the time and frequency domains. Clearly, the SPICE solver input language represents an excellent implementation of this fact since sources can be specified such that they apply both in the time and the frequency domains. This also helps the flexibility of the overall solution.

In Chapter 2, we give an introduction to circuit analysis necessary for a PEEC solution. Our solution approach and most SPICE circuit solvers today are based on the modified nodal analysis (MNA) method. Importantly, this approach also leads to the *dc* solution for PEEC. The fundamental idea is to set up the MNA circuit matrix for all our solutions. This is accomplished with so-called matrix stamps where each circuit element is added to the circuit matrix with a *matrix stamp*. This enters the contribution to the circuit matrix in a clear, systematic way.

Stamps for different circuit elements are given in Appendix B. We need to consider the circuit solutions for both the time and frequency domains. In addition, we present a circuit-oriented approach for which frequency-dependent elements can be included also in a time domain solution. This is done with a synthesis process. Another useful approach presented for adding macromodels to the circuit matrix is the recursive convolution approach. Section 2.11 considers circuits with delays, which is fundamental for full-wave PEEC models.

Chapter 3 introduces the underlying electromagnetic concepts necessary for the PEEC models presented in the book. Further details on the fundamentals for the EM concepts can be found in the above-mentioned texts. We have to assume that the reader of our book is knowledgeable in the solution of Maxwell's equations. The presentations are clearly oriented toward IE solutions. The necessary Green's functions that are used in this text are given. We only use scalar Green's functions for which we use the lowercase g. We also use the fundamental concepts necessary for surface IE-based PEEC equations. We end Chapter 3 with a short discussion of the numerical solution of IEs.

Chapter 4 is dedicated to the computation of capacitances – an integral part of PEEC models. The basics of IE-based solutions for capacitance problems is considered. The potential coefficients in the solutions are basic elements in the remaining chapters. For completeness, we also included computations using a differential equation approach. Further, the importance of projection meshing is considered since it may lead to errors in the results. Models for representing capacitances with retardation for PEEC equivalent circuits are presented. Finally, the computation of partial potential coefficient is given in Appendix D.

Chapter 5 considers important aspects of inductance calculations. The main topic is the development of partial inductance concepts since they are the fundamental building block of most PEEC models. The general concept of partial inductance is described and its application for general inductance computation is discussed. Concepts such as open-loop inductances are introduced and examples are given for the use of the open-loop concept.

The basic task of the computation of partial inductances is delegated to Appendix C. Other issues such as problems with the partial inductance for large conductors are shown to be of importance. Finally, the use of the PEEC concepts for the modeling of transmission lines and the efficient modeling of plane pairs is considered as a useful application of partial inductances.

In Chapter 6, we present the building details for PEEC models that utilize partial inductances as well as partial coefficients of potential and resistances. The presentation of these concepts is done systematically so that the derivations in the previous chapters can be used. Importantly, the conductor cells are described in detail and the PEEC equivalent circuits are given. Other aspects such as the implementation of the continuity equation as Kirchhoff's current law is derived in terms of the physical geometry. More details for the capacitance models used in PEEC are given. Finally, the circuit equations for the model are presented, which includes models with delays or retardation.

Chapter 7 gives a detailed development of the equations for the nonorthogonal case, which is the extension of the PEEC approach to nonorthogonal conductors and dielectrics. This necessary step will facilitate the computation problems with nonrectangular-shaped objects. Mainly, we found that a purely rectangular or Manhattan representation of the geometry is insufficient for some problems.

Manhattan modeling can do an excellent job for many problems, especially for on-chip problems. The additional complexity introduced by nonrectangular coordinates certainly leads to increases in compute time and additional challenges in the algorithms. Hence, we do address both the rectangular and the nonrectangular cases in a consistent way. Advances have also been made for both orthogonal and nonorthogonal geometries. We present formulations for the evaluation of the nonorthogonal partial elements in the appendices.

Chapter 8 is dedicated to the important topic of the geometrical description of the bodies and their subdivision or meshing. This subdivides the problem into the appropriate geometrical cells for which the partial elements are computed. Two fundamentally different approaches exist today. In many IE approaches, the geometry is subdivided into small triangular cells that can be computed at a relatively small cost. This approach is well suited, for example, in solving problems such as the scattering from airplanes.

For electronic systems, efficient subdivision may consist of cells with very large aspect ratios. This may result in a reduction in the number cells and ultimately unknowns. Large differences in cell sizes are allowed. However, a larger effort is needed for the meshing as well as for the computation of the partial elements. In this chapter, we discuss details for the implementation for this type of meshing.

Chapter 9 covers the inclusion of skin-effect in PEEC models. The issue is fundamentally very challenging since it can add considerably to the compute time needed to solve a problem. Fortunately, the PEEC method is inherently suitable for 3D skin-effect models. Several different models are included depending on the current flow and other geometrical features. The modeling of skin-effect problems in terms of circuits is a natural, convenient solution. This is shown to be the case in this chapter where several circuit-oriented models are presented. Also, a surface IE skin-effect model is presented. Examples are given where we also compare the integral equation results with the volume skin-effect models.

Chapter 10 considers PEEC with dielectric models. These models also include loss models for the dielectrics. It is clear that the inclusion of the dielectrics in PEEC is important today. Approaches such as the method-of-images are presented for the computation of the Green's functions for layered dielectric structures. In the gigahertz range, the dielectric losses can dominate skin-effect losses. Hence, the higher frequency made the inclusion of dielectric losses relevant.

The challenge is to provide loss models that operate equally well in both the time and the frequency domains. It is evident from this chapter that it is easy to include other circuit models in the implementation of PEEC circuits. The last topic in this chapter is inclusion of dielectrics in the time domain using recursive convolution. This allows the inclusion of relatively complex structures without excessive compute times.

In Chapter 11, we consider techniques for the modeling of problems that include magnetic materials. First, we consider the conventional magnetic reluctance concepts since they lead to circuit models, although they are not directly connected to PEEC. Concepts such as magnetic potentials and magnetic charge are considered next for the IE solution.

The key aspect of this part of the chapter is concerned with PEEC formulations that also includes magnetic bodies. The derivations are parallel to Chapter 6, where we are building conventional PEEC models. Finally, surface IE approaches are also suitable for the inclusion of magnetic materials. For this reason, we include a PEEC surface formulation in this chapter.

Chapter 12 includes incident and radiated fields, which represent an important part of EM models, especially for electromagnetic interference (EMI) and antenna problems. First, the incident field computation is considered, which results in additional independent voltage sources. Then a PEEC-specific, sensor-based approach is used to compute both the electric and magnetic fields. Finally, more conventional direct far field computations are included.

The concepts of stability and passivity are considered in Chapter 13. We introduce the fundamental circuit concepts needed for considering this complicated topic. Both the time and frequency domain solutions are included. Conventional PEEC models without delays can be treated in a relatively straightforward way. While the theory is needed, we also consider the more practical side where possible.

Interesting results have been obtained from running conventional circuits. However, PEEC circuits with delays lead to very challenging *descriptor* systems. This chapter also considers techniques to improve the stability and passivity for realistic problems. Examples are given to show the effectiveness and the implementation of some of these concepts.

REFERENCES

1. O. C. Zienkiewicz and Y. K. Cheung. Finite elements in the solution of field problems. *The Engineer*, 220:507–510, September 1965.
2. K. S. Yee. Numerical solution of initial boundary value problems involving Maxwell's equations in isotropic media. *IEEE Transactions on Antennas and Propagation*, 14(5):302–307, May 1966.
3. P. B. Johns and R. L. Beurle. Numerical solution of 2-dimensional scattering problems using a transmission-line matrix. *Proceedings of the IEEE*, 59(9):1203–1208, September 1971.
4. A. E. Ruehli. Electrical analysis of interconnections in a solid state circuit environment. In *ISSCD, IEEE International Solid-State Circuits Conference*, Volume XV, pp. 64–65, February 1972.
5. T. Weiland. Eine Methode zur Loesung der Maxwellschen Gleichungen fuer sechskompoentige Felder auf diskreter Basis. *Archiv der Elektrischen Uebertragung*, 31:116–120, 1977.
6. Z. D. Chen and M. M. Ney. The method of weighted residuals: a general approach to deriving time- and frequency-domain numerical methods. *IEEE Antennas and Propagation Magazine*, 51(1):51–70, February 2009.
7. J. M. Jin. *The Finite Element Method in Electromagnetics*. John Wiley and Sons, Inc., New York, 2nd edition, 2002.
8. R. F. Harrington. *Field Computation by Moment Methods*. Macmillan, New York, 1968.
9. B. A. Finlayson and L. E. Scriven. The method of weighted residual: a review. *Applied Mechanics Reviews*, 19:735–748, 1966.
10. M. F. Krawchuk. Application of the method of moments to the solution of linear differential and integral equations (in Ukrainian). *Kiev Soobshch, Akademii Nauk UkSSR*, 1:168, 1932.
11. L. Pillegi, R. Rohrer, and C. Visweswariah. *Electronic Circuits and System Simulation Methods*. McGraw-Hill Book Company, New York, 1995.
12. M. Celik, L. Pileggi, and A. Odabasioglu. *IC Interconnect Analysis*. Kluwer Academic Publishers, 2002.
13. A. E. Ruehli, A. D. Ling, and A. Ruehli Eds. *Circuit Analysis, Simulation and Design, Part 2*. Elsevier Science Publishers B.V., North Holland, 1987.
14. C. Paul. *Inductance, Loop and Partial*. John Wiley and Sons, Inc., New York, 2010.
15. Y. Zhu and A. Cangellaris. *Multigrid Finite Element Methods for Electromagnetic Field Modeling*. John Wiley and Sons, Inc., New York, 8th edition, 2006.
16. C.-K. Cheng, J. Lillis, S. Lin, and N. Chang. *Interconnect Analysis and Synthesis*. John Wiley and Sons, Inc., New York, 2000.
17. J. Nitsch, F. Gronnwald, and G. Wollenberg. *Radiating Nonuniform Transmission-Line Systems and the Partial Element Equivalent Circuit Method*. John Wiley and Sons, Inc., New York, 2009.
18. A. Ametani, Ed. *Numerical analysis of power system transients and dynamics*. The Institution of Engineering and Technology, London, United Kingdom, 2015.
19. C. A. Balanis. *Advanced Engineering Electromagnetics*. John Wiley and Sons, Inc., New York, 1989.

20. W. C. Chew. *Waves and Fields in Inhomogeneous Media*. IEEE Press, New York, 1995.
21. C. R. Collin. *Field Theory of Guided Waves*. IEEE Press, New York, 1991.
22. J. A. Kong. *Electromagnetic Wave Theory*. EMW Publishing, Cambridge, MA, 2008.
23. C. R. Paul. *Analysis of Multiconductor Transmission Lines*. John Wiley Interscience, IEEE Press, Hoboken, NJ, 2nd edition, 2008.
24. R. Plonsey and R. Collin. *Principles and Applications of Electromagnetic Fields*. McGraw-Hill, New York, 1961.
25. S. Ramo, J. R. Whinnery, and T. Van Duzer. *Fields and Waves in Communication Electronics*. John Wiley and Sons, Inc., New York, 1994.
26. A. E. Ruehli. Equivalent circuit models for three dimensional multiconductor systems. *IEEE Transactions on Microwave Theory and Techniques*, MTT-22(3):216–221, March 1974.
27. L. W. Nagel. SPICE: a computer program to simulate semiconductor circuits. Electronics Research Laboratory Report, ERL M520, University of California, Berkeley, CA, May 1975.

2

CIRCUIT ANALYSIS FOR PEEC METHODS

2.1 CIRCUIT ANALYSIS TECHNIQUES

The fundamental concepts in this book are based on circuit and electromagnetic (EM) theories. We formulate as much as possible in the circuit domain since it adds additional capabilities and flexibilities to the solution techniques.

In this chapter, we introduce the circuit techniques relevant to this book. As was stated in the introduction in Chapter 1, Maxwell's equations can be represented in terms of equivalent circuits. The circuit solution techniques presented in this chapter are clearly oriented toward EM and relevant circuits (Ckt) problems. We also show how circuit techniques have advantages for the solution of EM problems. Also, mathematical techniques such as the numerical solution of differential equations (DE) must be included in circuit approaches.

It is generally understood that learning EM techniques is not easy. However, we hope that our circuit-oriented approach facilitates the introduction of the new concepts. Today, many excellent books are available on circuit theory [1–6] and many SPICE-type solvers [7] are based on these techniques.

Another key issue is that a large part of the problems of interest include combinations of Ckt and EM problems. Hence, it is a very important issue that the bridge between Ckt and EM parts be seamless. For this reason, the approach we pursue in the partial element equivalent circuit (PEEC) method is to transform the entire problem in rigorous ways, into the circuit domain as much as possible.

2.2 OVERALL ELECTROMAGNETIC AND CIRCUIT SOLVER STRUCTURE

A fundamental question for an EM and an EM/Ckt solution is what should be the basis of the internal structure. It is clear that we can learn from the structure of SPICE-type [7] circuit solvers in several ways. First, a SPICE input language can be viewed as a flexible high-level

Circuit Oriented Electromagnetic Modeling Using the PEEC Techniques, First Edition.
Albert E. Ruehli, Giulio Antonini, and Lijun Jiang.

description of the EM/Ckt problem. This adds much flexibility to the problem solving. Using a high-level description of the EM structure has some advantages even in comparison to graphical model description only. The SPICE-type `.subckt` subcircuit macromodel description can be used to subdivide large problems. Today, many different alternatives exist for input specifications.

As indicated in the previous section, the language description of the geometry does not have to give all the details but rather can be at a macrolevel. As shown below, the input language statement is internally translated into the so-called *circuit element stamps* that directly enters the contribution of the element to the system or circuit matrix.

Since the topology of a PEEC Ckt is given by the geometry, we can use so-called physical macros or block element stamps as well at the conventional circuit level element by element stamps. Further, a macrocircuit consisting of several elements can be represented by a subcircuit. These overall stamps may be reused multiple times in a model.

No matter how we specify the circuits and geometries, in this chapter we can assume that the circuit topology is already available for the entire problem. We do assume that all the circuits are of a linear time-invariant nature and are to be solved in the time and/or the frequency domain. Hence, the approach will concentrate on the solution of the important class of problems described.

The fundamental technique we pursue here is based on the modified nodal analysis (MNA) technique [6, 8, 9], which is used in most SPICE-type circuit solvers. The approach has shown to apply to a wide range of applications, for example, Refs [10] and [11].

The MNA approach is based on the idea that we are *stamping* the circuit elements into the circuit matrix element by element to assemble the system of equations of interest. We have to supply all the stamps for the element type that our solver needs to provide. It is also evident that we easily can add new element stamps to the solver. Another issue in a SPICE structured solver is the switching between the time and frequency domain. In the original SPICE input language, it is very easy to switch between the two domains. We try to make the PEEC formulation to work in both domains as much as possible. Hopefully, very similar stamps are available for all elements such that the setup of the circuit matrix is not fundamentally different. This is a great advantage since it avoids duplications of large parts of the solver program implementation that is usually required for separate solvers.

For some problems, an alternative approach to switch between domains is through the fast Fourier transform (FFT) solution to convert the frequency domain data into the time domain results through the inverse fast Fourier transform (IFFT). The opposite is also true as it is done in the finite difference time domain (FDTD) method. However, a time domain solver is required for serious nonlinear problems.

Using the MNA formulation greatly enhances the capabilities of the PEEC method since a *dc* solution can be obtained without additional work [12, 13], a feature many EM formulations and solvers lack. This feature is very important for applications. Some EM techniques implemented sophisticated loop-star transformations to obtain a solution with an improved low-frequency response [14–16]. However, even though these techniques do improve the low-frequency behavior, they do not lead to a proper *dc* solution. Also, the implementation of these techniques is complex. The MNA approach pursued in this book results in the EM and Ckt solutions including *dc*.

We start the technical discussion of this chapter with some circuit analysis concepts that are necessary for the general PEEC approach. The fundamental observation is that neither *loop* nor *nodal* analysis is sufficient for the representation of all types of circuits. The MNA approach has both voltage and current unknowns. This leads to a very convenient, general solution approach. Importantly, graph algorithms are in principle not necessary to set up the MNA. However, we discuss the graph theoretic aspects needed since this helps the simultaneous setup of large geometrical objects with a known structure. This leads to stamps for sections of a problem.

2.3 CIRCUIT LAWS

2.3.1 Kirchoff's Current Law

We start the details of the circuit formulation with Kirchhoff's laws. One of the most important laws for MNA circuit analysis is Kirchhoff's current law (KCL). Another one is Kirchhoff's voltage law (KVL). In general, both Kirchhoff's current and voltage laws are of fundamental importance. All circuit solutions are based on these laws. *A simple form of the KCL states that the sum of all currents at a node is zero* [1, 5, 6].

To be specific, we consider the circuit shown in Fig. 2.1. If we apply the KCL to node 1, we obtain

$$-I_i + I_1 + I_2 + I_3 = 0. \tag{2.1}$$

Note that we assume as a convention that the current direction *flowing away from the node is positive*. Hence, we have assigned the direction of the current in each branch. We clearly do not know the actual direction of the current until we solve the circuit equations, but this is irrelevant since the sign of the current will be negative if we choose the sign in the other direction. Therefore, we do not need to worry about the choice of the arrow directions as long as we apply the KCL.

In all cases, we need to declare a single node as the ground or datum node. We do not label the ground node since we do not have to write the KCL for the ground nodes. After some thought, we recognize that writing KCL equation for all ground nodes would add redundant information to the circuit equations. This is the case since we already wrote the KCL equation for each of the components at other nonground nodes. Hence, adding the equation at the ground nodes would lead to a linearly dependent, singular circuit matrix. An example for this is Fig. 2.1 where only nodes 1–3 are labeled while all ground notes are not considered for the KCL.

2.3.2 Kirchoff's Voltage Law

In this section, we consider KVL. *The KVL states that the sum of voltages around a loop is zero*. We count the contribution of a voltage as positive from plus to minus. The application

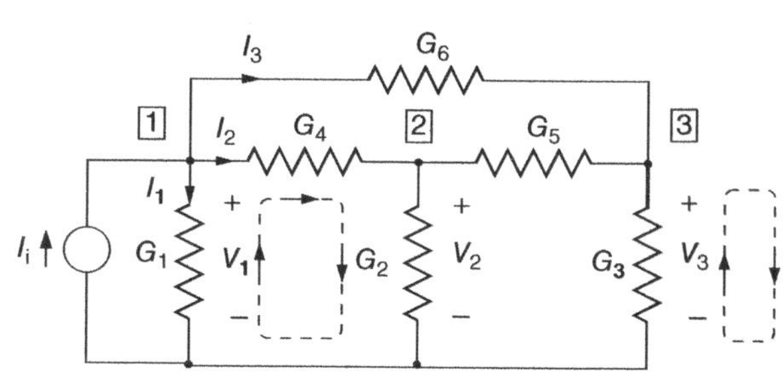

FIGURE 2.1 Resistance circuit example for Kirchhoff's voltage and current laws.

of the KVL is shown in the circuit in Fig. 2.1. Even for the MNA approach, we implicitly apply the KVL for each branch without being aware of it. We show the application for two different cases. If the KVL is applied to the left loop in the schematic for the branch resistor G_4, we get

$$-V_1 + V_{b4} + V_2 = 0, \tag{2.2}$$

where V_{b4} is the voltage across G_4 or the *branch voltage*. The second example loop is given for G_3 where only one nodal voltage is involved. For this case, we have the simpler equation

$$-V_3 + V_{b3} = 0. \tag{2.3}$$

It is clear that to fully describe the behavior of the circuit, we can apply the KVL implicitly to each of the components. As we see, the MNA circuit element stamps that we introduce here will automatically take care of both KCL and KVL.

2.3.3 Branch Impedances

At this point, we consider the equation that describes the *branch current and/or voltage relation* for each of the circuit elements. We start with a simple resistor branch. Again we use the circuit in Fig. 2.1. It is seen that each branch of the circuit consists of a resistor with a conductivity of $G_m = 1/R_m$. We can use the KVL with respect to ground to set up the branch relation. If we pick the branch with a conductance $G_4 = 1/R_4$, we can obtain the following branch relation:

$$-V_1 + \frac{I_2}{G_4} + V_2 = 0. \tag{2.4}$$

Hence, (2.4) can be solved for the current I_2 contribution to the KCL at node 1, which is

$$I_2 = G_4 V_1 - G_4 V_2. \tag{2.5}$$

It is known that the SPICE input language statement is also circuit element oriented. For example, the resistor in Fig. 2.1 is represented by its name, node indexes, and the resistance value:

```
R4   1   2   2 kOhms
```

This information can directly be used to identify and compute G_4 in the formulation of circuit matrices.

2.3.4 Incomplete Kirchhoff's Current Law

It is key for a general, flexible circuit solver to assemble the circuit matrix *element by element* rather than implementing the KCL such that all elements connecting to a node are handled simultaneously. For this reason, we account for the current at a node on an element basis [6, 8]. We call this process *element stamping* since for each type of circuit element we develop a so-called matrix stamp with its contribution. The contributions are then *stamped* or entered into the circuit matrix. This process is executed until

TABLE 2.1 Matrix element stamp for resistor R_4.

Row/column	V_1	V_2	RHS
1	G_4	$-G_4$	0
2	$-G_4$	G_4	0

all the current contribution for each node has been accounted for. As an example, the branch impedance stamp for the resistor R_4 can be derived from (2.5) and is shown in Table 2.1.

Its first row is obtained using the current through G_4 from node 1. Its second row is obtained using the current through G_4 from node 2. Hence, the stamps for multiple elements shall all be considered for all connected nodes except for the ground node. Given the matrix stamp for a resistor, we can set up the circuit equation for the circuit in Fig. 2.1 on an element-by-element basis. We add the new stamps to the matrix each time a new node voltage is required by an element. Hence, the circuit matrix for the circuit in Fig. 2.1 is given by

$$\begin{bmatrix} G_1+G_4+G_6 & -G_4 & -G_6 \\ -G_4 & G_2+G_4+G_5 & -G_5 \\ -G_6 & -G_5 & G_3+G_5+G_6 \end{bmatrix} \begin{bmatrix} V_1 \\ V_2 \\ V_3 \end{bmatrix} = \begin{bmatrix} I_i \\ 0 \\ 0 \end{bmatrix} \tag{2.6}$$

where I_i is the input current source current. Hopefully, this simple example helps to understand the setup of the basic circuit matrix by the stamping approach. In this example, we also used the current source stamp from Appendix B where stamps are given for a large class of circuit elements.

2.4 FREQUENCY AND TIME DOMAIN ANALYSES

It is clear that, for a general solution, we need to have matrix stamps for all types of circuit elements and relationships. A list of relevant elements, sources, and their relationships are given in Table 2.2 for both time and frequency domains. A list of stamps corresponding to this table is given in Appendix B. We consider several stamps in this section as examples.

TABLE 2.2 Small list of circuit elements for which we need MNA matrix stamps.

Circuit elements	Frequency domain	Time domain
R Resistance	$V(s) = R\,I(s)$	$v(t) = R\,i(t)$
C Capacitance	$I(s) = sCV(s)$	$i(t) = Cd\,v(t)/dt$
L Inductances	$V(s) = sLI(s)$	$v(t) = Ld\,i(t)/dt$
CCVS (Current controlled voltage source)	$V(s) = h\,I(s)$	$v(t) = h\,i(t)$
CCCS (Current controlled current source)	$I(s) = f\,I(s)$	$v(t) = f\,i(t)$
VCVS (Voltage controlled voltage source)	$V(s) = e\,V(s)$	$v(t) = e\,v(t)$
VCCS (Voltage controlled current source)	$I(s) = g\,V(s)$	$i(t) = g\,v(t)$

2.5 FREQUENCY DOMAIN ANALYSIS FORMULATION

In this section, we give a short introduction to frequency domain solver issues. Since the frequency domain analysis is widely used, we do not present basic concepts. Fundamentally, the circuit is solved at each frequency of interest. Therefore, at each frequency, we have to update the values corresponding to the frequency-dependent elements in the MNA matrix. For convenience, we express the frequency in terms of the Laplace variable $s = j\omega$. First, we want to point out that many SPICE language statements for elements such as capacitances and inductances are the same for the time and frequency domains. This adds simplification as well as generality to the overall approach.

We use a short-circuit capacitor example from Chapter 4 to show the application of a frequency domain stamp. For the two-conductor system in Fig. 2.2, we have

$$\begin{aligned} Q_1 &= Cs_{11}\,\Phi_1 + Cs_{12}\,\Phi_2 \\ Q_2 &= Cs_{21}\,\Phi_1 + Cs_{22}\,\Phi_2 \end{aligned} \tag{2.7}$$

where Q_1 and Q_2 are charges, and Φ_1 and Φ_2 are conductor potentials on the two conductors, respectively. Cs_{ij} is the self- or coupling capacitance of the two conductors. We also observe that self-coefficients reach to infinity. The fundamental observation is that the connection to infinity for the equivalent circuit is replaced with a single ground node. This will be the case for the PEEC models with capacitive models.

Equation (2.7) is a system of short-circuit capacitances. Its corresponding currents in the frequency domain can be obtained from (2.7) as

$$I_1(s) = s\,Cs_{11}\,V_1(s) + s\,Cs_{12}\,V_2(s) \tag{2.8a}$$

$$I_2(s) = s\,Cs_{21}\,V_1(s) + s\,Cs_{22}\,V_2(s). \tag{2.8b}$$

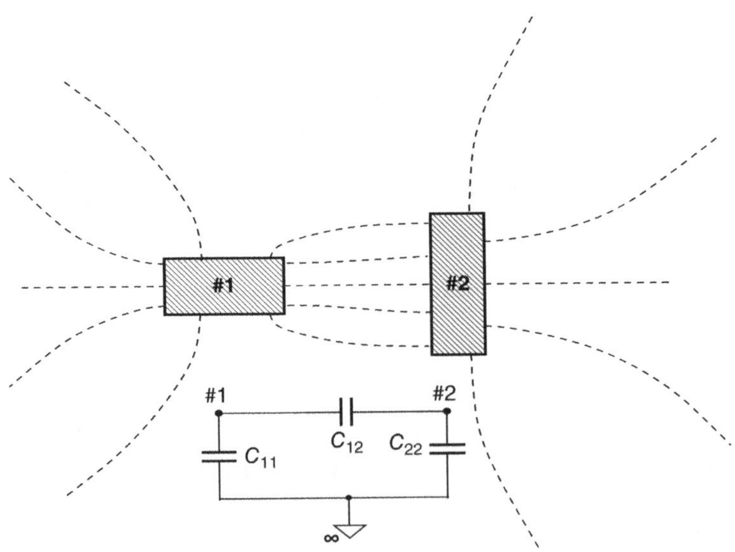

FIGURE 2.2 Flux patterns and equivalent circuit for two-conductor system.

TABLE 2.3 Matrix element stamp for capacitor C_{12} in frequency domain.

Row/column	V_1	V_2	RHS
1	$s\,C_{12}$	$-s\,C_{12}$	0
2	$-s\,C_{12}$	$s\,C_{12}$	0

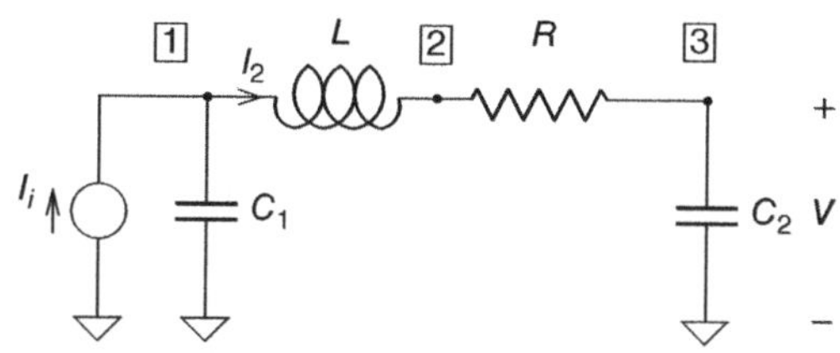

FIGURE 2.3 A small example circuit that includes an inductor.

However, short-circuit capacitances can be replaced with three two-terminal capacitances where the coupling capacitance is $C_{12} = -Cs_{12}$. The stamp that connects the capacitor between nodes 1 and 2 is given in Table 2.3. Hence, the 2×2 matrix is assembled in the same way.

Next, we show how to construct the stamping of an MNA matrix in the frequency domain for the small circuit in Fig. 2.3. For completeness, we include capacitances as well as an inductor in the example. We notice that the grounded capacitors contribute only one entry. Using the appropriate stamps from Appendix B, we obtain

$$\begin{bmatrix} s\,C_1 & 0 & 0 & +1 \\ 0 & 1/R & -1/R & 0 \\ 0 & -1/R & (1/R + s\,C_2) & -1 \\ 1 & -1 & 0 & -s\,L \end{bmatrix} \begin{bmatrix} V_1 \\ V_2 \\ V_3 \\ I_2 \end{bmatrix} = \begin{bmatrix} I_i \\ 0 \\ 0 \\ 0 \end{bmatrix}. \tag{2.9}$$

The MNA matrix can be simplified through combining the inductance and the resistance stamps to impedance formulations by eliminating node 2 in Fig. 2.3. This results in fewer unknowns shown in

$$\begin{bmatrix} s\,C_1 & 0 & +1 \\ 0 & +s\,C_2 & -1 \\ 1 & -1 & -(R + s\,L) \end{bmatrix} \begin{bmatrix} V_1 \\ V_3 \\ I_2 \end{bmatrix} = \begin{bmatrix} I_i \\ 0 \\ 0 \end{bmatrix}. \tag{2.10}$$

The resistance in series to an inductance is used in one new simplifying stamp.

We further choose this circuit to illustrate the basic structure of a PEEC circuit. As presented in Chapter 6, the topology of a PEEC circuit is very similar to the circuit in Fig. 2.3. In fact, the MNA matrix for a simple PEEC circuit is

$$\begin{bmatrix} \mathbf{G} + s\mathbf{C} & \mathbf{A} \\ \mathbf{A}^T & -(\mathbf{R} + s\mathbf{L}) \end{bmatrix} \begin{bmatrix} \mathbf{\Phi} \\ \mathbf{I} \end{bmatrix} = \begin{bmatrix} \mathbf{I}_i \\ \mathbf{V}_i \end{bmatrix}, \tag{2.11}$$

where $\mathbf{V}_i$ represents independent voltage sources and $\mathbf{\Phi}$ represents the potential or voltage to infinity. This comparison of the simple circuit in (2.10) and (2.11) may give some insights into how the PEEC matrix formulation is constructed. Again, details for the model are given in Chapter 6.

At this point, we want to insert another important observation regarding the condition of the MNA circuit matrix. As we see, the notation for MNA equations such as (2.11) is conventionally written in the general form

$$s\,\mathcal{C}\,\mathbf{X}(s) + \mathcal{G}\,\mathbf{X}(s) = \mathcal{B}\,\mathbf{X}_{in}(s) \tag{2.12a}$$

$$\mathbf{X}_{out}(s) = \mathcal{L}^T\mathbf{X}(s). \tag{2.12b}$$

We also observe that this corresponds to the general form of a descriptor system that is considered in (2.77).

It is also worth noticing that matrix $\mathcal{G}$ or/and $\mathcal{C}$ may be singular if there are nodes without resistive or capacitive elements connected. A small example where the $\mathcal{C}$ matrix part in (2.12) is singular can be constructed. Specifically, the removal of either capacitor to ground, C_1 or C_2 in the example in Fig. 2.3 will lead to a singular $\mathcal{C}$ matrix in (2.12).

Figure 2.4 shows an example circuit with both matrices $\mathcal{G}$ and $\mathcal{C}$ nonsingular.

The corresponding matrix is given in (2.13):

$$\begin{bmatrix} s\,C_1 + G_1 & 0 & +1 \\ 0 & +s\,C_2 + 1/R_3 & -1 \\ 1 & -1 & -(R_2 + s\,L) \end{bmatrix} \begin{bmatrix} V_1 \\ V_3 \\ I_2 \end{bmatrix} = \begin{bmatrix} I_i \\ 0 \\ 0 \end{bmatrix}. \tag{2.13}$$

For completeness, we add two additional formulations for the circuit equations besides the general descriptor form. First, we eliminate the currents $\mathbf{I}$ in (2.11) to obtain by matrix manipulations an admittance matrix form

$$\left[(\mathbf{G} + s\mathbf{C}) + \mathbf{A}(\mathbf{R} + s\mathbf{L})^{-1}\mathbf{A}^T\right]\mathbf{\Phi} = \mathbf{I}_i + \mathbf{A}(\mathbf{R} + s\mathbf{L})^{-1}\mathbf{V}_i. \tag{2.14}$$

Alternatively, we can eliminate the potential $\mathbf{\Phi}$ to obtain an impedance formulation again by matrix manipulations

$$\left[(\mathbf{R} + s\mathbf{L}) + \mathbf{A}^T(\mathbf{G} + s\mathbf{C})^{-1}\mathbf{A}\right]\mathbf{I} = -\mathbf{V}_i + \mathbf{A}^T(\mathbf{G} + s\mathbf{C})^{-1}\mathbf{I}_i. \tag{2.15}$$

We should note that both these formulations may have some applications. However, both the admittance and the impedance formulations are based on time-consuming matrix inversions. We also want to point out that the admittance (2.14) and impedance (2.15) formulations may be ill-conditioned at low frequencies. Both formulations include terms that scales with s and $1/s$.

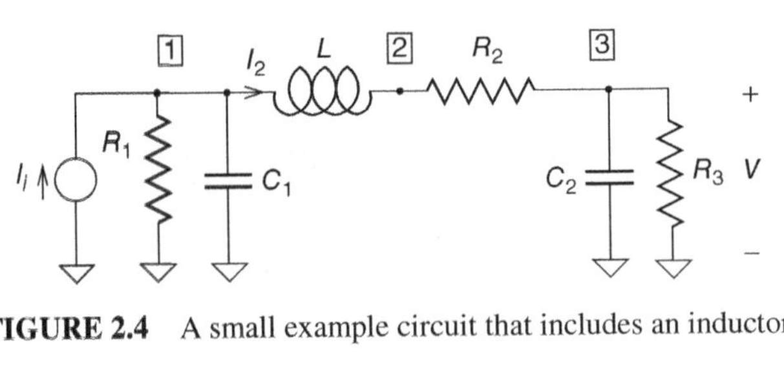

FIGURE 2.4 A small example circuit that includes an inductor.

More details can be found in Ref. [17]. It is demonstrated that the system matrix (2.11) is well behaved from a full-wave solution at high frequencies to a pure resistive circuit solution at *dc*, thereby enabling *dc*-to-daylight simulations. The better conditioning at low frequencies is based on including additional current unknowns. Such a limitation can be alleviated by exploiting the physical property of the divergence-free nature of current inside conductors and dielectrics. This allows adopting loop currents and removing potentials and branch currents inside the volumes, thus reducing the global number of unknowns [18].

The MNA equations (2.11) represents a general formulation of the circuit equations that can easily incorporate delays, leading to a system of delayed differential equations [19–21], as discussed in Section 2.11.1 and Chapter 6.

2.6 TIME DOMAIN ANALYSIS FORMULATIONS

In Section 2.4, we considered the frequency domain solver formulation. An important aspect of the solution approach is to directly solve the circuit equation in the time domain. As an introduction to this approach, we give the matrix stamp for a capacitor in the time domain. This is based on $i(t) = C_{12}\, d(v_1(t) - v_2(t))/dt$, and the so-called backward Euler (BE) integration method for the time derivative of the voltage that is given in Section 2.6.2. The corresponding capacitor stamp is given in Table 2.4, where Δt_p is the time step and p denotes the *p*resent time index, as it is explained in the following section. Enforcing the KVL to each circuit branch and the KCL to each independent node of the circuit illustrated in Fig. 2.1, results in the MNA [9] system of equations.

If we consider Table 2.2 and compare the elements for the time and the frequency domain, we observe that the multiplication by $s = j\omega$ can be replaced by the operator d/dt. Hence, the question is how we solve the circuit equations in the time domain. The basic approach is to subdivide the time axis into uniform or nonuniform time steps. As is commonly done, the time step at time t_p is called Δt_p, such that we can change the time step if necessary. We observe that, with uniform time steps, the MNA matrix will not change. Many time domain PEEC codes are based on uniform time steps due to the time saving that results.

Exception are solvers for nonlinear problems where a nonuniform time step is almost mandatory. The main issue is that the transitions during the fast changing nonlinearities usually require smaller time steps. The present time step is called Δt_p and the past steps are called Δt_{p-1}, Δt_{p-2}, etc.

As an example, the time derivative of a voltage v is approximated as

$$\frac{dv}{dt} = \dot{v} \simeq \frac{v_p - v_{p-1}}{t_p - t_{p-1}} = \frac{v_p - v_{p-1}}{\Delta t_p}. \tag{2.16}$$

TABLE 2.4 Matrix element stamp for capacitor in time domain.

Row/column	v_1	v_2	RHS
1	$C_{12}/\Delta t_p$	$-C_{12}/\Delta t_p$	$C_{12}/\Delta t_p\, v_1^{p-1} - C_{12}/\Delta t_p\, v_2^{p-1}$
2	$-C_{12}/\Delta t_p$	$C_{12}/\Delta t_p$	$C_{12}/\Delta t_p\, v_2^{p-1} - C_{12}/\Delta t_p\, v_1^{p-1}$

For a time domain PEEC-circuit solver, we need to consider the solution of the circuit DE formulations that are of the general form

$$\mathbf{f}\left[\mathbf{x}(t), \dot{\mathbf{x}}(t), \mathbf{u}(t)\right] = \mathbf{0}, \tag{2.17}$$

where $\mathbf{x}$ represents the vector of unknowns, which are in our formulations the nodal voltages and some branch currents. In general, the functions $\mathbf{f}$ could be nonlinear.

For general PEEC models, the formulation is somewhat more complicated. As we observe in Chapter 6, for a so-called *full-wave* solution some of the currents and voltages can be delayed of the form $i(t - \tau)$ or $v(t - \tau)$ where τ is the time delay.

As is considered in Section 2.11.1, this is the equivalent of the multiplication by $e^{-s\tau}$ in the complex frequency domain as indicated in the Section 2.5. Of course, the time delay is also related to distance by $\tau = R/v$, where R is the physical distance and v is again the propagation speed of the EM field in the material at hand. With this, we can take the time delay for the partial potential coefficients and the partial inductances into account. This leads to somewhat more complicated time domain form of (2.17), which is

$$\mathbf{f}\left[\mathbf{x}(t), \mathbf{x}(t-\tau), \dot{\mathbf{x}}(t), \dot{\mathbf{x}}(t-\tau), \mathbf{u}(t), \mathbf{u}(t-\tau)\right] = \mathbf{0} \tag{2.18}$$

which is called a delay differential equation (DDE) of the neutral type [19]. As we see in Chapter 6, the actual PEEC formulation leads to systems of this type with a multitude of delays.

Since we do consider only linear parts, for the PEEC model we can simplify the MNA equations as

$$\mathcal{C}\,\dot{\mathbf{x}}(t) + \mathcal{G}\mathbf{x}(t) = \mathcal{B}\,\mathbf{u}(t) \tag{2.19}$$

for the system without delays, which is a set of differential-algebraic equations (DAE).

The general case for the linear system for the case with delay can be written as an extensions of (2.19)

$$\mathcal{C}_0\,\dot{\mathbf{x}}(t) + \mathcal{G}_0\,\mathbf{x}(t) + \sum_i \mathcal{C}_i\,\dot{\mathbf{x}}(t-\tau_i) + \sum_i \mathcal{G}_i\,\mathbf{x}(t-\tau_i) = \sum_i \mathbf{u}_i(t-\tau_i). \tag{2.20}$$

Fortunately, we can use the same approach for setting up the circuit matrix in the time as well as the frequency domain. For the solution in the time domain, we need to numerically solve the resultant DE for general PEEC circuits.

2.6.1 Numerical Integration of Time Domain Differential Equations

In the time domain, we are required to solve the MNA equations in the form of DE, which can be represented in the general form of (2.17). In fact, they form DAE [6, 22]. Surprisingly, some time domain formulations used for time integral equation solutions do not take advantage of the results from the mathematical research. We should note that mathematicians have extensively treated the numerical integration of DE in detail, for example, Ref. [23].

Some circuit-oriented books present work on numerical integration techniques [1, 2, 5, 6]. Hence, we need to state that time integration is a well-developed area.

An implementation of a general circuit solution SPICE-type time domain solution is introduced in Ref. [24]. This work also includes the solution of nonlinear problems. This issue is also considered in most of the above-cited texts, where details are given in Refs [4, 6]. We are not further considering nonlinear problems in this text since, so far, we have not conducted extensive research in the area for PEEC circuits.

The time integration of the MNA equations for a full-wave PEEC circuit is even more challenging than for conventional circuits and quasistatic PEEC circuits. The difficulty is due to the delay between the circuit elements, which adds a new aspect to the conventional circuit theory. The delays have a major impact on the strategy for the time solution of these problems. First, the stability and the passivity in both the time and frequency domains are changed. These issues are considered in Chapter 13.

Also, the types of numerical integration methods that are suitable are different. First, it is best to use the so-called one-step methods rather than multistep methods such as the BD2 method considered below due to the rapid changes in the delayed signals. However, the so-called one-step methods such as the BE and equivalently BD1 are possible suitable methods [6].

Oscillatory solutions are very common for EM problems due to the large number of resonances in the responses. This issue, which needs to be considered for PEEC time domain solutions, requires the accurate representation of oscillatory waveforms. This is equivalent to the correct representations of resonance peaks in the frequency domain. However, the correct response reproduction without excessive damping is challenging. Importantly, different numerical integration methods exhibit strong or weak damping behaviors. To quantify this issue, we consider the following:

Definition 1 *A lossless resonance circuit stimulated with an initial condition leads to an oscillatory waveform which resonates at a frequency ω with a constant amplitude A for each frequency in the range of interest. A numerical solution must be valid in the solution interval up to the solve time interval t_{max}. We call the loss in amplitude of the numerical solution A_N of the oscillatory circuit $\Delta A = |A - A_N|$. This loss in amplitude, which is a function of the integration time step h, is called numerical damping of the integration method.*

Of course, if the amplitude A_N increases with time, the solution is unstable. Finally, the relative amplitude error due to numerical damping is given by

$$A_{error} = \frac{|A - A_N|}{A}. \tag{2.21}$$

Hence, it is important that the numerical damping error is sufficiently small especially for oscillatory waveforms.

For the time domain solution, we have discretized the time axis into small time segments Δt as is conventionally done. For a SPICE-type solver with nonlinear circuits, we have to take nonuniform time steps to deal with the nonlinearities. However, for linear PEEC-type circuits, we mostly use uniform time steps. The key issue is that the circuit matrix will not change between time steps, since the matrix stamps do not change if the time step is not changed. This leads to a fast solution since the same circuit matrix is used at each unchanged time step.

As mentioned above, one-step integration methods are preferable for the delayed circuit elements (2.20). However, the best method without the delayed elements is different from the case where we do have delays. For this reason, we have to consider both systems with

TABLE 2.5 Time integration methods of importance.

Integration method	k_p	k_{p-1}	k_{p-2}	α	β
Forward Euler	$1/\Delta t_p$	$-1/\Delta t_p$	0	0	1
BD1 (backward Euler)	$1/\Delta t_p$	$-1/\Delta t_p$	0	1	0
BD2 (Gear)	$1.5/\Delta t_p$	$-2/\Delta t_p$	$1/(2\Delta t_p)$	1	0
Trapezoidal	$1/\Delta t_p$	$-1/\Delta t_p$	0	0.5	0.5
Theta θ	$1/\Delta t_p$	$-1/\Delta t_p$	0	θ	$(1-\theta)$

and without delays. The methods that are used most for SPICE circuit solvers (nondelayed) are the BD2 and trapezoidal rule (TR) in Table 2.5. Dahlquist, a mathematician, was able to show that A-stable linear multistep methods are of order two [1]. This will be further considered in Chapter 13.

Basic delayed models are very different. It is impossible to guarantee stability due to the large number of poles where some of the very high-frequency poles are in the right half plane leading to eventual instability. PEEC systems with delay belong to the class of problems called *distributed*, which are more challenging than the *rational systems* that result without delays.

The stability issue is a very old one as is well known from integral equation formulations that originally used explicit integration methods [25]. Recently, most researchers switched to implicit integration methods that are much more stable, especially for larger time steps. A method that is used for PEEC models [26] is the so-called *theta* or Liniger–Willoughby method [27], where the derivative is represented as a combination of the present time step (p) and also at one time step back ($p-1$) with the time step Δt. Hence, this is still a one-step method. Of course, other one-step methods can be considered. Bellen suggested that Runge–Kutta methods such as the Lobatto IIIC [23] may be very suitable. However, this leads to a system with twice the number of unknowns.

2.6.2 List of Integration Methods for PEEC Solver

The choice of an appropriate integration method is important for any time domain solver. Fortunately, all the integration methods we consider here can be represented in the following general form for the case where uniform time steps are used. The time axis is represented by a time step Δt_p and the vector is used to represent t_i, where $i = 1, 2, 3, \ldots, P$, where P is the total number of time point given by the largest solution time. We can represent all the methods of interest in the general form

$$\alpha\, \dot{x}_p = k_p\, x_p + k_{p-1}\, x_{p-1} + k_{p-2}\, x_{p-2} - \beta\, \dot{x}_{p-1}. \tag{2.22}$$

The coefficients for each of the methods are given in Table 2.5.

The explicit forward Euler method is added to the table for completeness. Explicit solution requires very small time steps that are needed to guarantee the convergence. For many years, this type of explicit formulas has been used for time domain integral equations [25] that contributed to instabilities. Recently, more solvers started to use implicit numerical techniques, such as all the other approaches listed in Table 2.5.

We give two examples for the interpretation of the table. For example, the BD2 method is given by

$$1.5x_p - 2x_{p-1} + 0.5x_{p-2} = \Delta t\, \dot{x}_p, \tag{2.23}$$

while the θ method is given by

$$x_p = x_{p-1} + \Delta t_p[(1-\theta)\dot{x}_{p-1} + \theta \dot{x}_p], \tag{2.24}$$

where $\theta \in [0, 1]$ for all applications.

For nondelay circuits, the integration methods for PEEC are similar to the ones used in a SPICE circuit solver where BD1, BD2, and the trapezoidal method are among the dominant techniques. On the other hand, for the delayed full-wave PEEC models, a one-step method such as BD1 or the θ method is used. Of course, the θ method can represent three methods listed right above it. For $\theta = 0$, it reduces to the forward Euler method, while for $\theta = 1$, it is the backward Euler method.

Further for $\theta = 0.5$, it represents the TR. It was shown in Ref. [19] that the TR is not suitable for the problems at hand since it fails to have damping at very high frequencies. The method can drastically change its properties depending on the value of θ, which makes it a very flexible method. We show that for our applications, the important values are $\theta \in [2/3, 1]$. We should also note that the θ method is the only method besides the TR in this class that requires the storing of past values of the derivatives $\dot{x}_{p-1}$. With this, we can replace the derivative in the MNA circuit matrix with $\dot{x}_p$ for all the above equations.

We can see from Table 2.5 that all the methods besides the BD2 method are, in principle, suitable for the solution of DDEs since for one-step methods we have $k_{p-2} = 0$. More details on numerical integration methods is considered in Chapter 13.

We want to end this section with an example for a time domain stamp for a capacitor C_1 between nodes 1 and 2, which would be invoked with a SPICE statement of the form

```
C1  1  2  1pF.
```

We know the branch equation for the capacitor is

$$i_{C_1} = C_1 \frac{d(V_1 - V_2)}{dt}. \tag{2.25}$$

Also, if we want to set up a stamp that works for both the BD1 and BD2 methods according to Table 2.5, we can use

$$\frac{d(V_1 - V_2)}{dt} \simeq k_p \left(V_1^{(p)} - V_2^{(p)}\right) + k_{p-1}\left(V_1^{(p-1)} - V_2^{(p-1)}\right) + k_{p-2}\left(V_1^{(p-2)} - V_2^{(p-2)}\right). \tag{2.26}$$

With this, we have the formulation for a time domain stamp for a capacitor as given in Appendix B along with the stamps of other circuit elements.

We should compare the result in Table 2.6 to the simpler example for the BE one-step method given in Table 2.5.

TABLE 2.6 Matrix element stamp for capacitor in time domain.

Row/column	V_1	V_2	RHS
k	$k_p C_1$	$-k_p C_1$	$-C_1[k_{p-1}(V_1^{(p-1)} - V_2^{(p-1)}) + k_{p-2}(V_1^{(p-2)} - V_2^{(p-2)})]$
ℓ	$-k_p C_1$	$k_p C_1$	$-C_1[k_{p-1}(V_2^{(p-1)} - V_1^{(p-1)}) + k_{p-2}(V_2^{(p-2)} - V_1^{(p-2)})]$

2.6.3 Initial Conditions for Time Solver with Delays

More details are given in this section for time domain delay equations (2.20). We assume that the system to be solved has at least parts that are described by DDE. If the derivatives also have delays, they are called neutral DDEs or NDDEs. One of the problems considered here are the initial conditions for all the unknown variables in the solution vector **x**.

The circuit solution usually starts at $t = 0$. However, there are elements to evaluate with a time delay. Assume that such an element P has a delay τ_M and that we need to evaluate $P\,x(t - \tau_M)$. If τ_M is the largest delay in the system, it is reasonable to initialize all variables starting at $-\tau_M$. Usually, we initialize all variable from $t = -\tau_M$ to $t_0 = 0$. Hence, we have to provide initial waveform data for negative times such that first time step that the solver needs to know the value of the solution vector $\mathbf{x}(t_0 - \tau_1)$ as the initial conditions.

The task is accomplished by initializing all waveforms from $-\tau_M$ to $t_0 = 0$ with the initial values $\mathbf{x}(t = 0)$. Since the end of solution time is usually much larger than the largest delay τ_M, the additional storage required for the initial values is usually small. Of course, if the initial values are all zero, then this will be an array of zeros. Assuming a uniform time step of size Δt, then the number of negative time steps $N_{Neg} \geq \tau_M/\Delta t$.

Special techniques must be applied if the initial conditions of some elements of the initial voltage are not zero. Much compute time can be wasted if some input *dc* nonzero voltage sources are involved and if we then let the solution find the correct initial values for all elements of the solution vector **x**. An efficient solution to this problem can be found if we eliminate all the inductances in the PEEC model to avoid oscillatory solutions. Then, the remaining circuit can be integrated with a very large time step Δt such that the appropriate initial voltages in **x** are found.

2.7 GENERAL MODIFIED NODAL ANALYSIS (MNA)

As is evident in Chapter 6, the circuit unknowns for a simple PEEC model are the potential $\mathbf{\Phi}$ and inductive currents $\mathbf{I}_\ell$. Of course, potentials are voltages with respect to the ground node that represents infinity from an EM point of view. If charges **q** or magnetic flux variables $\mathbf{\Psi}$ are required for some other circuit elements, they can easily be included as additional unknowns (e.g., see [1, 5, 8, 9]). We next show how the MNA approach can be implemented rather easily in a general and robust way. This is certainly part of the reason that the majority of SPICE solvers use the MNA approach. Once we are familiar with the techniques, we can set up circuit matrices by inspection.

We next show an example to illustrate that the MNA approach leads to a relatively easy way to set up the time domain circuit matrix for the example circuit in Fig. 2.3.

An example for the SPICE description for this circuit would be

```
Ii   1   Gnd   pulse (0mA 10mA 0ns 0.1ns 0.1ns 0.2ns 10ns)
C1   1   Gnd   1pF
R    2     3   11Ohms
C2   3   Gnd   2pF
L    1     2   1nH
```

We observe that each of the statements represents a command for invoking an element stamp for a particular circuit element. The order in which the SPICE elements are listed can be used to set up the list of unknowns. Hence, let us assume the unknowns be listed as

V_1, V_2, V_3, I_L. By using this order of the unknowns, using the appropriate element stamps, we get the system

$$\begin{bmatrix} k_p\,C_1 & 0 & 0 & 1 \\ 0 & 1/R & -1/R & -1 \\ 0 & -1/R & 1/R + k_p\,C_2 & 0 \\ 1 & -1 & 0 & -k_pL \end{bmatrix} \begin{bmatrix} V_1 \\ V_2 \\ V_3 \\ I_L \end{bmatrix} = \begin{bmatrix} -C_1\,k_{p-1}\,V_1^{(p-1)} - C_1\,k_{p-2}\,V_1^{(p-2)} + I_i \\ 0 \\ -C_2\,k_{p-1}\,V_2^{(p-1)} - C_2\,k_{p-2}\,V_2^{(p-2)} \\ +L\,k_{p-1}\,I_L^{(p-1)} - L\,k_{p-2}\,I_L^{(p-2)} \end{bmatrix} \tag{2.27}$$

for both BD1 and BD2 integration methods. We observe two issues related to the circuits matrix. First, if we choose a resistor value of 0, then some entries go to infinity; this leads to problems in the solution. The solution accuracy may be impaired even if the resistance value is extremely small. However, we can use the current stamp for a resistor to reduce the system size and, at the same time, to solve the zero resistance problem. This leads to the following reduced system

$$\begin{bmatrix} k_p\,C_1 & 0 & 1 \\ 0 & k_p\,C_2 & 0 \\ 1 & -1 & -R - k_pL \end{bmatrix} \begin{bmatrix} V_1 \\ V_3 \\ I_L \end{bmatrix} = \begin{bmatrix} -C_1\,k_{p-1}\,V_1^{(p-1)} - C_1\,k_{p-2}\,V_1^{(p-2)} + I_i \\ -C_2\,k_{p-1}\,V_3^{(p-1)} - C_2\,k_{p-2}\,V_3^{(p-2)} \\ +L\,k_{p-1}\,I_L^{(p-1)} - L\,k_{p-2}\,I_L^{(p-2)} \end{bmatrix}. \tag{2.28}$$

We apply this kind of node reduction while setting up the PEEC circuit matrix since we have many situations with a partial inductance in series with a resistor. Hence, a reduction of the nodes and, importantly, unknowns is accomplished.

2.7.1 Matrix Kirchhoff's Current Law and Stamps

Instead of writing the KCL for each node, we can set up a KCL matrix $\mathbf{A}$ where each row represents a different node. Hence, the column vector $\mathbf{a}_n$ represents each node in the circuit, or

$$\mathbf{a}_n = \begin{bmatrix} n_1 & n_2 & n_3 & \cdots & n_n \end{bmatrix}^T, \tag{2.29}$$

where n_n is the number of nodes. To write the KCL, we have to acknowledge all the branches that are connected to a particular node n_k. This is done with a row vector where all the branches b are listed.

$$\mathbf{a}_b = \begin{bmatrix} b_1 & b_2 & b_3 & \cdots & b_b \end{bmatrix}, \tag{2.30}$$

where b_b is the number of branches. The matrix entries acknowledge the existence or absence of a connection of a branch to a node. If a branch is connected to a node and if it is assumed to leave the node, we acknowledge this with a $+1$ in the matrix. A branch that is incident to a node is taken as negative with a -1. If a branch is not connected to the node, then this results in a 0 in the KCL matrix $\mathbf{A}$.

To illustrate the case, we take the simple circuit in Fig. 2.1 and we replace it with a so-called directed graph G. Usually, we can assign to each branch an arbitrary direction, but for comparison purposes we give in Fig. 2.5 the branches attached to node 1 the same direction as in Fig. 2.1. Note that we do not follow the names of the resistors for the branch

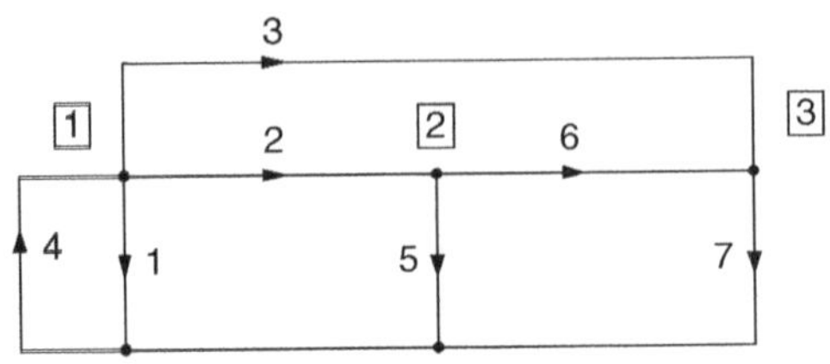

FIGURE 2.5 Directed graph G for the circuit in Fig. 2.1.

labels. Next, we want to write down the matrix KCL $\mathbf{A}$ for this circuit. From node 1, we find that

$$I_1 + I_2 + I_3 - I_4 = 0. \tag{2.31}$$

We assign the node voltage vector and the branch current vector in a simple way

$$\mathbf{v}_n = [1\ \ 2\ \ 3]^T; \qquad \mathbf{i}_b = \begin{bmatrix} I_1 & I_2 & I_3 & I_4 & I_5 & I_6 & I_7 \end{bmatrix}. \tag{2.32}$$

The branch vector $\mathbf{i}_b$ corresponds to the current directions assigned to them. To set up the matrix KCL equations for this case is quite simply given by

$$\mathbf{A} = \begin{bmatrix} 1 & 1 & 1 & -1 & 0 & 0 & 0 \\ 0 & -1 & 0 & 0 & 1 & 1 & 0 \\ 0 & 0 & -1 & 0 & 0 & -1 & 1 \end{bmatrix}. \tag{2.33}$$

It is clear from this that, for example, the first row of $\mathbf{A}$ multiplied by the branch current vector $\mathbf{i}_b$ represents the KCL for the first node. Hence, it is evident that

$$\mathbf{A}\mathbf{i}_b = \mathbf{0} \tag{2.34}$$

is the *matrix KCL*. We see in the following that we also want to consider an incomplete KCL for matrix stamping $\mathbf{A}$ for our applications.

2.7.2 Matrix Kirchhoff's Voltage Law

It is clear from the above derivation that another relation is of importance. In (2.2), we gave an example for the local KVL relationship between the branch and the nodal voltages. Also, in Fig. 2.1, the KVL loop is shown for this example to completely specify the relationships between all the branch voltages $\mathbf{v}_b$ for each element and the nodal voltages $\mathbf{v}_n$ for the graph or circuit. Hence, we would like to have another matrix relationship between these two voltage vectors. Surprisingly, at first this relation is

$$\mathbf{v}_b = \mathbf{A}^T\ \mathbf{v}_n, \tag{2.35}$$

where $\mathbf{A}$ is the same matrix as in the matrix KCL. So, we have to show that this is true. As we can see, this relation considers the node voltages that are connected to the branches. Hence, if a branch is connected to a node, it will contribute +1 or a −1 entry to the matrix $\mathbf{A}^T$. By inspection, we find the following relation directly from the circuit in Fig. 2.3

$$v_{b1} = v_{n1} \tag{2.36a}$$

$$v_{b2} = v_{n1} - v_{n2} \tag{2.36b}$$

$$v_{b3} = v_{n1} - v_{n3} \tag{2.36c}$$

$$v_{b4} = -v_{n1} \tag{2.36d}$$

$$v_{b5} = v_{n2} \tag{2.36e}$$

$$v_{b6} = v_{n2} - v_{n3} \tag{2.36f}$$

$$v_{b7} = v_{n3}. \tag{2.36g}$$

We now can observe why the $\mathbf{A}^T$ matrix works for the branch voltage to nodal voltage relationship. A branch results in a $+1$ entry in $\mathbf{A}^T$ if the branch is connected and the voltage drop is in the direction of the current for $\mathbf{A}$. Also, a branch voltage is directed with the minus sign at a node, then the voltage entry is a -1 in $\mathbf{A}^T$, which is the same as a current being directed toward a node in $\mathbf{A}$ that is acknowledged with a -1. Finally, if a branch is not connected to a node, a zero results in both cases. Hence, we understand why the $\mathbf{A}$ matrix is related to $\mathbf{A}^T$.

We also want to observe that an MNA stamp does, for each branch, automatically use the specific entries into $\mathbf{A}$ for the current equations and $\mathbf{A}^T$ for the nodal connections to the circuit. So, it takes over the functions of both since it implicitly uses the KCL and KVL.

2.7.3 Matrix KCL Solution of MNA Equations for PEEC

We consider a circuit that is relevant for the PEEC model in most chapters, which is shown in Fig. 2.6. This also serves as an example for the application of the matrix KCL for the circuits of interest to us in this text. We can treat the currents for the capacitance as $\mathbf{I}_c$, for the internal capacitive current sources $\mathbf{I}_s$, the inductances as $\mathbf{I}_\ell$, and finally the input current sources as $\mathbf{I}_i$. Note that an independent current source could be attached to each node in Fig. 2.6. However, possible independent current sources are not shown in this figure. We should note that inspecting the stamps given in Appendix B for resistors can either be included with the inductance currents as is the case in (2.13) or stamped similar to the capacitor admittances stamps using G.

The KCL matrix that is split up into submatrices corresponding to the different components, which is

$$\mathbf{A}_c\mathbf{I}_c + \mathbf{A}_s\mathbf{I}_s + \mathbf{A}_\ell\mathbf{I}_\ell + \mathbf{A}_i\mathbf{I}_i = \mathbf{0}, \tag{2.37}$$

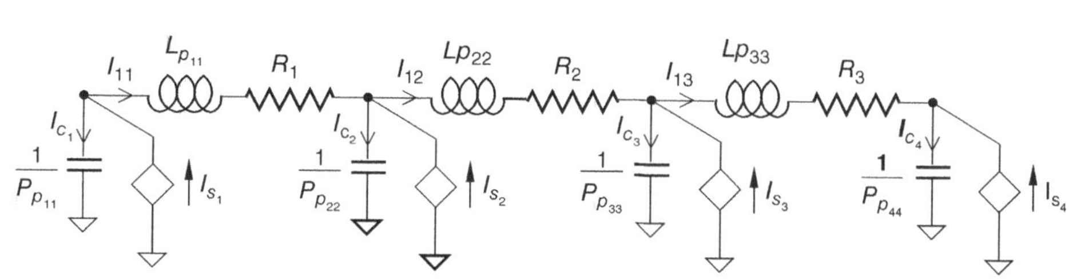

FIGURE 2.6 Example PEEC model for three fundamental loops, which includes the capacitive part.

where $\mathbf{A}_c$, $\mathbf{A}_s$, $\mathbf{A}_\ell$, and $\mathbf{A}_i$ are the KCL connectivity matrices corresponding to capacitive, controlled current sources, inductive, and input currents.

So, in a sense, we want to stamp all elements of the same type simultaneously to make the process more efficient. Of course, this works best for structured circuits such as a PEEC model. It is evident how we subdivide the matrix current law according to the element types shown in Fig. 2.6. This results in all the elements of the same type that can be stamped into the matrix together. We still can add additional elements to the circuit matrix such as load resistors, etc., by stamping them at an element-by-element basis.

The total capacitive current for each capacitive source current for each node is given by $I_T = I_c - I_s$. The details of the derivation have been assigned to Problem 2.3.

Since all capacitor nodes are connected to ground, $\mathbf{A}_c$ is an identity matrix and $\mathbf{A}_s$ is minus an identity matrix. Hence, what we call the total capacitive current is

$$\mathbf{I}_T = \mathbf{I}_c - \mathbf{I}_s \tag{2.38}$$

by using (2.37). As shown in Section 6.3.4, the PEEC current controlled current source (CCCS) models are given by a matrix of the type

$$\begin{bmatrix} I_{s1} \\ I_{s2} \\ I_{s3} \\ I_{s4} \end{bmatrix} = \begin{bmatrix} 0 & Pp_{12}/Pp_{11} & Pp_{13}/Pp_{11} & Pp_{14}/Pp_{11} \\ Pp_{21}/Pp_{22} & 0 & Pp_{23}/Pp_{22} & Pp_{24}/Pp_{22} \\ Pp_{31}/Pp_{33} & Pp_{32}/Pp_{33} & 0 & Pp_{34}/Pp_{33} \\ Pp_{41}/Pp_{44} & Pp_{42}/Pp_{44} & Pp_{43}/Pp_{44} & 0 \end{bmatrix} \begin{bmatrix} I_{T1} \\ I_{T2} \\ I_{T3} \\ I_{T4} \end{bmatrix} \tag{2.39}$$

which we write as $\mathbf{I}_s = \mathbf{M}\,\mathbf{I}_T$ and where the total capacitive current is $\mathbf{I}_T = \mathbf{I}_c - \mathbf{I}_s$. We borrow the following equation (6.46) from Chapter 6 as

$$\mathbf{Pd}^{-1}\frac{d\mathbf{\Phi}}{dt} = \left(\hat{\mathbf{I}} + \mathbf{M}\right)\mathbf{I}_T(t). \tag{2.40}$$

The branch relation translated to the frequency domain for the capacitors is $\mathbf{I}_c = \mathbf{Pd}^{-1} s\,\mathbf{\Phi}_n$ where $\mathbf{Pd}$ is a diagonal matrix with Pp_{mm} on the diagonal or $\mathbf{Pd} = \text{diag}(Pp_{11}, Pp_{12}, \ldots,$ $Pp_{nn})$ where $\mathbf{Pd}^{-1}$ is the diagonal matrix with the diagonal elements $1/Pp_{ii}$ for the ith row. Hence, the total capacitance currents are from (2.40)

$$\left(\hat{\mathbf{I}} + \mathbf{M}\right)\mathbf{I}_T = s\mathbf{Pd}^{-1}\mathbf{\Phi}. \tag{2.41}$$

From the matrix KCL equations (2.37) and (2.38), we obtain by premultiplying it with $\left(\hat{\mathbf{I}} + \mathbf{M}\right)$

$$s\mathbf{Pd}^{-1}\mathbf{\Phi} + \left(\hat{\mathbf{I}} + \mathbf{M}\right)\mathbf{A}_\ell \mathbf{I}_\ell = -\left(\hat{\mathbf{I}} + \mathbf{M}\right)\mathbf{A}_i \mathbf{I}_i. \tag{2.42}$$

This leads to the first row in the MNA circuit matrix where the unknowns are the node potentials $\boldsymbol{\phi} = \mathbf{\Phi}_n$ and the inductor currents $\mathbf{I}_\ell$.

Fortunately, the second row in the MNA matrix is much simpler to obtain where the inductive-resistive branch is given by

$$\mathbf{V}_\ell = (s\mathbf{Lp} + \mathbf{R})\,\mathbf{I}_\ell. \tag{2.43}$$

Here, **Lp** is the matrix of coupled partial inductances and **R** is a diagonal matrix of the resistances. Finally, the branch to nodal voltages are related to the potential as

$$\mathbf{V}_\ell = \mathbf{A}_\ell^T \mathbf{\Phi}_n, \tag{2.44}$$

which forms the second row in the MNA matrix by replacing $\mathbf{V}_\ell$ in the last two equations. This finally leads to the formulation for the matrix as

$$\begin{bmatrix} s\mathbf{Pd}^{-1} & (\hat{\mathbf{I}}+\mathbf{M})\mathbf{A}_\ell \\ \mathbf{A}_\ell^T & -(s\mathbf{Lp}+\mathbf{R}) \end{bmatrix} \begin{bmatrix} \mathbf{\Phi}_n \\ \mathbf{I}_\ell \end{bmatrix} = \begin{bmatrix} -(\hat{\mathbf{I}}+\mathbf{M})\mathbf{A}_i\mathbf{I}_i \\ \mathbf{0} \end{bmatrix}. \tag{2.45}$$

Hence, we can see that using the matrix KCL, we can stamp multiple elements of the same type in a matrix sense. Of course, this is only the case for given topologies such as the PEEC circuit considered here. Note that in a PEEC circuit with capacitors the *ground* node or the node potentials $\Phi_n = 0$ is referred to the node at ∞.

2.7.4 Matrix KCL for Conductor Example

To show how the matrix KCL will be applied to a PEEC-type geometry, we consider the example in Fig. 2.7. The lower conductor 1 is subdivided into three nodes along the z (upward) direction. Hence, the current vector, $\mathbf{I}_{1,1} = [I_{1,1}, I_{2,1}, \ldots, I_{20,1}]$, $\mathbf{V}_{1,1} = [V_1, V_2, \ldots, V_{12}]$. The second conductor 2 has the unknown currents $\mathbf{I}_{2,2} = [I_{1,2}, I_{2,2}, I_{3,2}, I_{4,2}]$, $\mathbf{V}_{1,2} = [V_1, V_2, V_3, V_4]$. Note that we labeled the branches according to the label for the impedances, which we assume to consist of resistances in series to partial inductances. Also, the positive direction for the impedances is always assumed to be in the x-front, y-right, and z-up directions.

Conductor 2 is assumed to be very thin such that only a single layer of nodes is assigned.

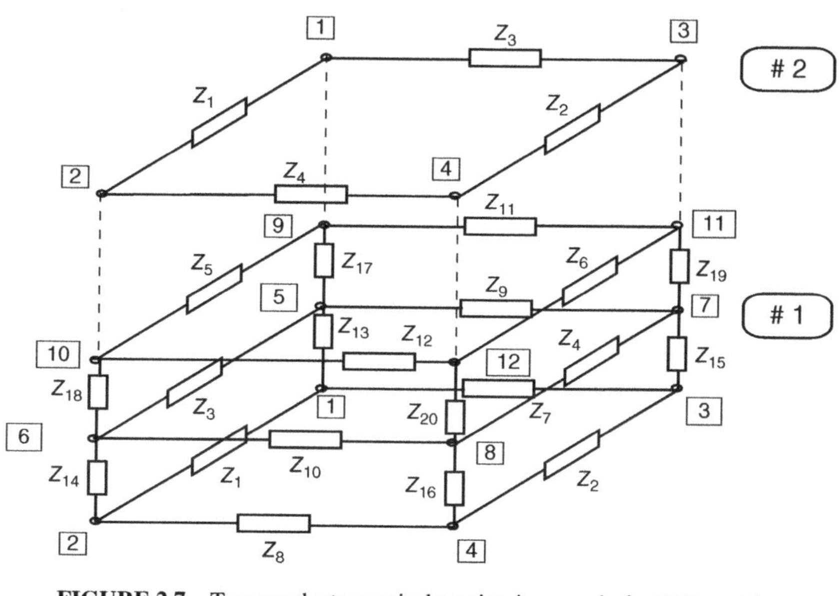

FIGURE 2.7 Two-conductor equivalent circuit example for KCL matrix.

We give as an example the KCL matrix for the lower part of the model that includes 20 currents. The lower part #1 is represented by 12 nodes. Hence, the 12×20 matrix is

$$
\mathbf{A}_\ell = \begin{bmatrix}
+1 & 0 & 0 & 0 & 0 & 0 & +1 & 0 & 0 & 0 & 0 & 0 & +1 & 0 & 0 & 0 & 0 & 0 & 0 & 0 \\
-1 & 0 & 0 & 0 & 0 & 0 & 0 & +1 & 0 & 0 & 0 & 0 & 0 & +1 & 0 & 0 & 0 & 0 & 0 & 0 \\
0 & +1 & 0 & 0 & 0 & 0 & -1 & 0 & 0 & 0 & 0 & 0 & 0 & 0 & +1 & 0 & 0 & 0 & 0 & 0 \\
0 & -1 & 0 & 0 & 0 & 0 & 0 & -1 & 0 & 0 & 0 & 0 & 0 & 0 & 0 & +1 & 0 & 0 & 0 & 0 \\
0 & 0 & +1 & 0 & 0 & 0 & 0 & 0 & +1 & 0 & 0 & 0 & -1 & 0 & 0 & 0 & +1 & 0 & 0 & 0 \\
0 & 0 & -1 & 0 & 0 & 0 & 0 & 0 & 0 & +1 & 0 & 0 & 0 & -1 & 0 & 0 & 0 & +1 & 0 & 0 \\
0 & 0 & 0 & +1 & 0 & 0 & 0 & 0 & -1 & 0 & 0 & 0 & 0 & 0 & -1 & 0 & 0 & 0 & +1 & 0 \\
0 & 0 & 0 & -1 & 0 & 0 & 0 & 0 & 0 & -1 & 0 & 0 & 0 & 0 & 0 & -1 & 0 & 0 & 0 & +1 \\
0 & 0 & 0 & 0 & +1 & 0 & 0 & 0 & 0 & 0 & +1 & 0 & 0 & 0 & 0 & 0 & -1 & 0 & 0 & 0 \\
0 & 0 & 0 & 0 & -1 & 0 & 0 & 0 & 0 & 0 & 0 & +1 & 0 & 0 & 0 & 0 & 0 & -1 & 0 & 0 \\
0 & 0 & 0 & 0 & 0 & +1 & 0 & 0 & 0 & 0 & -1 & 0 & 0 & 0 & 0 & 0 & 0 & 0 & -1 & 0 \\
0 & 0 & 0 & 0 & 0 & -1 & 0 & 0 & 0 & 0 & 0 & -1 & 0 & 0 & 0 & 0 & 0 & 0 & 0 & -1
\end{bmatrix} \tag{2.46}
$$

Of course, this is only one part of the derivation of the elements needed for the general MNA matrix in (2.45). However, for a problem with inductances and resistances, the formulation is given by

$$
\begin{bmatrix} 0 & \mathbf{A}_\ell \\ \mathbf{A}_\ell^T & -(s\mathbf{Lp} + \mathbf{R}) \end{bmatrix} \begin{bmatrix} \mathbf{\Phi}_n \\ \mathbf{I}_\ell \end{bmatrix} = \begin{bmatrix} -\mathbf{A}_i \mathbf{I}_i \\ \mathbf{0} \end{bmatrix}. \tag{2.47}
$$

as is evident from (2.45). In PEEC circuits without capacitances, we need to choose a common reference ground node in each independent subpart, which is otherwise not connected. Also, in this example that has two independent parts, we can choose one *common ground reference node* from the lower and upper subcircuits. Of course, for the reference node, we do not need to write or eliminate the potential or voltage for the node. Also, we do not need to write a KCL equation for the reference nodes, else the MNA matrix becomes singular.

Since voltages are potential differences, the choice of a node will only result in a potential shift while the voltages or potential differences are the same. It is clear that in the model without the capacitance the potential to infinity is not connected.

2.8 INCLUDING FREQUENCY DEPENDENT MODELS IN TIME DOMAIN SOLUTION

Important for any circuit solver, it is desirable to keep the frequency- and time-dependent representations as compatible as possible. As part of this, we would like to include frequency-dependent models in the time domain. At first, this seems to be a contradiction to include such elements. However today, the inclusion is very common [28, 29]. A general representation of macromodeling solutions is given in Ref. [30]. Loss models can be included in PEEC models using frequency-dependent elements.

It is clear that we can obtain more information if the results in both domains are available. If conventional linear equivalent circuit models are available, they are immediately suitable for both domains. No other work is required for this case. However, this is not always feasible. In this section, we consider the case where frequency-dependent models are available, which we would like to convert into models that are suitable for both time and frequency domain solvers.

Some conditions must be satisfied for a frequency domain model to be suitable for a time domain representation. For a linear time domain solver, the voltages and currents have an

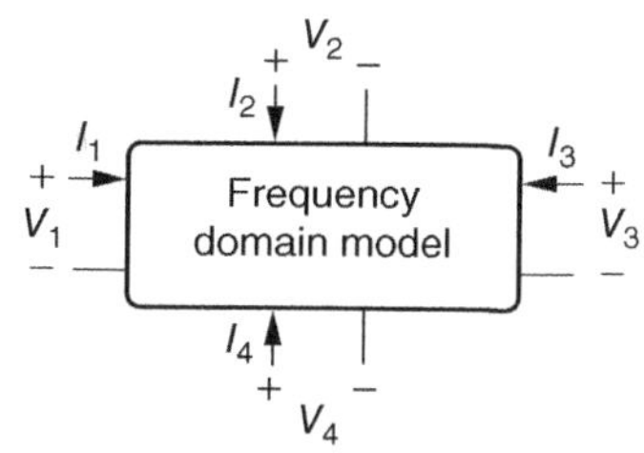

FIGURE 2.8 Example of frequency domain macromodel with four ports.

upper frequency content given by the fastest relevant input signal to the time domain solver. Hence, we assume that the frequency domain description of a model is sufficiently accurate up to the highest f_{Max} frequency required for the time domain model. We need to note that the same is true for the lowest frequency f_{Min} represented in the model. Especially for very long time domain input signals as they occur in channels, models must be provided down to very low frequencies. For example, if the time domain solve time $T_{solve} = 100\,\text{ns}$, then the lowest frequency in the data should be at least $f_{Min} \leq \alpha/T_{solve}$ where $\alpha = 0.1 \cdots 0.01$. This avoids a shift in the quasi *dc* level of the solution. Of course, ideal data models would also include a *dc* solution.

We first need to enter our frequency domain data into the circuit solver in the form made available from a model generator. An important aspect of the tables is the fact that a sufficient number of discrete frequency points must be available for the full frequency range of importance. Then, the final goal will be to come up with the appropriate MNA stamps so that we can include the models in our circuit solver.

An important question is how much of the initial description will be included in the available frequency domain model. One form of the data is to represent it in the form of impedances $\mathbf{Z}(s)$, admittance $\mathbf{Y}(s)$, or scattering parameter $\mathbf{S}(s)$. Also, the data may be the impedance of a single port or a scalar or, alternatively, a matrix for a multiport situation.

A partial description of the model is in terms of ports as illustrated in Fig. 2.8. In this example, only four port currents are available as well as four voltages or potential differences are specified at the ports. It is clear that in this case, only a fraction of the internal variables of the systems are available at the ports.

We know that for the complete description of all states of a circuit, we need to have the MNA equations that involves both the nodal voltages and additional currents. Hence, the models we consider here lead to a partial description of the system. We can assume that the models described represent the relevant aspects of the models at hand. However, aspects like a complete *dc* solutions may be missing. These issues are considered in Chapter 13.

We can assume that the other circuit descriptions are used. The data may be available in tables or as analytical expressions. It is important to include a sufficient number of frequency data points in a table format. The data can be given in a single or in multiple tables. This may require considerable storage especially for a large number of ports, etc. Here, we consider input/output representation, such as the impedance $\mathbf{Z}$, admittance $\mathbf{Y}$, and the scattering matrix form $\mathbf{S}$. The admittance formulation $\mathbf{I} = \mathbf{YV}$ is in detail given by

$$\begin{bmatrix} I_1(s) \\ I_2(s) \\ \vdots \\ I_N(s) \end{bmatrix} = \begin{bmatrix} y_{11}(s) & y_{12}(s) & \cdots & y_{1N}(s) \\ y_{21}(s) & y_{22}(s) & \cdots & y_{2N}(s) \\ \vdots & \vdots & \ddots & \vdots \\ y_{N1}(s) & y_{N2}(s) & \cdots & y_{NN}(s) \end{bmatrix} \begin{bmatrix} V_1(s) \\ V_2(s) \\ \vdots \\ V_N(s) \end{bmatrix} \tag{2.48}$$

for which a four-port example system is shown in Fig. 2.8. For this example, $\mathbf{Y}$ is a 4×4 matrix.

The impedance formulation in the form $\mathbf{V} = \mathbf{ZI}$ similarly is given by

$$\begin{bmatrix} V_1(s) \\ V_2(s) \\ \vdots \\ V_N(s) \end{bmatrix} = \begin{bmatrix} z_{11}(s) & z_{12}(s) & \cdots & z_{1N}(s) \\ z_{21}(s) & z_{22}(s) & \cdots & z_{2N}(s) \\ \vdots & \vdots & \ddots & \vdots \\ z_{N1}(s) & z_{N2}(s) & \cdots & z_{NN}(s) \end{bmatrix} \begin{bmatrix} I_1(s) \\ I_2(s) \\ \vdots \\ I_N(s) \end{bmatrix}, \tag{2.49}$$

where N is the number of ports.

Very often, the data is given in terms of scattering matrices. Hence, we need to be able to include them as well. We may have to input scattering matrix data into our solver, or in this case, we also would like to output scattering matrix data. One approach is based on converting $\mathbf{Y}$ or $\mathbf{Z}$ matrices into a scattering parameter form. We follow the notation in Ref. [31].

The data as scattering parameters are given as

$$\begin{bmatrix} V_1^-(s) \\ V_2^-(s) \\ \vdots \\ V_N^-(s) \end{bmatrix} = \begin{bmatrix} S_{11}(s) & S_{12}(s) & \cdots & S_{1N}(s) \\ S_{21}(s) & S_{22}(s) & \cdots & S_{2N}(s) \\ \vdots & \vdots & \ddots & \vdots \\ S_{N1}(s) & S_{N2}(s) & \cdots & S_{NN}(s) \end{bmatrix} \begin{bmatrix} V_1^+(s) \\ V_2^+(s) \\ \vdots \\ V_N^+(s) \end{bmatrix}. \tag{2.50}$$

In this formulation [31], the vector V_k^+ represents the voltage incident on port k and V_k^- is the reflected voltage on port $1 \leq k \leq N$.

The scattering matrix techniques are strongly related to the transmission and reflection concepts for transmission lines [31, 32]. It should be noted that an alternative notation is $\mathbf{V}^+ = \mathbf{a}$ and $\mathbf{V}^- = \mathbf{b}$. However, we think that the notation of [31] is more descriptive and we adhere to this notation. Similar to the transmission lines we have

$$\begin{aligned} \mathbf{V} &= \mathbf{V}^+ + \mathbf{V}^- \\ \mathbf{Z}_r\mathbf{I} &= \mathbf{V}^+ - \mathbf{V}^-, \end{aligned} \tag{2.51}$$

where the so-called *reference impedance* is given by $\mathbf{Z}_r = \mathrm{diag}[Z_{r1}, Z_{r2}, \ldots, Z_{rN}]$. Usually all reference impedances $\mathbf{Z}_r$ are 50 Ω.

Starting from a data set of measured or simulated S-parameters, the following relation can be established [31]:

$$\mathbf{Y} = \left\{ \left[\frac{Z_0^*}{\sqrt{R_0}} \right] + \mathbf{S} \left[\frac{Z_0}{\sqrt{R_0}} \right] \right\}^{-1} \{\hat{\mathbf{I}} - \mathbf{S}\} \left[\frac{1}{\sqrt{R_0}} \right], \tag{2.52}$$

where $\hat{\mathbf{I}}$ is again the identity matrix and where Z_0, Z_0^*, and R_0 are, respectively, the complex reference impedance, its conjugate part, and its real part. Hence, scattering data can always be converted into impedance or admittance data.

The data can also be specified as a model in an analytical form. For example, a scalar impedance can be represented as

$$Z(s) = \frac{B(s)}{A(s)} = \frac{b_{m+1}\, s^{m+1} + b_m\, s^m + \cdots + b_1\, s + b_0}{a_m\, s^m + a_{m-1} s^{m-1} + \cdots + a_1\, s + 1}. \tag{2.53}$$

This formulation could also be used to specify each entry of any of the above matrices. We can consider any one of the above forms. If the degree of the numerator is higher by one than the denominator, we can rewrite this as

$$Z(s) = R + sL + \frac{\gamma_{m-1}\, s^{m-1} + \gamma_{m-2}\,(s)^{m-2} + \cdots + \gamma_1\,(s) + \gamma_0}{\alpha_m\,(s)^m + \alpha_{m-1}\,(s)^{m-1} + \cdots + \alpha_1\,(s) + 1} \tag{2.54}$$

by division. In general, we would like to convert the frequency domain models in such a way that their contribution can be included into the time domain MNA matrix. Some models may already be available as frequency-independent circuits. Of course, they can directly be implemented in a circuit solver.

2.9 INCLUDING FREQUENCY DOMAIN MODELS IN CIRCUIT SOLUTION

We consider two different approaches to achieve the transformation from the matrix forms in (2.48), (2.49), or (2.50) to an approach that can be used in a SPICE-type circuit solver. In this section, we present an approach that aims at achieving a circuit representation, which can easily be used in an MNA matrix-based circuit solver [33].

The first part in a two-step approach is a pole-zero representation with both real and complex residue–pole pairs. As we see, this is accomplished in terms of branch admittances. In the second step, the branch admittances are represented with elements that can directly be used in a circuit representation for which we apply the conventional stamping approach to set up the circuit matrix in both domains.

2.9.1 Equivalent Circuit for Rational Approximation of Transfer Functions

In this section, we want to show how we can represent the **Y** matrix in terms of equivalent circuits. The approach was presented in Ref. [33]. In order to keep the example simple, we consider a four-port example as shown in Fig. 2.8. The admittance matrix (2.48) reduces for this case to

$$\mathbf{Y_4} = \begin{bmatrix} y_{11} & y_{12} & y_{13} & y_{14} \\ y_{21} & y_{22} & y_{23} & y_{24} \\ y_{31} & y_{32} & y_{33} & y_{34} \\ y_{41} & y_{42} & y_{43} & y_{44} \end{bmatrix}. \tag{2.55}$$

We start this section by considering the way an admittance matrix is stamped into the MNA matrix in the frequency domain. The stamp used is of the same form as the one for a conductance in (B.2), and we can interpret this matrix in terms of this approach. We specifically consider this issue in Appendix B, where the elements are given by the conductance stamp (B.2). In exactly the same way, we can relate the conductance by an admittance in the

TABLE 2.7 Matrix element stamp for admittance *Y*.

Row/column	V_k	V_ℓ	RHS
k	$Y_{k\ell}$	$-Y_{k\ell}$	0
ℓ	$-Y_{k\ell}$	$Y_{k\ell}$	0

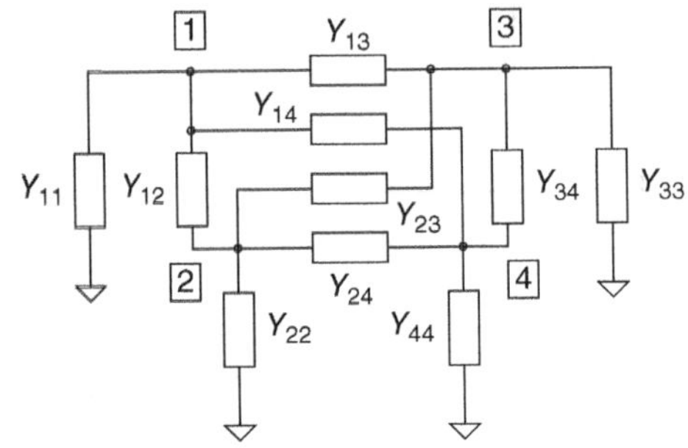

FIGURE 2.9 Equivalent circuit in terms of admittances for a four-terminal synthesis model.

frequency domain to get the matrix element stamp (Table 2.7), where the equivalent circuit is shown in Fig. 2.9. It is easy to verify that the stamping process leads to the diagonal elements

$$y_{kk} = \sum_{\ell=1}^{N} Y_{k\ell} = Y_{k1} + Y_{k2} + Y_{k3} + Y_{k4} \tag{2.56}$$

and the off-diagonal elements are given by the other stamp entry (2.57):

$$y_{k\ell} = -Y_{k\ell}. \tag{2.57}$$

We can conclude from this that we can make a circuit model for the **Y** matrix provided that we can model the individual branch impedances $Y_{k\ell}$. Hence, we know that if we have an expression for $y_{k\ell}$ for all entries, we also know $Y_{k\ell}$.

Fortunately, the vector fitting (VF) procedure [34–36] has been devised that can be used to convert the **Y** matrix into a set of rational approximations for each element $y_{k\ell}$, which are of the form

$$y_{k\ell}(s) = \sum_{m=1}^{N_p} \frac{r_{m,k\ell}}{(s - p_m)} + d_{k\ell} + s\ e_{k\ell}, \tag{2.58}$$

where $r_{m,k\ell}$ is a residue for a coefficient and p_m is a pole for a particular coefficient computed by VF. Also, $d_{k\ell}$ and $e_{k\ell}$ are results of VF. Finally, N_p is the number of poles for the approximation that can be chosen by the user. Passivity of the rational approximation can be enforced as outlined in Ref. [35]. Fortunately, it is well known that each term in (2.58) is associated with a lumped equivalent circuit [33, 37, 38].

The next and final step is to come up with an equivalent circuit that can then easily be entered into a circuit solver. Hence, the last step in this approach, given in this section, is the conversion of (2.58) into circuits.

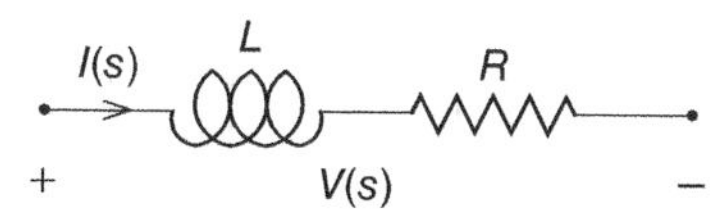

FIGURE 2.10 Equivalent RL circuit for real pole synthesis.

To start the circuit representation, we consider the series connection of a resistor R with an inductance L as shown in Fig. 2.10.

The admittance corresponding to this circuit is

$$y(s) = \frac{I(s)}{V(s)} = \frac{1}{R + sL} = \frac{\frac{1}{L}}{s + \frac{R}{L}}. \tag{2.59}$$

For this circuit with one real pole, we can by comparison find the circuit elements as

$$L = \frac{1}{r_{RL}}, \quad R = -p_{RL}L = -\frac{p_{RL}}{r_{RL}}, \tag{2.60}$$

where r_{RL} is the residue and p_{RL} is the pole of the circuit in Fig. 2.10.

Next, we consider the lossy LRC series circuit in Fig. 2.11 that leads to a complex pole/residue pair.

It is straightforward to compute the admittance for this circuit as

$$y(s) = \frac{1}{L} \frac{s + \frac{1}{R_2C}}{s^2 + \left(\frac{R_1}{L} + \frac{1}{R_2C}\right)s + \left(\frac{R_1}{L}\frac{1}{R_2C} + \frac{1}{LC}\right)}. \tag{2.61}$$

In this case, the two poles and two terms in (2.61) are involved in the representation. Fortunately, we can find an equivalent circuit for (2.61) that matches this behavior where the circuit elements are given by

$$L = \frac{1}{r_1 + r_2} \tag{2.62a}$$

$$R_1 = \frac{1}{r_1 + r_2}\left[-(p_1 + p_2) + \frac{1}{r_1 + r_2}(r_1p_2 + r_2p_1)\right] \tag{2.62b}$$

$$C = \frac{r_1 + r_2}{p_1p_2 + \left[-(p_1 + p_2) + \frac{1}{r_1+r_2}(r_1p_2 + r_2p_1)\right]\frac{(r_1p_2+r_2p_1)}{(r_1+r_2)}} \tag{2.62c}$$

$$R_2 = -\frac{1}{C}\frac{r_1 + r_2}{r_1\,p_2 + r_2\,p_1}, \tag{2.62d}$$

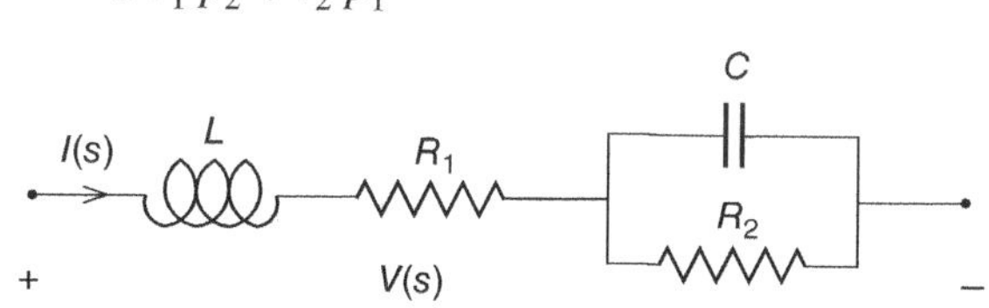

FIGURE 2.11 First equivalent circuit for complex pole/residue pair synthesis.

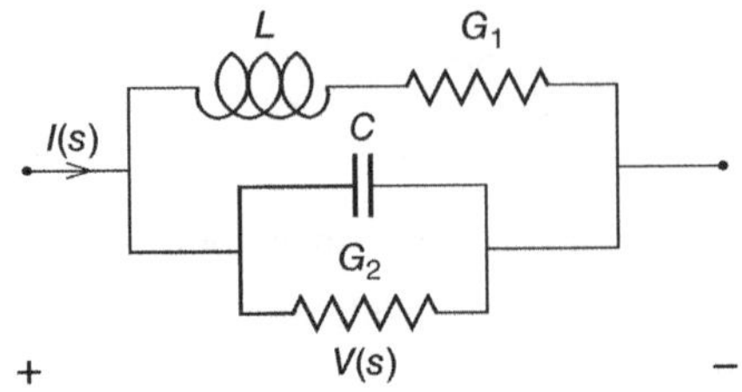

FIGURE 2.12 Second equivalent circuit for complex pole/residue pair synthesis.

where r_1, r_2 denote the complex conjugate residues and p_1, p_2 are the complex conjugate poles. For the same complex conjugate pole/residue pair, another dual equivalent circuit can be synthesized, as shown in Fig. 2.12.

The corresponding impedance is

$$z(s) = \frac{1}{C}\frac{s + \frac{1}{G_1 L}}{s^2 + \left(\frac{G_2}{C} + \frac{1}{G_1 L}\right)s + \left(\frac{G_2}{C}\frac{1}{G_1 L} + \frac{1}{LC}\right)}. \tag{2.63}$$

The circuit parameters can be evaluated by duality with respect to the admittance equivalent circuit previously described, providing

$$C = \frac{1}{r_1 + r_2} \tag{2.64a}$$

$$G_1 = \frac{1}{r_1 + r_2}\left[-(p_1 + p_2) + \frac{1}{r_1 + r_2}(r_1 p_2 + r_2 p_1)\right] \tag{2.64b}$$

$$L = \frac{r_1 + r_2}{p_1 p_2 + \left[-(p_1 + p_2) + \frac{1}{r_1+r_2}(r_1 p_2 + r_2 p_1)\right]\frac{(r_1 p_2 + r_2 p_1)}{(r_1 + r_2)}} \tag{2.64c}$$

$$G_1 = -\frac{1}{C}\frac{r_1 + r_2}{r_1 p_2 + r_2 p_1}. \tag{64d}$$

In summary, in this two-step approach, we first use the VF [34–36] technique to come up with a rational approximation for the original input system. Then, it is shown that the resultant elements can be approximated with equivalent circuits. As the last step, we can enter the equivalent circuit as input to a SPICE-type circuit solver. Again, these models can be used in both the time and the frequency domain.

2.9.2 Inclusion of Frequency Domain Models in a Time Domain Circuit Solver

Very similar to the approach in Section 2.9.1, where we started from some data in tables, we also can use different techniques such as VF to come up with a rational model. At this point, the approach differs from the equivalent circuit model in the last section by directly stamping the contribution into the MNA matrix. Hence, we show how the stamps are constructed.

We start with a simple impedance model. The small circuit that we would like to include in an MNA matrix is shown in Fig. 2.13. To be clear, this small circuit could also be included

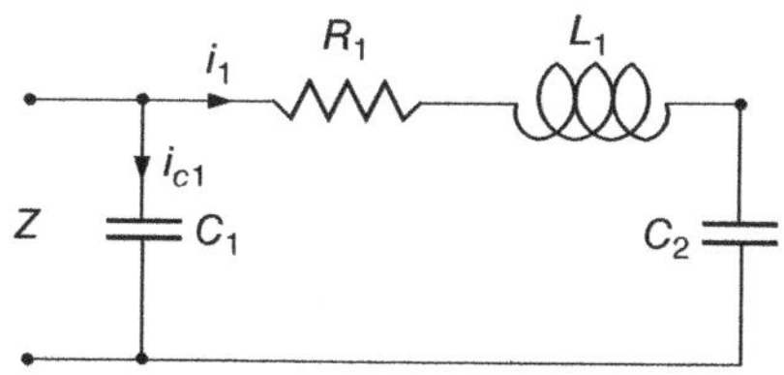

FIGURE 2.13 Example for impedance circuit.

TABLE 2.8 Matrix element stamp for single pole.

Row/column	$x_m^{(p)}$	$i^{(p)}$	RHS
m	$k_p - p_m$	$-r_m$	$k_{p-1}\, x_m^{(p-1)} + k_{p-2}\, x_m^{(p-2)}$
aux (v)	1	0	0

with the conventional MNA solution approach. However, we want to take the rational function in the frequency domain form

$$Z(s) = \frac{s^2 + s\frac{R_1}{L_1} + \frac{1}{L_1\,C_2}}{sC_1\left[s^2 + s\frac{R_1}{L_1} + \frac{1}{L_1}\left(\frac{C_1+C_2}{C_1C_2}\right)\right]} \tag{2.65}$$

into account. Importantly, the circuit representation in Fig. 2.13 is not necessary and it only serves as an example. The residue/pole representation for a general circuit can be obtained from VF in the above rational form

$$Z(s) = d + \sum_{m=1}^{M_r} \frac{r_m}{s - p_m} + \sum_{m=1}^{M_c} \left[\frac{r_{m,r} + j\, r_{m,i}}{s - p_{m,r} - j\, p_{m,i}} + \frac{r_{m,r} - j\, r_{m,i}}{s - p_{m,r} + j\, p_{m,i}}\right], \tag{2.66}$$

where M_r are the real poles and M_c are the complex pole pairs. The constant d can be interpreted as a series resistance, while we can write the equivalent differential equation for the real poles and residues, or $dx_m/dt - p_m x_m = r_m i(t)$ where the current $i(t)$ is the same for all the terms in series. We note that the state variable x_m is equivalent to a voltage. Hence, we can augment the conventional MNA circuit matrix by using a stamp that corresponds to the differential equation with the BD2 discretization (see Table 2.5). This is reported in Table 2.8 for this specific case, where $i^{(p)} = i(t_p)$. Note that we use (p) to indicate the present time step. Complex poles are best treated as pole pairs with the poles $p_{m,r} \pm jp_{m,i}$ and the residues $r_{m,r} \pm jr_{m,i}$ leading to a differential equation [39, 40]. Here, the solution is derived by assuming a complex pair of solutions $x_{m,r} \pm jx_{m,i}$ leading to the stamp (Table 2.9) where $\text{RHS}_1 = +k_{p-1}\, x_{m,r}^{(p-1)} + k_{p-2}\, x_{m,r}^{(p-2)}$ and $\text{RHS}_2 = +k_{p-1}\, x_{m,i}^{(p-1)} + k_{p-2}\, x_{m,i}^{(p-2)}$.

Finally, the stamps connect the impedance part of the MNA to the state matrix equations as shown in Fig. 2.14. As is the case for an inductance stamp, the current through all the elements and the voltage rows are shared. Hence, it is clear that the circuit matrix shares a single column for the current i and a single row for all the voltages aux(v).

TABLE 2.9 Matrix element stamp for a complex pole pair.

Row/column	$x_{m,r}^{(p)}$	$x_{m,i}^{(p)}$	$i^{(p)}$	RHS
$m1$	$k_p - p_{m,r}$	$p_{m,i}$	$-r_{m,r}$	RHS_1
$m2$	$-p_{m,i}$	$k_p - p_{m,r}$	$-r_{m,i}$	RHS_2
aux (v)	2	0	0	0

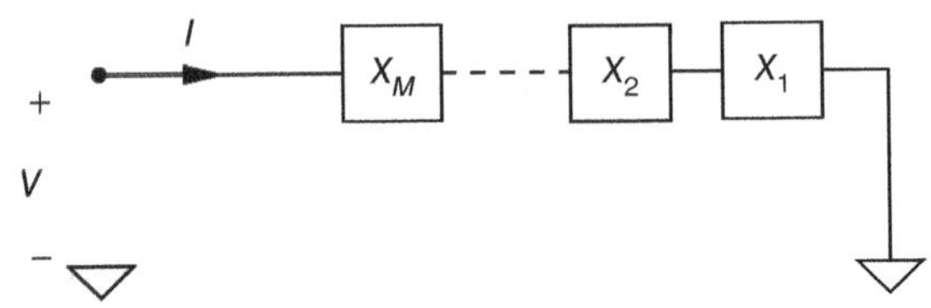

FIGURE 2.14 Macromodel for impedance circuit.

2.9.3 General Inclusion of Frequency Domain Admittance Models

As the last step, we want to show how a stamp model can be constructed for an admittance input matrix. Hence, we start out with a model or macromodel in an admittance residue–pole form given in Section 2.9.1. An expression for the matrix elements is in the general form

$$y_{k\ell} = g_{k\ell} + \sum_{m=1}^{M_r} \frac{r_{k\ell}^m}{s - p_m} + \sum_{m=1}^{M_c} \left[\frac{r_{k\ell}^{(m,r)} + j\, r_{k\ell}^{(m,i)}}{s - p_{m,r} - j\, p_{m,i}} + \frac{r_{k\ell}^{(m,r)} - j\, r_{k\ell}^{(m,i)}}{s - p_{m,r} + j\, p_{m,i}} \right], \tag{2.67}$$

where the same set of poles is used for all admittances. For use in an MNA matrix, the next step is to convert the pole–residue models into the time domain matrix stamps for the contribution of admittance models into the time domain circuit equation. We rewrite the Y-matrix system using (2.67) in the residue/pole form

$$\begin{bmatrix} I_1 \\ I_2 \\ I_3 \end{bmatrix} = \begin{bmatrix} g_{11} & g_{12} & g_{13} \\ g_{21} & g_{22} & g_{23} \\ g_{31} & g_{23} & g_{33} \end{bmatrix} \begin{bmatrix} V_1 \\ V_2 \\ V_3 \end{bmatrix} + \frac{1}{s - p_1} \begin{bmatrix} r_{11}^1 & r_{12}^1 & r_{13}^1 \\ r_{21}^1 & r_{22}^1 & r_{23}^1 \\ r_{31}^1 & r_{23}^1 & r_{33}^1 \end{bmatrix} \begin{bmatrix} V_1 \\ V_2 \\ V_3 \end{bmatrix} + \frac{1}{s - p_2} \begin{bmatrix} r_{11}^2 & r_{12}^2 & r_{13}^2 \\ r_{21}^2 & r_{22}^2 & r_{23}^2 \\ r_{31}^2 & r_{32}^2 & r_{33}^2 \end{bmatrix} \begin{bmatrix} V_1 \\ V_2 \\ V_3 \end{bmatrix} + \cdots, \tag{2.68}$$

TABLE 2.10 Matrix stamp for two ports of circuit in Fig. 2.8.

Row/column	$I_{m,1}^{(p)}$	$I_{m,2}^{(p)}$	$v_1^{(p)}$	$v_2^{(p)}$	RHS
v_1	$k_p - p_m$	0	$-r_{12}^m$	$-r_{13}^m$	RHS_1
v_2	0	$k_p - p_m$	$-r_{21}^m$	$-r_{22}^m$	RHS_2
I_1	1	0	0	0	0
I_2	0	1	0	0	0

where we only considered the real poles for space reasons. The first term can be interpreted as a resistive circuit while the mth pole can be written in matrix form as

$$\mathbf{I}_m = \frac{1}{s - p_m} \mathbf{R}_m \mathbf{V}_\ell \tag{2.69}$$

with the contributions to the currents as

$$\begin{bmatrix} I_{k,1} \\ I_{k,2} \\ I_{k,3} \end{bmatrix} = \mathbf{G}\,\mathbf{V}_\ell + \begin{bmatrix} I_{1,1} \\ I_{1,2} \\ I_{1,3} \end{bmatrix} + \cdots + \begin{bmatrix} I_{m,1} \\ I_{m,2} \\ I_{m,3} \end{bmatrix} + \cdots . \tag{2.70}$$

Each of the individual currents in (2.70) must be entered into the MNA matrix of the time domain solver. Hence, it is clear that poles can be included by the addition of residue/pole stamps.

The next step is to represent the currents in the time domain. This is accomplished from (2.69) by writing it as

$$\frac{d\mathbf{I}_m}{dt} - p_m \mathbf{I}_m = \mathbf{R}_m \mathbf{V}_\ell \tag{2.71}$$

and by using a numerical integration method such as BD2 (2.22). The combined currents are of the form

$$(k_p - p_m)\mathbf{I}_m^{(p)} - \mathbf{R}_m \mathbf{V}_\ell^{(p)} = k_{p-1}\mathbf{I}_m^{(p-1)} + k_{p-2}\mathbf{I}_m^{(p-2)}. \tag{2.72}$$

Finally, we are ready to consider the matrix stamps so that we can easily add the frequency domain macromodel to an MNA circuit matrix. From (2.72), we recognize that the stamp has to involve all the voltages corresponding to the port voltages for the interconnect model. However, we only include two of the four ports in the matrix stamp from the example in Fig. 2.8. Each pole m is added to circuit matrix using the stamp given in Table 2.10 where $\text{RHS}_1 = k_{p-1} I_{m,1}^{(p-1)} + k_{p-2} I_{m,1}^{(p-2)}$, $\text{RHS}_2 = k_{p-1} I_{m,2}^{(p-1)} + k_{p-2} I_{m,2}^{(p-2)}$. From (2.72) and (2.70), it is evident that adding more poles to the model is rather simple.

2.9.4 State-Space and Descriptor Representations

If poles and residues of a system are known, it can be useful to obtain its *state-space* representation, meaning a system of first-order ordinary differential equations (ODEs). This approach is known as *realization* [41]. Then, the equations can be incorporated in a circuit solver or solved numerically.

In general, a system of ODEs can be written as

$$\frac{d}{dt}\mathbf{x}(t) = \mathbf{A}\mathbf{x}(t) + \mathbf{B}\mathbf{u}(t) \tag{2.73a}$$

$$\mathbf{y}(t) = \mathbf{C}\mathbf{x}(t) + \mathbf{D}\mathbf{u}(t), \tag{2.73b}$$

where $\mathbf{A} \in \mathrm{R}^{p\times p}$ is the state-space matrix, $\mathbf{B} \in \mathrm{R}^{p\times n}$ is the selection matrix mapping the inputs onto the ODEs, $\mathbf{C} \in \mathrm{R}^{n\times p}$ is the matrix that relates the state variables and the outputs $\mathbf{y}$, $\mathbf{D} \in \mathrm{R}^{n\times n}$ is the matrix relating the inputs and the outputs, $\mathbf{x}$ is the vector of the state variables of size p, and $\mathbf{u}$ is the vector containing the inputs of size n (where n represents the number of ports).

For the sake of simplicity, let us assume we want to find the realization of an admittance 2×2 matrix and a two-pole rational model has already been identified where

$$\mathbf{Y} = \begin{bmatrix} \mathbf{Y}_{11} & \mathbf{Y}_{12} \\ \mathbf{Y}_{21} & \mathbf{Y}_{22} \end{bmatrix} = \begin{bmatrix} \sum_{i=1}^{q} \frac{K_i^{1,1}}{s-p_i} & \sum_{i=1}^{q} \frac{K_i^{1,2}}{s-p_i} \\ \sum_{i=1}^{q} \frac{K_i^{2,1}}{s-p_i} & \sum_{i=1}^{q} \frac{K_i^{2,2}}{s-p_i} \end{bmatrix}. \tag{2.74}$$

In the following, the Jordan canonical realization [41] is adopted. It requires four state variables and the system can be represented as

$$\frac{d}{dt}\begin{bmatrix} x_1 \\ x_2 \\ x_3 \\ x_4 \end{bmatrix} = \begin{bmatrix} p_1 & 0 & 0 & 0 \\ 0 & p_2 & 0 & 0 \\ 0 & 0 & p_3 & 0 \\ 0 & 0 & 0 & p_4 \end{bmatrix} \begin{bmatrix} x_1 \\ x_2 \\ x_3 \\ x_4 \end{bmatrix} + \begin{bmatrix} 1 & 0 \\ 0 & 1 \\ 1 & 0 \\ 0 & 1 \end{bmatrix} \begin{bmatrix} u_1 \\ u_2 \end{bmatrix} \tag{2.75a}$$

$$\mathbf{y} = \begin{bmatrix} i_1 \\ i_2 \end{bmatrix} = \begin{bmatrix} k_1^{1,1} & k_1^{1,2} & k_2^{1,1} & k_2^{1,1} \\ k_1^{2,1} & k_1^{2,2} & k_2^{2,1} & k_2^{2,2} \end{bmatrix} \begin{bmatrix} x_1 \\ x_2 \\ x_3 \\ x_4 \end{bmatrix} + \begin{bmatrix} 0 & 0 \\ 0 & 0 \end{bmatrix} \begin{bmatrix} u_1 \\ u_2 \end{bmatrix}. \tag{2,75b}$$

Conjugate complex pairs of poles need to be treated differently in the time domain. If two poles form a complex conjugate pair $\lambda_{1,2} = \omega \pm jz$, such that $C_{i,j} = (r \pm jv)_{i,j}$, then the corresponding residues for each port in state-space form becomes

$$\frac{d}{dt}\begin{bmatrix} x_1 \\ x_2 \\ x_3 \\ x_4 \end{bmatrix} = \begin{bmatrix} \omega & 0 & z & 0 \\ 0 & \omega & 0 & z \\ -z & 0 & \omega & 0 \\ 0 & -z & 0 & \omega \end{bmatrix} \begin{bmatrix} x_1 \\ x_2 \\ x_3 \\ x_4 \end{bmatrix} + \begin{bmatrix} 2 & 0 \\ 0 & 2 \\ 0 & 0 \\ 0 & 0 \end{bmatrix} \begin{bmatrix} u_1 \\ u_2 \end{bmatrix} \tag{2.76a}$$

$$\mathbf{y} = \begin{bmatrix} i_1 \\ i_2 \end{bmatrix} = \begin{bmatrix} r_{1,1} & r_{1,2} & v_{1,1} & v_{1,2} \\ r_{2,1} & r_{2,2} & v_{2,1} & v_{2,2} \end{bmatrix} \begin{bmatrix} x_1 \\ x_2 \\ x_3 \\ x_4 \end{bmatrix}. \tag{2.76b}$$

These equations are completed by the terminal conditions. They can include equivalent linear circuit synthesis, or can be described with additional linear/nonlinear equations.

The *descriptor* form of the ODE equation is important that is similar to the state form and reads

$$\mathbf{E}\,\frac{d}{dt}\mathbf{x}(t) = \mathbf{A}\mathbf{x}(t) + \mathbf{B}\,\mathbf{u}(t) \tag{2.77}$$

$$\mathbf{y}(t) = \mathbf{C}\mathbf{x}(t) + \mathbf{D}\,\mathbf{u}(t).$$

Often, the MNA equations result in the so-called DAE [6] in descriptor form. For these mixed circuits, **E** is singular in the MNA equations such that it is difficult to convert the descriptor system (2.77) into a state-space system (2.73) by inverting **E**. Issues such as this have made the descriptor form important. In this text, we employ a notation that is conventionally used for circuit equation. An example represents the MNA equation in (13.10). This notation is also more compatible with the one used for the EM part of the book.

2.10 RECURSIVE CONVOLUTION SOLUTION

2.10.1 Conventional Convolution

Convolution is another approach for the inclusion of frequency-dependent elements in the time domain. It is important to point out that this section is the same for conventional convolution and recursive convolution. The time domain response corresponding to the transfer function $H(s)$ is the output response $h(t)$ to a stimulus with a so-called impulse $\delta(t)$ at its input [32, 40, 42]. Hence, the key application of this approach is suited for series-connected input–output connected blocks.

A clean derivation of the recursive convolution approach is lengthy as will be apparent from our derivation that is based on [42]. We start with a conventional convolution

$$x(t) = \int_0^t h(t-\tau)y(\tau)d\tau, \tag{2.78}$$

where $y(t)$ is the input signal, $h(t)$ is the impulse response of the system, and $x(t)$ is the output response. We notice that in this approach we need to know the input signal $y(t)$ in time. The Laplace transform of $h(t)$ is denoted as $H(s)$ and we assume that it is a rational function of s.

First, and we assume that it is a rational derivation that all time steps Δt are the same. Then, the discrete form of the integral is

$$x(t_p) = \Delta t \sum_{k=0}^{K_p} h(t_p - t_k)\, y(t_k), \tag{2.79}$$

where t_p means the *present* time and K_p is the number of time steps Δt up to t_p.

The transfer function could be generated with any one of the above considered approaches in the form of (2.58). Here, we only assume noncomplex poles to keep the derivation simpler where the transfer function $H(s)$ can be written as

$$H(s) = d + \sum_{m=1}^{M_r} \frac{r_m}{s - p_m}, \tag{2.80}$$

where M_r is the number or residue–pole pairs. The inverse Laplace transform for $H(s)$ is

$$h(t) = d\delta(t) + \sum_{m=1}^{M_r} r_m \, e^{p_m t} u(t), \tag{2.81}$$

where $\delta(t)$ is the delta function and $u(t)$ is a unit step at $t = 0$. If we insert (2.81) into the discrete convolution (2.79), we get

$$x(t_p) = \Delta t \sum_{k=0}^{K_p} \left[d\ \delta(t_p - t_k) + \sum_{m=1}^{M_r} r_m \, e^{p_m(t_p - t_k)} \right] y(t_k). \tag{2.82}$$

It is clear that we can find the output $x(t_p)$ for the transfer function (2.82) in response to the known input function $y(t)$. The time-consuming issue of this formulation is the double summation that has to be executed at each new present time t_p.

2.10.2 Recursive Convolution

It is clear that the evaluation of (2.82) at each time step t_p is computationally expensive. The speedup technique for convolution by the *recursive* approach is due to [43].

The task of the recursive convolution is to reduce the number of operations performed. We start by rewriting (2.82) in two parts as

$$x(t_p) = \Delta t\ d \sum_{k=0}^{K_p} \delta(t_p - t_k)\, y(t_k) + \Delta t \sum_{k=0}^{K_p} \sum_{m=1}^{M_r} r_m \, e^{p_m(t_p - t_k)}\, y(t_k). \tag{2.83}$$

Looking at the first term, we find that the only time $\delta(t_p - t_k)$ is nonzero for $t_p = t_k$, which simplifies the first term to

$$\Delta t \sum_{k=0}^{K_p} d\ \delta(t_p - t_k)\, y(t_k) = \Delta t\ d\ y(t_p). \tag{2.84}$$

The second term of (2.83) can also be simplified by using

$$e^{p_m(t_p - t_k)} = e^{p_m(t_p - t_{p-1} + t_{p-1} - t_k)} = e^{p_m \Delta t} e^{p_m(t_{p-1} - t_k)}. \tag{2.85}$$

With (2.85), the second part is

$$\Delta t \sum_{k=0}^{K_p} \sum_{m=1}^{M_r} r_m \, e^{p_m(t_p - t_k)}\, y(t_k) = \Delta t \sum_{m=1}^{M_r} r_m \, e^{p_m \Delta t} \sum_{k=0}^{K_p} e^{p_m(t_{p-1} - t_k)}\, y(t_k). \tag{2.86}$$

Using the results from (2.84), (2.83), and (2.86) result in

$$x(t_p) = \Delta t\ d\ y(t_p) + \Delta t \sum_{m=1}^{M_r} r_m\ e^{p_m \Delta t} \sum_{k=0}^{K_p} e^{p_m(t_{p-1}-t_k)} y(t_k), \tag{2.87}$$

since $t_p - t_{p-1} = \Delta t$.

The computationally expensive part is due to the double summation. Next, we subdivide (2.87) into parts that are used to update the present value and the part that only depends on past values

$$w(t_{p-1}) := \sum_{k=0}^{K_p} e^{p_m(t_{p-1}-t_k)} y(t_k). \tag{2.88}$$

We clearly can update (2.88) recursively by rewriting it

$$w(t_{p-1}) = e^{p_m \Delta t} \sum_{k=0}^{K_p} e^{p_m(t_{p-2}-t_k)}\ y(t_k) = e^{p_m \Delta t}\ w(t_{p-2}), \tag{2.89}$$

again by using (2.85) for the time step t_{p-1} where $w(t_{k-2})$ is evident from (2.88) and where we recursively compute $w(t_0), w(t_1), \ldots, w(t_{k-2})$.

Using this approach, we can compute the solution as

$$x(t_p) = \Delta t\ y(t_p)\ d + \Delta t \sum_{m=1}^{M_r} r_m\ e^{p_m \Delta t} w(t_{p-1}), \tag{2.90}$$

where we save computing time with the recursive update.

2.11 CIRCUIT MODELS WITH DELAYS OR RETARDATION

Time-delayed signals are not common in the conventional circuit theory. However, delay or retardation plays an ever-increasing role in the circuit theory. An exception is conventional transmission lines that were first introduced by Branin [44]. Sources of the form of $v_2(t) = v_1(t - \tau)$ are delaying a signal by the time τ. The equivalence in the frequency domain is given by (2.93). The theory of delays in transmission lines is extensively covered by Paul [32]. However, three different regimes exist for PEEC models that are *dc*, *quasistatic*, and *full-wave*. Delay is of importance only for the full-wave regime.

The delay becomes important if the size of the object or the distance between two objects becomes large. The time domain criterion for the onset of the need for a full-wave solution with delay can simply be specified if the size of a problem D or the distance between objects, which we also call D for simplicity is

$$D = c\ T_r, \quad c = \frac{c_0}{\sqrt{\mu_r \epsilon_r}}, \tag{2.91}$$

where c_0 is the speed of light in the vacuum, c is the velocity in the material at hand, μ_r is the relative permeability, and ϵ_r is the relative permittivity. T_r is the smallest rise time of the signals in the system.

Similarly, in the frequency domain, the problem size D has to be a fraction of the smallest wavelength λ_{min}. Hence, we can estimate that for the full-wave regime we need

$$D < \frac{\lambda_{min}}{k}, \quad \lambda_{min} = \frac{c}{f_{max}}, \tag{2.92}$$

where f_{max} is the largest frequency in the system and k a pertinent factor ranging from 10 to 100. Of course, in any system there may be subsystems where we do not need to include delays or retardation.

2.11.1 Inclusion of Delays in the Circuit Domain

It is clear from the previous section that a full-wave model includes delays or retardation in the PEEC circuit. Delay is simple in the time domain in which a current $f(t) = i(t)$ or voltage $f(t) = v(t)$ is delayed by including $f(t - \tau)$. For PEEC models, the delay is $\tau = R/c$, where R is the physical distance and c is the velocity in the material according to (2.91).

In the frequency domain, if a current or voltage $F(\omega)$ is delayed by a delay τ, then the Laplace transform of $f(t - \tau)$ is given by $e^{-s\tau}$. Hence, in the general case, we just multiply the delayed elements by the $e^{-s\tau}$ in the MNA matrix. Of course, we can also write the delay term as $\tau = R/c$. We also want to introduce another notation, $\beta = \omega/c$, such that

$$e^{-s\tau} = e^{-j\beta R} \tag{2.93}$$

with $\beta = \omega/c$, is also a widely used form. In the EM text, the notation used is $\beta = k$, while in the circuit domain and the transmission line work using β is more common. In the following, we use β instead of k.

Next, we use a small inductance example to illustrate the impact of relatively small delays on the impedance. In Fig. 5.16, a bar is shown that we subdivide into three subsections or partial inductances. Also, in (5.32) the open-loop inductance of the bar can be computed as

$$L_o = 3\, Lp_{11} + 4\, Lp_{12} + 2\, Lp_{13}, \tag{2.94}$$

since the three sections are assumed to have the same length.

A physical delay between the coupled segments results in a delayed mutual coupling reaction for the segments. To take this into account, we introduce the delayed currents for the coupling elements only where τ is the delay approximated as due to the conductor center distances. This results in

$$L_o = 3\, Lp_{11} + 4\, Lp_{12}\, e^{-s\,\tau_{12}} + 2\, Lp_{13}\, e^{-s\,\tau_{13}}, \tag{2.95}$$

where $\tau_{12} < \tau_{13}$ and where τ_{13} is the delay between conductors 1 and 3, etc. Since all three conductor segments are physically close to each other, we can expand the exponentials in a Taylor series. Considering one of the couplings, we get

$$Lp_{12}\, e^{-s\,\tau_{12}} = Lp_{12}\,(1 - s\,\tau_{12} + \cdots). \tag{2.96}$$

If we compare the inductive coupling inductance $j\omega\, Lp_{12}$ with the delayed part, we find since $s = j\omega$ that

$$j\omega\, Lp_{12}\, e^{-s\,\tau_{12}} = j\omega\, Lp_{12} + \omega^2\, \tau_{12}\, Lp_{12}. \tag{2.97}$$

We can interpret the term $\omega^2\, \tau_{12}\, Lp_{12}$ as frequency-dependent damping resistor introduced by the retardation. It is interesting to observe that the retardation results in damping for physically close conductors for sufficiently high angular frequencies ω. We conclude from this that full-wave models considered in Section 6.3.6 exhibit more damping in the resonance behavior of models.

We hope that this clearly shows that delays play an important part of the circuit theory for EM circuits. It is apparent that the so-called full-wave PEEC models with delay result in other new challenging issues.

PROBLEMS

2.1 *Resistance circuit*

Verify that you understand that equation (2.6) corresponds to the resistor circuit in Fig. 2.1 using the matrix stamp in Table 2.1. Write a small Matlab program to find the resistance between the current source terminal given that all resistors are 1 kΩ. Hint: If you apply a current of 1 mA, the voltage will be proportional to the voltage at node 1. Please use all microelectronics-type units in Table A.2 in Appendix A for all circuit elements and variables.

2.2 *Time domain solution*

For the circuit in Fig. 2.3 implement the MNA equations in a small Matlab problem with the data given at the beginning of section 2.7 by the SPICE input language as shown above the equation. Choose different time steps Δt_p to find the largest time step that results in a good solution. Note that the input current pulse has an amplitude of 10 mA with a rise time and a fall time of 0.1 ns. The width of the pulse roof is 0.2 ns. Hence, it is a trapezoidal pulse shape.

2.3 *Capacitance model*

Derive in detail that the MNA equation (2.45) that is based on the special capacitance current relations (2.39) and (2.38). Start from the relation for the potential (4.10), or $\mathbf{\Phi}_n = \mathbf{Pp\, Q}$ for a two-conductor capacitance problem and take the time derivative on both sides of the equation since $i = dQ/dt$. Also, verify the capacitance equivalent circuit in Fig. 2.6 is correct.

2.4 *Equivalent circuit synthesis*

Synthesize the equivalent circuit of the open-ended input impedance of a transmission line whose residues and poles are listed in Table 2.11.

TABLE 2.11 Poles and residues of the open-ended input impedance of a transmission line.

Residues [$\Omega\,s^{-1}$]	Poles [s^{-1}]
9.09e+10	−4.54e+07
9.09e+10 ± 5.20e+08i	−2.38e+07 ± 3.77e+09i
9.09e+10 ± 2.60e+08i	−2.38e+07 ± 7.54e+09i
9.09e+10 ± 1.73e+08i	−2.38e+07 ± 1.13e+10i
9.09e+10 ± 1.30e+08i	−2.38e+07 ± 1.50e+10i

2.5 *State-space model identification*
For the pole–residue set of the previous problem, find the state-space matrices according to Section 2.9.4.

REFERENCES

1. J. Vlach and K. Singhal. *Computer Methods for Circuit Simulation.* Van Nostrand Reinhold Co., New York, 1983.
2. A. Ruehli Ed. *Circuit Analysis, Simulation and Design, Part 1.* Elsevier Science Publishers B.V., North-Holland, 1986.
3. L. O. Chua, C. A. Desoer, and E. S. Kuh. *Linear and Nonlinear Circuits.* McGraw-Hill Book Company, New York, 1987.
4. J. Ogrodzki. *Circuit Simulation Methods and Algorithms.* CRC Press, Boca Raton, FL, 1994.
5. L. Pillegi, R. Rohrer, and C. Visweswariah. *Electronic Circuits and System Simulation Methods.* McGraw-Hill Book Company, New York, 1995.
6. F. N. Najm. *Circuit Simulation.* John Wiley and Sons, Inc., New York, 2010.
7. L. W. Nagel. SPICE: a computer program to simulate semiconductor circuits. Electronics Research Laboratory Report, ERL M520, University of California, Berkeley, CA, May 1975.
8. C. Ho, A. E. Ruehli, and P. Brennan. Interactive circuit analysis and design using APL. In *Proceedings of the IEEE International Symposium on Circuits and Systems*, May 1975.
9. C. Ho, A. Ruehli, and P. Brennan. The modified nodal approach to network analysis. *IEEE Transactions on Circuits and Systems*, 22(6):504–509, June 1975.
10. A. M. Rushdi. Development of modified nodal analysis into a pedagogical tool. *IEEE Transactions on Education*, 28(1):17–25, February 1985.
11. R. Escarela-Perez, E. Melgoza, and J. Alvarez-Ramirez. Multi-slice modeling in circuit-field coupled systems using finite-element and modified nodal analyses. *IEEE Transactions on Magnetics*, 46(1):67–74, January 2010.
12. D. Gope, A. E. Ruehli, and V. Jandhyala. Solving the low frequency EM-CKT problems using the PEEC method. In *Digest of Electrical Performance of Electronic Packaging*, Austin, TX, pp. 351–354, October 2005.
13. D. Gope, A. Ruehli, and V. Jandhyala. Solving low-frequency EM-CKT problems using the PEEC method. *IEEE Transactions on Advanced Packaging*, 30(2):313–320, May 2007.
14. G. Vecchi. Loop-star decomposition of basis functions in the discretization of the EFIE. *IEEE Transactions on Antennas and Propagation*, 51(2):339–346, February 1999.
15. J. S. Zhao and W. C. Chew. Integral equation solution of Maxwell's equation from zero frequency to microwave frequencies. *IEEE Transactions on Antennas and Propagation*, 48:1635–1645, October 2000.

16. V. I. Okhmatowski, J. D. Morsey, and A. Cangellaris. Loop-tree implementation of the adaptive integral method (AIM) for numerically-table, broadband, fast electromagnetic modeling. *IEEE Transactions on Antennas and Propagation*, 52(8):2130–2140, August 2004.
17. D. Gope, A. E. Ruehli, and V. Jandhyala. Speeding up PEEC partial inductance computations using a QR-based algorithm. *IEEE Transactions on Very Large Scale Integration (VLSI) Systems*, 15(1):60–68, January 2007.
18. G. Antonini, G. Miscione, and D. Frigioni. Hybrid formulation of the equation systems of the 3-D PEEC model based on graph algorithms. *IEEE Transactions on Circuits and Systems I*, 57(1):249–261, January 2010.
19. A. Bellen, N. Guglielmi, and A. Ruehli. Methods for linear systems of circuit delay differential equations of neutral type. *IEEE Transactions on Circuits and Systems*, 46(1):212–216, January 1999.
20. P. J. Restle, A. E. Ruehli, S. G. Walker, and G. Papadopoulos. Full-wave PEEC time-domain method for the modeling of on-chip interconnects. *IEEE Transactions on Computer-Aided Design of Integrated Circuits and Systems*, 20(7):877–886, July 2001.
21. G. Antonini and P. Pepe. Input-to-state stability analysis of partial element equivalent circuit models. *IEEE Transactions on Circuits and Systems I*, 56(3):673–684, March 2009.
22. K. E. Brenan, S. L. Campell, and L. R. Petzold. *Numerical Solution of Initial-Value Problems in Differential-Algebraic Equations*. Elsevier Science Publishers, New York, 1989.
23. E. Hairer and G. Wanner. *Solving Ordinary Differential Equations II, Stiff and Differential Algebraic Problems*. Springer-Verlag, New York, 1991.
24. R. K. Brayton, F. G. Gustavson, and G. D. Hachtel. A new efficient algorithm for solving differential-algebraic systems using implicit backward differential formulas. *Proceedings of the IEEE*, 60(1):98–108, 1972.
25. B. P. Rynne. Comments on a stable procedure in calculating the transient scattering by conducting surfaces of arbitrary shape. *IEEE Transactions on Antennas and Propagation*, APP-41(4):517–520, April 1993.
26. W. Pinello, A. C. Cangellaris, and A. E. Ruehli. Hybrid electromagnetic modeling of noise interactions in packaged electronics based on the partial-element equivalent circuit formulation. *IEEE Transactions on Microwave Theory and Techniques*, MTT-45(10):1889–1896, October 1997.
27. L. Lapidus and J. H. Seinfeld. *Real and Complex Analysis, Numerical Solution of Ordinary Differential Equations*. ACADPR, Ljubljana, 1971.
28. R. Achar, P. K. Gunupudi, M. Nakhla, and E. Chiprout. Passive interconnect reduction algorithm for distributed/measured networks. *IEEE Transactions on Circuits and Systems II*, 47(4):287–301, April 2000.
29. W. H. Kao, C.-Y. Lo, M. Basel, and R. Singh. Parasitic extraction: current state of the art and future trends. *Proceedings of the IEEE*, 89(5):729–739, May 2001.
30. S. Grivet-Talocia and B. Gustavsen. *Passive Macromodeling: Theory and Applications*. John Wiley and Sons, Inc., New York, 2016.
31. D. M. Pozar. *Microwave Engineering*. John Wiley and Sons, Inc., New York, 1998.
32. C. R. Paul. *Analysis of Multiconductor Transmission Lines*. John Wiley Interscience, IEEE Press, Hoboken, NJ, 2nd edition, 2008.
33. G. Antonini. Spice compatible equivalent circuits of rational approximation of frequency domain responses. *IEEE Transactions on Electromagnetic Compatibility*, 45(3):502–512, August 2003.
34. B. Gustavsen and A. Semlyen. Rational approximation of frequency domain responses by vector fitting. *IEEE Transactions on Power Apparatus and Systems*, 14(3):1052–1061, July 1999.
35. B. Gustavsen. Enforcing passivity for admittance matrices approximated by rational functions. *IEEE Transactions on Power Delivery*, 16(1):97–104, February 2001.

36. B. Gustavsen. Computer code for rational approximation of frequency dependent admittance matrix. *IEEE Transactions on Power Delivery*, 17(4):97–104, October 2002.
37. T. Mangold and P. Russer. Full-wave modeling and automatic equivalent-circuit generation of millimeter-wave planar and multilayer structures. *IEEE Transactions on Microwave Theory and Techniques*, 47(6):851–858, June 1999.
38. G. Antonini, A. C. Scogna, and A. Orlandi. Equivalent network synthesis for via holes discontinuities. *IEEE Transactions on Advanced Packaging*, 25(4):528–535, November 2002.
39. R. Achar and M. Nakhla. Simulation of high-speed interconnects. *Proceedings of the IEEE*, 89(5):693–728, May 2001.
40. M. Celik, L. Pileggi, and A. Odabasioglu. *IC Interconnect Analysis*. Kluwer Academic Publishes, Boston, MA, Dordrecht, London, 2002.
41. T. Kailath. *Linear Systems*. Prentice Hall, Englewood Cliffs, NJ, 1980.
42. M. Swaminathan and A. E. Engin. *Power Integrity Modeling and Design for Semiconductors and Systems*. Prentice Hall, Upper Saddle River, NJ, 2007.
43. A. Semlyen and A. Dabuleanu. Fast and accurate switching transient calculations on transmission lines with ground return using recursive convolutions. *IEEE Transactions on Power Apparatus and Systems*, 94(2):561–571, March 1975.
44. F. H. Branin. Transient analysis of lossless transmission lines. *Proceedings of the IEEE*, 55(11):2012–2013, November 1967.

3

MAXWELL'S EQUATIONS

3.1 MAXWELL'S EQUATIONS FOR PEEC SOLUTIONS

One set of fundamental techniques necessary for partial element equivalent circuit (PEEC) models is the circuit techniques in Chapter 2. Furthermore, we require EM techniques that are based on Maxwell's equations presented in this chapter. We give a general overview of the different possible forms of Maxwell's equations and the resultant differential and integral equation formulations. Hopefully, this leads to the understanding of various appropriate solutions. In this chapter, we emphasize existing results and conclusions based on Maxwell's equations instead of derivations of the fundamentals on which they are based. Our aim is to provide essential tools for connecting lumped circuit analysis and electromagnetic modeling. We concentrate on solutions that have practical relevance for PEEC modeling. As an example, we can state that the type of problems we want to solve also requires a *dc* solution since such connections exist in these problems. Hence, it is important that our approaches include *dc* solutions. Hopefully, we include sufficient references for the reader who wants to study the origin of some of the concepts in more details.

3.1.1 Maxwell's Equations in the Differential Form

To start, we choose a general form of Maxwell's equation [1, 2] that includes fictitious magnetic charges q_m as well as fictitious magnetic currents $\boldsymbol{J}_m$. We show that this can lead to important solution techniques while it is clear that both quantities do not exist in real-life physical systems. Importantly, these quantities result in a dual formulation in the electric field $\boldsymbol{E}$ and the magnetic field $\boldsymbol{H}$, which has advantages. We need to consider time and frequency domain formulations since they both are important. For this reason, we use the Laplace transform variable $s = j\omega$ in the frequency domain and the time derivative $\partial/\partial t$ in

Circuit Oriented Electromagnetic Modeling Using the PEEC Techniques, First Edition.
Albert E. Ruehli, Giulio Antonini, and Lijun Jiang.

the time domain. With this, Maxwell's equations can be written as

$$\nabla \times \boldsymbol{E} = -\boldsymbol{J}_{vm} - s\,\boldsymbol{B} \tag{3.1a}$$

$$\nabla \times \boldsymbol{H} = \boldsymbol{J}_{ve} + s\,\boldsymbol{D} \tag{3.1b}$$

$$\nabla \cdot \boldsymbol{D} = q_{ve} \tag{3.1c}$$

$$\nabla \cdot \boldsymbol{B} = q_{vm}. \tag{3.1d}$$

We use the notation $q_{ve}\,(\boldsymbol{r}, s)$ for the electric charge volume density in $\mathrm{C/m^3}$. However, if a solution does not involve magnetic charges, we simplify the notation to $q_v\,(\boldsymbol{r}, s)$ since the electric charge is obvious. The magnetic volume charge density is $q_{vm}\,(\boldsymbol{r}, s)$ in Weber/m^3. If the electric charge is distributed only on the surface, we use the surface charge representation $q_{se}\,(\boldsymbol{r}, s)$ for the surface charge in $\mathrm{C/m^2}$. Of course, a magnetic surface charge density is represented by $q_{sm}\,(\boldsymbol{r}, s)$ in Weber/m^2.

Names used for the Maxwell's equation (3.1a) is the Faraday's law and (3.1b) represents Ampere's circuital law. Equation (3.1c) is the Gauss' law for electric field and (3.1d) is the Gauss' law for magnetic field. It can be shown that for the above-mentioned four equations, only two are independent. Gauss' law (3.1d) and (3.1c) can both be derived from (3.1b) and (3.1a), respectively.

The notation for current densities is analogous. The volume current density $\boldsymbol{J}_{ve}\,(\boldsymbol{r}, s)$ is the impressed electric current volume density in $\mathrm{A/m^2}$. The surface current density $\boldsymbol{J}_{se}\,(\boldsymbol{r}, s)$ is in A/m. The notation for magnetic current densities is similar. The magnetic volume current density is $\boldsymbol{J}_{vm}\,(\boldsymbol{r}, s)$ in $\mathrm{V/m^2}$. The magnetic surface current density is $\boldsymbol{J}_{sm}(\boldsymbol{r}, s)$ in V/m. We should note that for magnetic quantities we *always* use the magnetic subscript index m for clarity, such as in q_{sm} and $\boldsymbol{J}_{vm}$. We always need to be reminded that fictional magnetic quantities can be involved.

The electric field intensity $\boldsymbol{E}\,(\boldsymbol{r}, s)$ is in V/m and $\boldsymbol{H}\,(\boldsymbol{r}, s)$ is the magnetic field intensity in A/m. Further, the displacement flux density is $\boldsymbol{D}\,(\boldsymbol{r}, s)$ in $\mathrm{C/m^2}$. The magnetic flux density $\boldsymbol{B}\,(\boldsymbol{r}, s)$ is in Weber/m^2.

The following constitutive relations are satisfied by $\boldsymbol{H}, \boldsymbol{B}, \boldsymbol{E}$, and $\boldsymbol{D}$ for an isotropic media:

$$\boldsymbol{B} = \mu\,\boldsymbol{H} \tag{3.2a}$$

$$\boldsymbol{D} = \epsilon\,\boldsymbol{E} \tag{3.2b}$$

$$\boldsymbol{J}_v = \sigma \boldsymbol{E}, \tag{3.2c}$$

where μ is the permeability in H/m, ϵ is the permittivity in F/m, and σ is the conductivity of the material in S/m. Again, we call the conductive current $\boldsymbol{J}_v$ rather than $\boldsymbol{J}_{ve}$ for the case without magnetic currents.

Other extremely important relations for our work are the so-called continuity equations. They are based on charge conservation principle and consider the time-dependent charge density variation. The relation between the electric current density and electric charge density is

$$\nabla \cdot \boldsymbol{J}_{ve} + s\,q_{ve} = 0. \tag{3.3}$$

Similarly, for the magnetic current density and magnetic charge density, the charge conservation law is given by

$$\nabla \cdot \boldsymbol{J}_{vm} + s\, q_{vm} = 0. \tag{3.4}$$

3.1.2 Maxwell's Equations in the Integral Form

For completeness we also list Maxwell's equations in an integral form. This formulation has many applications. An important one is that it forms the basis for the finite integration technique (FIT), which is a differential equation solution technique [3].

The approach is based on the integration of Maxwell's equations (3.1a) and (3.1b) over local surfaces and volumes. We present the Maxwell's equations in the time domain, which is similar to the frequency domain formulation. We use the egg-shaped half volume in Fig. 3.1 to illustrate the formulas.

We start with the first two equations (3.1a) and (3.1b) to obtain

$$\int_S (\nabla \times \boldsymbol{E}) \cdot \hat{\boldsymbol{n}}\, dS = \oint_C \boldsymbol{E} \cdot dl = -\int_S \boldsymbol{J}_{vm} \cdot \hat{\boldsymbol{n}}\, dS - \frac{\partial}{\partial t} \int_S \boldsymbol{B} \cdot \hat{\boldsymbol{n}}\, dS \tag{3.5a}$$

$$\int_S (\nabla \times \boldsymbol{H}) \cdot \hat{\boldsymbol{n}}\, dS = \oint_C \boldsymbol{H} \cdot dl = \int_S \boldsymbol{J}_{ve} \cdot \hat{\boldsymbol{n}}\, dS + \frac{\partial}{\partial t} \int_S \boldsymbol{D} \cdot \hat{\boldsymbol{n}}\, dS. \tag{3.5b}$$

The surface integrals are over the surface S in Fig. 3.1. The second form over the closed contour C is obtained by applying Stokes' theorem given in the following section (3.34).

The second part of the surface formulation of Maxwell's equations is based on the Gauss' theorem in (3.33), which is

$$\oint_S \boldsymbol{D} \cdot \hat{\boldsymbol{n}}\, dS = \int_{\mathcal{V}} q_{ve}\, d\mathcal{V} \tag{3.6a}$$

$$\oint_S \boldsymbol{B} \cdot \hat{\boldsymbol{n}}\, dS = \int_{\mathcal{V}} q_{vm}\, d\mathcal{V}, \tag{3.6b}$$

where $\hat{\boldsymbol{n}}$ is the outward normal in Fig. 3.1 to the surface S of volume $\mathcal{V}$. To simplify the notation of the above equations, we did not indicate that the field variables in (3.5a)–(3.6b) depend on the spatial position $\boldsymbol{r}$ and time t.

The left-hand side of (3.5a) represents the electromotive force (*emf*). Further, the left-hand side of (3.5b) is the magnetomotive force (*mmf*). Both equations clearly indicate the coupling between the time-varying electric field and the time-varying magnetic field.

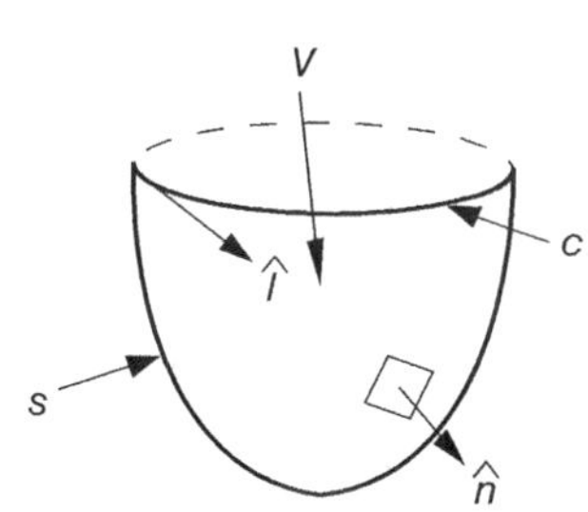

FIGURE 3.1 Example for surface and volume for the integrals.

The electric continuity equation in the form of charge conservation is for the electric volume current given by

$$\oint_S \boldsymbol{J}_{ve} \cdot \hat{\boldsymbol{n}}\, dS + \frac{\partial}{\partial t}\int_{\mathcal{V}} q_{ve}\, d\mathcal{V} = 0, \tag{3.7}$$

and for the magnetic volume current as

$$\oint_S \boldsymbol{J}_{vm} \cdot \hat{\boldsymbol{n}}\, dS + \frac{\partial}{\partial t}\int_{\mathcal{V}} q_{vm}\, d\mathcal{V} = 0. \tag{3.8}$$

3.1.3 Maxwell's Equations and Kirchhoff's Circuit Laws

The fundamental idea of PEEC is to model Maxwell's equations to the circuit domain. Here, we give a general intuitive view. We pursue this issue in great detail in most chapters of this book. Chapter 1 of [1] gives a very detailed derivation and explanation for the contents of this section.

Equation (3.5a) can be interpreted as time-varying magnetic flux Ψ through a closed loop generating an emf, which can be represented as [1]

$$\sum V = \oint_C \boldsymbol{E} \cdot d\boldsymbol{l} = -\frac{\partial}{\partial t}\int_S \boldsymbol{B} \cdot \hat{\boldsymbol{n}}\, dS = -\frac{\partial \Psi}{\partial t} = -L_s \frac{\partial i}{\partial t}, \tag{3.9}$$

where L_s is an inductance due to a loop area. The relation between magnetic flux and the current i is given by $\Psi = L_s\, i$. Inductance concepts are considered in detail in Chapter 5 since they are important elements in the PEEC circuit equations.

As we learned from Chapter 2, we can write the Kirchhoff's voltage law (KVL) for a resistance, inductance, capacitance (RLC) lumped circuit loop for a simple situation as

$$V_i + V_R + V_C + V_L = 0, \tag{3.10}$$

where V_i may be due to a voltage source, V_R is the voltage drop due to a lumped resistor, V_C is the voltage drop from a lumped capacitor, and V_L is the voltage drop along the lumped inductance. We also can consider that the variation of the magnetic field through a finite circuit loop area represents the KVL as

$$V_i + V_R + V_C = -L_s \frac{\partial i}{\partial t} = -\frac{\partial}{\partial t}\int_S \boldsymbol{B} \cdot \hat{\boldsymbol{n}}\, dS. \tag{3.11}$$

From (3.7), the connection with a capacitance can be established by the following derivation

$$\oint_S \boldsymbol{J}_{ve} \cdot \hat{\boldsymbol{n}}\, dS = -\frac{\partial}{\partial t}\int_{\mathcal{V}} q_{ve}\, d\mathcal{V} = -\frac{\partial}{\partial t} Q = -C_s \frac{\partial V}{\partial t}, \tag{3.12}$$

where C_s is a capacitance due to the capacitive coupling between a conducting node to the ground.

As an example, a general circuit node may have parallel connected lumped R,L,C elements as well as a source current source i_i in parallel. Then, the conventional Kirchhoff's current law (KCL) will yield

$$i_i + i_R + i_C + i_L = 0. \tag{3.13}$$

Hence, we can see from this example how the continuity equation can be connected to the KCL in the circuit domain. The continuity equation for PEEC is considered in Section 6.3.1.

3.1.4 Boundary Conditions

Boundaries with different properties play an important role in many relevant problems. Hence, it is important to understand how to deal with such boundary conditions in the solution of electromagnetic problems. We consider an interface between two materials as shown in Fig. 3.2, where the normal vector points from the material 2 into material 1. This leads to the following boundary conditions for different materials.

$$-\hat{\boldsymbol{n}} \times (\boldsymbol{E}_1 - \boldsymbol{E}_2) = \boldsymbol{J}_{sm} \tag{3.14a}$$

$$\hat{\boldsymbol{n}} \times (\boldsymbol{H}_1 - \boldsymbol{H}_2) = \boldsymbol{J}_{se} \tag{3.14b}$$

$$\hat{\boldsymbol{n}} \cdot (\boldsymbol{D}_1 - \boldsymbol{D}_2) = q_{se} \tag{3.14c}$$

$$\hat{\boldsymbol{n}} \cdot (\boldsymbol{B}_1 - \boldsymbol{B}_2) = q_{sm}. \tag{3.14d}$$

Equation (3.14a) states that the tangential components of the electric field intensity are discontinuous in the presence of a surface magnetic current. Further, (3.14b) states that the tangential components of the magnetic field intensity are discontinuous due to the interface electric current. The third equation (3.14c) states that the normal components of the displacement flux density are discontinuous due to the presence of an interface electric charge density. Equation (3.14d) states that the normal components of the magnetic flux density are discontinuous in the presence of an interface magnetic charge density.

We next consider the case where material 2 in Fig. 3.2 is a perfect electric conductor (PEC) with infinite conductivity σ. Then, the electric field $\boldsymbol{E}_2$ is zero inside the conductor.

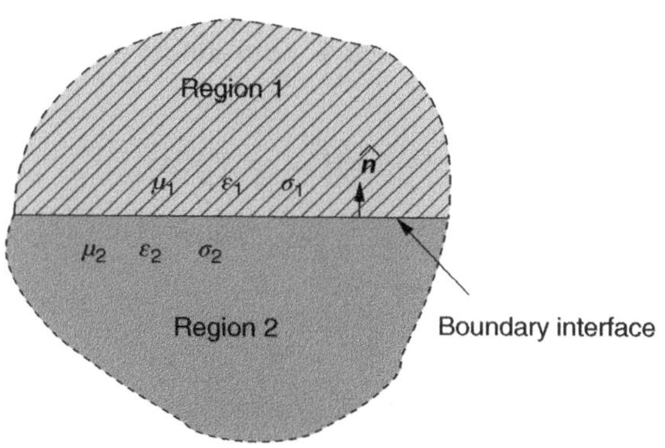

FIGURE 3.2 Example interface between two different materials.

For a solution where we set the nonphysical magnetic current and magnetic charge to zero, the boundary condition at the interface between material 1 and a PEC region 2 reduces to

$$-\hat{n} \times \boldsymbol{E}_1 = 0 \tag{3.15a}$$

$$\hat{n} \times \boldsymbol{H}_1 = \boldsymbol{J}_{se} \tag{3.15b}$$

$$\hat{n} \cdot \boldsymbol{D}_1 = q_{se} \tag{3.15c}$$

$$\hat{n} \cdot \boldsymbol{B}_1 = 0. \tag{3.15d}$$

Essentially, (3.15a) states that the tangential component of the electric field at the interface is 0. Further, (3.15d) states the normal component of the magnetic field at the interface is 0.

We also want to consider the important case where material 2 is a good but nonperfect conductor. For this case, another boundary condition connecting the tangential electric field and magnetic fields is frequently used. This leads to what is called the impedance boundary condition defined as

$$\boldsymbol{E}_t(s) = Z(s)\, \hat{n} \times \boldsymbol{H}_t(s), \tag{3.16}$$

where $Z(s)$ is called the surface impedance relating to the tangential components of the electrical and magnetic fields, $\boldsymbol{E}_t(s)$ and $\boldsymbol{H}_t(s)$, respectively. This condition has an assumption that the incident filed is close to orthogonal to the surface.

3.2 AUXILIARY POTENTIALS

3.2.1 Magnetic Vector Potential A and Electric Scalar Potential Φ_e

An important indirect approach for the solution of Maxwell's equations is based on the definition of auxiliary vector potentials. From (3.1d) we have $\nabla \cdot \boldsymbol{B} = 0$. Then from the vector identity

$$\nabla \cdot \nabla \times \boldsymbol{A} = 0, \tag{3.17}$$

we find that we can define $\boldsymbol{A}$ such that it obeys the above identity. From this, it is evident that we can have

$$\boldsymbol{B} = \mu\, \boldsymbol{H} = \nabla \times \boldsymbol{A}, \tag{3.18}$$

where $\boldsymbol{B}$ is derived from the vector potential $\boldsymbol{A}$. Substituting (3.18) into (3.1a), we obtain for the source-free single harmonic case

$$\nabla \times \boldsymbol{E} = -\mu\, s\, \boldsymbol{H} = -s\, \nabla \times \boldsymbol{A}. \tag{3.19}$$

Collecting the terms with the cross product, we get

$$\nabla \times (\boldsymbol{E} + s\, \boldsymbol{A}) = 0. \tag{3.20}$$

We note that both terms inside the parentheses should be electric fields. Next, we use the vector identity

$$\nabla \times \nabla\, \Phi = 0. \tag{3.21}$$

Comparing (3.20) and (3.21), we see that we can write

$$\boldsymbol{E} = -s\,\boldsymbol{A} - \nabla\,\Phi_e, \tag{3.22}$$

where $\boldsymbol{E}$ is the scattered field due to all sources. Φ_e is the electric scalar potential. If an external incident field $\boldsymbol{E}^{inc}$ is enforced, equation (3.22) can be rewritten as

$$\boldsymbol{E} = \boldsymbol{E}^{inc} - s\,\boldsymbol{A} - \nabla\,\Phi_e. \tag{3.23}$$

Inside a lossy medium, the total field is related to the conducting current by (3.2c). Hence,

$$\boldsymbol{E} = \frac{\boldsymbol{J}_c}{\sigma}. \tag{3.24}$$

For this case, (3.23) becomes

$$\boldsymbol{E}^{inc} = \frac{\boldsymbol{J}_c}{\sigma} + s\,\boldsymbol{A} + \nabla\,\Phi_e. \tag{3.25}$$

This is also called the electric field integral equation (EFIE). We note that this is one of the foundation formulas for the PEEC method.

3.2.2 Electric Vector Potential F and Magnetic Scalar Potential Φ_m

Analogous to the last section, the definition of an electrical vector potential $\boldsymbol{F}$ helps the formulation of additional solution methods. This is part of the efforts in providing a dual formulation.

We start with electric displacement flux density $\boldsymbol{D}$ that is solenoidal in a source-free region. We again use the vector identity

$$\nabla \cdot (-\nabla \times \boldsymbol{F}) = 0. \tag{3.26}$$

Hence, we can in the same way define the electric vector potential $\boldsymbol{F}$ with $\boldsymbol{D} = -\nabla \times \boldsymbol{F}$, or

$$\boldsymbol{E} = -\frac{1}{\epsilon}\nabla \times \boldsymbol{F}. \tag{3.27}$$

At the sourceless region, (3.1b) can be written as

$$\nabla \times \boldsymbol{H} = \epsilon\, s\, \boldsymbol{E}. \tag{3.28}$$

Using the definition of $\boldsymbol{F}$, we can reduce this to

$$\nabla \times (\boldsymbol{H} + s\,\boldsymbol{F}) = 0. \tag{3.29}$$

This leads to the important equation

$$\boldsymbol{H} = -\nabla\Phi_m - s\,\boldsymbol{F}, \tag{3.30}$$

where Φ_m is the magnetic scalar potential.

In (3.30), $\boldsymbol{H}$ is the scattered magnetic field due to all sources. If an external incident magnetic field $\boldsymbol{H}^{inc}$ is applied, the total field is the sum of the incident field $\boldsymbol{H}^{inc}$ and the scattered one. Hence, it is

$$\boldsymbol{H} = \boldsymbol{H}^{inc} - \nabla\Phi_m - s\,\boldsymbol{F}. \tag{3.31}$$

In the absence of an external applied magnetic field, the final formulation for the magnetic field is

$$\boldsymbol{H} = -s\,\boldsymbol{F} - \nabla\,\Phi_m. \tag{3.32}$$

This is the basic magnetic field integral equation (MFIE) that is directly corresponding to the EFIE in (3.25).

3.2.3 Important Fundamental Relationships

Several important theorems can be used to facilitate the derivation and the solution of the electromagnetic wave equations. It is important to understand the mathematical formulation as well as physical meaning. Our summary of these electromagnetic theorems does not include details since many excellent textbooks exist today on the fundamentals of electromagnetics, e.g., [1].

The important Gauss' law states that the integration of the divergence of a field variable is equal to the total flux of the field variable through the surface enclosing that volume. If $\boldsymbol{f}$ is a field vector, the mathematical representation of this theorem is

$$\int_{\mathcal{V}} \nabla\cdot\boldsymbol{f}\,d\mathcal{V} = \oint_{s} \boldsymbol{f}\cdot\hat{\boldsymbol{n}}\,dS. \tag{3.33}$$

A second important law is Stokes' theorem, which states that the total curl flux of a field through a given surface is equal to the field's loop integration over the closed boundary of that surface. Its mathematical representation is

$$\int_{S} (\nabla\times\boldsymbol{f})\cdot\hat{\boldsymbol{n}}\,dS = \oint_{\ell} \boldsymbol{f}\cdot d\boldsymbol{\ell}. \tag{3.34}$$

We note that we will extensively use Stokes' law in Chapter 5.

A form of Gauss' theorem considers the relationship between two scalar functions ψ and ϕ in terms of the Laplacian operator ∇^2. It is generally represented as

$$\int_{\mathcal{V}} (\phi\,\nabla^2\psi\; - \;\psi\,\nabla^2\phi)\,d\mathcal{V} = \oint_{S} (\phi\,\nabla\psi\; - \;\psi\,\nabla\phi)\cdot\,\hat{\boldsymbol{n}}\,dS. \tag{3.35}$$

This theorem is frequently used for wave equations as well as integral equations.

3.3 WAVE EQUATIONS AND THEIR SOLUTIONS

3.3.1 Wave Equations for *E* and *H*

The vector wave equations can be derived from Maxwell's equation for the electric field $\boldsymbol{E}$ and the magnetic field $\boldsymbol{H}$. In an isotropic region, using (3.1a) and applying the curl operation on both sides results in

$$\nabla \times \nabla \times \boldsymbol{E} = -\nabla \times \boldsymbol{J}_{vm} - \mu\, s\, \nabla \times \boldsymbol{H}. \tag{3.36}$$

Using the fundamental vector Laplacian identity [4], we get

$$\nabla \times \nabla \times \boldsymbol{F} = \nabla(\nabla \cdot \boldsymbol{F}) - \nabla^2 \boldsymbol{F}, \tag{3.37}$$

and replacing $\nabla \times \boldsymbol{H}$ in (3.36) with (3.1b), we obtain

$$\nabla(\nabla \cdot \boldsymbol{E}) - \nabla^2 \boldsymbol{E} = -\nabla \times \boldsymbol{J}_{vm} - \mu\, s\, \boldsymbol{J}_{ve} - \epsilon\, \mu\, s^2 \boldsymbol{E}. \tag{3.38}$$

Hence, using (3.1c), the wave equations in terms of the electrical field in an isotropic region results in

$$\nabla^2 \boldsymbol{E} - \epsilon\, \mu\, s^2 \boldsymbol{E} = \nabla \times \boldsymbol{J}_{vm} + \mu\, s\, \boldsymbol{J}_{ve} + \frac{\nabla q_{ve}}{\epsilon}. \tag{3.39}$$

Using the Duality Principle of Maxwell's equations, we can directly obtain the wave equation for the magnetic field to be

$$\nabla^2 \boldsymbol{H} - \mu\, \epsilon\, s^2 \boldsymbol{H} = -\nabla \times \boldsymbol{J}_{ve} + \epsilon\, s\, \boldsymbol{J}_{vm} + \frac{\nabla q_{vm}}{\mu}. \tag{3.40}$$

Of course, a derivation similar to (3.39) can be employed to obtain (3.40).

3.3.2 Wave Equations for A, F, and Φ_e

Based on the Laplacian identity (3.37), the magnetic vector potential satisfies the following expression:

$$\nabla \times \nabla \times \boldsymbol{A}(\boldsymbol{r}, t) = \nabla(\nabla \cdot \boldsymbol{A}(\boldsymbol{r}, t)) - \nabla^2 \boldsymbol{A}. \tag{3.41}$$

Based on (3.18) and taking curl operation on both sides, we have

$$\nabla \times \nabla \times \boldsymbol{A} = \mu \nabla \times \boldsymbol{H} = \mu \boldsymbol{J}_{ve} + \mu \epsilon s \boldsymbol{E}_A. \tag{3.42}$$

Because of (3.30), by combining (3.41) and (3.42), we have

$$\nabla^2 \boldsymbol{A} - \mu \epsilon s^2 \boldsymbol{A} = -\mu \boldsymbol{J}_{ve} + \nabla(\nabla \cdot \boldsymbol{A}(\boldsymbol{r}, t) + \mu \epsilon s \Phi_e). \tag{3.43}$$

We can enforce the Helmholtz equation, if the Lorenz gauge (or Lorenz condition) [1] is used, which is

$$\nabla \cdot \boldsymbol{A} + \mu \epsilon s \Phi_e = 0 \tag{3.44}$$

for the magnetic vector potential

$$\nabla^2 \boldsymbol{A} - \mu \epsilon s^2 \boldsymbol{A} = -\mu \boldsymbol{J}_{ve}. \tag{3.45}$$

It is possible to express the potential Φ_e in terms of the charge density following the same steps, which leads to the Helmholtz equation for the electric scalar potential, or

$$\nabla^2\Phi_e - \mu\epsilon s^2\Phi_e = -\frac{q_{ve}}{\epsilon}. \tag{3.46}$$

Using a similar process, we could derive the Helmholtz equation for the electric vector potential $\boldsymbol{F}$

$$\nabla^2\boldsymbol{F} - \mu\epsilon s^2\boldsymbol{F} = -\epsilon\boldsymbol{J}_{vm}, \tag{3.47}$$

and also

$$\nabla^2\Phi_m - \mu\epsilon s^2\Phi_m = -\frac{q_{vm}}{\mu}. \tag{3.48}$$

3.3.3 Solution of the Helmholtz Equation

In the time domain, the Helmholtz equation for the magnetic vector potential $\boldsymbol{A}$ is

$$\nabla^2\boldsymbol{A}(\boldsymbol{r},t) - \mu\epsilon\frac{\partial^2\boldsymbol{A}}{\partial t^2}(\boldsymbol{r},t) = -\mu\boldsymbol{J}_{ve}(\boldsymbol{r},t). \tag{3.49}$$

For a homogeneous medium, (3.49) has a closed-form solution for the magnetic vector potential $\boldsymbol{A}(\boldsymbol{r},t)$ due to an electric current $\boldsymbol{J}_{ve}(\boldsymbol{r},t)$ in the volume $\mathcal{V}'$ as

$$\boldsymbol{A}(\boldsymbol{r},t) = \frac{\mu}{4\pi}\int_{\mathcal{V}'}\frac{\boldsymbol{J}_{ve}(\boldsymbol{r}',t')}{|\boldsymbol{r}-\boldsymbol{r}'|}d\mathcal{V}'. \tag{3.50}$$

Equation (3.46) can be written for a homogeneous material in the time domain as

$$\nabla^2\Phi_e(\boldsymbol{r},t) - \mu\epsilon\frac{\partial^2\Phi_e(\boldsymbol{r},t)}{\partial t^2} = -\frac{q_{ve}(\boldsymbol{r},t)}{\epsilon}. \tag{3.51}$$

Another important closed-form solution can be obtained for an electric scalar potential $\Phi_e(\boldsymbol{r},t)$ due to the charge distribution. This formula takes into account the charge $q_{se}(\boldsymbol{r}',t)$ residing on the exterior surface of the conductors. For a homogeneous medium the solution of (3.46) is given by

$$\Phi_e(\boldsymbol{r},t) = \frac{1}{4\pi\epsilon}\int_{S'}\frac{q_{se}(\boldsymbol{r}',\boldsymbol{t}')}{|\boldsymbol{r}-\boldsymbol{r}'|}dS'. \tag{3.52}$$

In (3.50) and (3.52), t' denotes the time at which the current and charge distributions, $\boldsymbol{J}_{ve}$ and q_{ve}, act as sources for $\boldsymbol{A}$ and Φ_e, respectively. The source of the difference between t and t' is the finite value of the speed of light in a homogeneous medium, which is $v = 1/\sqrt{\mu\epsilon}$. Hence, they can be related as

$$t = t' + \frac{|\boldsymbol{r}-\boldsymbol{r}'|}{v}, \tag{3.53}$$

as considered in Section 2.11. The derivation of the relations (3.50) and (3.52) are based on Maxwell's equations as well as the Lorentz gauge (3.44).

We should note that both potentials (3.52) and (3.50) are of importance throughout this book as is evident from the following section.

3.3.4 Electric Field Integral Equation

A formula that is fundamental to the PEEC method is the integral equation derived in this section. If we write an equation for the sum of electric fields in a conductor based on (3.50) and (3.52) into (3.25) and add a possible external incident electric field $\boldsymbol{E}^{inc}(\boldsymbol{r}, t)$, we obtain the important EFIE

$$\boldsymbol{E}^{inc}(\boldsymbol{r}, t) = \frac{\boldsymbol{J}_{ve}(\boldsymbol{r}, t)}{\sigma} + \frac{\partial}{\partial t}\frac{\mu}{4\pi}\int_{\mathcal{V}'} \frac{\boldsymbol{J}_{ve}(\boldsymbol{r}', \boldsymbol{t}')}{|\boldsymbol{r} - \boldsymbol{r}'|}\, d\mathcal{V}' + \nabla\Phi_e(\boldsymbol{r}, t) \tag{3.54}$$

where we consider a lossy, nonperfect conductor. The electric charge is assumed to be on the surfaces of the conductors. It is represented by the potential (3.52).

Hence, the integral equation includes (3.54) and (3.52). In addition to (3.52), we also have to enforce the conservation of charge with the continuity equation (3.7), or

$$\nabla \cdot \boldsymbol{J}_{ve}(\boldsymbol{r}, t) = -\frac{\partial q_{ve}(\boldsymbol{r}, t)}{\partial t}. \tag{3.55}$$

Since the charge is located only on the surface of conductors, inside conductors equation (3.55) reduces to

$$\nabla \cdot \boldsymbol{J}_{ve}(\boldsymbol{r}, t) = 0. \tag{3.56}$$

However, for the charge on the surfaces on a conductor, we find using the surface divergence that

$$\hat{\boldsymbol{n}} \cdot \boldsymbol{J}_{ve}(\boldsymbol{r}, t) = -\frac{\partial q_{se}(\boldsymbol{r}, t)}{\partial t}, \tag{3.57}$$

where $\hat{\boldsymbol{n}}$ is the outward normal to the surface S'. Finally, we can summarize the relevant equations as

$$\boldsymbol{E}^{inc}(\boldsymbol{r}, t) = \frac{\boldsymbol{J}_{ve}(\boldsymbol{r}, t)}{\sigma} + \frac{\partial}{\partial t}\frac{\mu}{4\pi}\int_{\mathcal{V}'} \frac{\boldsymbol{J}_{ve}(\boldsymbol{r}', \boldsymbol{t}')}{|\boldsymbol{r} - \boldsymbol{r}'|}\, d\mathcal{V}' + \nabla\Phi_e(\boldsymbol{r}, t) \tag{3.58a}$$

$$\Phi_e(\boldsymbol{r}, t) = \frac{1}{4\pi\epsilon}\int_{S'} \frac{q_{se}(\boldsymbol{r}', \boldsymbol{t}')}{|\boldsymbol{r} - \boldsymbol{r}'|}\, dS', \qquad \boldsymbol{r} \in S' \tag{3.58b}$$

$$\nabla \cdot \boldsymbol{J}_{ve}(\boldsymbol{r}, t) = 0, \boldsymbol{r} \in \mathcal{V}' \tag{3.58c}$$

$$\hat{\boldsymbol{n}} \cdot \boldsymbol{J}_{ve}(\boldsymbol{r}, t) = -\frac{\partial q_{se}(\boldsymbol{r}, t)}{\partial t}, \qquad \boldsymbol{r} \in S'. \tag{3.58d}$$

The unknowns of such a problem are represented by the current density $\boldsymbol{J}_{ve}(\boldsymbol{r}, t)$ in the interior of the conductors, the charge density $q_{se}(\boldsymbol{r}, t)$ on the surface of the conductors, and the electric scalar potential distribution $\Phi_e(\boldsymbol{r},t)$ of conductors, which can be directly expressed as a function of the charge density for $\boldsymbol{r} \in S'$.

Equations (3.58a)–(3.58d) can be rewritten using the Laplace variable s as

$$\boldsymbol{E}^{inc}(\boldsymbol{r},s) = \frac{\boldsymbol{J}_{ve}(\boldsymbol{r},s)}{\sigma} + \frac{s\mu}{4\pi}\int_{\mathcal{V}'} \frac{\boldsymbol{J}_{ve}(\boldsymbol{r}',s)e^{-s\tau}}{|\boldsymbol{r}-\boldsymbol{r}'|} d\mathcal{V}' + \nabla\Phi_e(\boldsymbol{r},s) \tag{3.59a}$$

$$\Phi_e(\boldsymbol{r},s) = \frac{1}{4\pi\epsilon}\int_{S'} \frac{q_{se}(\boldsymbol{r}',s)e^{-s\tau}}{|\boldsymbol{r}-\boldsymbol{r}'|} dS', \qquad \boldsymbol{r} \in S' \tag{3.59b}$$

$$\nabla \cdot \boldsymbol{J}_{ve}(\boldsymbol{r},s) = 0, \qquad \boldsymbol{r} \in \mathcal{V}' \tag{3.59c}$$

$$\hat{\boldsymbol{n}} \cdot \boldsymbol{J}_{ve}(\boldsymbol{r},s) = -s\, q_{se}(\boldsymbol{r},s), \qquad \boldsymbol{r} \in S', \tag{3.59d}$$

where $\tau = |\boldsymbol{r}-\boldsymbol{r}'|/c$ and s is the Laplace variable. Again, both the time and frequency domain formulations for the EFIE are fundamental to the PEEC method.

3.4 GREEN'S FUNCTION

The Green's function is the basic solution of the wave equation for a local source, which is a point source for 3D. A solution using Green's function with respect to space variables such as x, y, z can best be compared with the equivalent in the time domain convolution with an impulse response considered in the circuit analysis in Chapter 2. The Green's function can be compared to an impulse response in the time domain.

Two different classes of Green's functions can be defined – static and dynamic – depending on the type of wave equations for which they are a solution. Further, the homogeneous and inhomogeneous Green's functions are defined depending upon the medium environment. Specifically, inhomogeneous Green's function represents different media.

Observe that homogeneous Green's function can be used to solve inhomogeneous problems. For example, the Poggio-Miller-Chang-Harrington-Wu-Tsai (PMCHWT) integral equation method in Section 11.4.2 can be based on homogeneous Green's functions while the solution applies to nonhomogeneous problems. Also, it is important to observe that delayed static Green's functions are used in full-wave time domain PEEC models.

It is usually difficult to find closed-form solutions for inhomogeneous Green's functions for complex structures. As an example, the Green's function for the full-wave Helmholtz equation in the homogeneous medium is defined by the following wave equation for a point source:

$$(\nabla^2 + \beta^2)g(\boldsymbol{r},\boldsymbol{r}') = -\delta(\boldsymbol{r}-\boldsymbol{r}'), \tag{3.60}$$

where $\boldsymbol{r}$ is the field point location and $\boldsymbol{r}'$ is the source point position. The wave number $\beta > 0$ for nonzero frequencies is considered in the following section.

3.4.1 Notation Used for Wave Number and Fourier Transform

It is noted that for the regular physical wavenumber $k^2 = \omega^2\mu\epsilon$ in the frequency domain and $k^2 = -s^2\mu\epsilon$ using the Laplace transform. In many books and papers, β is also used to represent the wave number. Further, β is also used for the phase constant, which is different

TABLE 3.1 Notation used for propagation parameters.

Representation	Transmission line notation	Physics notation
Lossless material, propagation constant	β	k
Lossy material, propagation constant	γ	jk
Lossy material, attenuation constant	α	$j\alpha$
Lossy material, phase constant	β	$j\beta$ or jk

from the wave number that is determined by the medium property and the frequency. The phase constant is determined by the guided wavelength. To be specific, $k = 2\pi/\lambda_0$, where λ_0 is the wavelength of the electromagnetic wave in an infinite homogeneous medium for the given frequency. $\beta = 2\pi/\lambda_g$ where λ_g is the guided wavelength of the electromagnetic wave in the specific structure at the given frequency.

To make the issue more complicated, for a lossy medium, k is complex. For this case, it is usually written as $k = \beta - j\alpha$. The real part β corresponds to the wavelength while the imaginary part corresponds to the loss. The different usages of k and β can cause confusions if they are not properly distinguished. The comparison between the two notations is shown in Table 3.1. In this book, we use β as the wave number as is done in Ref. [1] since it is a clear notation. It is also the conventional transmission line notation.

In the electromagnetic analysis, Fourier transform and its inverse are frequently used. The Fourier transform using the wave number β from the space domain y for $f(y)$ is

$$\mathcal{F}(\beta) = \int_{-\infty}^{\infty} f(y)\, e^{-j\beta y}\, dy. \tag{3.61}$$

Correspondingly, the inverse Fourier transform is

$$f(y) = \frac{1}{2\pi}\int_{-\infty}^{\infty} \mathcal{F}(\beta)\, e^{j\beta y}\, d\beta. \tag{3.62}$$

Other forms are also in use besides the conventional notations. The Fourier transform is used in Section 5.8.2.

3.4.2 Full Wave Free Space Green's Function

The solution of (3.60) is called the full-wave free-space Green's function

$$g(\boldsymbol{r},\boldsymbol{r}') = \frac{e^{-j\beta|\boldsymbol{r}-\boldsymbol{r}'|}}{4\pi|\boldsymbol{r}-\boldsymbol{r}'|}. \tag{3.63}$$

When the frequency is zero, Helmholtz equation degrades to Poisson's equation.

$$\nabla^2 g(\boldsymbol{r},\boldsymbol{r}') = -\delta(\boldsymbol{r}-\boldsymbol{r}'). \tag{3.64}$$

Its solution is called the static Green's function:

$$g(\boldsymbol{r},\boldsymbol{r}') = \frac{1}{4\pi|\boldsymbol{r}-\boldsymbol{r}'|}. \tag{3.65}$$

It is noted that the lowercase g is used exclusively for scalar Green's functions. But other types of Green's functions in this book employ the uppercase G as the Green's function symbol.

If the vector wave equation for the electric field is established based on Maxwell's equations, we have

$$\nabla\times\nabla\times\boldsymbol{E}(\boldsymbol{r}) - \beta^2\boldsymbol{E}(\boldsymbol{r}) = -j\omega\mu\boldsymbol{J}_{ve}(\boldsymbol{r}), \tag{3.66}$$

then the corresponding Green's function is defined as the dyadic Green's function by the following equation:

$$\nabla\times\nabla\times\overline{\boldsymbol{G}}(\boldsymbol{r},\boldsymbol{r}') - \beta^2\overline{\boldsymbol{G}}(\boldsymbol{r},\boldsymbol{r}') = -\overline{\boldsymbol{I}}\delta(\boldsymbol{r}-\boldsymbol{r}'), \tag{3.67}$$

where $\overline{\boldsymbol{G}}(\boldsymbol{r},\boldsymbol{r}')$ is the dyadic Green's function and $\overline{\boldsymbol{I}}$ the unitary dyad. As a result, the electric field solution is

$$\boldsymbol{E}(\boldsymbol{r}) = j\omega\mu\int_{\mathcal{V}}\overline{\boldsymbol{G}}(\boldsymbol{r},\boldsymbol{r}')\cdot\boldsymbol{J}_{ve}(\boldsymbol{r}')d\mathcal{V}', \tag{3.68}$$

where the dyadic Green's function is

$$\overline{\boldsymbol{G}}(\boldsymbol{r},\boldsymbol{r}') = \left[\overline{\boldsymbol{I}} + \frac{\nabla\nabla}{\beta^2}\right]g(\boldsymbol{r},\boldsymbol{r}'). \tag{3.69}$$

In the implementation, the dyadic Green's function is expanded as a 3×3 matrix. Its element composition is shown as follows:

$$\begin{bmatrix} G_{xx}(\boldsymbol{r},\boldsymbol{r}') & G_{xy}(\boldsymbol{r},\boldsymbol{r}') & G_{xz}(\boldsymbol{r},\boldsymbol{r}') \\ G_{yx}(\boldsymbol{r},\boldsymbol{r}') & G_{yy}(\boldsymbol{r},\boldsymbol{r}') & G_{yz}(\boldsymbol{r},\boldsymbol{r}') \\ G_{zx}(\boldsymbol{r},\boldsymbol{r}') & G_{zy}(\boldsymbol{r},\boldsymbol{r}') & G_{zz}(\boldsymbol{r},\boldsymbol{r}') \end{bmatrix}. \tag{3.70}$$

Substituting it into (3.68), we have

$$\begin{bmatrix} E_x(\boldsymbol{r}) \\ E_y(\boldsymbol{r}) \\ E_z(\boldsymbol{r}) \end{bmatrix} = j\omega\mu\int_{\mathcal{V}}\begin{bmatrix} G_{xx}(\boldsymbol{r},\boldsymbol{r}') & G_{xy}(\boldsymbol{r},\boldsymbol{r}') & G_{xz}(\boldsymbol{r},\boldsymbol{r}') \\ G_{yx}(\boldsymbol{r},\boldsymbol{r}') & G_{yy}(\boldsymbol{r},\boldsymbol{r}') & G_{yz}(\boldsymbol{r},\boldsymbol{r}') \\ G_{zx}(\boldsymbol{r},\boldsymbol{r}') & G_{zy}(\boldsymbol{r},\boldsymbol{r}') & G_{zz}(\boldsymbol{r},\boldsymbol{r}') \end{bmatrix}\cdot\begin{bmatrix} J_{vex}(\boldsymbol{r}') \\ J_{vey}(\boldsymbol{r}') \\ J_{vez}(\boldsymbol{r}') \end{bmatrix}d\mathcal{V}'. \tag{3.71}$$

3.5 EQUIVALENCE PRINCIPLES

Equivalence principles are very useful for solving problems with regions of different materials. They will be used for both volume- and surface-type integral equation formulations.

Here, we present the so-called *volume equivalence principle* that we use in Chapter 10 to solve combined conductor and dielectric problems. The fundamental derivation of the method is given in the following section, while the derivation for the PEEC method is given in Section 10.4.5.

The second approach is based on the *surface equivalence principle*. Its basic principle is presented in Section 3.5.2 and a PEEC implementation is presented in Section 11.4.2. The formulation includes geometries with dielectric and magnetic materials where $\mu_r > 1$ and $\epsilon_r > 1$.

3.5.1 Volume Equivalence Principle

For a region that can consist of dielectric materials with position-dependent permittivity ϵ and permeability μ, the field can be solved using the volume equivalence principle. In general, an external field excitation can come from a current density $\boldsymbol{J}$ somewhere in the region of interest. Maxwell's equations for $\boldsymbol{E}$ (3.1a) and $\boldsymbol{H}$ (3.1b) are

$$\nabla \times \boldsymbol{E} = -\mu \frac{\partial \boldsymbol{H}}{\partial t} \tag{3.72a}$$

$$\nabla \times \boldsymbol{H} = \boldsymbol{J}_{ve} + \epsilon \frac{\partial \boldsymbol{E}}{\partial t}. \tag{3.72b}$$

If the same source is placed at the same position of a free space with ϵ_0 and μ_0, we have

$$\nabla \times \boldsymbol{E}_0 = -\mu_0 \frac{\partial \boldsymbol{H}_0}{\partial t} \tag{3.73a}$$

$$\nabla \times \boldsymbol{H}_0 = \boldsymbol{J}_{ve} + \epsilon_0 \frac{\partial \boldsymbol{E}_0}{\partial t}. \tag{3.73b}$$

The scattered field can be computed by subtraction, or $\boldsymbol{E}_s = \boldsymbol{E} - \boldsymbol{E}_0$ and $\boldsymbol{H}_s = \boldsymbol{H} - \boldsymbol{H}_0$. The results are

$$\nabla \times \boldsymbol{E}_s = -\boldsymbol{J}^v_{vm} - \mu_0 \frac{\partial \boldsymbol{H}_s}{\partial t} \tag{3.74a}$$

$$\nabla \times \boldsymbol{H}_s = \boldsymbol{J}^v_{ve} + \epsilon_0 \frac{\partial \boldsymbol{E}_s}{\partial t}, \tag{3.74b}$$

where $\boldsymbol{J}^v_{ve}$ and $\boldsymbol{J}^v_{vm}$ are the electric and magnetic volume current densities.

$$\boldsymbol{J}^v_{vm} = (\mu - \mu_0) \frac{\partial \boldsymbol{H}_0}{\partial t} \tag{3.75a}$$

$$\boldsymbol{J}^v_{ve} = (\epsilon - \epsilon_0) \frac{\partial \boldsymbol{E}_0}{\partial t}. \tag{3.75b}$$

In this general form, additional equivalent volume magnetic current and equivalent volume electric current sources are used. We only consider the electrical integration formulation with the electric current density $\boldsymbol{J}^v_{ve}$. This current density can be interpreted as the source of the scattered field generated in this formulation. Adding this equivalent source into the electric field equation (3.59a), the result is

$$\boldsymbol{E}^{inc}(\boldsymbol{r},s) = \frac{\boldsymbol{J}_{ve}(\boldsymbol{r},s)}{\sigma} + \frac{s\mu}{4\pi}\int_{\mathcal{V}'} \frac{\boldsymbol{J}_{ve}(\boldsymbol{r}',s)e^{-s\tau}}{|\boldsymbol{r}-\boldsymbol{r}'|}d\mathcal{V}' + \nabla\Phi_e(\boldsymbol{r},s)$$
$$+ \frac{s\mu}{4\pi}\int_{\mathcal{V}'} \frac{\boldsymbol{J}^{v}_{ve}(\boldsymbol{r}',s)e^{-s\tau}}{|\boldsymbol{r}-\boldsymbol{r}'|}d\mathcal{V}'. \tag{3.76}$$

If we insert the source current density into the equation, we get

$$\boldsymbol{E}^{inc}(\boldsymbol{r},s) = \frac{\boldsymbol{J}_{ve}(\boldsymbol{r},s)}{\sigma} + \frac{s\mu}{4\pi}\int_{\mathcal{V}'} \frac{\boldsymbol{J}_{ve}(\boldsymbol{r}',s)e^{-s\tau}}{|\boldsymbol{r}-\boldsymbol{r}'|}d\mathcal{V}' + \nabla\Phi_e(\boldsymbol{r},s)$$
$$+ \frac{s^2\mu}{4\pi}\int_{\mathcal{V}'} \frac{(\epsilon-\epsilon_0)\boldsymbol{E}_0(\boldsymbol{r}',s)e^{-s\tau}}{|\boldsymbol{r}-\boldsymbol{r}'|}d\mathcal{V}'. \tag{3.77}$$

We should mention that we apply this volume equivalence formulation in Chapter 10.

3.5.2 Huygens' Equivalence Principle

The Huygens' equivalence principle is one of the key surface equivalence principles of interest. All surface equivalence principles originate from the uniqueness theory, which states that the matching tangential field components are sufficient to guarantee a unique solution of the field. The tangential field at the surface of the region boundary must be matched inside a closed dielectric or conductor region as shown in Fig. 3.3. Potential external incident electric and magnetic fields are indicated by E^{inc} and H^{inc}, respectively. Surprisingly, the fields inside the closed boundary can be arbitrarily chosen since they do not impact the tangential fields.

A set of equivalent surface currents with any choice of the internal field $\boldsymbol{E}_{in}$ and $\boldsymbol{H}_{in}$ will appear based on the fundamental boundary conditions.

$$\boldsymbol{J}_{se} = \hat{\boldsymbol{n}} \times (\boldsymbol{H} - \boldsymbol{H}_{in}) \tag{3.78a}$$

$$\boldsymbol{J}_{sm} = -\hat{\boldsymbol{n}} \times (\boldsymbol{E} - \boldsymbol{E}_{in}). \tag{3.78b}$$

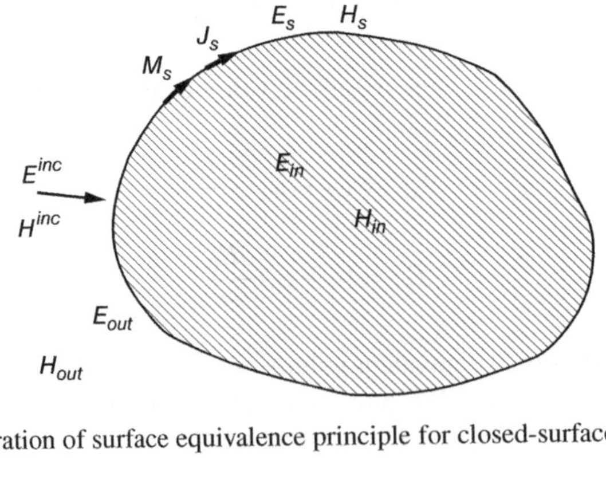

FIGURE 3.3 Illustration of surface equivalence principle for closed-surface two-region problem.

We observe that the flexible choice of the internal field will result in different equivalent surface currents. These equivalent currents reproduced the field outside the closed region. This is Huygens' equivalence principle.

We set the internal field $\boldsymbol{E}_{in} = 0$ and $\boldsymbol{H}_{in} = 0$. The resultant equivalent current is named the *Love's equivalent current*. Importantly, the outside field is regenerated by the equivalent electric current $\boldsymbol{J}_{se}$ and magnetic current $\boldsymbol{J}_{sm}$. Due to (3.45) and (3.22), the contribution from $\boldsymbol{J}_{se}$ to the $\boldsymbol{E}$ field can be taken into account by the magnetic vector potential $\boldsymbol{A}$ and the scalar potential Φ_e.

Because of (3.47) and (3.30), the contribution from $\boldsymbol{J}_{sm}$ to the $\boldsymbol{E}$ field can be counted by the magnetic vector potential $\boldsymbol{F}$ and scalar potential Φ_m. As a result, the $\boldsymbol{E}$ field can be represented by the equivalent sources as follows:

$$\boldsymbol{E}_s = -s\boldsymbol{A}(\boldsymbol{J}_{se}) - \nabla\Phi_e - \frac{1}{\epsilon}\nabla \times \boldsymbol{F}(\boldsymbol{J}_{sm}). \tag{3.79}$$

Consequently, the magnetic field can be derived to be represented using the equivalent sources as follows:

$$\boldsymbol{H}_s = -s\boldsymbol{F}(\boldsymbol{J}_{sm}) - \nabla\Phi_m + \frac{1}{\mu}\nabla \times \boldsymbol{A}(\boldsymbol{J}_{se}). \tag{3.80}$$

Thus, $\boldsymbol{A}(\boldsymbol{J}_{se})$ can be obtained from (3.49). Also, $\boldsymbol{F}(\boldsymbol{J}_{sm})$ can be obtained from the similar representations. Φ_e shall be obtained from (3.51) using the electric charge distribution on the surface. Φ_m shall be obtained similarly using the magnetic charge distribution on the surface.

3.6 NUMERICAL SOLUTION OF INTEGRAL EQUATIONS

A fundamental step in the practical solution of Maxwell's equations is the discretization into finite size cells. The numerical technique used for the approximations represents an important issue. In this section, we give a general outline for integral equation techniques. PEEC is a member of this class.

All solution techniques must result in a finite but sufficiently large number of unknowns such that an accurate numerical solution can be obtained. Of course, we also want to keep the number of unknowns and the compute time as small as possible. Basically, all solution methods are based on the so-called *Weighted Residuals Method* (WRM), as discussed in Chapter 1 [5]. It was originally called the *Method of Error Distribution Principles* [6], which is also a good descriptive name since we have to make approximations in several places that introduce errors in comparison to the exact Maxwell's equations.

How the general WRM approach is employed for different formulations was illustrated in Ref. [7]. It was also illustrated how the WRM approach is applied to solve different formulations based on differential as well as on integral equations. Sometimes, the *method of moments* (MoM) name is also considered to be a WRM [8] technique. The problem with this is that different meanings have been associated with this name. Some researchers assume that the name is limited to integral equations, while others assume that MoM is generally same as WRM [7, 9]. Also, we should point out that MoM has no relation to the moment matching method in the macromodeling area [10, 11].

Fundamentally, the WRM solution technique consists of several steps. The first step is the discretization of objects, including conductors and dielectrics, into finite blocks, bars, or surfaces cells for which we locally preassign fixed current directions or charge distributions areas. For example, one conductor or dielectric block can be partitioned into a collection of cells. The fixed current or field distribution assigned within the cell is called the *basis function*. Hence, each cell has its own fixed basis function and a conductor is represented by a set of fixed basis functions for each of the cell. With this, the current direction is fixed on each cell. However, the strength of the current on each cell is unknown.

To be specific about the second approximation used, we consider the computation of capacitances where we need to match the potential on a surface as illustrated in Fig. 4.1. Similar to the basis function, the equipotential on a conductor is matched with an approximating function usually called a *testing function*. For example, it is quite common to match the potential at a single point only on a cell leading to the so called collocation method. However, as we see, in PEEC we want to have symmetric circuit elements. All elements such as resistors, capacitors, and inductors are fundamentally symmetrical. This requires that both the basis function and the testing functions to be the same for PEEC circuits. This condition is also called a Galerkin-type method, which has its general origin in Ref. [12].

The third key step of the WRM method pertains to the way the resultant matrix system of equations is solved [7]. In the PEEC method, this consists of solving circuit equations based on the approximated system from the first two steps. Other members of the WRM family include techniques such as least squares solutions. Further, techniques such as the Generalized Multipole Technique [13] and multipole approaches [14] should be included in the class of solutions. Since the PEEC solution is circuit oriented, its connection to WRM may not be obvious. The solution step uses a circuit solver that is more like a SPICE tool [15]. Today, numerous techniques are used to solve the resultant matrix solution faster.

We next formulate the above discussion in a symbolic operator form. In general, we can write both differential and integral equations this way.

$$\mathcal{L}\, \boldsymbol{x}(\boldsymbol{r}) = \boldsymbol{b}(\boldsymbol{r}), \tag{3.81}$$

where $\mathcal{L}$ is a differential or integral equation operator. An example of a differential equation in this form is given by (3.64), and an integral equation operator example is given in (9.44). Here, $\boldsymbol{x}(\boldsymbol{r})$ is the unknown function of the spatial coordinate $\boldsymbol{r}$. We assume that the time or frequency for which the equations are solved and also $\boldsymbol{b}(\boldsymbol{r})$ is known. The solution with the WRM method is given next. If we subdivide the problem into N cells, we can name the fixed known basis function assigned to the jth cell to be f_j. Then, if we subdivide the problem into N cells, we can represent the solution as

$$\boldsymbol{x}(\boldsymbol{r}) = \sum_{j=1}^{N} \alpha_j f_j. \tag{3.82}$$

We now can substitute the solution (3.82) into the operator equation (3.81) to get

$$\mathcal{L}\, \boldsymbol{x}(\boldsymbol{r}) = \sum_{j=1}^{N} \alpha_j\, \mathcal{L} f_j = \boldsymbol{b}(\boldsymbol{r}), \tag{3.83}$$

where the unknowns now are the N coefficients α_j. As an example, in most PEEC models the function f_j is a constant pulse function.

We can view (3.83) as a row of a matrix representing the problem. The next step with the testing function is to set up a system of equation based on an averaging integration process of the equation. Mathematically, the averaging or testing function for the ith row of the matrix yields

$$\sum_{j=1}^{N} \alpha_j \frac{1}{\mathcal{V}} \int_{\mathcal{V}} f_i \, \mathcal{L} f_j \, d\mathcal{V} = \frac{1}{\mathcal{V}} \int_{\mathcal{V}} f_i \, \boldsymbol{b}(\boldsymbol{r}) \, d\mathcal{V}, \tag{3.84}$$

while $\mathcal{V}$ represents a volume, but it can also be an area $\mathcal{S}$ or even a length ℓ. Here, f_i is a weighting function that is the same as f_j for the circuit and the Galerkin-type solution.

Once the basis and weighting or expansion functions are chosen, we can proceed to the third WRM step, which will solve the resultant system of equations.

Since the book is based on PEEC circuit techniques, we do not use the above mathematical formulation for the solutions. Fortunately, the circuit-oriented approach used yields a clear way to present the techniques.

PROBLEMS

3.1 *The independence of Maxwell's equations*

Derive Gauss' law for the electric field and also Gauss' law for the magnetic field from Ampere's law (3.1b) and Faraday's law (3.1a), respectively.

3.2 *The static and dynamic fields*

Assume the field is static (no time-dependent variation), rewrite Maxwell's equations and the corresponding wave equations. Point out key differences between the static field you derived and the dynamic field based on the Helmholtz equation.

3.3 *Green's function*

Let us assume the wave equation

$$(\nabla^2 + \beta^2) g(\boldsymbol{r}, \boldsymbol{r}') = -\delta(\boldsymbol{r} - \boldsymbol{r}'),$$

where $\delta(\boldsymbol{r} - \boldsymbol{r}')$ is the Dirac delta function with zero value everywhere except at point $\boldsymbol{r}'$ and where its volume integral over the whole space equal to 1. Show that, for three-dimensional (3D) space, the solution of $g(\boldsymbol{r}, \boldsymbol{r}')$ is

$$g(\boldsymbol{r}, \boldsymbol{r}') = \frac{e^{-j\beta|\boldsymbol{r}-\boldsymbol{r}'|}}{4\pi|\boldsymbol{r} - \boldsymbol{r}'|}.$$

Use the conventional $e^{j\omega t}$ as the time dependency reference.

3.4 *Magnetic field generated by currents*

Assume a square conducting loop has the side length ℓ. The conductor is assumed to have a diameter d where ($d \ll \ell$). Assume that a uniform *dc* current I is applied. Calculate the total magnetic flux created at the center of the loop.

3.5 *The electric field generated by charges*
Two large zero thickness square plates with a side length ℓ are totally overlapping. They are spaced by a small distance d. A dielectric material is placed between the two plates with permittivity ϵ. We assume that the top plate is charged with a (nonphysical) uniformly distributed positive charge and an equal and opposite uniformly distributed negative charge is on the bottom plate. Assume that the nonuniform field at the plate edges can be ignored. Calculate the capacitance between the plates.

REFERENCES

1. C. A. Balanis. *Advanced Engineering Electromagnetics*. John Wiley and Sons, Inc., New York, 1989.
2. J. A. Stratton. *Electromagnetic Theory*. McGraw-Hill Book Company, New York, 1941.
3. T. Weiland. Eine Methode zur Loesung der Maxwellschen Gleichungen fuer sechskompoentige Felder auf diskreter Basis. *Archiv der Elektrischen Ubertragung*, 31:116–120, 1977.
4. S. Ramo, J. R. Whinnery, and T. Van Duzer. *Fields and Waves in Communication Electronics*. John Wiley and Sons, Inc., New York, 1994.
5. B. A. Finlayson. The method of weighted residuals and its relation to certain variational principles for the analysis of transport processes. *Chemical Engineering Science*, 20:395–404, May 1965.
6. L. Collatz. *The Numerical Treatment of Differential Equations*. Springer-Verlag, Berlin and New York, 1960.
7. Z. D. Chen and M. M. Ney. The method of weighted residuals: a general approach to deriving time- and frequency-domain numerical methods. *IEEE Antennas and Propagation Magazine*, 51(1):51–70, February 2009.
8. R. F. Harrington. *Field Computation by Moment Methods*. Macmillan, New York, 1968.
9. J. M. Jin. *The Finite Element Method in Electromagnetics*. John Wiley and Sons, Inc., New York, 2nd edition, 2002.
10. L. Pillegi, R. Rohrer, and C. Visweswariah. *Electronic Circuits and System Simulation Methods*. McGraw-Hill Book Company, New York, 1995.
11. M. Celik, L. Pileggi, and A. Odabasioglu. *IC Interconnect Analysis*. Kluwer Academic Publishers, Boston, MA, Dordrecht, London, 2002.
12. B. G. Galerkin. Sterzhni i plastiny. ryady v nekotoykh voprosakh uprogogo ravnovesiya sterzhnei i plastin. *Vestb. Inzhen. i Tekh. Petrograd*, 19(2):897–908, 1915.
13. C. Hafner. *The Generalized Multipole Technique for Computational Electromagnetics*. Artech House, Boston, MA, 1990.
14. W. C. Chew, J.-M. Jin, E. Michielssen, and J. Song. *Fast and Efficient Algorithms in Computational Electromagnetics*. Artech House, Boston, MA, 2001.
15. L. W. Nagel. SPICE: a computer program to simulate semiconductor circuits. Electronics Research Laboratory Report, ERL M520, University of California, Berkeley, CA, May 1975.

4

CAPACITANCE COMPUTATIONS

We consider capacitance modeling in this chapter. Capacitance calculations have a long history from Maxwell [1] to early precomputer work, for example, Ref. [2]. However, the majority of the important work is based on the availability of computers, for example, Refs [3–5].

Capacitance computations also represent a key component of partial element equivalent circuit (PEEC) models. This is evident from Chapter 6. Subdividing the conductors into cells is fundamental to PEEC. Two different issues exist. As shown in Fig. 4.1, the charge on the conductors is not uniform. However, the potential of a conducting plate is constant as shown. As we will see, at high frequencies conducting surfaces are no longer equipotentials. In both cases, we need to subdivide the plate into cells such that the charge can assume a different value at each cell.

We start out with the computation of capacitance as a stand-alone solution. This expands the scope of this chapter and it includes aspects that are important for capacitance modeling in general. We hope that this gives the reader a good understanding on how to compute and approximate capacitance values.

To start with, we assume that we want to compute the capacitance for a thin conducting plate or a set of plates. For the single plate, the capacitance C is governed by the well-known relation between the charge Q and potential Φ given by $Q = C\Phi$. The example of a single plate is shown in Fig. 4.1. As shown in Fig. 4.1a, the capacitance for the single plate in free space is computed with respect to infinity. It is important to understand that infinity is the only "true" ground where the potential $\Phi = 0$ V. This plays an important role in PEEC models in general. From a SPICE circuit point of view in Chapter 2, this is the true *ground* node for three-dimensional problems.

In general, the voltage V is defined as a potential difference, or $V = \Phi_1 - \Phi_2$ where Φ_1 and Φ_2 are two different potentials of two different conductors. In Fig. 4.1b , we give an

Circuit Oriented Electromagnetic Modeling Using the PEEC Techniques, First Edition.
Albert E. Ruehli, Giulio Antonini, and Lijun Jiang.

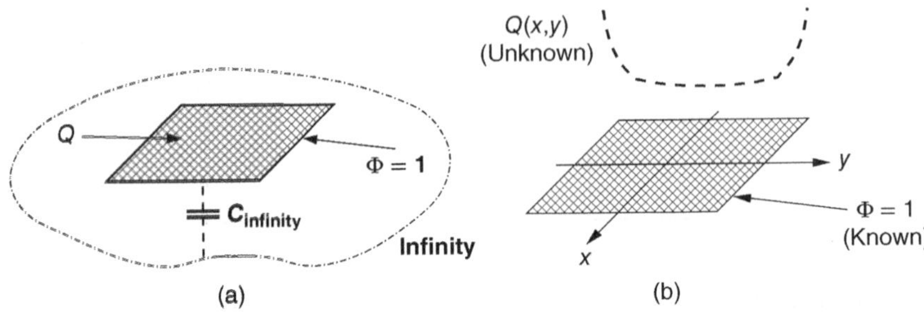

FIGURE 4.1 Single plate. (a) Capacitance to infinity. (b) Charge distribution.

example of a square plate. We can assume that the potential of the plate is $\Phi = 1$ V. This leads to $C = Q$. Hence, we have to find the total charge first before getting the capacitance. As is indicated schematically in Fig. 4.1b, the charge density is the smallest at the center while it goes to infinity toward the edges of the plate due to the forces exerted on the charges at any point on the plate, in opposite directions. Charge is limited to the plate surface.

The challenging aspect of the capacitance determination is the computation of the total charge on the surface of the plate. This is accomplished by subdividing the charge plate and by adding up the charges of all subcells on the plate. The techniques for computing the charges are given in Section 4.3.

4.1 MULTICONDUCTOR CAPACITANCE CONCEPTS

The capacitance concept for multiconductor geometries is considered in this section. In Refs [6] and [7], these issues are considered in more detail since important applications are in the interconnect modeling area.

In general, we have a system with K conductors for which we compute multicapacitances. An example is given in Fig. 4.2 where we have $K = 4$ conductors. In this example, we assume that conductors 1 and 2 are electrically connected. The same is assumed for conductors 3 and 4. Obviously, for high frequencies, the connection between the conductors would also have to be modeled, while for a capacitance problem we can assume that the result of the connections is that conductors are at the same potential and charges are shared. The properties of the small connecting wires are ignored in this example. Hence, the resultant system can be represented with a 2×2 capacitance matrix since the conductor pairs are of the same Φ. Of course, the conductors are in the presence of the zero potential reference at ∞ that has to be taken into account.

Two–dimensional (2D) models are extensively used for transmission lines and other applications. However, since the transmission line conductors in a 2D system extend from $-\infty$ to $+\infty$ in length, the potential reference cannot be at ∞ anymore. Hence, we *artificially*

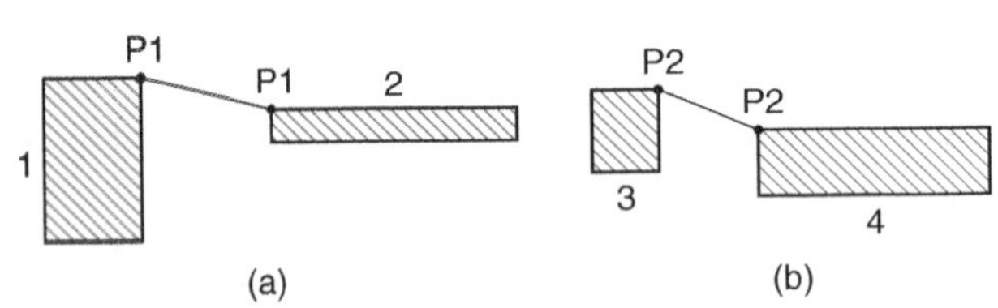

FIGURE 4.2 Multiconductor example where some of the conductors are electrically connected.

need to declare one conductor in a multiconductor system to be the zero voltage reference $\Phi = 0\ V$. It is evident from this that 2D models consist of at least two conductors. The capacitance of a single wire that extends to infinity is ∞.

Of course, there are also three-dimensional situations where we assume a 0 V reference at a finite point in a geometry. However, this is a *local* reference which, as we show in Section 2.7.4, is independent of the ground reference at infinity. It is evident from the above considerations that the mixture of two- and three-dimensional structures leads to challenging reference situations. Surprisingly, workable solutions can be found for many arrangements.

A three-dimensional multiconductor situation with the conventional zero potential at ∞ is considered next. The capacitance model for this case, shown in Fig. 4.3, is an obvious extension of the single capacitance model in Fig. 4.2. We need to establish the connection between this capacitance model and the capacitance matrix formulation. Actually, for real-life applications, an amazingly large number of different cases can be relevant.

4.2 CAPACITANCE MODELS

We start with the concept of multiconductor capacitance that is important for many different aspects of modeling of real situations. Most of the techniques apply to two-dimensional as well as three-dimensional geometries. The units for the two-dimensional case, where the conductors are theoretically infinitely long, are necessarily per unit length (F/m or pF/mm) in our normalized units since the infinite structure has an infinite total capacitance. For the three-dimensional case (see Appendix A), units are simply pF or other units that lead to conveniently scaled numbers. Examples are given in Appendix A.

4.2.1 Capacitance Models for Multiconductor Geometries

Figure 4.3 shows a two-conductor geometry. Intuitively, the equivalent circuit shown seems to correctly represent the two conductors and the coupling between them.

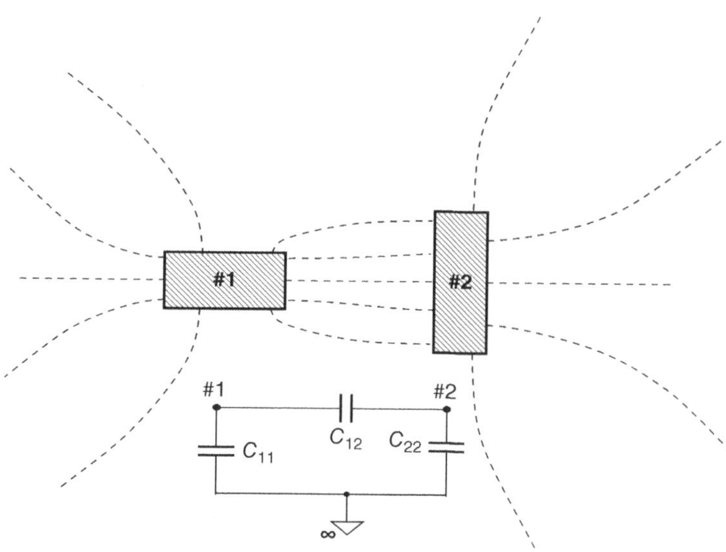

FIGURE 4.3 Flux patterns and equivalent circuit for a two-conductor system.

An interesting issue is that both capacitances that connect to the zero potential at ∞ can be considered to connect to a *common node* representing infinity. As is shown in Fig. 4.3, the charge on conductor 1 is shared between the two-terminal or circuit capacitances C_{11} and C_{12}. The definition of a *two-terminal* capacitance is given by

$$C_{ij} = \frac{Q_{ij}}{(\Phi_i - \Phi_j)}, \quad i \neq j, \tag{4.1}$$

where Q_{ij} is the charge associated with C_{ij} on both conductors i and j. Note that the voltage across the capacitors is $V = \Phi_i - \Phi_j$. In the specific case of the two-conductor model in Fig. 4.3 by applying (4.1) to conductor 1 the charge on conductor is given by

$$Q_1 = Q_{11} + Q_{12} = (C_{11} + C_{12})\Phi_1 - C_{12}\Phi_2. \tag{4.2}$$

Similarly, the charge on conductor 2 is given by

$$Q_2 = Q_{21} + Q_{22} = -C_{21}\Phi_1 + (C_{22} + C_{21})\Phi_2. \tag{4.3}$$

4.2.2 Short Circuit Capacitances

We want to relate this result to the so-called *short circuit capacitances* **Cs**, which is the fundamental capacitance model given by $\mathbf{Q} = \mathbf{Cs}\,\boldsymbol{\Phi}$. The name is simple to remember since the elements are computed under the condition that all the other potentials are shorted or set to zero.

If we apply this equation to a two-conductor system, we have

$$\begin{aligned} Q_1 &= Cs_{11}\Phi_1 + Cs_{12}\Phi_2 \\ Q_2 &= Cs_{21}\Phi_1 + Cs_{22}\Phi_2. \end{aligned} \tag{4.4}$$

We observe that the charge induced on conductor 1 by setting $\Phi_2 = 1$ V and shorting conductor 1 or $\Phi_1 = 0$ V is equal to Cs_{12} in terms of the value. Hence, we obtain Cs_{12} from the charge computation with a special short circuit setup. For the general case of K conductors, it is obvious that all other $K - 1$ conductors need to be shorted to ground or infinity. Hence, the name *short circuit capacitances* is appropriate for them. As we show below, the short circuit capacitances are usually obtained first by numerical capacitance solutions.

Of course, we need to relate the short circuit capacitances (4.4) to the two-terminal or circuit capacitances (4.2). By comparing the two systems, we find that

$$\begin{bmatrix} Cs_{11} & Cs_{12} \\ Cs_{21} & Cs_{22} \end{bmatrix} = \begin{bmatrix} C_{11} + C_{12} & -C_{12} \\ -C_{21} & C_{21} + C_{22} \end{bmatrix}. \tag{4.5}$$

From this comparison, it is obvious that the off-diagonal terms are in general where the circuit capacitances are given by $C_{ij} = -Cs_{ij}$. We can also generalize the result for the diagonal entries that can be represented as the sum of all terminal capacitances in each row. This can in general be written as

$$Cs_{ii} = \sum_{j=1}^{K} C_{ij}. \tag{4.6}$$

TABLE 4.1 Matrix element stamp for two terminal capacitances into short circuit capacitance matrix.

$C_{k\ell}$	$-C_{k\ell}$	$-C_{\ell k}$	$C_{\ell k}$
k	$C_{k\ell}$	$-C_{k\ell}$	0
ℓ	$-C_{\ell k}$	$C_{\ell k}$	0

This is a somewhat unconventional approach to show how the short circuit capacitance matrix is assembled using the two-terminal capacitances. In fact, we can generalize the approach by using so-called matrix stamps that are used for circuit analysis as part of the MNA approach in Chapter 2. Starting from (4.1), we see how the charge associated with each circuit capacitor C_{ij} is related to the potentials Φ_i and Φ_j. So, the contribution of each of the capacitances to the **Cs** matrix is represented with the matrix stamp in Table 4.1.

The short circuit matrix stamping pattern becomes clear if we consider a 4×4 matrix describing a four-conductor system

$$\begin{bmatrix} C_{11}+C_{12}+C_{13}+C_{14} & -C_{12} & -C_{13} & -C_{14} \\ -C_{21} & C_{21}+C_{22}+C_{23}+C_{24} & -C_{23} & -C_{24} \\ -C_{31} & -C_{32} & C_{31}+C_{32}+C_{33}+C_{34} & -C_{34} \\ -C_{41} & -C_{42} & -C_{43} & C_{41}+C_{42}+C_{43}+C_{44} \end{bmatrix}. \tag{4.7}$$

Each off-diagonal term of the short circuit matrix is given by $Cs_{mk} = -C_{mk}$ that connects conductors m and k.

As a last step, we give the short circuit capacitance matrix using the matrix Kirchhoff's current law (KCL) equation or (2.34), which is an incidence matrix **A**. In the approach for forming the **Cs** matrix, we first form a branch capacitance matrix that is similar to a branch impedance matrix for circuits. This large $K \times K$ matrix is simply a collection of all circuit elements

$$\mathbf{Cb} = \text{diag}[C_{11}, C_{12}, \ldots\ C_{K(K-1)}, C_{KK}], \tag{4.8}$$

where K is the number of conductors. With this, the short circuit matrix can be formed as

$$\mathbf{Cs} = \mathbf{A}\ \mathbf{Cb}\ \mathbf{A}^{\mathbf{T}}. \tag{4.9}$$

One of the **Cs** formulation given may be best for a specific application.

4.2.3 Coefficient of Potential Matrix Pp

As shown in Section 6.3.4, short circuit capacitances are originally computed from the so-called partial coefficient of potential Pp_{km}. The coefficient of potential matrix that relates potential to charge is simply the inverse of the short circuit capacitances, or

$$\mathbf{\Phi} = (\mathbf{Cs})^{-1}\mathbf{Q} = \mathbf{Pp}\ \mathbf{Q}. \tag{4.10}$$

An important aspect that needs to be pointed out is that the size of this coefficient of potential matrix is given by the number of conductors K. Hence, we deal with the total charge Q_m on each conductor m and the potential Φ_m of each conductor. This is quite different from the subdivided partial conductors considered below for capacitance computations.

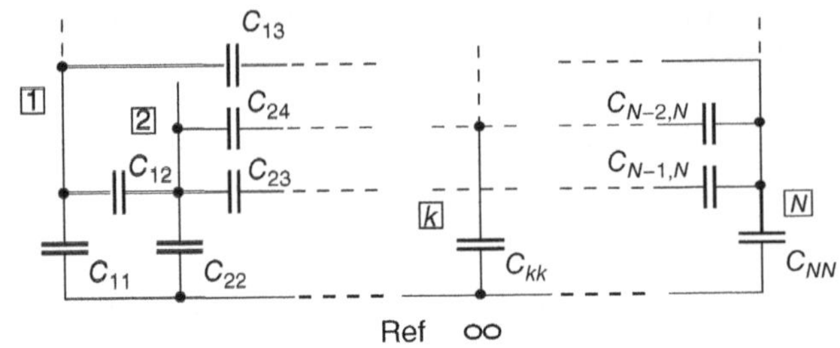

FIGURE 4.4 Capacitance model for *N - conductor* system.

4.2.4 Capacitance of Conductor Systems

The model in Fig. 4.4 gives an example of the capacitance equivalent circuit for a problem which includes N conductors and where the reference **Ref** is the potential at infinity. As is shown, in a complete model, the two-terminal coupling capacitor exists between all conductors.

We can consider two examples for the capacitance of a conductor in the presence of other conductors. As a first situation, we assume that all the other conductors are grounded. Hence, in Fig. 4.4 we want to know the capacitance of conductor k where all the other nodes are ground. It is easy to see that the solution we are looking for is

$$\mathbf{Q} = \mathbf{Cs}\ \mathbf{\Phi} \tag{4.11}$$

where only $\Phi_k = 1$ and all the other conductors are at $\Phi = 0$. Then, the capacitance of conductor k is given by Q_k.

A second situation leads to a more complicated evaluation. If some of the conductors are floating, then the potential (voltage) they assume becomes unknown. The same is the case for the charge.

Hence, in this case, both the charge as well as the potential of some conductors is not known. Of course, the exceptions are the grounded conductors with a potential $\Phi = 0$ and the conductor for which we apply exceptions, the potential $\Phi = 1$. Then, the system to be solved for the charges is

$$\mathbf{P}\ \mathbf{Q} = \mathbf{\Phi} \tag{4.12a}$$

$$\mathbf{Cs}\ \mathbf{\Phi} = \mathbf{Q} \tag{4.12b}$$

where the unknows consist of the potential of the floating conductors and their charge. We give examples for the elimination of floating capacitance nodes for some simple examples in the next section.

4.2.5 Elimination of a Floating Conductor Node

We assume that we have computed the **P** or **Cs** by the technique presented in 4.3. Eliminating a shorted node is given in the previous section. Assuming that all shorted conductors or nodes are eliminated we assume that the size of the resultant capacitance matrix is not too large. Here we provide an example for a three node system where we want to eliminate

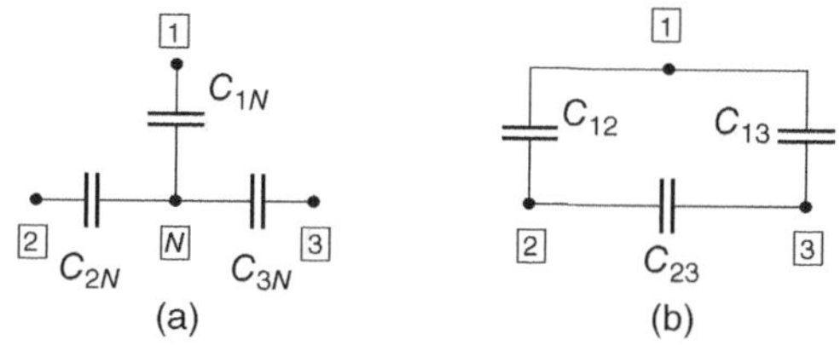

FIGURE 4.5 Example for star node N (a) and Delta elimination of node N (b).

a floating node. We note that it as well could be the reference or ground node of the system.

In this example, we want to compute the capacitance C_2 from node 2 to ground by eliminating the node where nodes 1 and 3 are grounded in Fig. 4.5. From circuit analysis we find the capacitance at node 2 as

$$C_2 = [C_{2N}(C_{1N} + C_{3N})]/(C_{1N} + C_{2N} + C_{3N}) \tag{4.13}$$

where the circuit now has only 2 nodes where one of them is the reference node.

As a second example, we give the 4 node problem in Fig. 4.5 where we want to eliminate the floating node N from the equivalent circuit.

The reduced 3 node model is shown in Fig. 4.5(b). The computation of the reduced capacitor model is given in terms of the 4 node model as

$$C_{12} = \frac{C_{1N}}{1 + C_{1N}/C_{2N} + C_{3N}/C_{2N}} \tag{4.14}$$

$$C_{13} = \frac{C_{3N}}{1 + C_{3N}/C_{1N} + C_{2N}/C_{1N}} \tag{4.15}$$

$$C_{23} = \frac{C_{2N}}{1 + C_{1N}/C_{3N} + C_{2N}/C_{3N}} \tag{4.16}$$

More general cases, where the capacitance matrix is larger than two, is more difficult for the elimination of floating nodes. For this case, the so-called Y - Δ (Y-delta) method is used.

4.2.6 Floating or Reference Free Capacitances

In Fig. 4.4 the capacitances for N conductors are shown. This assumes that all conductors are not connected to other objects etc. So, it is somewhat of a theoretical question.

A specific question is related to the reference node. Usually, to avoid a singular circuit solution we do not write an equation for the reference node in Fig. 4.4. However, for the equivalent circuit, we can choose any node as the reference node which is more appropriate for a reference, like for example a small mobile product. Hence, any node can be chosen from the equivalent circuit as a reference. However, the circuit equations must be changed by writing an equation for the reference node and by deleting the row and column for the new reference node.

4.3 SOLUTION TECHNIQUES FOR CAPACITANCE PROBLEMS

The previous sections are based on the assumption that we know the short circuit capacitance matrix **Cs** for the conductor configuration. For the quasistatic case, the conductors represent equipotential surfaces. As illustrated in Fig. 4.1, the challenge is how to compute the nonuniform charge distributions on conductor surfaces.

To find a solution, we have to use a WRM approximation. In the example, we use the model shown in Fig. 4.6 for a plate where a conducting sheet is subdivided into rectangles of constant charge distribution. We should note that the rectangles are assumed to be isolated by an infinitesimal gap so that the charge is constrained to the rectangular cells or

$$q(\boldsymbol{r}) = \frac{Q}{S}, \tag{4.17}$$

where q is the charge density and Q is the total charge on the cell, $\boldsymbol{r}$ is located on the individual cells and S is the surface area of the cell. Hence, $q(\boldsymbol{r})$ is a uniform charge density on the cell whose location is indicated by the vector $\boldsymbol{r}$.

We next consider the integral equation (IE) to find the charges. Equation (3.52) is a solution to Maxwell's equations of the form

$$\Phi(\boldsymbol{r}) = \frac{1}{\epsilon}\int_{S'} g(\boldsymbol{r},\boldsymbol{r}')\, q(\boldsymbol{r}')dS', \tag{4.18}$$

where S' is the surface where the charge is located, q is the surface charge density, and g is the free space quasistatic Green's function given by (3.65)

$$g(\boldsymbol{r},\boldsymbol{r}') = \frac{1}{4\pi}\frac{1}{|\boldsymbol{r}-\boldsymbol{r}'|}, \tag{4.19}$$

which essentially is the potential generated by a point source in the free space. Here, $\boldsymbol{r}$ is the vector of the observation point and $\boldsymbol{r}'$ is the vector of the source point.

The above discretization is the key part for the transformation of the IE for the charge (4.18) into the solution of a matrix problem. The unknowns clearly are the charges on the cells. Hence, we can apply (4.18) in terms of integrals over all cells, or

$$\Phi(\boldsymbol{r}) = \frac{1}{\epsilon}\int_{S_1} g(\boldsymbol{r},\boldsymbol{r}')\, q_1(\boldsymbol{r}')dS' + \frac{1}{\epsilon}\int_{S_2} g(\boldsymbol{r},\boldsymbol{r}')\, q_2(\boldsymbol{r}')dS' + \cdots, \tag{4.20}$$

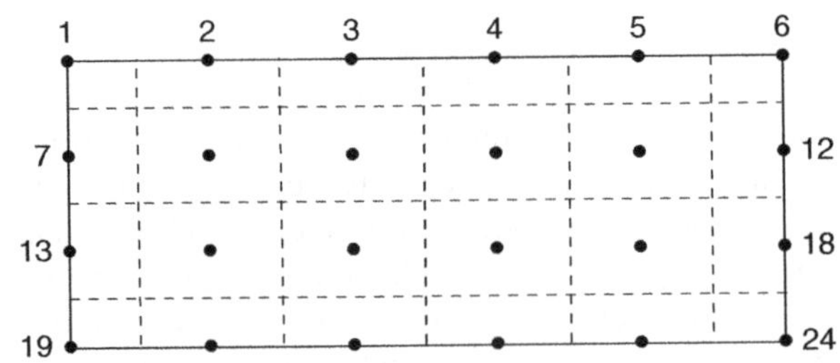

FIGURE 4.6 Mesh cells for charge distribution on the conductor.

where S_ℓ represents the surface of cell ℓ. Assuming the uniform charge density on each cell, we can rewrite (4.20) as

$$\Phi(\boldsymbol{r}) = \frac{Q_1}{\epsilon S_1}\int_{S_1} g(\boldsymbol{r},\boldsymbol{r}')dS' + \frac{Q_2}{\epsilon S_2}\int_{S_2} g(\boldsymbol{r},\boldsymbol{r}')dS' + \cdots, \tag{4.21}$$

where $Q_i = S_i\, q_i$. We notice that the integrals over the Green's function in (4.21) can be evaluated once the locations of the cell ℓ and the observation point $\boldsymbol{r}_k$ are known:

$$Pp_{k\ell} = \frac{1}{\epsilon S_\ell}\int_{S_\ell} g(\boldsymbol{r}_k,\boldsymbol{r}')\, dS'. \tag{4.22}$$

The last step in forming a system of equations can again be done with different WRM approximations.

A more obvious approach for the solution is to place the vector $\boldsymbol{r}$ at the center of each cell yielding a row in the matrix system of the form

$$\boldsymbol{\Phi} = \mathbf{Pp}\ \mathbf{Q}. \tag{4.23}$$

The solution for the charges is simply $\mathbf{Q} = \mathbf{Pp}^{-1}\ \boldsymbol{\Phi}$. This approach is called a *collocation* technique.

For the example, in Fig. 4.6 we have 24 unknown charges. The potential can be normalized to 1 on all cells. Hence, the potential vector is $\boldsymbol{\Phi} = [1\ 1\ 1 \cdots 1\ 1]^T$ of the length of the number of cells on the conductor, which is 24 in this example. This results in the vector of the 24 unknown charges $\mathbf{Q}$. Finally,

$$C_{11} = Q_T = \sum_{k=1}^{24} Q_k \tag{4.24}$$

results in an approximation to the capacitance C_{11} or the total charge Q_T.

We can revisit the above solution approach by comparing the result of (4.23) evaluated on any cells to find the difference or errors as

$$Error = |\boldsymbol{\Phi} - \mathbf{Pp}\ \mathbf{Q}|, \tag{4.25}$$

where *Error* is the residual potential error, which would be zero for an exact solution. In the collocation approach, we match the potential at conductor surfaces at the center of each cell shown in Fig. 4.6.

In the process of finding a circuit solution, we note that capacitances must be symmetric. Hence, we tried to make the coefficient in (4.22) symmetric with an additional averaging process, or

$$Pp_{k\ell} = \frac{1}{\epsilon S_k S_\ell}\int_{S_k}\int_{S_\ell} g(\boldsymbol{r}_k,\boldsymbol{r}_\ell)\, dS_k\, dS_\ell. \tag{4.26}$$

This corresponds to averaging the potential over each cell as

$$\Phi_{average} = \frac{1}{S}\int_S \Phi(\boldsymbol{r})\, d\,S, \tag{4.27}$$

where S is the cell surface. This corresponds to a Galerkin approximation since we use a constant approximation for both the charge and the potential. This is part of the WRM approximation technique. The evaluation of the integrals in (4.26) is considered in Appendix D. It is clear that the efficient evaluation of the integrals has a considerable impact on the efficiency of the solution.

As a reference, we give the capacitance of a 1 m^2 zero thickness square plate, which is about 41 pF [8]. In the cited work, this result is compared to other techniques such as the collocation method. Very often, it is useful to have such a result for the debugging of new numerical codes.

4.3.1 Differential Equation (DE) Methods for Capacitance Computations

In the previous section, an IE solution was presented, which is part of the PEEC approach. An alternate interesting approach for capacitance computation is based on differential equations (DEs) formulations. Further, a combination of the different techniques works best for some applications. We also can combine some models based on DE techniques together with partial PEEC model inductances. For this reason, we also consider simple DE circuit-based models. Graphical techniques were used to approximate such solutions in the precomputer era [9].

Simple DE models are not well suited for single plate capacitance computations. The IE methods in the previous section yield relatively simple solutions for this type of problem. Problems for which DE method yields good approximate solutions are related to a parallel plate capacitor in Fig. 4.7.

Starting from one of Maxwell's equations (3.1 c), we get

$$\nabla \cdot \epsilon_0 \epsilon_r \boldsymbol{E} = q, \tag{4.28}$$

which can also be regarded as the differential form of the Gauss theorem. It is assumed that between the plates there we only have a uniform z component of the electric field. Applying the Gauss theorem to a surface S enclosing one of the plates, we obtain

$$E = \frac{Q}{\epsilon_0 \epsilon_r S}, \tag{4.29}$$

where $Q = \int q\, dS = qS = q\, \ell\, w$ is the total charge on the plates and we assume that a uniform dielectric with a relative dielectric constant ϵ_r exists in the area between the plates.

Furthermore, integrating $\boldsymbol{E} = -\nabla\Phi$ along the thickness of the capacitor, we obtain

$$V = E\, d, \tag{4.30}$$

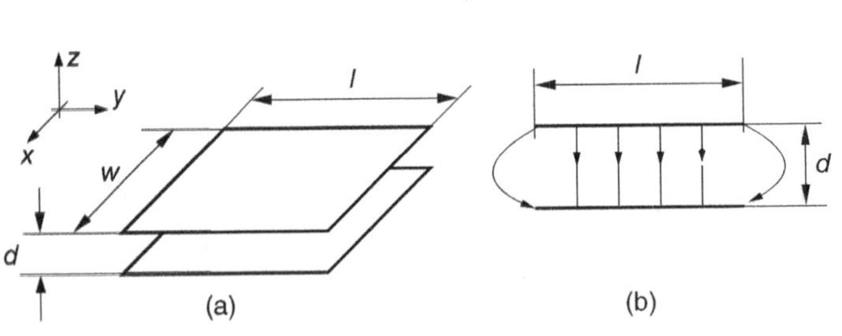

FIGURE 4.7 (a) Parallel plate conductor. (b) Side view and field lines.

where V is the voltage across the capacitor and d its thickness. Combining (4.29) and (4.30) the result becomes

$$C = \frac{Q}{V} = \frac{\epsilon_0\,\epsilon_r\,\ell\,w}{d}. \tag{4.31}$$

As an example with $\epsilon_r = 10$, a capacitor with an area of 1 mm^2 and $d = 0.1$ mm has a capacitance $C = 0.8854$ pF.

The capacitance C obtained with (4.31) is reasonably accurate for closely located plates $d \ll \min(\ell, w)$ since the field between the plates is much larger than the external fields. However, (4.31) becomes less accurate for large spacings d since the fringe fields outside of the plates are becoming significant. The fringe factor for general squares is given in Ref. [4], which was computed using the IE solution in the previous section.

The key question is how we can take the contribution of the field lines outside the plates into account. They are illustrated in Fig. 4.3. For this, we want to consider an approximate circuit solution for the fringing fields. Unfortunately, also the free space must be subdivided into capacitive cells at least in the neighborhood of the conductors. It is clear that the model can become very complex.

As an example, we construct a two-dimensional (2D) capacitive models. Figure 4.8 shows a subdivision mesh with nodes where the potential is specified. For the two-dimensional case, we assume that the structure is infinitely long in the z direction such as a transmission line structure. Hence, the result will be per unit length capacitances in the z direction. In this model, plates of the local capacitors are always between two nodes. The incomplete equivalent circuit corresponding to the mesh in Fig. 4.8 is shown in Fig. 4.9.

A simple application of the equivalent circuit in Fig. 4.9 for a two-dimensional parallel capacitance problem is shown in Fig. 4.7. The upper plate is represented by the connected nodes 7–9 and the lower plate is represented by the connected nodes 12–14. Hence, the nodes and capacitances such as 7–8 are shorted. For the symmetric problem, we can assume that the upper plate is +0.5 V and the lower plate is −0.5 V as shown in Fig. 4.10. For clarity, we left all capacitances between the node out.

Since we apply 1 V between the plates, we get $Q = C_{pp}V = C_{pp}$. Hence, we need to evaluate the total charge induced on one of the plates. We set up the equations for the circuit represented by the capacitive mesh in Fig. 4.9. For example, for node 1 we set up an equation of the form

$$Q_1 = C_{12}(\Phi_1 - \Phi_2) + C_{16}(\Phi_1 - \Phi_6) = 0. \tag{4.32}$$

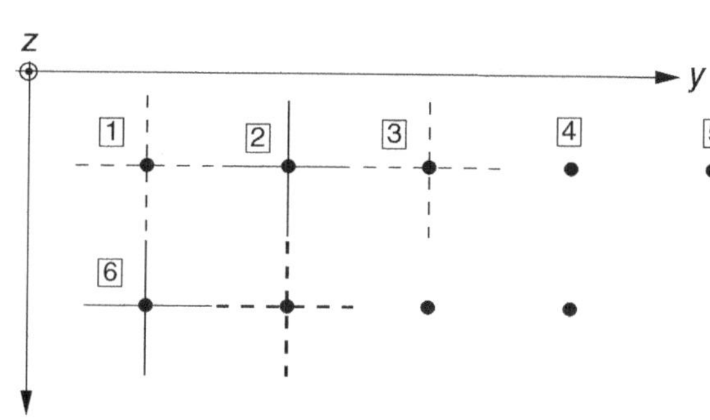

FIGURE 4.8 Mesh for two-dimensional capacitance model.

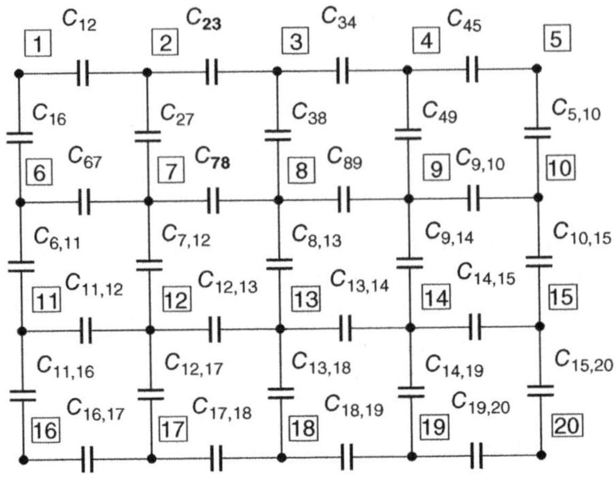

FIGURE 4.9 Two-dimensional capacitance model for free space or dielectric region.

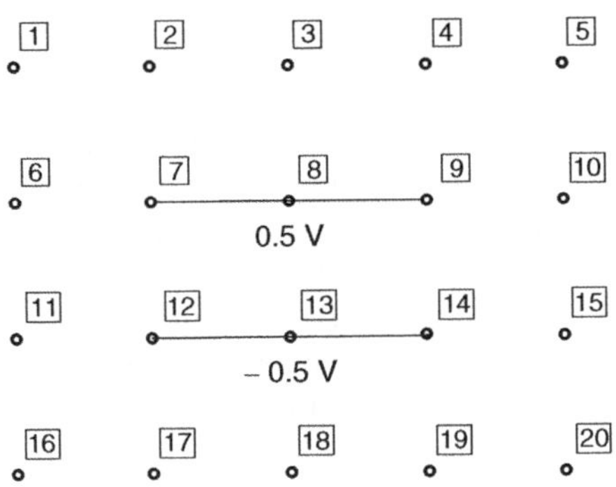

FIGURE 4.10 Two 2D plates embedded in capacitance mesh.

As another example, the equation for node 7 on the upper plate is given by

$$Q_7 = C_{27}(\Phi_7 - \Phi_2) + C_{7,12}(\Phi_7 - \Phi_{12}) + C_{67}(\Phi_7 - \Phi_6). \tag{4.33}$$

From this, it is evident how to set up a circuit matrix system of the form $\mathbf{Q} = \mathbf{C\Phi}$. We can choose any one of the nodes on the lower plate as the zero voltage reference for all the other nodes. If we choose node 13, the symmetry is preserved. Finally, the capacitance is given by $C_{pp} = Q_7 + Q_8 + Q_9$. Further, symmetry can be used to simplify the system.

The modeling approach is evident from the above explanation and the extension to three dimensions is structurally complex but otherwise straightforward. For this reason, we consider only one local three- dimensional model in more detail. Figure 4.11a shows a capacitance plates for x and y-directions for the node located at the origin. For simplicity only one plate for the z-direction is shown.

The plates in the dash-dot lines are the capacitor plates in the x- and z-directions. The solid line plates are for capacitors in the y-direction. Only one dashed line plate is for capacitors in the z-direction. Finally, the equivalent circuit is shown in Fig. 4.11b. The circuit solution of the problem represented by the capacitance model in Fig. 4.11 is evident from Chapter 2.

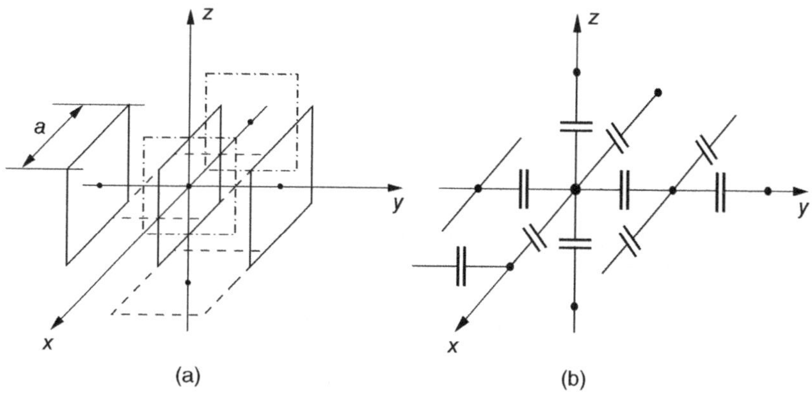

FIGURE 4.11 Three-dimensional capacitance model for dielectric region. (a) Subdivision of space. (b) Equivalent circuit.

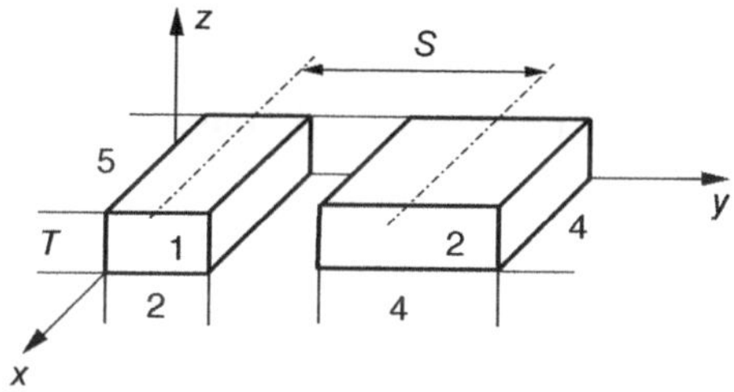

FIGURE 4.12 Two conductor example for the capacitance sensitivity.

4.4 MESHING RELATED ACCURACY PROBLEMS FOR PEEC MODEL

We only consider in this text the capacitance aspect of the meshing accuracy problem. However, the meshing sensitivity issue presented here can also be related to the meshing for inductance cells in Chapter 5. The PEEC solution approach for capacitances is given in Section 4.3. Also, the construction of meshes is presented in Chapter 8. Here, we consider important issues regarding the influence of the mesh subdivision on the solution accuracy. Since we consider sensitivities only, all dimensions in Fig. 4.12 can be viewed as normalized units.

It is clear that the accuracy of the computed capacitance values depend on the quality of the mesh. An important question is how fine the mesh subdivisions have to be for a sufficiently accurate answer without using too many cells or subdivisions. The sensitivity in (4.34) evaluates the quality of a mesh

$$S_{k\ell}(N) = \left| \frac{|C_{k\ell}(2N) - C_{k\ell}(N)|}{C_{k\ell}(2N)} \right|, \tag{4.34}$$

where N refers to the number of cells in (4.23) and k and ℓ are the two different conductors.

Here, $C_{k\ell}(2N)$ means the capacitance calculated using $2N$ mesh cells while $C_{k\ell}(N)$ refers to the capacitance calculated using N cells. It is clear that the accuracy of an off-diagonal capacitance values is usually more sensitive to the cell division. Therefore,

TABLE 4.2 Sensitivity factor for C_{12} capacitance to meshing.

Spacing	$S_{12}(20)$ for $T = 0$	$S_{12}(20)$ for $T = 0.2$	$S_{12}(50)$ for $T = 0$	$S_{12}(50)$ for $T = 0.2$
4	0.051	0.03	0.031	0.018
7	0.033	0.019	0.02	0.011
11	0.031	0.017	0.018	0.01

we use an off-diagonal coupling capacitance value for this measure. Practical experience with this sensitivity measure has shown that $S_{k\ell}(N) \leq 0.02$ indicates that the number of cells N leads to a sufficiently accurate answer.

The sensitivities are given for the mutual capacitance C_{12} for the two-conductor geometry in Fig. 4.12 for different conductor thicknesses T and conductor-to-conductor spacing S. The sensitivities in Table 4.2 are for 20 and 50 cells, respectively. We can observe that the sensitivity with respect to the number of cells is smaller for thin conductors. Of course, the sensitivity also decreases with larger conductor-to-conductor spacings S. In a problem with many conductors, the sensitivity of conductors with a small mutual capacitance tends to be less accurate.

4.4.1 Impact of Meshing on Capacitances and Stability and Passivity

In this section, we consider a very important meshing issue pertaining to the location of the cells on conductors. Essentially, the mesh subdivision or cells on an object can greatly impact the capacitance results.

We consider the impact of badly placed cells and on the partial capacitances and partial inductance values. The impact on the accuracy of capacitance calculations was first presented in Ref. [10] and the potential impact on the quality of the PEEC model solution is considered in Chapter 13.

The basic problem is that close conductors that are not well meshed can lead to poor results. In Fig. 4.13, we show a three- conductor problem where the lower plate, after meshing, consists of cells numbered 3–5. We chose the geometry of the thin conductors such that two top plates are exactly on top of the lower plate in the x direction with the same length $l_x = 100$ cm. All five cells are subdivided into two cells along y. Both top plates are $l_y = 50$ cm wide with a small spacing $S = 10$ cm in the middle. The total width of all bottom conductors is $y = 110$ cm.

The lower plate is subdivided into three width cells, 3–5, as shown in Fig. 4.13. To emphasize the problem, we chose the width for cells 3 and 5 to be 27.5 cm and cell 4 to be 55 cm. Hence, the top plates clearly do not project onto the lower plate. Finally, the plate-to-plate spacing in the z-direction is chosen to be 2.5 cm to emphasize the projection issue.

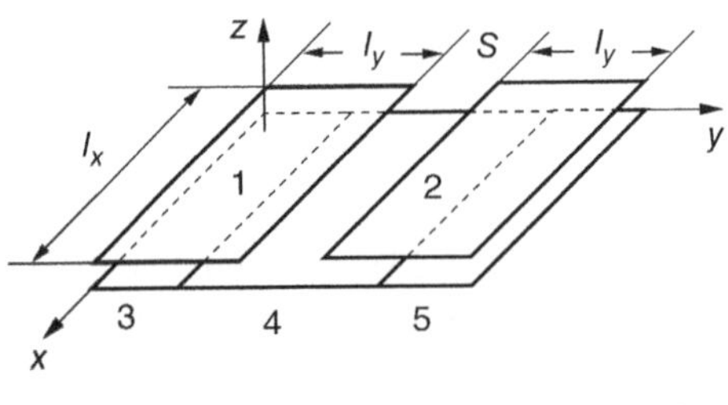

FIGURE 4.13 Three-conductor problem with overlapping cells.

If we compute the capacitances using the approach in Section 4.3, then the matrix solution will be in the form (4.23)

$$\mathbf{Q} = \mathbf{Pp}^{-1}\,\boldsymbol{\Phi} \tag{4.35}$$

for the subdivided system at the cell subdivision level. To reduce the system to the conductor plate level, we simply have to set all the partial potentials Φ on a plate, such as conductor 1 in Fig. 4.13 to be the same, Φ_1. Then by adding all the partial charges Q_k from all the cells on each conductor plate, we get the reduced system of short circuit capacitances:

$$\begin{bmatrix} Q_1 \\ \vdots \\ Q_4 \\ Q_5 \end{bmatrix} = \begin{bmatrix} Cs_1 & \cdots & Cs_{14} & Cs_{15} \\ \vdots & \vdots & \vdots & \vdots \\ Cs_{41} & \cdots & Cs_{44} & Cs_{45} \\ Cs_{51} & \cdots & Cs_{54} & Cs_{55} \end{bmatrix} \begin{bmatrix} \Phi_1 \\ \vdots \\ \Phi_4 \\ \Phi_5 \end{bmatrix}. \tag{4.36}$$

Finally, the capacitance of interest $C_{12} = -Cs_{12}$ is found from $Cs_{12} = Q_1/\Phi_2$ for the case when all the other plates are at potentials 0 and $\Phi_2 = 1$.

It is clear that for computing C_{12}, in Fig. 4.14 with the non overlapping cells the potentials of all conductors except for conductor 2 are set to zero. The potential on cell 2 is $\Phi = 1$, resulting in positive charges. Note that we created a bad overlap between cells 2 and 4 as shown in Fig. 4.13. This results in an excessive negative charge that is induced to cell 4. This in turn couples from cell 4 to cell 1 with the wrong amount of charge. Hence, $C_{12} = -Cs_{12}$ is computed incorrectly.

The capacitances computed for this case is a 3×3 symmetric matrix since all the lower plates are at the same potential. Then, the computed capacitances for Fig. 4.13 are in picofarad

$$C_{11} = C_{22} = 9.53, C_{33} = C_{55} = 21.09, C_{13} = C_{25} = 186.79, C_{12} = -37.90. \tag{4.37}$$

where we notice the wrong negative coupling capacitance between the two top conductors.

We compare this result with the case where we use the so- called *projection meshing*. This meshing is presented in Chapter 8. Using projection meshing, we have to subdivide the lower plate such that the cells of the top conductor exactly project onto the lower conductor as shown in Fig. 4.14 where the width of conductor 4 is also $l_y = 10$ cm, which corresponds to S in the top plate in both Figs. 4.13 and 4.14. The new computed symmetric capacitances for the new meshing in pF are

$$C_{11} = C_{22} = 8.97, C_{33} = C_{55} = 23.21, C_{13} = C_{25} = 219.68, C_{12} = 6.92. \tag{4.38}$$

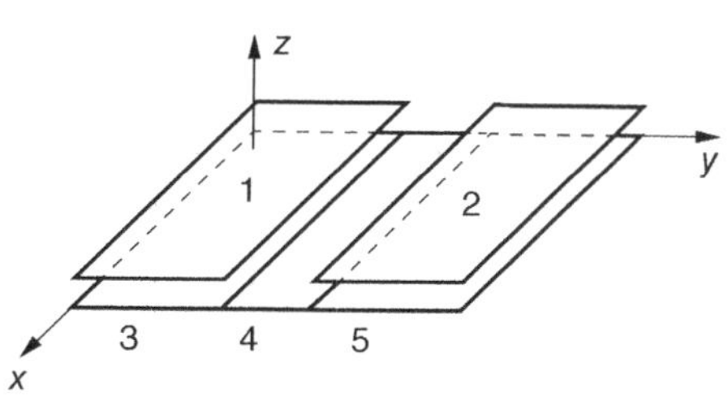

FIGURE 4.14 Three-conductor problem with nonoverlapping projection cells.

We observe that the coupling capacitance is now positive such that the negative C problem is solved, in addition to the improvement in the accuracy that is achieved with the better meshing without an excessive number of meshing cells or unknowns.

In a larger real problem, this small example geometry part with negative elements could be embedded in a model with otherwise correct partial elements. Then, the impact of negative capacitances could create stability or passivity problems that we consider in Chapter 13. Even for quasistatic PEEC models, all circuit elements should be positive for stability.

4.5 REPRESENTATION OF CAPACITIVE CURRENTS FOR PEEC MODELS

Three different models can be used to represent the capacitance part of a PEEC model. Two different types of models are based either on partial potential coefficients or on partial capacitances. These models are presented in the following section.

4.5.1 Quasistatic Capacitance–based Model

Conventional two-terminal capacitances (4.1) can be used only for quasistatic PEEC models. Section 6.3.4 presents the development of these models. However, we notice that this approach is quite expensive since the capacitances have to be precomputed. The $\mathbf{Pp}^{-1}$ matrix (4.23) is derived by the technique in Section 4.3. We want to point out that the approaches in the following two sections avoid the solution of the partial potential coefficient matrix system.

We consider a small example for three cells shown in Fig. 4.15 that corresponds to three cells or half-cells. We assume that they are obtained with meshing covered in Chapters 6 and 8. To make it slightly more general, we assume that cells 1 and 2 are located on a single plate or conductor while cell 3 may be on another conductor. However, cells 1 and 2 are also electrically isolated from each other by an infinitesimal gap, so that in reality the three cells are of the same type for our capacitance problem. Again, using the techniques in Section 6.3.4, we can obtain the capacitance model shown in Fig. 4.16. The simple model is very clear. Each node in the PEEC model will have the capacitance model attached. Hence, this model leads to coupling among all the nodes without delays. Still, this model has some applications especially smaller quasistatic models where the compute time is not an issue.

4.5.2 Current Source-Based Model for the Capacitances

We start with the same small three cell example in Fig. 4.15 to present this model. For full-wave models, we have to consider the fact that the couplings have a delay $\tau_{k\ell}$ as

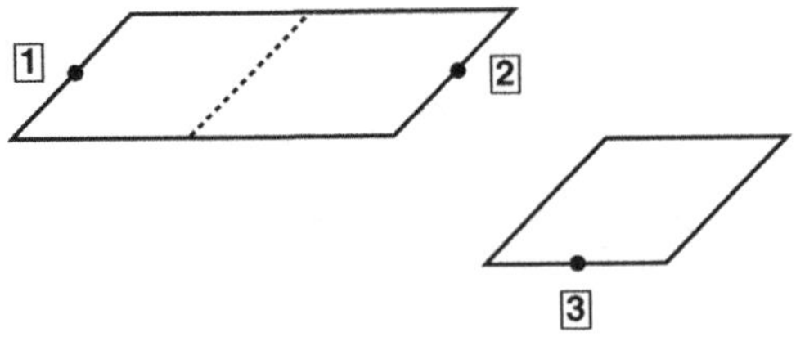

FIGURE 4.15 An example of three capacitive cells for a PEEC circuit.

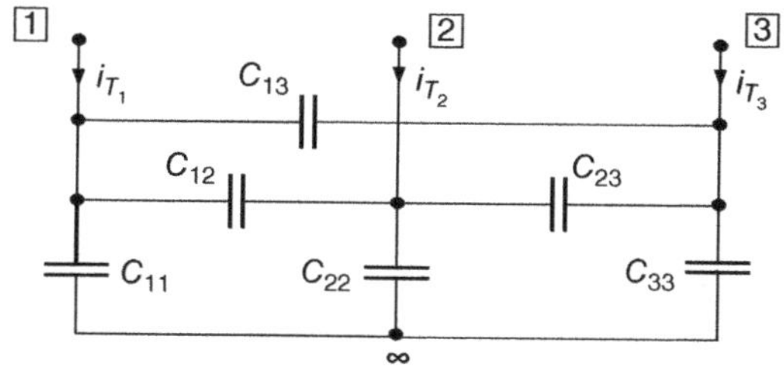

FIGURE 4.16 Example PEEC capacitance model for small three cell example.

discussed in Section 2.11.1, which corresponds to multiplication by $e^{-s\tau_{k\ell}}$ in the frequency domain.

Importantly, the capacitance models in this and the following section are applicable for quasistatic or full-wave models in the time and frequency domains. The model for the capacitive currents in PEEC models is directly based on the partial potential coefficients since it avoids capacitance precomputations.

The potential coefficients for the three-conductor systems in Fig. 4.15 lead to the system of equations

$$\Phi_1 = Pp_{11}\,Q_1 + Pp_{12}\,Q_2 + Pp_{13}\,Q_3 \tag{4.39a}$$

$$\Phi_2 = Pp_{21}\,Q_1 + Pp_{22}\,Q_2 + Pp_{23}\,Q_3 \tag{4.39b}$$

$$\Phi_3 = Pp_{31}\,Q_1 + Pp_{32}\,Q_2 + Pp_{33}\,Q_3, \tag{4.39c}$$

where Q_k is the total charge on cell k and Pp_{km} is the subdivided, normalized potential coefficient for conductors k and m. We start the derivation with the computation of currents from charges, which is given by

$$i(t) = \frac{dQ(t)}{dt} \quad \text{or} \quad I(s) = sQ(s), \tag{4.40}$$

in the time domain or in the frequency domain, respectively. If we take (4.39a) and divide it by Pp_{11}, we get

$$\frac{1}{Pp_{11}}\Phi_1 = Q_1 + \frac{Pp_{12}}{Pp_{11}}\,Q_2 + \frac{Pp_{13}}{Pp_{11}}\,Q_3. \tag{4.41}$$

Taking the time derivative on both sides of (4.41), we get

$$\frac{1}{Pp_{11}}\frac{d\Phi_1}{dt} = \frac{dQ_1}{dt} + \frac{Pp_{12}}{Pp_{11}}\frac{dQ_2}{dt} + \frac{Pp_{13}}{Pp_{11}}\frac{dQ_3}{dt}. \tag{4.42}$$

Rearranging (4.42) and by using (4.40) we get the final equation of interest in terms of currents

$$i_{T1} = \frac{1}{Pp_{11}}\frac{d\Phi_1}{dt} - \frac{Pp_{12}}{Pp_{11}}i_{T2} - \frac{Pp_{13}}{Pp_{11}}\,i_{T3}. \tag{4.43}$$

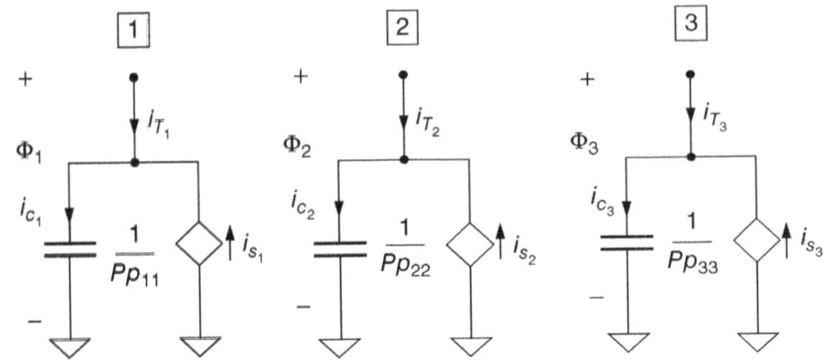

FIGURE 4.17 Current source capacitance equivalent circuit model for retarded PEEC model.

Finally, the last two terms of (4.43) can be separated and can be called i_{s1}, or

$$i_{s1} = \frac{Pp_{12}}{Pp_{11}} i_{T2} + \frac{Pp_{13}}{Pp_{11}} i_{T3}. \tag{4.44}$$

By inspection of (4.43) and (4.44), we can find an equivalent circuit for the situation. For example, the first term in (4.43) is clearly a capacitance to ground, which is a circuit node at ∞. This is shown in Fig. 4.17. We see that the value of the current source i_{s1} is given by (4.44). We notice that the sources i_{s1}, i_{s2}, and i_{s3} are current controlled current sources, or CCCS in the SPICE language. The concept of a common node at ∞ may be hard to accept from an intuitive point of view. However, it is the result of the analysis. A concept that the node at ∞ is the ground in a three-dimensional environment may be more intuitive.

4.5.3 Potential-Based Model for the Capacitances

In this section, we consider a second capacitance model that is based on a series connection of a voltage source with the self-capacitance in series. The resultant equivalent circuit for the three-plate model is shown in Fig. 4.18.

It is clear that the result is the same for both types of models. The series connection model leads to fewer branches connected to the node. The choice depends on the so-called degree d_N of the connecting node. The degree is determined by the number of branches

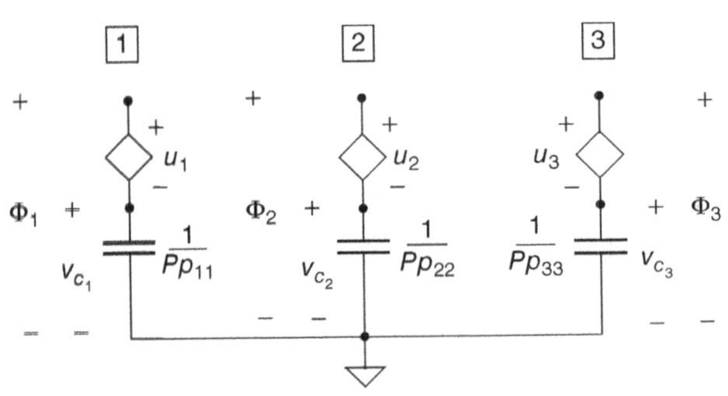

FIGURE 4.18 Voltage source capacitance equivalent circuit model for PEEC model.

that are incident to the nodes that the capacitors connect. If the degree or the number of connections $d_N < 4$, the voltage source model in this section is more efficient. If the degree of connections is larger than four on the connection nodes, the current source model should be used.

It is clear that the self-term $1/Pp_{mm}$ in the voltage source circuit model will be the same as for the current source model in Fig. 4.17. To derive the coupling term, we start with (4.44). As an example, the *km* single coupling term is

$$u_{km} = Pp_{km}\, Q_m = \frac{Pp_{km}}{Pp_{mm}}\, Vc_m, \tag{4.45}$$

since $Q_m = Vc_m/Pp_{mm}$ where Vc_m is the voltage across the self-capacitance $1/Pp_{mm}$. This is a contribution to the voltage source in series with the capacitance in Fig. 4.17.

We finally want to show how this model applies to the example in Fig. 4.15. The self-capacitances are obvious and the series voltage sources in (4.18). The controlled voltage sources for this model are given by

$$u_1 = \frac{Pp_{12}}{Pp_{22}}\, Vc_2 + \frac{Pp_{13}}{Pp_{33}}\, Vc_3 \tag{4.46a}$$

$$u_2 = \frac{Pp_{21}}{Pp_{11}}\, Vc_1 + \frac{Pp_{23}}{Pp_{33}}\, Vc_3 \tag{4.46b}$$

$$u_3 = \frac{Pp_{31}}{Pp_{11}}\, Vc_1 + \frac{Pp_{32}}{Pp_{22}}\, Vc_2. \tag{4.46c}$$

This voltage source model was first introduced in Ref. [11] and is also used in Ref. [12] and in other works.

PROBLEMS

4.1 *Capacitance matrices*

Verify that you understand (2.6). Replace the conductances with capacitances. Relate the resistance/conductance problem to the capacitance problem in Section 4.2.

4.2 *Capacitance computation*

The purpose of this problem is to make a small capacitance computation Matlab program. The approach for capacitance computations is given in Section 4.3. Compute the capacitance for a problem with a 1×1 m zero thickness conductor plate. Set up a 9×9 partial potential coefficient matrix which results in 9 unknown charges. Consider the meshing in Fig. 4.6. In this case, the only full mesh cell is at the center of the zero thickness metal sheet. All edge cells are usually half the width of the full cells. As a second case, solve the same problem with nonuniform cells where all the cells at the edge of the plate are of a width of only 0.1 m. Hence, the center cell will be 0.8×0.8 m. Compare the capacitances obtained from the two cases to show how more accurate the model is with the narrow edge cells. As a hint, the capacitance of the plate is somewhat less than 41 pF.

4.3 *Capacitance equivalent circuits*
Convert a 3×3 capacitance matrix into capacitance equivalent circuits as they are shown in both Figs. 4.17 and 4.18 for four conductors. Show that the external behavior of the equivalent circuits is the same as for the capacitance matrix formulation. You can use either a 3×3 short circuit capacitance matrix or a two-terminal capacitance model.

4.4 *Elements of the short circuit capacitance matrix*
The off diagonal elements of a correctly computed short circuit capacitance matrix are negative. Please explain why the mutual elements are negative. We assume that all conductors are perfect electrical conductors (PEC) in a free space environment for the model. Explain the relationship between the capacitance model of the structure and component of the capacitance matrix.

4.5 *Capacitance modeling*
Assume there are two congruent square perfect electrical conductor (PEC) plates as is shown in Fig. 4.7a. Each one has again a dimension of 1×1 m. The two plates are separated by a distance $d = 0.2$ m. Extend your program from problem 4.2 to this two-plate case. Apply the technique from Section 4.2.4 to compute the total capacitance of the two-conductor structure. What does this mean? To earn more points, try to apply some of the other techniques you did learn in this chapter.

REFERENCES

1. J. C. Maxwell. *A Treatease of Electricity and Magnetism*. Dover Publications, New York, 1954.
2. D. K. Reitan and T. J. Higgins. Accurate determination of the capacitance of a thin rectangular plate. *Transactions of the American Institute of Electrical Engineers, Part I: Communication and Electronics*, 75(6):761–766, January 1957.
3. P. D. Patel. Calculation of capacitance coefficients for a system of irregular finite conductors on a dielectric sheet. *IEEE Transactions on Microwave Theory and Techniques*, MTT-19(11):862–869, November 1971.
4. A. E. Ruehli and P. A. Brennan. Accurate metallization capacitances for integrated circuits and packages. *IEEE Journal Solid-State Circuits*, 4(4):289–290, August 1973.
5. K. Nabors and J. White. FastCap: a multipole accelerated 3D capacitance extraction program. *IEEE Transactions on Computer-Aided Design of Integrated Circuits and Systems*, 10(11):1447–1459, November 1991.
6. A. E. Ruehli. Electrical analysis of interconnections in a solid state circuit environment. In *ISSCD, IEEE International Solid-State Circuits Conference*, Volume XV, pp. 64–65, February 1972.
7. A. E. Ruehli Ed. *Circuit Analysis, Simulation and Design, Part 2*. Elsevier Science Publishers B.V., North-Holland, 1987.
8. A. E. Ruehli and P. A. Brennan. Efficient capacitance calculations for three-dimensional multiconductor systems. *IEEE Transactions on Microwave Theory and Techniques*, 21(2):76–82, February 1973.
9. J. D. Kraus. *Electromagnetics*. McGraw-Hill Book Company, New York, 1953.

10. A. E. Ruehli, P. A. Brennan, and H. W. Young. Recent progress in capacitance computation methods. In *Proceedings of the IEEE International Symposium on Circuits and Systems, Phoenix, AZ*, pp. 135–138, 1975.

11. H. Heeb and A. E. Ruehli. Retarded models for PC board interconnects or how the speed of light affects your SPICE circuit simulation. In *Proceedings of International Conference on Computer Aided Design*, pp. 70–73, November 1991.

12. J. Nitsch, F. Gronnwald, and G. Wollenberg. *Radiating Nonuniform Transmission-Line Systems and the Partial Element Equivalent Circuit Method*. John Wiley and Sons, Inc., New York, 2009.

5

INDUCTANCE COMPUTATIONS

The availability of computers had a large impact on inductance computations. We should recognize that the computations of inductances were a tedious task in the precomputer era due to the fact that they may involve integration over volumes in space. Hence, a very important aspect of early inductance work was to make the computation with a slide rule as simply as possible. References to the early work on inductance computations are given in Ref. [1] and also in Ref. [2].

In contrast, the computation of 10^6 partial inductances is not a large job today. Inductance computation has been a key part of the partial element equivalent circuit (PEEC) method from its beginning [3]. Of course, the concepts have evolved over the recent years as apparent from this chapter. Much work has been done by many researchers, for example, Refs [2, 4–7]. The satisfying aspect of the approach presented in this chapter is that a large class of important problems can be solved using this technique. For this reason, we treat this subject in detail.

The PEEC solution method is very flexible as is apparent. The approach makes inductance computations possible for many conductors in different configurations. Related approaches for PEEC inductance computations have been developed over the years. Such an example is a set of equal length parallel wires as they may occur on chips [8] and other on-chip approximations [9]. A form of PEEC where the coupling is done at the vector potential level has been presented in Ref. [10]. Further, speedup techniques have been devised [11].

In this chapter, we present an approach where we solve inductance problems for a multitude of configurations and applications. We also show that the computation of partial inductances is an important part of this process and we treat the general case for partial inductance computations for nonorthogonal structures in Chapter 7 and Appendix C.

Circuit Oriented Electromagnetic Modeling Using the PEEC Techniques, First Edition.
Albert E. Ruehli, Giulio Antonini, and Lijun Jiang.

Included in Appendix C is a collection of formulas for partial inductances and some other integrals.

5.1 LOOP INDUCTANCE COMPUTATIONS

As presented in Ref. [3], we start out with the basic concept of the computation of loop inductances for N loops. Figure 5.1 shows an example of such a loop structure for which we want to compute the inductances. We assume that they all have a small gap where we can inject a current and measure the induced voltage. In a realistic example, we can ignore the missed coupling due to the gaps since they are small compared to the loop size. The example in Fig. 5.1 does not have multibranch connected loops, a subject which is considered in this chapter.

In this example, we show N coupled loops. The desired inductance equivalent circuit is shown in Fig. 5.2. We give the definition of inductance for this simple case.

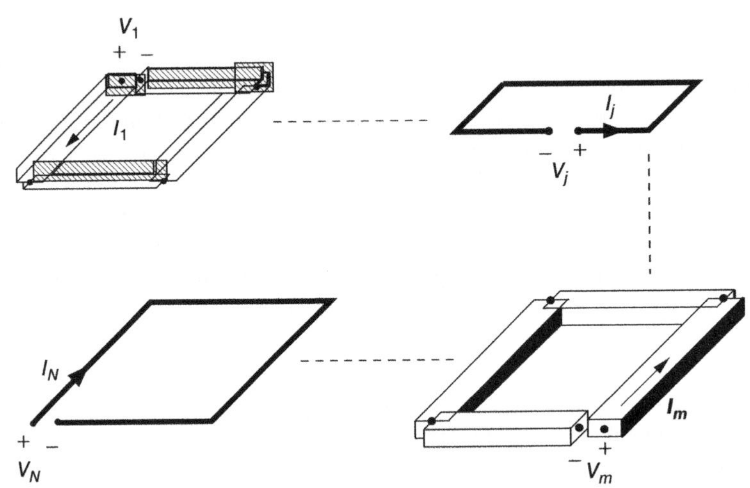

FIGURE 5.1 System of coupled conductor loops.

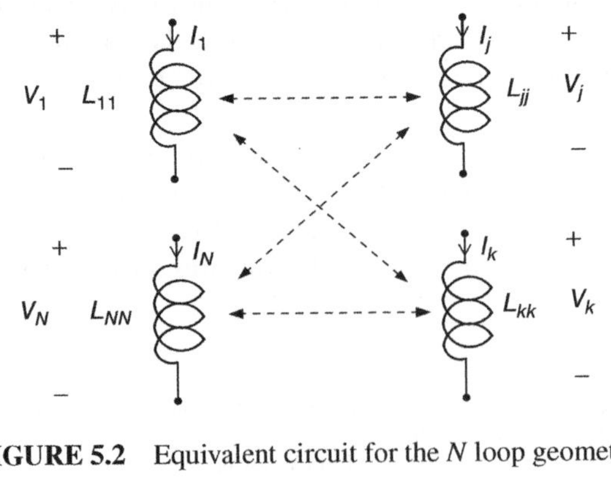

FIGURE 5.2 Equivalent circuit for the N loop geometry.

Definition 1 ***(Inductance for N loops)*** *The N^2 inductances for a system of N loop are defined by*

$$L_{km} = \frac{\psi_{km}}{I_m} = \frac{V_k}{s\, I_m}. \tag{5.1}$$

Where I_m is the current flowing in the mth loop and V_K is the voltage induced in the kth loop.

5.1.1 Loop Inductance Computation in Terms of Partial Inductances

As done in Ref. [3], we start the inductance computation by considering the flux ψ generated by loop m and picked up by a loop k similar to the ones in (5.1). The example geometry is the loop in Fig. 5.3. The unit vector $\hat{\boldsymbol{n}}$ is perpendicular to the loop area, and the magnetic flux through loop k is

$$\psi_{km} = \int_{S_k} \boldsymbol{B}_m \cdot \hat{\boldsymbol{n}}\, dS_k, \tag{5.2}$$

where S_k is the area of the kth loop. We would like to remind the reader again that we use S for a surface area or a flux area and $\mathcal{A}_c$ for a conductor current carrying cross section. Hence, we integrate the normal component of the magnetic field over the loop area. Using $(\int_{\mathcal{A}_c} d\mathcal{A}_c)/\mathcal{A}_c$ for the finite cross section of the conductors, we also average the integral over the conductor cross-sectional area depicted by $\mathcal{A}_{ck}$ in Fig. 5.3.

We should note that the same result can also be obtained from an energy point of view (see Chapter 11, Refs [12] and [2]).

The integration results in the flux in loop k

$$\psi_{km} = \frac{1}{\mathcal{A}_{ck}} \int_{\mathcal{A}_{ck}} \int_{S_k} \boldsymbol{B}_m \cdot \hat{\boldsymbol{n}}\, dS_k\, d\mathcal{A}_{ck}, \tag{5.3}$$

where again $\hat{\boldsymbol{n}}$ is the unit vector normal to the loop surface.

We continue the derivation by using (3.2a) for a homogeneous material with a relative permeability, which we assume to be $\mu_r = 1$ such that $\boldsymbol{B} = \mu_0 \boldsymbol{H}$. We start from Maxwell's

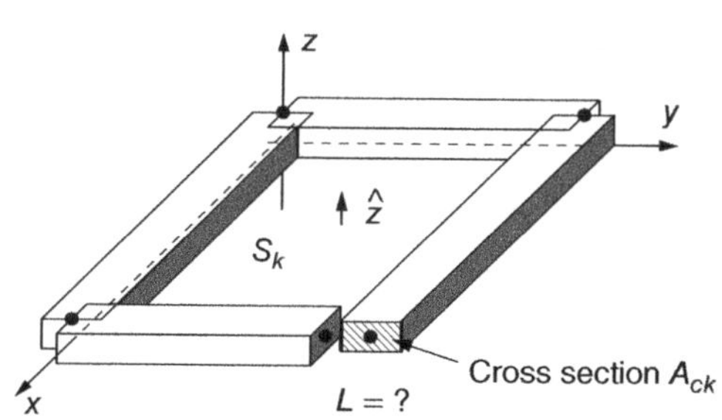

FIGURE 5.3 Example loop k for flux and vector potential computations.

equations (3.1b) for the quasistatic case, or

$$\nabla \times \boldsymbol{B} = \mu_0 \boldsymbol{J} \tag{5.4}$$

and, without magnetic charges, (3.1d) is

$$\nabla \cdot \boldsymbol{B} = 0. \tag{5.5}$$

Hence, we use the magnetic vector potential (3.18), such that

$$\boldsymbol{B} = \nabla \times \boldsymbol{A}. \tag{5.6}$$

Using (5.6), we can replace $\boldsymbol{B}_m$ with the vector potential $\boldsymbol{A}_m$ to get

$$\psi_{km} = \frac{1}{\mathcal{A}_{ck}} \int_{\mathcal{A}_{ck}} \int_{S_k} (\nabla \times \boldsymbol{A}_m) \cdot \hat{\boldsymbol{n}}\, dS_k\, d\mathcal{A}_{ck}. \tag{5.7}$$

By applying Stokes' theorem (3.34) to convert the area integral into a line integral, we obtain a line integral along the loop to compute the induced flux

$$\psi_{km} = \frac{1}{\mathcal{A}_{ck}} \int_{\mathcal{A}_{ck}} \int_{\ell_k} \boldsymbol{A}_m \cdot d\boldsymbol{\ell}_k\, d\mathcal{A}_{ck}. \tag{5.8}$$

We observe that the resultant line integral extends over the four sides of the loop. This leads to the important result that we can subdivide the integral into four integrals or segments, where each of them is over one side of the loop, or

$$\psi_{km} = \sum_{i=1}^{4} \frac{1}{\mathcal{A}_{ck}} \int_{\mathcal{A}_{ck}} \int_{\ell_{k,i}} \boldsymbol{A}_m \cdot d\boldsymbol{\ell}_{k,i}\, d\mathcal{A}_{ck}. \tag{5.9}$$

We want to set up the integrals over the sides separately. For this, we represent the orientation of the segments as $d\boldsymbol{\ell}_k = \hat{\boldsymbol{t}}_k\, d\ell_k$, where $\hat{\boldsymbol{t}}_k$ is the tangential unit vector. This individually represents each segment of the loop over which we integrate the flux area

$$\psi_{km} = \sum_{i=1}^{4} \frac{1}{\mathcal{A}_{ck}} \int_{\mathcal{A}_{ck}} \int_{\ell_{k,i}} \boldsymbol{A}_m \cdot \hat{\boldsymbol{t}}_{k,i}\, d\ell_{k,i}\, d\mathcal{A}_{ck}, \tag{5.10}$$

where, in general, the cross section of each conductor $\mathcal{A}_{ck}$ for the four sides can be different.

A second important step in the derivation is to replace the vector potential $\boldsymbol{A}_m$, in (5.10) by (3.50), given in Chapter 3 as

$$\boldsymbol{A}_m(\boldsymbol{r}, t) = \frac{\mu_0}{4\pi} \int_{V'} \frac{\boldsymbol{J}_{ve}(\boldsymbol{r}', t')}{|\boldsymbol{r} - \boldsymbol{r}'|}\, d\mathcal{V}'. \tag{5.11}$$

We want to apply (5.11) to a current filament in the cross section in the conductors in Fig. 5.3. In this example, both the observation point location $\boldsymbol{r}$ and the source point $\boldsymbol{r}_m$ are located on loop conductors.

We assume that the current in the loop conductors is uniform in the cross section and perform the averaging integration over the cross section to get the total current. This leads to the vector potential

$$\boldsymbol{A}_m(\boldsymbol{r}) = \frac{\mu_0}{4\pi}\frac{I_{cm}}{\mathcal{A}_{cm}}\int_{\mathcal{A}_{cm}}\int_{\ell_m}\frac{\hat{\boldsymbol{t}}_{m,j}}{R_{rm}}\,d\ell_m\,d\mathcal{A}_{cm}, \tag{5.12}$$

where $R_{rm} = |\boldsymbol{r} - \boldsymbol{r}_m|$ and I_{cm} is the total current in the bar m. Note that the vector potential $\boldsymbol{A}_m$ is in the same direction as the current I_{cm}. This is indicated by the tangential unit vector $\hat{\boldsymbol{t}}_m$.

It is clear that we again can break up the integral over the loop m into four parts, one for each side, or for $j = 1$ to 4:

$$\boldsymbol{A}_m(\boldsymbol{r}) = \sum_{j=1}^{4}\frac{\mu_0}{4\pi}\frac{I_{cm}}{\mathcal{A}_{cm}}\int_{\mathcal{A}_{cm}}\int_{l_{m,j}}\frac{\hat{\boldsymbol{t}}_m}{R_{rm}}\,d\ell_m\,d\mathcal{A}_{cm}. \tag{5.13}$$

The next step is to substitute $\boldsymbol{A}_m$ from (5.13) into (5.10) resulting in

$$\psi_{km} = \sum_{i=1}^{4}\sum_{j=1}^{4}\frac{I_m}{\mathcal{A}_{ck}\,\mathcal{A}_{cm}}\frac{\mu_0}{4\pi}\int_{\mathcal{A}_{ck}}\int_{\mathcal{A}_{cm}}\int_{l_{k,i}}\int_{l_{m,j}}\frac{\hat{\boldsymbol{t}}_{m,j}\cdot\hat{\boldsymbol{t}}_{k,i}}{R_{rm}}\,d\ell_m\,d\ell_k\,d\mathcal{A}_{cm}\,d\mathcal{A}_{ck}. \tag{5.14}$$

Finally, using (5.1) to determine the inductance between the loops $L_{km} = \psi_{km}/I_m$, or using (5.14), we get

$$L_{loop} = \sum_{i=1}^{4}\sum_{j=1}^{4}\frac{1}{\mathcal{A}_{ck}\,\mathcal{A}_{cm}}\frac{\mu_0}{4\pi}\int_{\mathcal{A}_{ck}}\int_{\mathcal{A}_{cm}}\int_{l_{k,i}}\int_{l_{m,j}}\frac{\hat{\boldsymbol{t}}_{m,j}\cdot\hat{\boldsymbol{t}}_{k,i}}{R_{rm}}\,d\ell_m\,d\ell_k\,d\mathcal{A}_{cm}\,d\mathcal{A}_{ck}. \tag{5.15}$$

We observe that we compute the partial self-inductance if we apply this result to only one part where $m = k$. With this, we have a form of the equations that leads to the partial inductances in the following section.

5.1.2 Circuit Model for Partial Inductance Loop

We use the loop inductance derived in the previous section as a starting point for a much more powerful circuit-oriented general approach for inductance computations using partial inductances. By considering Fig. 5.1, we see that we approximate each section of the loops by a rectangular bar.

As a first example to the general case for inductance computations, we consider the loop in Fig. 5.3. We recognize that we can simplify the loop inductance equation in (5.15) into

$$L_{loop} = \sum_{i=1}^{4}\sum_{j=1}^{4}Lp_{ij}, \tag{5.16}$$

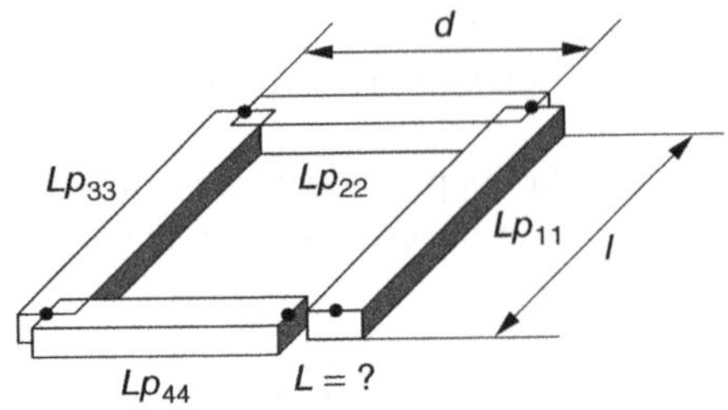

FIGURE 5.4 Single loop approximated by four bars.

where

$$Lp_{ij} = \frac{1}{\mathcal{A}_k\,\mathcal{A}_m}\frac{\mu_0}{4\pi}\int_{\mathcal{A}_k}\int_{\mathcal{A}_m}\int_{\ell_{k,i}}\int_{\ell_{m,j}}\frac{\hat{t}_{k,i}\cdot\hat{t}_{m,j}}{R_{km}}\,d\ell_m\,d\ell_k\,d\mathcal{A}_m\,d\mathcal{A}_k. \qquad (5.17)$$

From this example, we find the following definition of a partial inductance

Definition 2 ***(Partial inductance)*** *A partial self-inductance Lp_{ii} is defined for a single piece of conductor as $V_i = Lp_{ii}\, sI_i$, where V_i is the voltage along the conductor i and I_i is the current in the conductor i. Similarly, the partial mutual inductance is defined between two pieces of conductors i and j as*

$$V_i = Lp_{ij}\; s\; I_j \qquad (5.18)$$

with the formula for the partial inductance given by (5.17). Partial inductances are symmetrical as is evident from equation (5.18).

We recognize that in principle each segment can have its own cross section, such as $\mathcal{A}_m$ for segment m. But this seems to create a redistribution of the currents. Mainly, the local current at the corner of the conductors is approximated. However, as it is apparent from Chapter 6, the general current flow can be modeled, but the model will be more complex.

Probably the most important aspect of the partial inductance concept is the fact that it obeys all the laws of circuit analysis [3]. This is shown to be the case for our simple loop

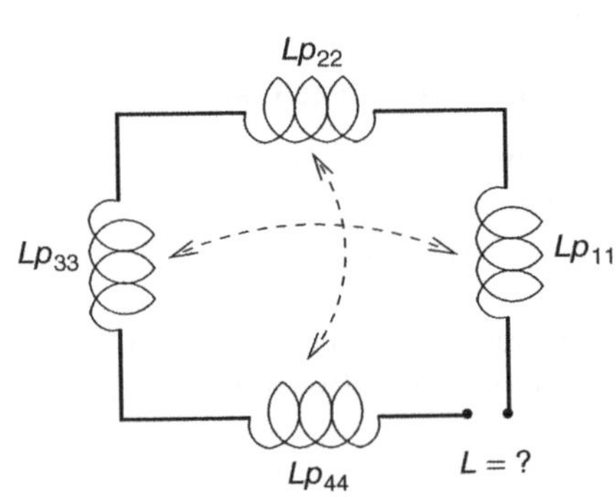

FIGURE 5.5 PEEC equivalent circuit for the single loop.

formed by four cells or bars that we show in Fig. 5.4. It is clear that the equivalent circuit for this loop is as shown in Fig. 5.5.

The approach works for arbitrary connected bars that eventually form closed circuits.

5.2 INDUCTANCE COMPUTATION USING A SOLUTION OR A CIRCUIT SOLVER

Inductance estimations can be performed for simple geometries using approximate models. The loop inductance in (5.16) is a good example. This type of model can often be used to understand fundamental issues.

Fortunately, all the modified nodal analysis (MNA) circuit solver techniques in Chapter 2 can be applied for a problem with arbitrary conductor arrangements. Any problem can be solved using these circuit solution techniques. Importantly, we also can use any SPICE-type circuit solver to solve general inductance problems. However, in many situations the assignment of the current direction to the inductive branches seems to be confusing at first. But using a simple rule will yield correct results. The rule can be stated as follows:

Rule ***(Assignment of conductor current directions).*** *All inductive currents in branches of a PEEC circuit need to be labeled as pointing in the +x-, +y-, and +z-directions, respectively. The correct coupling signs are obtained if we also label all the partial inductances* Lp_{km} *in the matrix as positive.*

This rule automatically leads to the correct solution. After the solution is obtained, the actual sign of the elements indicates whether the current flows in the assigned direction. As usual, a positive current indicates that the current flows in the assigned direction, whereas a negative sign indicates current flows in the opposite direction.

5.3 FLUX LOOPS FOR PARTIAL INDUCTANCE

The flux that contributes to the inductance is confined inside the loop as shown in Fig. 5.3. We need to show how the flux is also confined using partial inductances. This issue is investigated in Refs [2, 3, 13]. In Fig. 5.6a, we consider two black, short parallel wires that we assume to be two parts of the same loop.

We next want to answer the question how the flux is confined to the interior of the loop that is between the two wires. We can associate a flux area to the partial inductance perpendicular to the ends such that the perpendicular sides of the area do not contribute to the partial inductance loop. This is given by the crosshatched area for Lp_{11}. We can see from the definition of a partial inductance (5.17) that the partial mutual inductance Lp_{12} between the two wires with the same current in the opposite direction simply cancels the flux area outside the loop, leaving the crosshatched flux inside the loop.

We notice that the partial mutual inductance formula Lp_{12} is the same as Lp_{11} for the integration area outside of the flux area inside the loop. This keeps the flux to the inside of the loop between the conductors as we know is the case.

In Fig. 5.6b, we show the case where the dashed conductor is not parallel to the one at $x = 0$ along the y-axis. As shown, we can approximate the tiled conductor with staircases consisting of parallel and perpendicular conductors. We see that the flux areas for each

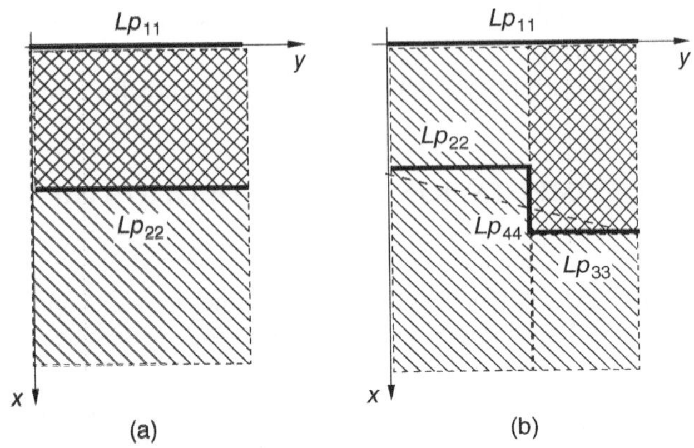

FIGURE 5.6 Two examples for the flux loops for partial inductances.

subsegment corresponding to the tilted wire will generate its own flux situation as Fig. 5.6a. However, we also see that it adds a contribution in the y-direction in exactly the same way as the x direction. This part is not shown in Fig. 5.6b. Furthermore, it is clear that the same rule applies for loops with a nonplanar surface. With the staircase solution, we can show that any loop geometry can be represented.

Finally, we need to consider situations for a PEEC circuit that also includes capacitances in the loops. It is clear that we cannot apply a voltage or current source to loops that does not form a closed circuit. We observe that a PEEC circuit for a dipole antenna includes two conductors that seem to be open circuited at the ends. It is apparent in Chapter 6 that these loops are closed by capacitances such that a return current exists. Practical situations are usually more complicated.

In many practical problems, the flux associated with a PEEC circuit with multiple branches can be quite complicated and it may not be obvious why the circuit laws apply for the general case. However, we know that PEEC models satisfy Maxwell's equations.

5.4 INDUCTANCES OF INCOMPLETE STRUCTURES

5.4.1 Open-Loop Inductances

Perhaps the most important concept for inductance computation, besides the partial inductances, is the *open-loop* inductance. We illustrate the issue with a practical example. Numerous electronic components are sold to be placed on a printed circuit board (PCB). Such an attached component may have two or more connections to the PCB. We cannot determine the inductance of some loops that connect through the pins since the exact closure of the paths at the PCB are not known to the manufacturer of the components. They are only known once the component is placed on the specific PCB with known connection wires.

How can we specify the inductance of such loops if only part of the connections is known? To give an example of an open-loop inductance, we consider the simple example in Fig. 5.7. This simplified figure may represent a typical component where the vertical pins are connected (soldered) to the board. Therefore, the question is whether it makes sense to compute the inductance of the component without the onboard connections. Importantly,

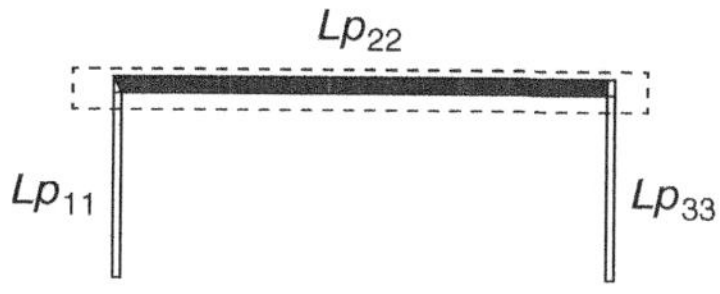

FIGURE 5.7 Example of a device with two pins and an open loop.

we can compute the incomplete *open-loop* inductance Lo for the rest of the loop on the component as

$$Lo_{11} = Lp_{11} + Lp_{22} + Lp_{33} - 2\,Lp_{13}. \tag{5.19}$$

We need to remember that loops will always be closed. In this example, the open inductance Lo_{11} is smaller or at least inaccurate compared to the loop inductance due to the missing PCB inductances.

Another specific example is the inductance of a decoupling capacitor called ESL. Its geometry for a PCB is not too different from Fig. 5.7. Importantly, the manufacturer of the component should specify values for the inductances of these components. An interesting question is if the inductance value for such a component specified is the open-loop inductance or a value computed with some presumed closed-loop path?

To deal with the issue, we need to formally define the concept of the *open-loop* inductance.

Definition 3 ***(Open-loop inductance)*** *Open-loop inductance(s) of a component or part must be defined such that we eventually can compute the closed loop inductance. However, sufficiently small gaps are permitted.*

Remember that partial inductances are defined for single conductors only.

5.4.2 Open-Loop Macromodels

We need to consider another practical approximation for a PEEC model based on the open-loop issue. Assume that this brings up the very challenging question on how we should characterize an entire component such that they can be used in conjunction with or embed them into other PEEC models, once the rest of the geometry is specified.

To be specific, we again take the example of a component on a PCB that could be a resistor or a capacitor. Hence, such a component has connections similar to Fig. 5.7. However, we need a circuit model for the component. An example of an approximate macromodel for a capacitor is shown in Fig. 5.8a and an equivalent circuit is shown in Fig. 5.8b.

This example model for the capacitor deserves some explanation. Most of the partial inductance models for the wires are self-evident. However, Lp_{33} may be an approximation for the hidden equivalent conductors in the capacitor, which the manufacturer of the component could determine, and C is the value of the capacitor given by the manufacturer. Hence, with all these values we almost could add this component model to any PEEC model for the capacitor on the PCB.

Importantly, we need one more piece of information about the component. We also need to take into account the partial mutual inductances Lp_{k3} or the coupling from the component to the element of the board interconnect. For example, we call the interconnect element k

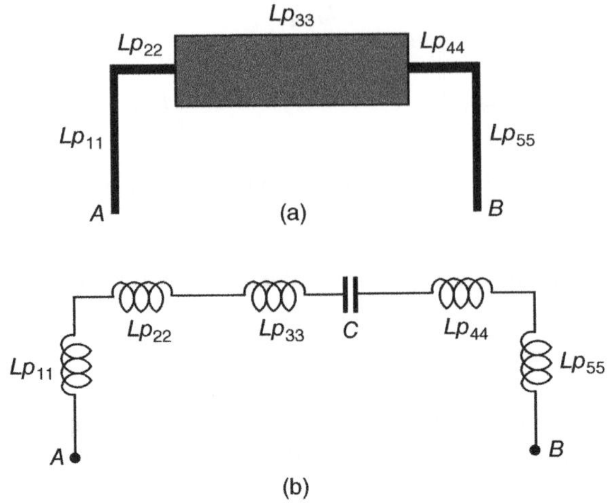

FIGURE 5.8 Example geometry and equivalent circuit for capacitor macromodel.

which is part of the structure to which the component is attached to. Then, we also need to compute the coupled partial mutual inductances. One of the elements is the partial mutual inductance Lp_{3k} between the capacitor model in Fig. 5.8 and the interconnect k element. Hence, it is clear that we need to know more detailed information about the approximate shapes in the component to compute these partial mutual inductances.

From this example, we can deduce the following information for an open-loop macromodel.

Definition 4 ***(Open-loop macromodel)*** *An open-loop macromodel consists of self- and mutual partial inductances for a part of a geometry. Its purpose is to characterize a geometry consisting of a set of conductor pieces that can be characterized with an inductance macromodel. The model may also include the geometrical specification of parts for the evaluation of partial mutual inductances to specify the properties of loops that are not closed.*

It is apparent that this is a flexible approach for the inductance modeling for part of component or geometry.

We also want to observe that by comparing the best result for the loop inductance (5.16) and the open-loop inductance (5.19), we can get an estimate of the error in using the incomplete open-loop inductance value as a lower bound. Usually, some approximate closing path(s) can be used to give an indication of the closed-loop inductances.

5.4.3 Examples for Open-Loop Inductances

We consider the impact of the open-loop inductance in comparison to the closed loop for the example shown in Fig. 5.9. For the comparison, we only allow an infinitesimal gap g in Fig. 5.9 for the closed-loop comparative inductance. In Fig. 5.10, we show the computed open-loop inductance as the example for an increasing gap width g. The entire front half side is eliminated for the largest gap width. For some problems, where we need an

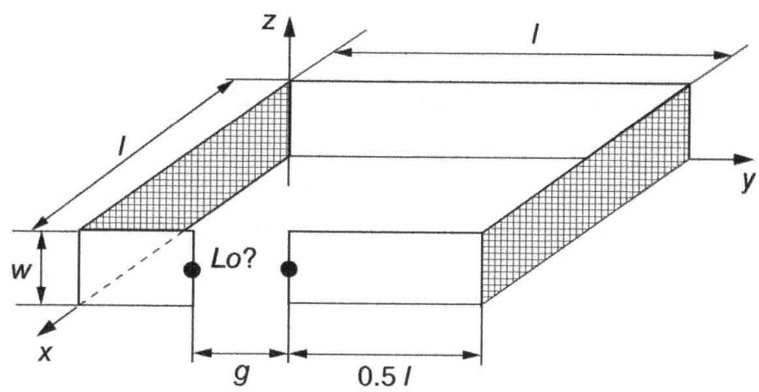

FIGURE 5.9 Example geometry for open-loop inductance computation.

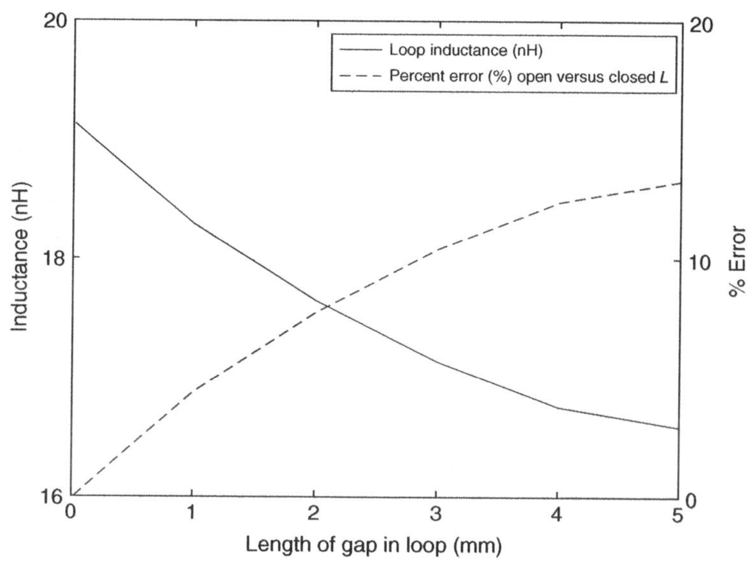

FIGURE 5.10 Open-loop inductance example for geometry in Fig. 5.9.

approximate value for the inductance of an incomplete situation, we still may use an approximate closed-loop value for an estimate.

5.5 COMPUTATION OF PARTIAL INDUCTANCES

It is evident that the PEEC method leads to an efficient approach for the computation of inductances for any structure. Unfortunately, the evaluation of the partial inductances is not straightforward. The computation of *exact* partial inductances can be time consuming, while some *approximate* formulas designed before computers were available [1] may not be accurate enough for closely spaced conductors. A recent text makes a valuable contribution to partial inductance computations [2].

The aim of our Appendix C is to provide a list of partial inductances that are convenient to use for the evaluation of partial inductances. The approach is to use analytic expressions wherever possible. However, this does not exclude the use of numerical integration where necessary. Especially for some nonorthogonal and other complex geometries, we have to apply numerical integration concepts where analytic formulas are hard to find. Fortunately, continuous progress is being made by many researchers, for example, Refs [3, 14–16].

Other important details must be observed to obtain efficient, accurate partial inductances. Hence, we focus on this issue since it is key for practical computation. The evaluation of the partial inductance Lp (5.17) requires the integration over two volumes or six coordinates, x, y, z, x', y', z'. Clearly, obtaining reasonable inductances without computers was at best tedious in the precomputer era [1]. Still, the exact partial inductance computations can be too time consuming since the speed of the evaluation of the partial inductances can limit the size of problems we can solve.

Many researchers aimed at reducing the compute time for each Lp given by (5.17). A simple implementation requires N_L^2 partial inductances to be evaluated. However, due to the symmetry of the **Lp** matrix, only $N_L(N_L + 1)/2$ partial inductances must be evaluated. A general approach for the efficient evaluation of partial inductances is the *multifunction* technique, which is a mixture of many different approaches considered in [15, 16] and Appendix E. A small example of the integration simplification is given in the following section. Today, other techniques can be applied to speed up the computation of partial inductance matrices. They basically take advantage of the reduced rank [17] of the **Lp** matrix [11]. However, we do not consider these techniques in this text.

5.5.1 Approximate Formulas for Partial Inductances

As an example, volume filament (VFI) models for the skin effect considered in Chapter 9 need very accurate partial inductance evaluations due to the close spacing of the filament conductors. However, carefully applied approximate partial inductance formulas [1] and [2] are sufficient for some simple PEEC models. To make sure that they are applied appropriately, we labeled the *approximate* Lp's as such in Appendix C. Also, we need to consider numerical integration methods for different approximations. The high accuracy needed for some partial inductances may result in relatively large compute time if numerical techniques are used. Often, very good results can be obtained with sufficient accuracy if careful mixed analytic and numerical integration methods are used.

Next, we give an example for an approximate partial mutual inductance formula for distant conductors. Consider the case where two relatively short conductors are widely spaced. It is clear that approximations can be used if the distance R between too small conductors is large compared to the conductor size. We use the exact partial inductances as

$$\begin{aligned} Lp_{km} &= \frac{1}{\mathcal{A}_k\, \mathcal{A}_m} \frac{\mu_0}{4\pi} \int_{\mathcal{A}_k} \int_{\mathcal{A}_m} \int_{\ell_k} \int_{\ell_m} \frac{\hat{\imath}_k \cdot \hat{\imath}_m}{R_{km}}\, d\ell_m\, d\ell_k\, d\mathcal{A}_m\, d\mathcal{A}_k \\ &\cong \frac{\mu_0}{4\pi R_{km}} \frac{1}{\mathcal{A}_k\, \mathcal{A}_m} \int_{\mathcal{A}_k} \int_{\mathcal{A}_m} \int_{\ell_k} \int_{\ell_m} (\hat{\imath}_k \cdot \hat{\imath}_m)\, d\ell_m\, d\ell_k\, d\mathcal{A}_m\, d\mathcal{A}_k. \end{aligned} \tag{5.20}$$

This formula works for both orthogonal and nonorthogonal coordinates. A convenient further approximation yields

$$Lp_{km} \simeq \frac{\mu_0}{4\pi} \frac{\ell_k \cdot \ell_m}{R_{km}} = \frac{\mu_0}{4\pi} \frac{\ell_k\, \ell_m}{R_{km}} \cos\theta, \tag{5.21}$$

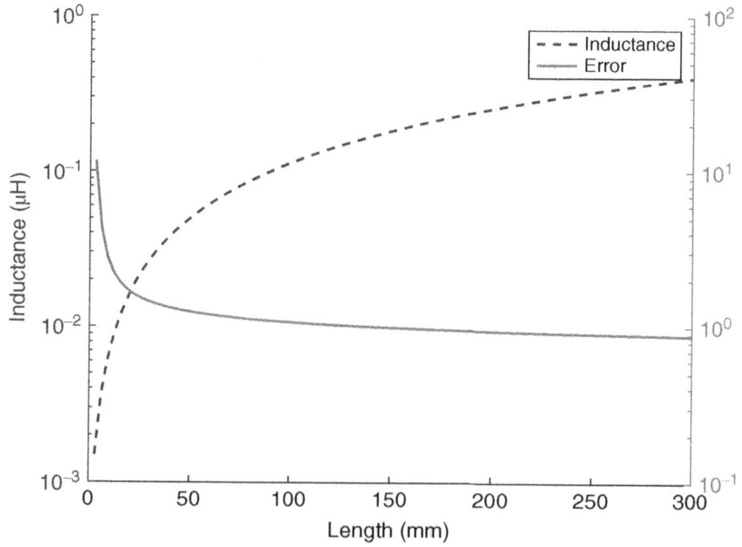

FIGURE 5.11 Error in Lp_{kk} for approximate formula.

where ℓ_k and ℓ_m are the length of filament approximation segments that are placed at the center of the cell cross sections. Here, θ is the angle between the filaments.

It is clear that this approximate formula can save compute time for distant conductors. In fact, the majority of couplings are far distance for large problems. In Section 5.7, we use (5.20) for the case of rectangular cross section, transmission line (TL) type, and ground plane geometry.

Other precomputer formulas for partial inductance have been given in Refs [18] and [1]. For example, the partial self- inductance of a rectangular bar is simplified to

$$Lp_{kk} \simeq \frac{\mu_0 \ell}{2\pi}\left[\log\left(\frac{2\,\ell}{W+T}\right) + 0.5 + \frac{2}{9}\,\frac{W+T}{\ell}\right], \tag{5.22}$$

where ℓ is the variable length and W is the width, and T is the thickness of the rectangular bar.

We compare the exact rectangular bar formula (C.35) from Appendix C with the result of the formula (5.22) in Fig. 5.11 using $W = 1$ mm and $T = 0.2$ mm as a function of the length. It is evident that such an error of several percent in the partial inductances can lead to poor overall inductance results.

5.5.2 Inductance Computations for Large Aspect Ratio Conductors

An important advantage of the PEEC method is that we often can make models with cells that are very long or wide in comparison to other dimensions. This is in part due to the special meshing that is employed by the approach as is presented in Chapter 8. This is in contrast to other methods such as the finite difference time domain (FDTD), finite integration (FIT), transmission line matrix (TLM), or finite element (FEM) and other integral equation (IE) methods where the ratio between the sides of a cell is limited. In fact, a desired ratio of length to width close to one is not unusual. PEEC cells of large aspect ratio are desirable for the type of problem we want to solve since a large gain in efficiency is possible for some geometries.

We should note that the evaluation of the partial inductances is difficult for extreme geometries. For this reason, we want to understand the limit of the side aspect ratios for which the partial inductance computation may become inaccurate. It should be clear that each analytic formula or inductance shape has its own limit. Other researchers have also considered this issue [14]. An extreme ratio for the dimensions may be a ratio of 10,000:1 for the length to width or thickness of a conductor. The limiting factor is that some of the analytic formulas for a partial self- or mutual inductance are very elaborate. It is this complexity that leads to numerical errors even if we use double precision arithmetic in the evaluation.

Several papers assert that the PEEC approach is not suitable for very long, narrow, parallel line-type structures and that the computation of loop inductance should be used to solve this problem. We show that, in fact, this is not the case. However, correctly solving the aspect ratio problem is not trivial. The proper solution of this problem involves two aspects. One aspect that relates to the PEEC model is considered in this section, whereas the second issue that relates to the evaluation of the partial inductances is considered in the following section.

Not surprisingly, it is difficult to evaluate partial inductance with a high accuracy for extreme geometries. As an example geometry, it is sufficient to consider a typical loop shown in Fig. 5.4. Here, we consider a physically long version shown in Fig. 5.12. The circuit solution for the loop inductance is given in (5.16). We regroup the terms of the equation in pairs in the form

$$L = (Lp_{11} - Lp_{13}) + (Lp_{22} - Lp_{24}) + (Lp_{33} - Lp_{31}) + (Lp_{44} - Lp_{42}). \tag{5.23}$$

It is apparent from Fig. 5.12 that we assume that the length ℓ of the loop is large compared to the spacing d. The length consists of the partial self inductances Lp_{11} and Lp_{33}, while the orthogonal end branches have the partial self-inductances Lp_{22} and Lp_{44}. Note that due to the orthogonality of the conductors, the only nonzero partial mutual inductances are $Lp_{13} = Lp_{31}$ and $Lp_{24} = Lp_{42}$.

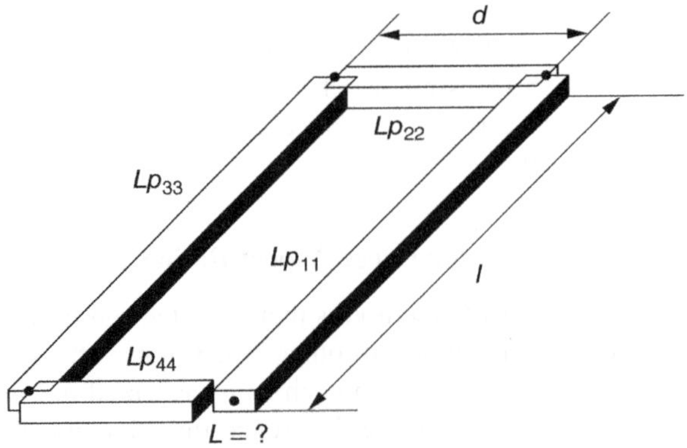

FIGURE 5.12 Long loop represented by four bars.

Now, we want to consider the case where the two bars with Lp_{11} and Lp_{33} are very long, or $\ell \gg d$. Approximations can be used for the end branch inductances since they are very short compared to the long branches of length ℓ. Clearly, the partial mutual inductances Lp_{24} and Lp_{42} will become very small due to the large coupling distance that is approximately ℓ and the lengths are d. In fact, we can approximate $Lp_{24} = Lp_{42} \simeq \mu_0 d^2/(4\pi\ell)$ using (5.21).

Further, the end partial self-inductances are small. Hence, we can either ignore them or just add the partial end self-inductances that are Lp_{22} and Lp_{44} to the loop inductance. Of course, for very large lengths, their contributions can also be approximated to be zero without impacting the results we want to show.

The key problem is the treatment of partial inductances where the length ℓ is much larger than all other conductor dimensions. Also, a very interesting limiting case is given by $\ell \rightarrow \infty$. Another important observation is that such conductors must be close to each other like a TL, else the inductance reaches such large values that this would not represent a reasonable physical situation. In the loop example, this implies that d is small. If we ignore the small end inductances, the loop inductance (5.23) simplifies to

$$L \simeq (Lp_{11} - Lp_{13}) + (Lp_{33} - Lp_{31}). \tag{5.24}$$

Taking the partial self-inductance alone, we get $\lim_{\ell\rightarrow\infty} Lp_{11} = \infty$. Hence, it would be a mistake to independently evaluate each partial inductance in (5.24). Instead, for long complex geometries, we take pairs of partial inductances such that each partial self-inductance has a closing part. Then, we define each pair a *difference inductance* where $Ld_{11} = Lp_{11} - Lp_{13}$ and $Ld_{33} = Lp_{33} - Lp_{31}$. Then the approximate inductance for the loop with very long conductors 1 and 3 given by $L \simeq Ld_{11} + Ld_{33}$. We consider the careful computation of the difference inductances and the partial inductances in the following section. The important observation is that per unit length difference inductances are given by

$$\lim_{\ell\rightarrow\infty} \frac{Ld(\ell)}{\ell} = k, \tag{5.25}$$

where k is a constant finite asymptotic value. This behavior strongly depends on the formulas used for the partial inductance. This aspect is considered in the following section.

The difficulties in the evaluation of partial inductance for extreme aspect ratios of the conductors were known from the start of computer- based solutions [3]. It was found that the evaluation of complex formulas for the partial inductances may lead to very unreliable results for extreme dimensions [14, 19]. The essence of the problem can be reduced to two closely located very long conductors with equal and opposite currents that we assume to be of the same size.

It is clear from (5.24) that the difference inductances converge according to (5.25) if both partial inductance formulas in Ld_{kk} have the same asymptotic behavior. This means that the partial mutual inductance formulation for Lp_{km} must, in the limit, give exactly the same result as the partial self-inductances Lp_{kk}. Practically, five digits of accuracy in the partial inductances is sufficient for extreme geometries.

Next, we want to consider the evaluation of the difference inductances for a realistic example. Important is the asymptotic behavior with respect to a long length ℓ. In the first example, a loop of round conductors with a radius r is used. We need to emphasize that these formulas found in Refs [1, 2] are approximations [20].

The key reasons that we use these approximate formulas are they can be used for both the partial self- and mutual and that they have the same asymptotic behavior. However, this does not give any indication about the exactness of the result. Using this, the difference between the partial self- and mutual inductances can be written as

$$\begin{aligned} Ld_1 &= Lp_{11} - Lp_{13} \\ &\simeq \left(\frac{\mu_0}{2\pi}\ell\left[\log\left(\frac{\ell}{r} + \sqrt{\left(\frac{\ell}{r}\right)^2 + 1}\right) + \frac{r}{\ell} - \sqrt{\left(\frac{r}{\ell}\right)^2 + 1}\right]\right) \\ &\quad - \left(\frac{\mu_0}{2\pi}\ell\left[\log\left(\frac{\ell}{d} + \sqrt{\left(\frac{\ell}{d}\right)^2 + 1}\right) + \frac{d}{\ell} - \sqrt{\left(\frac{d}{\ell}\right)^2 + 1}\right]\right). \end{aligned} \tag{5.26}$$

For very long lengths where $\ell \gg d$, the partial inductances can be approximated as

$$Lp_{11} \simeq \frac{\mu_0}{2\pi}\ell\left[\log\left(\frac{2\ell}{r}\right) - 1\right] \tag{5.28a}$$

$$Lp_{13} \simeq \frac{\mu_0}{2\pi}\ell\left[\log\left(\frac{2\ell}{d}\right) - 1\right]. \tag{5.28b}$$

Hence, for the case where both conductors have the same radius r and the same length ℓ, the approximate loop inductance is

$$\begin{aligned} Ld_1 &= Lp_{11} - Lp_{13} \\ &= \frac{\mu_0}{2\pi}\ell\left[\log\left(\frac{2\ell}{r}\right) - 1 - \log\left(\frac{2\ell}{d}\right) + 1\right] \\ &= \frac{\mu_0}{2\pi}\ell\left[\log\left(\frac{2\ell}{r}\right) - \log\left(\frac{2\ell}{d}\right)\right] \\ &= \frac{\mu_0}{2\pi}\ell\log\left(\frac{d}{r}\right). \end{aligned} \tag{5.29}$$

Finally, again under the condition that $\ell \gg d$, the per-unit length inductance is approximated as

$$\begin{aligned} L/\ell &= \frac{Ld_1 + Ld_3}{\ell} \\ &= \frac{\mu_0}{\pi}\log\frac{d}{r}. \end{aligned} \tag{5.30}$$

We note that this corresponds to the usual approximate per-unit length inductance of a two round conductor TL for dimensions (ℓ, d), which can be found in textbooks [21]. This derivation clearly shows partial and subinductance approaches can effectively be used

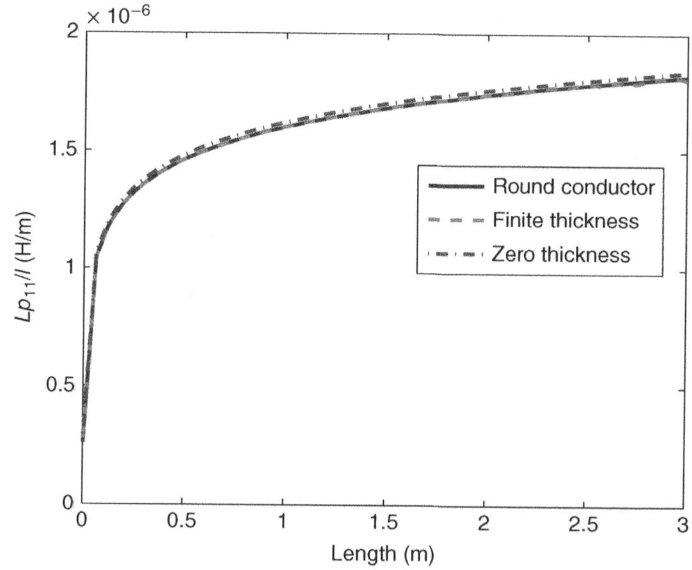

FIGURE 5.13 Partial self-inductance divided by length for different cross sections.

to compute the inductance of long loops. Importantly, the same approach also works for multiple-loop situations.

We next consider the same loop where the conductors consist of zero thickness sheets. The exact partial self- and mutual inductances are computed with the formula (C.24). We use the equation (C.19) for both the partial self- and the partial mutual inductances with the same asymptotic behavior for Lp_{ij} as for Lp_{ii}. The zero thickness loop problem using formula (5.24) for two conductors of the same width also yields the same difference inductance $Ld_{11} = Lp_{11} - Lp_{13}$, where $I_1 = -I_3$.

Results are presented for the long loop example with the three different cross sections corresponding to Fig. 5.12, where the short return conductors are ignored. The cross section of the round conductors is 252.3 μm. The zero thickness conductors have a width $w =$ 1 mm. Finally, the thickness for the rectangular conductors is $t = 50$ μm and the length for all conductors is increased over a wide range from $\ell = 1$ mm to $\ell = 3$ m. Hence, the smallest to largest dimension is 1 : 3000. We have the zero and finite thickness conductors touching each other on the width side as would be the case for a VFI skin-effect model. For this case, Lp_{11} will be only slightly larger than Lp_{13}. Hence, this is the most critical case to consider for the accuracy issue.

Figure 5.13 shows a comparison for the partial self- inductance divided by the length for the three different cases using the approximate round wire formula (5.28a) "(Round)," the zero thickness formula (C.19) "(Zero)," and also the exact volume formula for an equivalent area square shape given in Appendix C (C.35) as "(Finite)."

Similarly, in Fig. 5.14 partial mutual inductance formulas are compared where we see inaccuracies for large lengths for the "(Finite)" result in Appendix C for two rectangular bars.

Figure 5.15 shows the subtraction of the self and the mutual Lp. This subtraction takes place in most formulas. For lengths larger than 2 m, inaccuracies appear in the *rectangular* formulas using double precision due to their increased complexity. Hence, for this case,

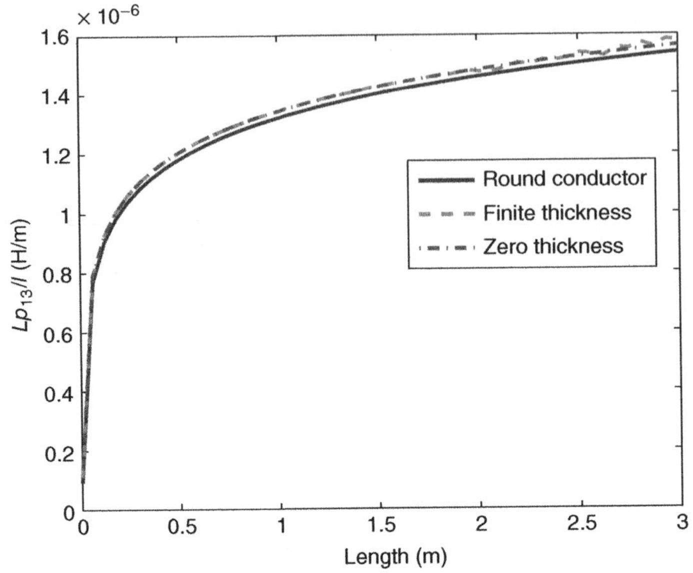

FIGURE 5.14 Partial mutual inductance divided by length for different cross sections.

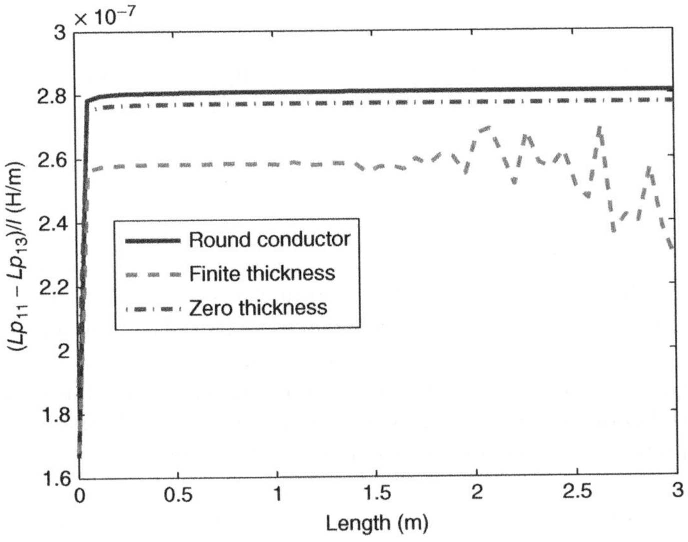

FIGURE 5.15 Subinductances divided by length for different cross sections.

other practical techniques such as breaking up the length of the conductor into segments are used.

This comparison shows that even for large aspect ratios, the simpler partial inductance formulas give consistent results. However, for the finite thickness conductor loop, the difference inductance becomes very inaccurate for long length. This is due to the more complicated analytic formulas for the partial inductances. This example illustrates how very complex formulas for the partial inductances can lead to inaccuracies in the inductance evaluation. We should mention that the example of extreme aspect ratios for the dimensions does not only apply to conductor length, but cells or conductors with other extreme dimension ratios can also lead to similar problems.

One of the advantages of the partial inductance concept is the ability to break a complicated three-dimensional problem into its constituent interactions resulting for general inductance calculations. In this section, we use a small PEEC model to address the accuracy issue that is encountered in *dense* problems. This is done using the two- conductor example from Fig. 5.12. In this approximate model, we represent each conductor by a partial inductance. If we ignore the vertical connections at the two ends, its inductance is approximated as $L = 2(Lp_{11} - Lp_{12}) = 2\,Ld_{11}$.

The inherent source of errors is due to the subtraction of the partial inductances for the closely located conductors. For close neighbor conductors, the first two digits in the partial inductances may be the same for Lp_{11} and Lp_{12} and only the third digit may contain information relevant for the result. Hence, a key source of problems is the exact computation of the partial inductances. Many formulas are not robust at all, especially for large conductor aspect ratios. Robust Lp computations for large aspect ratios are mandatory for PEEC.

We should use the formulas in Appendix C that are labeled *exact*. As we exemplified in the previous section, exact analytic formulas can also lead to errors even using double precision arithmetic for the computations. For this reason, we sometimes resort to the mixed analytic and numerical calculations. The analytic and the numerical evaluation of the integrals can be complementary. The weakness of analytic formulas can be very long conductors as we illustrated. However, the numerical integration along the long direction is also a poor idea due to the long compute times.

To give an example of the analytic-numerical evaluation, we use long filament partial inductances Lpf_{km} for a very long small cross-sectional wire. For filaments, we can apply the partial inductance formula for long conductors due to the simplicity of the formula (C.2). In this case, the filament m carries the entire subsection current. To compute the partial inductance for a conductor with an arbitrary cross section, we use

$$Lp_{km} = \frac{1}{\mathcal{A}_k \mathcal{A}_m} \int_{\mathcal{A}_k} \int_{\mathcal{A}_m} Lpf_{km}\, d\mathcal{A}_k\, d\mathcal{A}_m. \tag{5.31}$$

A variety of approximations for the cross-sectional integrals can be used. For small cross sections, the evaluation of the integrals over the cross sections $\mathcal{A}_k$ and $\mathcal{A}_m$ can efficiently be evaluated numerically using Gaussian integration, which is presented in Appendix E.1.1.

5.6 GENERAL INDUCTANCE COMPUTATIONS USING PARTIAL INDUCTANCES AND OPEN LOOP INDUCTANCE

It should be apparent at this point that partial inductances provide a flexible approach for the computation of the inductance for arbitrary geometries. In this section, we give simple examples to illustrate the use of some of the concepts given earlier.

The simple example includes three metal strip cells in series as shown in Fig. 5.16. Since they clearly do not form a loop, we can only compute the *open-loop* inductance Lo for the three metal bars connected to each other.

The partial inductance matrix for this problem is given in (5.32), where all the partial inductances are computed as reported in the previous section.

$$\begin{bmatrix} V_1 \\ V_2 \\ V_3 \end{bmatrix} = \begin{bmatrix} Lp_{11} & Lp_{12} & Lp_{13} \\ Lp_{21} & Lp_{22} & Lp_{23} \\ Lp_{31} & Lp_{32} & Lp_{33} \end{bmatrix} \begin{bmatrix} sI_1 \\ sI_2 \\ sI_3 \end{bmatrix}. \tag{5.32}$$

We recognize that all the currents in the bars are same. Hence, we have

$$I = I_1 = I_2 = I_3 \tag{5.33}$$

and the total voltage along all three bars is

$$V = V_1 + V_2 + V_3. \tag{5.34}$$

Finally, to compute the open-loop inductance, $Lo = V/sI$. We recognize that the result is obtained by adding all the partial inductances in the matrix in (5.32). We assume that all three conductors have the same length $\ell = \ell_1 = \ell_2 = \ell_3$. For this case, solving the above equations can be reduced to the open loop inductance

$$Lo = 3\ Lp_{11} + 4\ Lp_{12} + 2\ Lp_{13}. \tag{5.35}$$

Of course, the symmetry of the **Lp** matrix is used to simplify the result.

5.6.1 Closing the Loop for Open-Loop Problems

We next set up a closed-loop example based on the open-loop example considered in the previous section. Figure 5.16 gives the example in Fig. 5.17 with a possible closing loop. We notice that the two sides are orthogonal to the horizontal cells. Hence, the only couplings between the sides are Lp_{64} and Lp_{46}. However, Lp_{55} is coupled to the three horizontal cells. Again, the current in all the cells is same, $I_1 = I_2 = \cdots = I_6$.

As reported in Section 5.4.2, we use the open-loop inductance Lo for three lower bars (5.35) as part of solution. The three lower conductor cells have the same size. This is also

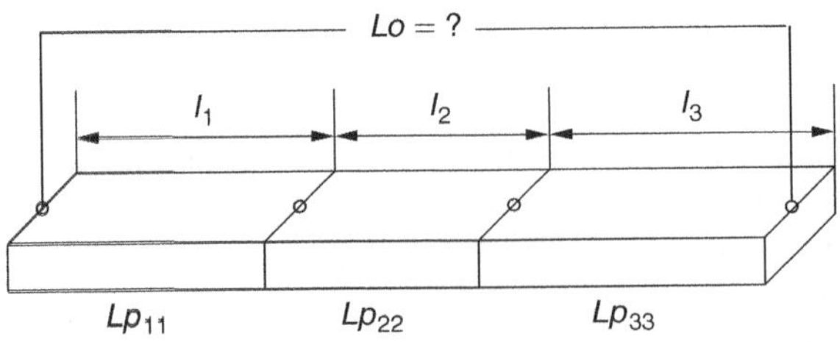

FIGURE 5.16 Open-loop inductance for three bars in series.

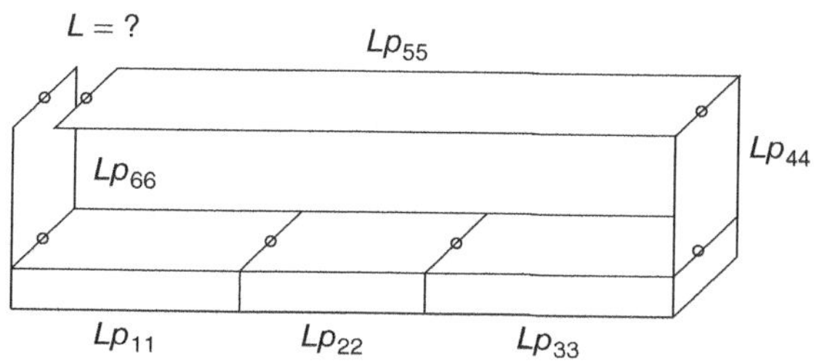

FIGURE 5.17 Loop inductance for three bars in series.

true for the two side conductors. We can set up the partial inductance matrix for the larger problem. However, the result can also be constructed by inspection as

$$L = Lo - 2Lp_{15} - 2Lp_{25} - 2Lp_{35} + Lp_{55} + 2Lp_{44} - 2Lp_{46}, \tag{5.36}$$

where the expression takes advantage of the symmetry of the **Lp** matrix and $Lp_{66} = Lp_{44}$.

A general observation is that in many models, partial or incomplete inductances are not marked as such. Errors may result since we know that open-loop inductances in Fig. 5.16, without a closing path like the one shown in Fig. 5.17, represent an incomplete macromodel.

It is important for us to deal with this situation since too many models, with incomplete open loops, lead to wrong results.

5.7 DIFFERENCE CELL PAIR INDUCTANCE MODELS

In Section 5.5.2, we observed the behavior of the difference inductance between

$$\frac{Ld}{\ell} = \frac{(Lp_{11} - Lp_{12})}{\ell}, \tag{5.37}$$

where ℓ is again the parallel conductor length. This yields finite results for long structures where $\ell \rightarrow \infty$ is in contrast to the partial self-inductance; where as the length ℓ goes to ∞, we have

$$\lim_{\ell \rightarrow \infty} \frac{Lp_{11}(\ell)}{\ell} = \infty. \tag{5.38}$$

An interesting set of difference models are based on symmetrical conductor TL-type models with conventional voltage difference ports with equal and opposing currents. For these structures, the difference inductance is finite. This is considered in the following section.

5.7.1 Inductances for Transmission Line-Type Geometries

We consider the inductance for TL-type inductance models [22, 23] in this section. We consider finite models for TL-type structures that cannot be obtained with conventional transverse electromagnetic (TEM) models. The models usually consist of long uniform cross-sectional structures. To generalize the models, we could also apply the VFI skin-effect models in Section 9.3 to all the conductors in the models. However, in this section we do not consider skin-effect issues. TEM TL conductors are theoretically infinite in length. Here we investigate real-life finite length lines and limit the models to inductances only. However, capacitance models that are not considered can be modeled similarly.

An interesting observation is that the conventional TL model is based on a differential equation representation. Still, it is important to understand that the per-unit length TEM-TL cross-sectional capacitances and inductances are usually computed with a 2D IE solver. The difference between the 2D TL models is the finite length of the 3D PEEC models that are considered in this section.

Partial inductances for wires that are long compared to the cross section are considered in Section 5.5.2. This is very useful for some very long wire model. By now it is clear that PEEC section models are based on coupling between all cells in the model. This is much different from the conventional TEM-TL models where the inductances models are not coupled along the length. These 1-D type models are based on the assumption that the line is infinitely long with a uniform cross-section.

Clearly for real-life TLs, practical errors result if the finite conductor length is relatively short compared to the cross-sectional conductor to ground spacing. For conventional TL models to give a reasonably accurate answer, a realistic length ℓ to wire-to-wire cross-sectional spacing d, ratio $Ratio = \ell/d$ should be larger than 10–20.

Fortunately, for finite length TL structures we can use partial inductance-based models to eliminate the length error. A PEEC model based on partial inductances that represents a section of a TL model is shown in Fig. 5.18. For this model, we chose the range for the section number to be from $m = -M_1$ to a positive $m = M_2$, so that the line can be extended to ∞ in both directions. For a semi-infinite line, we can cut the model anywhere, for instance with $m \geq 0$.

Compute the inductance for a single section k in Fig. 5.18. The voltage drop along the section is

$$V_k - V_{k-1} = \sum_{m=-M_1}^{M_2} Lp_{km}\, sI - \sum_{m'=-M_1}^{M_2} Lp_{km'}\, sI + \sum_{m'=-M_1}^{M_2} Lp_{k'm'}\, sI - \sum_{m=-M_1}^{M_2} Lp_{k'm}\, sI, \tag{5.39}$$

which is the voltage induced in the kth section in (5.39) including all the partial mutuals. We note that if our model extends from $M_1 = -\infty$ to $M_2 = +\infty$, then the section inductance is the same as that of a conventional TEM-TL model. The inductance for section k in (5.39)

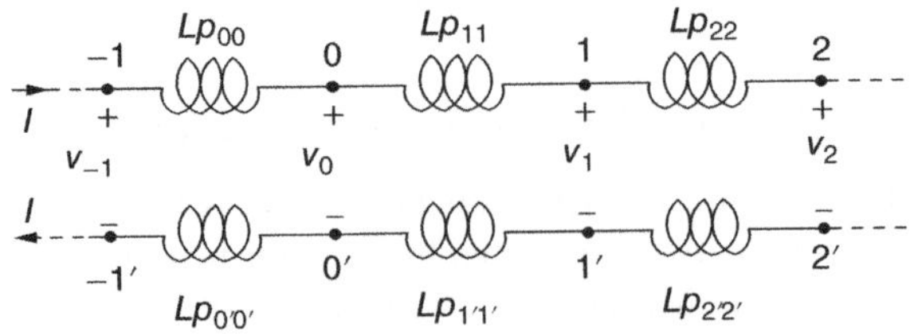

FIGURE 5.18 Finite two-wire line represented in terms of partial inductances.

can be rewritten in a more simple form as

$$L_k = \frac{V_k - V_{k-1}}{sI} = 2 \sum_{m=-M_1}^{M_2} (Lp_{km} - Lp_{k'm}) = 2 \sum_{m=-M_1}^{M_2} Ld_{km} \tag{5.40}$$

for an infinitely long line. If both conductors have the same cross sections for the symmetric situation where $Lp_{km} = Lp_{k'm'}$, then we can apply what we call a difference inductance as

$$Ld_{km'} = (Lp_{km} - Lp_{km'}), \tag{5.41}$$

to simplify the (5.40).

The concept of difference inductances is also efficient for other applications such as multiple TLs or other double-layer plane inductances as is considered in the following section. Of course, the cell pairs must be located directly above each other with opposing currents. The general case for coupling L_{km} between two cell pairs k and m is shown in Fig. 5.19. The coupling inductance can be computed as

$$L_{km} = Lp_{km} - Lp_{km'} + Lp_{k'm} - Lp_{k'm'} = 2\,(Lp_{km} - Lp_{km'}) = 2\,Ld_{km'} \tag{5.42}$$

with the difference inductance given in (5.41).

5.7.2 Approximate Inductive Coupling Calculation Between Difference Cell Pairs

The evaluation of all the partial or difference inductances can be very time consuming. We consider simplification and speedup of the evaluation of partial inductances for the dense sections of the MNA circuit matrix. First, we observed that the coupling from section to section decays much faster than the $1/R$ behavior of the conventional partial inductances where R is the center-to-center distance between the conductor cell pairs shown in Fig. 5.19.

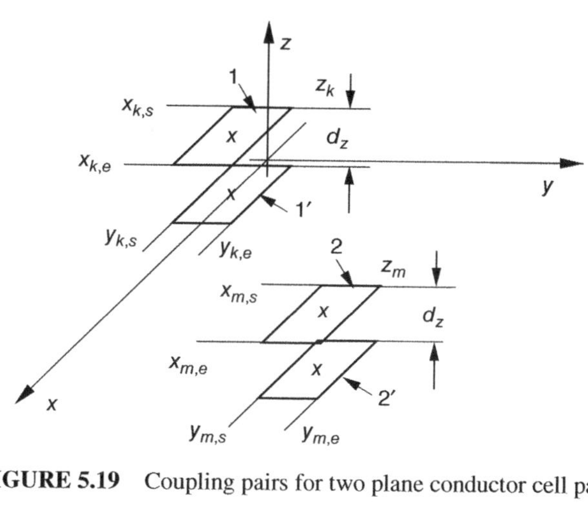

FIGURE 5.19 Coupling pairs for two plane conductor cell pair.

Hence, the difference inductance-based matrices can be sparsified, while this is not the case for matrices of conventional sparse matrices.

We want to apply the simplification of the **Ld** matrices for the general case shown in Fig. 5.19.

The geometry illustrated does not only consider the single TL inductance case where both cell pairs are along the x-axis but also the more general case for plane pairs that is treated in Section 5.7.4. The evaluation of the exact formulas for Lp_{km} is based on using conventional partial inductance formulas given in Appendix C. The basic inductance formulation for the two coupled cell pairs is given in (5.41).

A second, fast approximate evaluation of (5.41) is based on the filament approximation given in (5.21) for the partial inductance, or

$$Lp_{km} = \frac{\mu_0}{4\pi} \frac{\ell_{x_k} \ell_{x_m}}{R_{km}}, \tag{5.43}$$

where ℓ_{x_k} and ℓ_{x_m} are the cell lengths parallel in the x-direction. Further, the cell center-to-center distance between the cell pairs is

$$R_{km'} = \sqrt{(xc_k - xc_{m'})^2 + (yc_k - yc_{m'})^2 + d_z^{\,2}}, \tag{5.44}$$

where we assume that d_z is the same vertical cell distance for both pairs and that all cells are in the same planes such that $z_k = z_m$. For the TL case, we have $yc_k = yc_m$ and for the cells at the same level or $R_{km} = R_{k'm'}$.

We can rewrite the distance metric for the k to m' case as

$$R_{km'} = R_{km} \sqrt{1 + \frac{d_z^2}{R_{km}^2}}. \tag{5.45}$$

This leads to the approximate partial inductance using (5.21) for distant conductors that couples between the planes at the two different levels

$$Lp_{km'} \simeq \frac{\mu_0}{4\pi} \frac{\ell_{x_k} \ell_{x_m}}{R_{km}\sqrt{1 + \frac{d_z^2}{R_{km}^2}}} \simeq \frac{\mu_0}{4\pi} \frac{\ell_{x_k} \ell_{x_m} \left(1 - \frac{d_z^2}{2\,R_{km}^2}\right)}{R_{km}}, \tag{5.46}$$

where in the last step we approximate the square root with a series expansion under the condition that $d_z \ll R_{km}^2$. This leads to the approximate section-to-section coupling

$$L_{km} = 2\,Ld_{km} = 2\,(Lp_{km} - Lp_{km'}) = \frac{\mu_0}{4\pi} \frac{\ell_{x_k} \ell_{x_m}\, d_z^2}{R_{km}^3}, \tag{5.47}$$

where the center-to-center distance R_{km} in Fig. 5.19 is large compared to the plane-to-plane distance d_z.

TABLE 5.1 Comparison of section-to-section coupling inductance decay.

Section Distance (mm)	L_{12} (pH)	1 Fil. appr. L_{12} (pH)	Error (%)	2 Fil. appr. L_{12} (pH)	Error
1	21.978	4.00	81.80	5.0095	77.21
2	0.62408	0.5000	19.88	0.54407	12.82
3	0.16161	0.14815	8.33	0.15418	4.60
5	0.032963	0.03200	2.92	0.032477	1.47
8	0.0079019	0.0078125	1.31	0.0078582	0.55
10	0.0040291	0.0040000	0.72	0.0040150	0.35
15	0.00118980	0.00111852	0.32	0.0011872	0.15

We give an example for the reduction of the error with cell distance in the approximate section-to-section inductance L_{km}. In this example, the cells are square with $\ell_x = \ell_y =$ 1 mm since square cells are representing a worst case, which is also important for the plane-pair models in the following section. In this example, the vertical cell spacing is 0.2 mm. The fast in-plane spacing decay is evident from Table 5.1. The error is the normalized difference between the exact and approximated formula (5.47). We observe that this is the worst case in comparison to narrower cells. Also, results are given in Table 5.1, where each cell is represented by a two- wire Gaussian integration approximation, with a reduction in error of about a factor 2 for sufficient spacing. This leads to an even faster decay of the coupling.

5.7.3 Inductance of Finite and Semi-Infinite Length TL

The coupling computations in the previous paragraph seem to work also for the coupling between length sections of a TL. The fundamental coupling equation used (5.21) gives the partial inductance conductors for two distant filaments. Now, the equation derived in the previous section (5.47) L_{km} gives the coupling for conducting pairs with opposite currents as is the case in TLs, which is based on the difference inductances.

We apply the difference coupling equations to the TL formulation in Section 5.7.1 to study the coupling between TL sections. Very often, we do not take the coupling between length sections into account. For this reason, we study impact of the lack of length inductive coupling between the segments. We cut the TLs into N independent sections as shown in Fig. 5.19. In the comparison, we construct two models, one with couplings between all sections and one where the couplings between the sections are ignored. The error between the two solutions with and without couplings is computed. In summary, the section coupling between sections 1 and 2 is evaluated according to (5.40) as

$$L_{12} = 2\,Ld = 2\,(Lp_{12} - Lp_{12'}), \tag{5.48}$$

where the approximate formula (5.47) is used, while the case of all coupled N sections is the same as the difference inductance for the entire two-wire line.

In Table 5.2, L_{1-N} represents the inductance of the calculation of the wires with all couplings. Again, the section model in Fig. 5.20 consists of N uncoupled parts.

It is not surprising that the error increases as the number of noncoupled ports increases.

TABLE 5.2 Comparison of TL inductance v.s. the number of sections.

Sections N	L_{1-N}	L_{1-N} section model
1	0.27765	0.27765
2	0.582049	0.555299
4	1.196842	1.110598
6	1.813189	1.665898
8	2.429942	2.221197
10	3.046943	2.776496

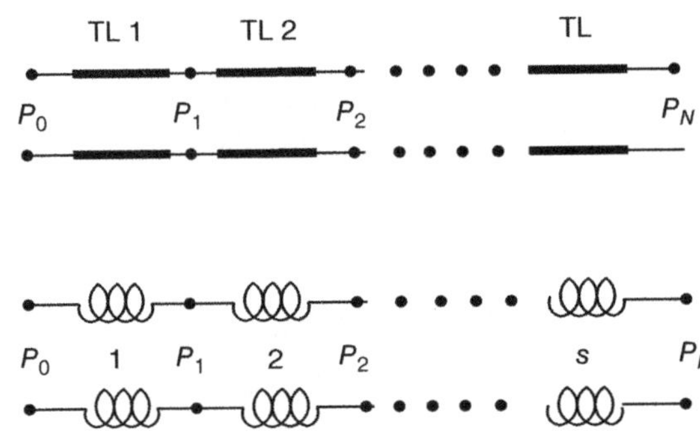

FIGURE 5.20 Transmission line model with line sections and equivalent circuit.

5.7.4 Plane Pair PEEC Models Based on Difference Currents

In the previous section, we showed that the inductive coupling between distant cell pairs drops off fast due to the canceling effect of the opposing currents in the partial inductance pairs. The same concepts lead to interesting application to the modeling for power plane pairs [24–26]. The purpose of the so-called power distribution network (PDN) is to supply voltage and ground to the integrated circuits (ICs).

We apply the current difference approach defined in Section 5.5.2 to plane pairs for printed circuit board [27, 28]. The approach also benefits from the rapid coupling decay that leads to a sparse circuit matrix. We need to mesh the planes according to Chapter 8 by laying out the surfaces with congruent conventional mesh cells. Hence, the cell structure and the node current structure are the same on both surfaces similar to the TL model in Fig. 5.18. The equal and opposing currents in the plane cells as well as the voltage differences at congruent nodes at the same x, y location lead to half the unknowns.

To keep the model simple, we consider a strictly inductance model. For the inductance-only model, the solution is fast since all the matrices are real. More complex parallel plane partial element equivalent circuit (PPP) models can include capacitances as well as resistances.

5.7.5 Parallel Plane PEEC Modeling

Realistic PC plane problems have specific connection and structures. The purpose of the decoupling capacitor used is to prevent the supply voltages from dropping due to the

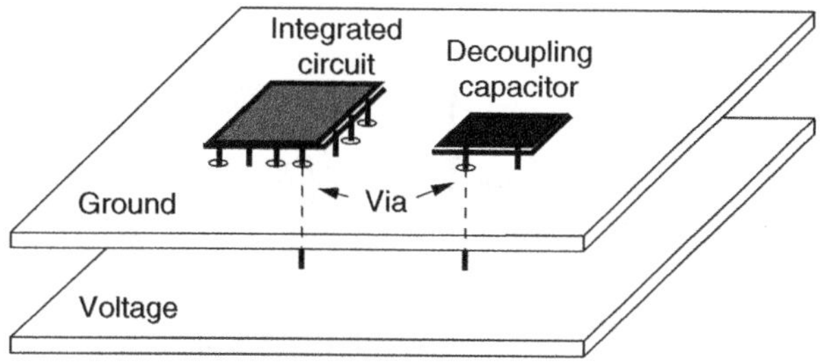

FIGURE 5.21 Parallel power planes example with IC and decoupling capacitor.

inductance of the path. The source of the voltage drop is the switching integrated circuits. The objective of the technique presented in this section is the efficient evaluation of the inductance between a pair of power planes that is part of the PDN circuit. The most simple example of such a plane pair distribution system is shown in Fig. 5.21.

The power plane system supplies the voltages to the appropriate pins at an IC. Usually, large decoupling capacitors keep the impedance (inductance) as low as possible between the supplied voltage and ground. An assumption made is that the frequency is high enough such that the skin depth (9.1) compared to the thickness of the conductors is small and all current flows between the inner surfaces of the plane pair shown in Fig. 5.21. Hence, the thickness of the planes isolates the currents to the inside the planes.

More generally, PPP cases are considered in Ref. [29] for both frequency and the time domain solutions. In this section, we are interested in inductance computation only under the condition that the impedance of the large decoupling capacitor(s) is inductive. Such a plane pair in general has multiple connections that can be capacitors, voltage, and ground via connections. Via connections are inductive at mid frequencies.

The approximate inductance for a decoupling capacitor is called ESL which is usually specified by the manufacturer. In fact, we consider the inductance modeling of a capacitor in Section 5.4.1. Fortunately, we can use a simple partial inductance model for the via connections, similar to the pin partial inductance in Fig. 5.8. Due to the orthogonality, the partial mutual inductances between the planes and vias are zero or very small. Hence, we can separate the plane inductances from the via inductances.

5.7.6 PEEC Inductance Plane Model with Orthogonal Meshing

We use them to represent the plane inductance in this model. We use the most simple example to explain the PPP formulation for the parallel plane model. The example shown in Fig. 5.22 does include a current source for the IC shown in Fig. 5.21 and a short circuit between the planes that represents a perfect decoupling capacitor. To keep the example simple, a minimum number of circuit elements are used. Each plane is represented by only four partial inductances shown in Fig. 5.22. We make the connection to the IC at node N_4 with the current source. The decoupling capacitor is located at node N_1 represented by a simple short circuit. Of course, the node pairs at each node represent ports between the plane pairs. For more realistic plane models, multiple capacitors ICs can be connected at available ports. At this point, we also assume that the difference inductances have been computed as reported in Section 5.7.2. We use them for inductances in this model.

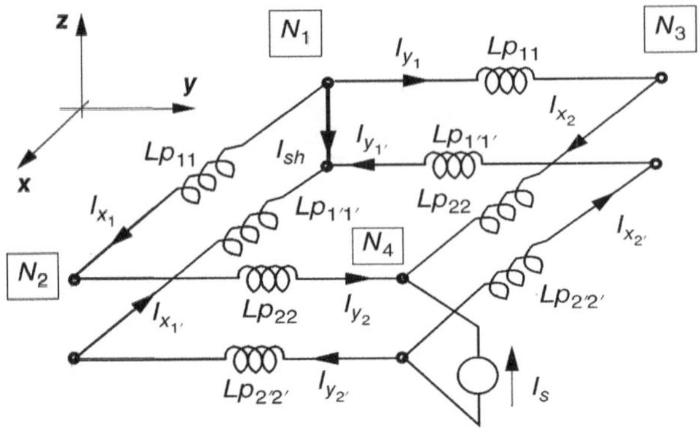

FIGURE 5.22 Small example geometry for two parallel plane PEEC model.

We set up the MNA equations for the simple test geometry in Fig. 5.22 according to Chapter 2. Here, the equations include voltage differences at the nodes and the opposing difference currents are between the planes leading to difference inductances Ld as given by (5.41). The reduction of unknowns is clear from this. The fundamental aspects of the MNA matrix are very similar, and we can use the circuit elements stamps in Appendix B. The MNA equations for the difference quantities is given by

$$\begin{bmatrix} 0 & 0 & 0 & 0 & 1 & 0 & 1 & 0 & 1 \\ 0 & 0 & 0 & 0 & -1 & 0 & 0 & 1 & 0 \\ 0 & 0 & 0 & 0 & 0 & 1 & -1 & 0 & 0 \\ 0 & 0 & 0 & 0 & 0 & -1 & 0 & -1 & 0 \\ 1 & -1 & 0 & 0 & -sLd_{x11} & -sLd_{x12} & 0 & 0 & 0 \\ 0 & 0 & -1 & 1 & -sLd_{x12} & -sLd_{x22} & 0 & 0 & 0 \\ 1 & 0 & -1 & 0 & 0 & 0 & -sLd_{y11} & -sLd_{y12} & 0 \\ 0 & 1 & 0 & -1 & 0 & 0 & -sLd_{y12} & -sLd_{y22} & 0 \\ 1 & 0 & 0 & 0 & 0 & 0 & 0 & 0 & 0 \end{bmatrix} \begin{bmatrix} V_{N1} \\ V_{N2} \\ V_{N3} \\ V_{N4} \\ I_{x1} \\ I_{x2} \\ I_{y1} \\ I_{y2} \\ I_{sh} \end{bmatrix} = \begin{bmatrix} 0 \\ 0 \\ 0 \\ I_s \\ 0 \\ 0 \\ 0 \\ 0 \\ 0 \end{bmatrix}. \tag{5.49}$$

At the node port N_4, we can compute the port inductance as V_{N4}/sI_s. For example, we inject a unit current of 1 mA and we measure the voltage leading to the inductive impedance in kΩ. The system would lead to a complex solution if we include the plane resistance in the model.

The basic ordering of the equations is given in (5.49). To make it clear, we also show the matrix structure in Fig. 5.23. The entries corresponding to the Kirchhoff's current law (KCL) **A** matrix parts are set up automatically by the inductance stamps. They can also be set up using matrix KCL stamps as is done in Section 2.7.1. We observe that the difference inductance Ld part is represented in decoupled x and y parts avoiding $x - y$ couplings. Further, adding the equations for short circuits is done at the last rows and columns. This facilitates changes such as adding other elements, for example, the addition of more decoupling capacitors to the solution. As is shown, the computationally expensive inductance part of the PEEC model and part of the MNA matrix are not impacted by adding elements to the model. The matrix equation ordering leads to flexibility in changes of the circuit topology.

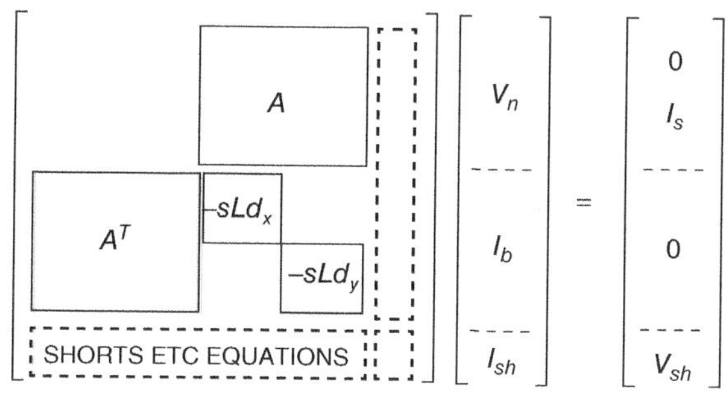

FIGURE 5.23 General schematic MNA matrix with additional equations.

5.7.7 Mesh Reduction Without Couplings of Nonparallel Inductances

So far, the formulation presented here avoids the coupling between the orthogonal matrix parts. We presented in this section how this is accomplished even for a mesh-size reduction. Several levels of mesh reduction are used to speed up the solution. Smaller mesh sizes near the contacts increase the accuracy of the solution. Also, the mesh size is increased for open areas to reduce the compute time. The advantage of a variable mesh size for the compute time can be considerable as is shown in Ref. [29]. Conventionally, the change in mesh size requires nonorthogonal cells. However, this results in the undesirable coupling between the Ld_x and Ld_y submatrices in the MNA matrix in Fig. 5.23. The approximate decoupled model in Fig. 5.24 shows nodes corresponding to larger cells and the refined mesh to half the cell size. The connection between smaller cells and larger cells must be established. We accomplish this with a so-called weighted KCL approach, which is presented next.

The PEEC inductance model corresponding to the meshing in Fig. 5.24 is shown in Fig. 5.25. We see from both figures that only half the refined mesh cells are connected directly at the boundary. Hence, a discontinuity results between the two regions. To take care of this situation, we locally modify the KCL equations.

Fundamentally, we redirect some of the current to the unconnected nodes of the fine mesh. We observe that we set up different weighted KCL equation near the corners depending on the location of the local nodes. Different weights have to be used for a convex and a concave corner as shown in Fig. 5.24. The weights are chosen such that the current distributed is as close as possible to the one without the mesh size reduction. Note that a factor 2 in mesh size is obtained at the interface as is evident from Fig. 5.24.

The nodes at the concave corner are considered first. As usual, we write the equation for currents leaving the nodes. The KCL for the corner node **G6** is

$$0.75\, I_{G6,G2} + I_{G6,S6} + I_{G6,S3} + 0.75\, I_{G6,G5} = 0. \tag{5.50}$$

The KCL for the node **S3** next to the concave corner is

$$0.25\, I_{G6,G2} + 0.25\, I_{G7,G3} + I_{S3,S7} + I_{S3,G7} + I_{S3,G6} = 0. \tag{5.51}$$

At the node **G7**, the approximate KCL for the noncorner node is

$$0.5\, I_{G7,G3} + I_{G7,S8} + I_{G7,S4} + I_{G7,S3} = 0. \tag{5.52}$$

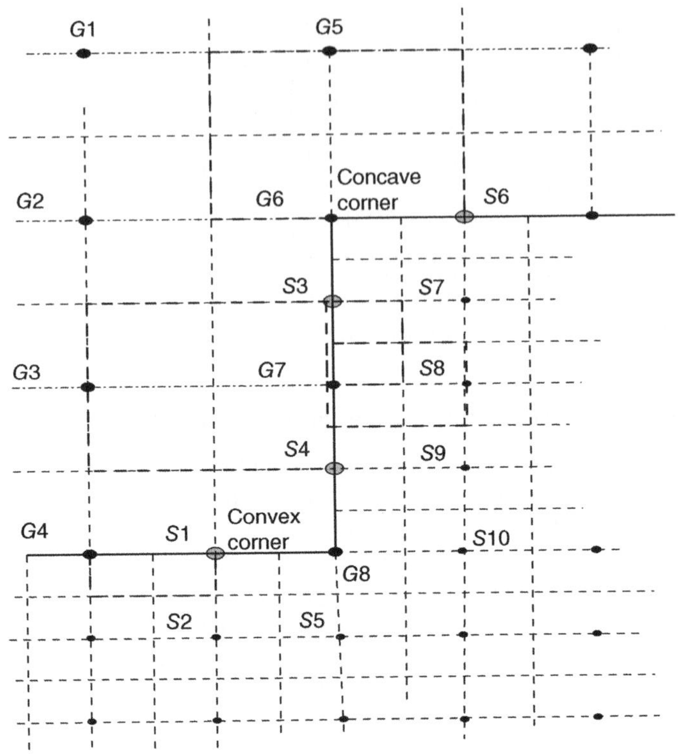

FIGURE 5.24 Mesh at corner interface between coarse and fine mesh.

The last two equations are used for all appropriate noncorner nodes.

The concave corner equations are similar with understandably somewhat different weighting factors. The KCL equation at the corner **G8** is given by

$$\frac{2}{3} I_{G8,S1} + I_{G8,S10} + I_{G8,S5} + \frac{2}{3} I_{G8,S4} = 0. \tag{5.53}$$

Finally, the KCL for neighbor node **S1** next to the convex corner is

$$I_{S1,G4} + I_{S1,G8} + I_{S1,S2} + 0.25\, I_{G4,G3} + \frac{1}{3} I_{G8,S4} = 0. \tag{5.54}$$

These weighted KCL equations cover all possible node configurations at the interface. The weights were found by comparing local currents with and without the interface such that it results in a good approximation of the local current density.

The mesh increase and reduction has a large impact on compute time and accuracy. This has been shown in Ref. [29]. The test problem consists of a simple 8 × 8 cm parallel plane configuration with a spacing of 0.2 mm. A symmetric source and short situation is used for this simple example with the source and the short 20 mm from each side. The source and short are placed in the middle 40 mm from each side. Hence, the short to source spacing is 60 mm.

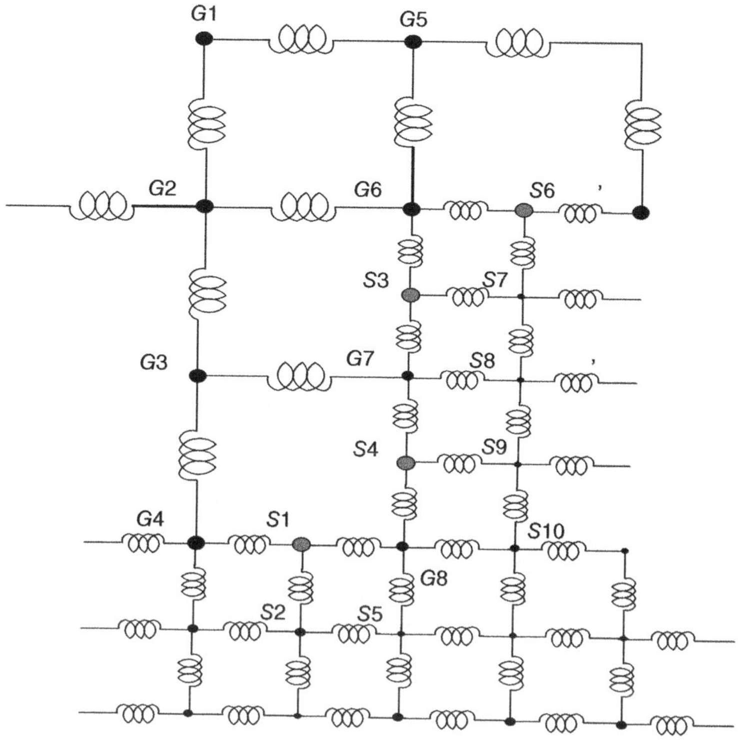

FIGURE 5.25 Equivalent circuit for reduction of meshing example for PPP inductance model.

TABLE 5.3 Some results for the PPP method for 10 × 10 cm planes.

Solution details	Number of unknowns	Inductance (pH)	Comp. time (s)
CST	Not available	345.7	Not available
Uniform mesh	77,404	349.3	172.0
1 Level submesh	20,108	355.7	13.6
2 Level submesh	6,836	357.5	2.5
3 Level submesh	3,852	358.5	1.5

The results of the experiment are given in Table 5.3. It is evident that the reduction in compute time due to the submeshing is excellent for the experimental Matlab program used.

5.8 PARTIAL INDUCTANCES WITH FREQUENCY DOMAIN RETARDATION

The evaluation of partial elements with retardation is an important issue for high frequencies where the quasistatic solutions are inadequate. Delay or retardation from a circuit point of view is addressed in Section 2.11.1. Retardation for full-wave (FW) solutions is considered in Chapter 3. Furthermore, full-wave (FW)PEEC models are developed in Chapter 6. In this section, we consider the retardation issue from a partial inductance perspective. The retarded potential coefficient issues are quite similar, but they are not considered.

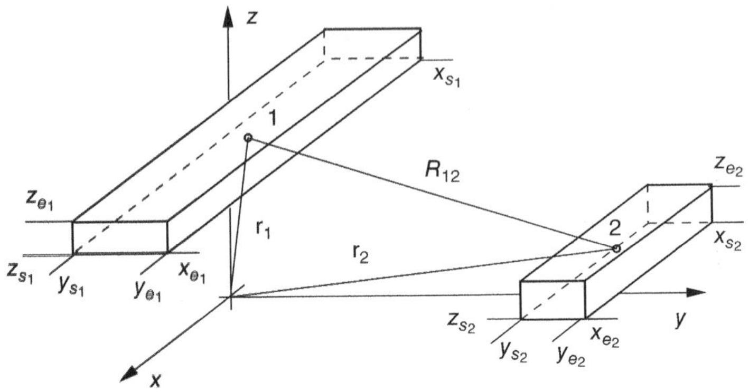

FIGURE 5.26 Two-bar example for a retarded partial inductance computation.

As an intermediate solution between the quasistatic and the high- frequency solution, it may be sufficient to include the retardation from cell center to cell center. The maximum cell size of $\lambda_{min}/20$ may limit the error to be sufficiently small if the f_{max} is small or if the conductors are small in size, unlike the example shown in Fig. 5.26.

In the time domain, we mostly utilize the center-to-center delay. For the two conductors example in Fig. 5.26, we may have to further subdivide the conductors such that the cell size is small enough. The retardation time t_d in the time domain is simply a delay between two points in space $\boldsymbol{r}_1$ and $\boldsymbol{r}_2$ as given in (2.11.1) as

$$\tau_d = R_{12}\,/v, \tag{5.55}$$

where v is the velocity in the material and where in Fig. 5.26

$$R_{12} = \sqrt{(x_1 - x_2)^2 + (y_1 - y_2)^2 + (z_1 - z_2)^2}. \tag{5.56}$$

In air $v = c$, the speed of light, and in a lossless dielectric it is $v = c/\sqrt{\epsilon_r}$, where ϵ_r is the relative dielectric constant. The time domain solver will have to take the delay of the couplings into account, a feature that is missing in a conventional SPICE circuit solver in both the time and the frequency domains.

For FW frequency domain solutions, the accuracy is improved if we also take the retardation inside the integral into account. The retardation is distributed over the areas of the cells. This leads to the following integrals for partial inductances. The formulation including the retardation term according to Section 3.4.2 is given by

$$Lp_{12}^R = \frac{\mu_0}{4\pi}\frac{1}{A_k A_m}\int_{z_{s1}}^{z_{e1}}\int_{y_{s1}}^{y_{e1}}\int_{x_{s1}}^{x_{e1}}\int_{z_{s2}}^{z_{e2}}\int_{y_{s2}}^{y_{e2}}\int_{x_{s2}}^{x_{e2}}\frac{e^{-s\,\tau_d}}{R_{12}}\,dx_2\,dy_2\,dz_2\,dx_1\,dy_1\,dz_1, \tag{5.57}$$

where $Lp_{12}^R = Lp_{12}^R(s)$ and $\tau_d = \tau_d(\boldsymbol{r}_1, \boldsymbol{r}_2)$. To simplify, we call $R_{12} = R$.

We first consider the situation for the partial self-inductance with delay, which is $Lp_{11}(s)$. In this case, both integrations will be over the same body. Again, the maximum physical dimensions of the conductor is limited to λ_{min}. This can be utilized to evaluate the delay term

$$e^{-s\tau_d} = e^{-j\beta R}, \tag{5.58}$$

where $\beta = \omega/v$. For the partial self-inductance, the exponent remains relatively small even for the highest angular frequency ω_{max}. For this reason, we can expand it in a Taylor series, or

$$e^{-j\beta R} \simeq 1 - j\,\beta\,R + (\beta\,R)^2/2 - \cdots = 1 - (\beta^2\,R^2)/2 - j\,\beta\,R - \cdots . \tag{5.59}$$

If we substitute the first three terms into the retarded partial inductance formula (5.57), then we will get three parts, or

$$\begin{aligned} Lp_{11}^R(s) \simeq{}& \frac{\mu_0}{4\pi\mathcal{A}_1^2}\int_{\mathcal{A}_1}\int_{\mathcal{A}_{1'}}\int_{\ell_1}\int_{\ell_{1'}} \frac{1}{R}\, d\ell_1'\, d\ell_1\, d\mathcal{A}_1'\, d\mathcal{A}_1 \\ &- \frac{\beta^2}{2}\frac{\mu_0}{4\pi\mathcal{A}_1^2}\int_{\mathcal{A}_1}\int_{\mathcal{A}_{1'}}\int_{\ell_1}\int_{\ell_{1'}} R\, d\ell_1'\, d\ell_1\, d\mathcal{A}_1'\, d\mathcal{A}_1 \\ &- j\beta\frac{\mu_0}{4\pi\mathcal{A}_1^2}\int_{\mathcal{A}_1}\int_{\mathcal{A}_{1'}}\int_{\ell_1}\int_{\ell_{1'}} d\ell_1'\, d\ell_1\, d\mathcal{A}_1'\, d\mathcal{A}_1. \end{aligned} \tag{5.60}$$

Interestingly, this simplifies to

$$Lp_{11}^R(s) \simeq Lp_{11} - \frac{\beta^2}{2}\frac{\mu_0}{4\pi\mathcal{A}_1^2}\int_{\mathcal{A}_1}\int_{\mathcal{A}_{1'}}\int_{\ell_1}\int_{\ell_{1'}} R\, d\ell_1\, d\ell_1'\, d\mathcal{A}_1\, d\mathcal{A}_1' - j\beta\frac{\mu_0}{4\pi}{\ell_1}^2. \tag{5.61}$$

Note that this results in a frequency domain impedance Lp^R whose imaginary part is

$$Re(Lp^R) \simeq Lp_{11} - \frac{\beta^2}{2}\frac{\mu_0}{4\pi\mathcal{A}_1^2}\int_{\mathcal{A}_1}\int_{\mathcal{A}_{1'}}\int_{\ell_1}\int_{\ell_{1'}} R\, d\ell_1'\, d\ell_1\, d\mathcal{A}_1'\, d\mathcal{A}_1. \tag{5.62}$$

The resistive part that results is evident from (5.60). Since this part is part of the partial inductance, we have to premultiply it with $j\omega$ that results in

$$\begin{aligned} Im(Lp^R) \simeq{}& -j\omega\frac{\beta^2}{2}\frac{\mu_0}{4\pi\mathcal{A}_1^2}\int_{\mathcal{A}_1}\int_{\mathcal{A}_{1'}}\int_{\ell_1}\int_{\ell_{1'}} R\, d\ell_1'\, d\ell_1\, d\mathcal{A}_1'\, d\mathcal{A}_1 \\ ={}& j\omega\left(-j\beta\frac{\mu_0}{4\pi}{\ell_1}^2\right) = \frac{\omega^2}{v}\frac{\mu_0}{4\pi}{\ell_1}^2 \end{aligned} \tag{5.63}$$

where the last step makes use of the fact that $\beta = \omega/v$. The conductor cross-sectional term cancels in the positive loss resistor. We observe a reduction in the partial inductance with frequency and a frequency-dependent series resistor that introduces damping especially at high frequency. However, this approximation cannot be applied if the structure or the delay is too large compared to the wavelength. This issue is addressed in the next section.

5.8.1 Thin Wire Example for Retarded Partial Inductances

We evaluate as an example the retarded partial inductance evaluation for relatively thin round wires using the derivations from the previous section, which uses the approximation of the exponential term. The partial inductance (5.17) is repeated in a simplified form with retardation in (5.64) which assumes that the current flows in the axial direction

$$Lp_{12}^R = \frac{1}{\mathcal{A}_1\,\mathcal{A}_2}\frac{\mu_0}{4\pi}\int_{\mathcal{V}_1}\int_{\mathcal{V}_2}\frac{e^{-j\beta R}}{R}\,d\mathcal{V}_2\,d\mathcal{V}_1, \tag{5.64}$$

where we evaluate only one term in the inductance part in (5.61). We assume that the length of the wire $(x_e - x_s)$ in Fig. 5.27 is at least two times the diameter d. We can either use the approximate formula (C.15) or the tube formula (C.17) for the quasi-static self term. The tube formula (C.17) is exact if $(x_e - x_s)$ is much larger than the tube diameter d. From the previous section, we get the impedance model for a retarded partial self-inductance as

$$Z_{Lp,11}^R = j\omega Lp_{11} + \frac{\omega^2}{v}\frac{\mu_0}{4\pi}(x_e - x_s)^2, \tag{5.65}$$

where Lp_{11} is the quasistatic partial inductance computed with the above formula and the resistive part is given by (5.63).

The evaluation of the retarded partial mutual inductance between two distant wire segments is more challenging. Here, we give a small example for the simpler case when the two wire segments are lined up as shown in Fig. 5.28.

A question is how the large phase factor associated with a mutual inductance is evaluated for the coupling term. The integral (5.64) reduces for this case to

$$Lp_{12}^R = \frac{\mu_0}{4\pi}\int_{x_1=0}^{e_1}\int_{x_2=b_2}^{e_2}\frac{e^{-j\beta|x_2-x_1|}}{|x_2-x_1|}\,dx_2\,dx_1. \tag{5.66}$$

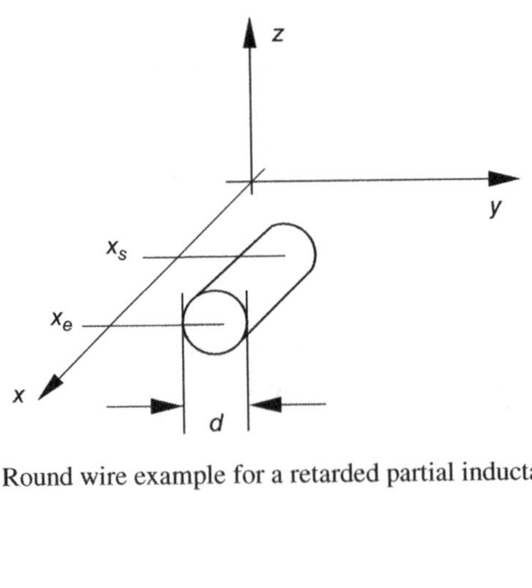

FIGURE 5.27 Round wire example for a retarded partial inductance evaluation.

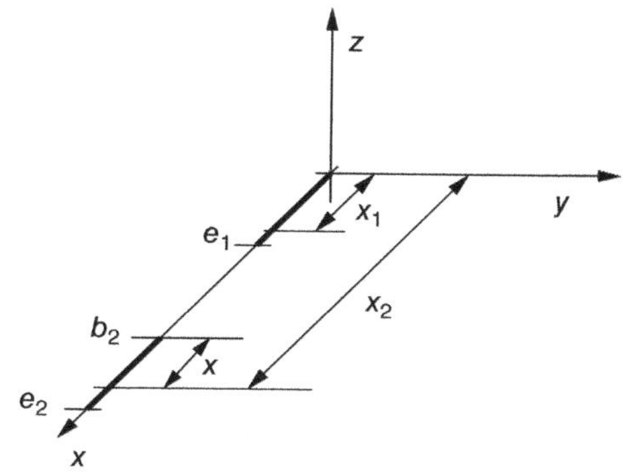

FIGURE 5.28 Two wires in series for computation of retarded partial inductance.

We use the following change in the variables where we replace x_2 by setting $x = x_2 - b_2$ in (5.66). This leads to the following result:

$$Lp_{12}^{RM} = \frac{\mu_0 e^{-j\beta b_2}}{4\pi} \int_{x_1=0}^{e_1} \int_{x=0}^{(e_2-b_2)} \frac{e^{-j\beta|x-x_1|}}{|b_2+x-x_1|} \, dx \, dx_1, \tag{5.67}$$

where we take the constant delay factor $e^{-j\beta b_2}$ outside of the integral. Finally, if we add and subtract $Lp_{12}\, e^{-j\beta b_2}$, the result will be

$$Lp_{12}^{R} = e^{-j\beta b_2} Lp_{12} - e^{-j\beta b_2} Lp_{12} + Lp_{12}^{RM}. \tag{5.68}$$

Different approaches can be used to approximate the second and third part on the right hand side of (5.68) since the delay term under the third term is now smaller such that series expansions can be used. We observe from Fig. 5.28 that the starting points of the two wires are in general much farther apart, leading to a large delay factor $e^{-j\beta b_2}$, which is for most cases much larger than the small ones for the evaluation due to the wires segments with a length of $\lambda/20$ or smaller. Also, we can view the second term as a correction factor to the conventional solution where the quasistatic solution is multiplied by the delay factor $e^{-j\beta b_2} Lp_{12}$. An example where this approach is used is given in Section 13.5.6.

5.8.2 General Case for Separated Conductor Partial Inductances with Retardation

We again refer to the example in Fig. 5.26, where we want to compute the partial inductance with retardation. If the two conductors are sufficiently separated, another integration method based on local expansion and far-field translation can be employed. In general, rewrite the partial inductance in the following simplified format:

$$Lp_{1,2}^{R}(s) = \frac{\mu_0}{4\pi \mathcal{A}_1 \mathcal{A}_2} \int_{z_{s1}}^{z_{e1}} \int_{y_{s1}}^{y_{e1}} \int_{x_{s1}}^{x_{e1}} \int_{z_{s2}}^{z_{e2}} \int_{y_{s2}}^{y_{e2}} \int_{x_{s2}}^{x_{e2}} \frac{e^{-j\beta R_{12}}}{R_{12}} \, dx_2 \, dy_2 \, dz_2 \, dx_1 \, dy_1 \, dz_1. \tag{5.69}$$

If the center of conductor 1 is $\boldsymbol{r}_{o1}$, the center of conductor 2 is $\mathbf{r}_{o2}$, the vector $\boldsymbol{D} = \mathbf{r}_{o2} - \boldsymbol{r}_{o1}$. Let the vectors to the center $\boldsymbol{r}_{L1} = \boldsymbol{r}_1 - \boldsymbol{r}_{o1}$ and $\mathbf{r}_{L2} = \mathbf{r}_2 - \mathbf{r}_{o2}$, then

$$\mathbf{r}_2 - \mathbf{r}_1 = \mathbf{r}_{L2} + \boldsymbol{D} - \mathbf{r}_{L1}, \tag{5.70}$$

where $||\boldsymbol{D}||$ is very large compared to the wavelength. Hence,

$$R_{12} = ||\boldsymbol{r}_2 - \boldsymbol{r}_1|| = ||\boldsymbol{r}_{L2} + \boldsymbol{D} - \boldsymbol{r}_{L1}||. \tag{5.71}$$

So, according to the addition theorem, the retarded coupling can be expressed as

$$\frac{e^{-j\beta R_{12}}}{R_{12}} = \frac{-j\beta}{4\pi} \int_{\theta=0}^{\pi} \int_{\phi=0}^{2\pi} e^{-j\boldsymbol{\beta}\cdot(-\mathbf{r}_{L1}+\mathbf{r}_{L2})} T_L(\hat{\boldsymbol{\beta}} \cdot \hat{\boldsymbol{D}}) \sin\theta d\theta \, d\phi, \tag{5.72}$$

where T_L is the multipole translator for high frequencies and is defined as

$$T_L(\hat{\boldsymbol{\beta}} \cdot \hat{\boldsymbol{D}}) = \sum_{l=0}^{L} (-1)^l (2l+1) j_l(\beta d) h_l^{(1)}(\beta D) P_l(\cos\theta), \tag{5.73}$$

where $h_l^{(1)}(kD)$ is the lth-order spherical Hankel function of the first type. P_l is the lth-order Legendre polynomial. Substituting (5.72) into (5.69), we have

$$\begin{aligned} Lp_{12}(s) = {} & \frac{-j\beta\mu_0}{16\pi^2} \frac{1}{\mathcal{A}_1\mathcal{A}_2} \int_{\theta=0}^{\pi} \int_{\phi=0}^{2\pi} \{T_L(\hat{\boldsymbol{\beta}} \cdot \hat{\boldsymbol{D}}) \\ & \cdot \int_{-L_1/2}^{L_1/2} \int_{-W_1/2}^{W_1/2} \int_{-T_1/2}^{T_1/2} e^{-j\boldsymbol{\beta}\cdot(-\boldsymbol{r}_{L1})} \, dx_1 \, dy_1 \, dz_1 \\ & \cdot \int_{-L_2/2}^{L_2/2} \int_{-W_2/2}^{W_2/2} \int_{-T_2/2}^{T_2/2} e^{-j\boldsymbol{\beta}\cdot(\mathbf{r}_{L2})} \, dx_2 \, dy_2 \, dz_2\} \sin\theta \, d\theta \, d\phi, \end{aligned} \tag{5.74}$$

where

$$\begin{aligned} L_1 &= x_{e1} - x_{s1}, \quad L_2 = x_{e2} - x_{s2} \\ W_1 &= y_{e1} - y_{s1}, \quad W_2 = y_{e2} - y_{s2} \\ T_1 &= z_{e1} - z_{s1}, \quad T_2 = z_{e2} - z_{s2}. \end{aligned} \tag{5.75}$$

Then, the analytic solution of the inner integrals can be achieved by considering the fact that $\boldsymbol{\beta} = \hat{\boldsymbol{x}}\beta_x + \hat{\boldsymbol{y}}\beta_y + \hat{\boldsymbol{z}}\beta_z$. Then

$$\begin{aligned} Lp_{12}(s) = {} & \frac{-j\beta\mu_0}{16\pi^2} \frac{1}{\mathcal{A}_1\mathcal{A}_2} \int_{\theta=0}^{\pi} \int_{\phi=0}^{2\pi} \Big\{ T_L\,(\hat{\boldsymbol{\beta}} \cdot \hat{\mathbf{D}}) \\ & \cdot [L_1 \cdot \mathrm{sinc}(\beta_x L_1/2) \cdot W_1 \cdot \mathrm{sinc}(\beta_y W_1/2) \cdot T_1 \cdot \mathrm{sinc}(\beta_z T_1/2)] \end{aligned}$$

$$\cdot [L_2 \cdot \mathrm{sinc}(\beta_x L_2/2) \cdot W_2 \cdot \mathrm{sinc}(\beta_y W_2/2) \cdot T_2 \cdot \mathrm{sinc}(\beta_z T_2/2)]$$

$$\Big\} \sin\theta \, d\theta \, d\phi, \tag{5.76}$$

where $\mathrm{sinc}(x) = \sin(x)/x$. Further simplifications give the following reduced result

$$Lp_{1,2}(s) = \frac{-j\beta\mu_0}{16\pi^2} L_1 L_2 \int_{\theta=0}^{\pi} \int_{\phi=0}^{2\pi} \Big\{ d^2 \, \beta T_L(\hat{\beta} \cdot \hat{\mathbf{D}})$$

$$\cdot [L_1 \cdot \mathrm{sinc}(\beta_x L_1/2) \cdot W_1 \cdot \mathrm{sinc}(\beta_y W_1/2) \cdot T_1 \cdot \mathrm{sinc}(\beta_z T_1/2)]$$

$$\cdot [L_2 \cdot \mathrm{sinc}(\beta_x L_2/2) \cdot W_2 \cdot \mathrm{sinc}(\beta_y W_2/2) \cdot T_2 \cdot \mathrm{sinc}(\beta_z T_2/2)]$$

$$\Big\} \sin\theta \, d\theta \, d\phi. \tag{5.77}$$

Then, numerical quadrature rules can be applied to the outer spectral integration to obtain the final results.

One numerical example is shown in Fig. 5.29, where conductor cells are coupled to each other. There, central distances are varied to check how accurate the complex inductance computation could be based on the fast multipole algorithm when it is compared with the direct computation results. Figures 5.30 and 5.31 show that excellent results are obtained. And the method could be used for close-coupling computation. The truncation number in this case is $L = 6$.

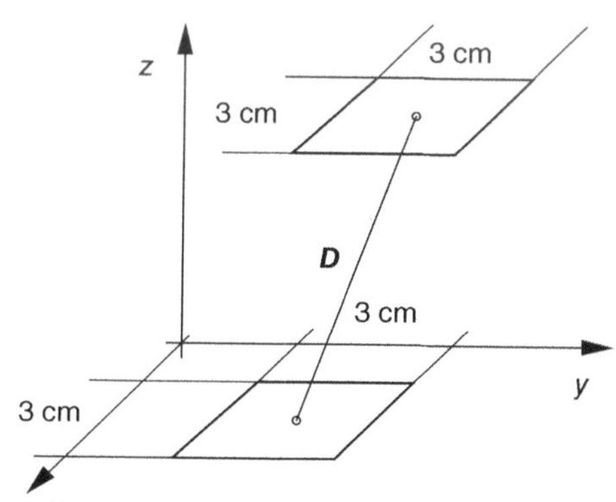

FIGURE 5.29 Geometry of two cells with a certain distance.

PROBLEMS

5.1 *Partial mutual inductance for two short wires*
Derive the formula for the partial mutual inductance for the two-wire geometry (C.2) in Appendix C. Make sure that your result agrees with the formula in (C.8).

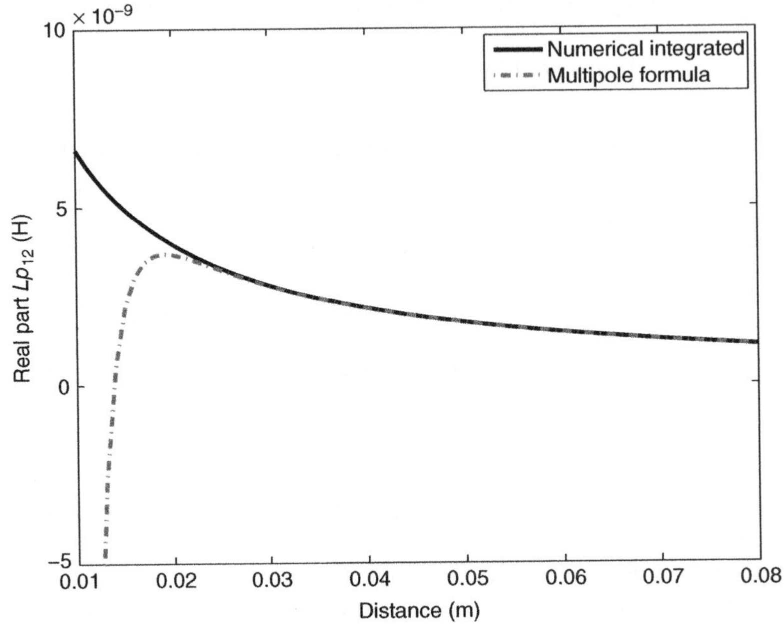

FIGURE 5.30 Mutual inductance real part of the coupled cells by the fast multipole algorithm compared with the theoretical result.

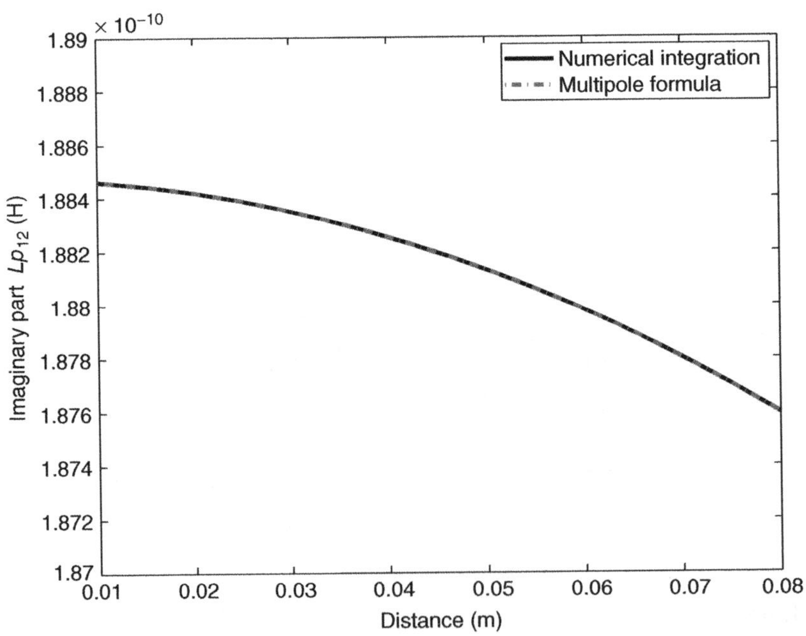

FIGURE 5.31 Mutual inductance imaginary part of the coupled cells by the fast multipole algorithm compared with the theoretical result.

5.2 *Single-loop inductance*
Compute the inductance of the loop in Fig. 5.32 in terms of the following partial inductance matrix. All the orthogonal partial inductances are zero as indicated. Compute the inductance $L = V_{AB}/(sI(s))$ of the loop by solving the circuit equations. Note that the $V_{AB} = V_1 + V_2 + V_3 + V_4$. Use the sign rule given in Section 5.2.

$$\begin{bmatrix} V_1 \\ V_2 \\ V_3 \\ V_4 \end{bmatrix} = \begin{bmatrix} Lp_{11} & 0 & +Lp_{13} & 0 \\ 0 & Lp_{22} & 0 & +Lp_{24} \\ +Lp_{31} & 0 & Lp_{33} & 0 \\ 0 & +Lp_{42} & 0 & Lp_{44} \end{bmatrix} \begin{bmatrix} sI_1 \\ sI_2 \\ sI_3 \\ sI_4 \end{bmatrix}. \tag{5.78}$$

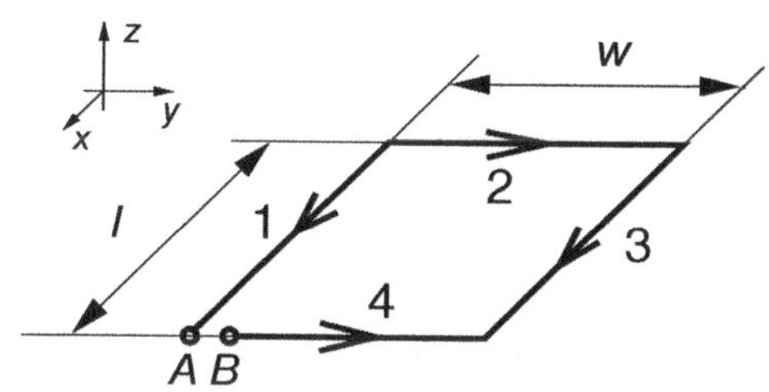

FIGURE 5.32 Single-loop inductance example.

5.3 *Inductance for loop with a gap*
For Problem 5.2, introduce a gap by reducing the length of branch 4 of the loop in the previous problem by half. Calculate the loop inductance for a 5 cm square loop with a square cross section of 1 mm. Compare the closed-loop inductance with the open-loop inductance with the 2.5-cm branch 4. For such an open geometry, different formulas can be used for the partial self- and mutual inductances. Round wires or rectangular ones can be used.

5.4 *Long loop*
If the loop in Fig. 5.32 is oblong, or nonsquare, the opposite partial self-inductances are equal resulting in $Lp_{11} = Lp_{33}$ and $Lp_{22} = Lp_{44}$. Therefore, (5.23) is reduced to

$$L = 2Lp_{11} - 2Lp_{13} + 2Lp_{22} - 2Lp_{24}. \tag{5.79}$$

However, if the loop is very long in the ℓ direction and the wires are almost touching, then we can ignore the small values of $2Lp_{22} - 2Lp_{24}$. Make comparisons for which aspect ratio of ℓ/w this contribution becomes small compared to the length inductances.

5.5 *Two-loop transformer*
Compute the inductances **L** of the two-loop transformer using the following partial inductance matrix:

$$\begin{bmatrix} V_{AB} \\ V_{CD} \end{bmatrix} = \begin{bmatrix} L_{11} & L_{12} \\ L_{21} & L_{22} \end{bmatrix} \begin{bmatrix} sI_1 \\ sI_2 \end{bmatrix}. \tag{5.80}$$

As an alternate approach use the loop to branch matrix **B** for the two set of wires. In Fig. 5.33, we use the dashed loop with the same direction for both loops. The **B** sets up the branch direction for each loop where each row indicates the branches in the loop by a + or − sign depending on the branch direction. For this example, we have

$$\mathbf{B_{ab}} = \begin{bmatrix} -1 & 1 & 1 & -1 & 0 & 0 & 0 & 0 \end{bmatrix}$$
$$\mathbf{B_{cd}} = \begin{bmatrix} 0 & 0 & 0 & 0 & -1 & 1 & 1 & -1 \end{bmatrix}.$$

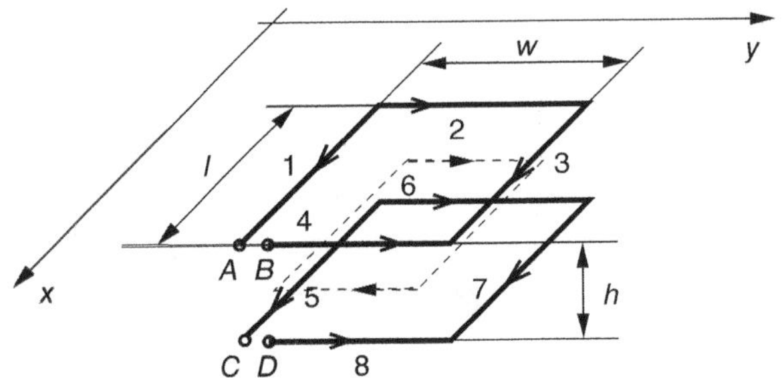

FIGURE 5.33 Two-loop example.

Show that the answer can be obtained as

$$\mathbf{V} = \mathbf{B L_p B}^T \, s\mathbf{I},$$

where the appropriate $\mathbf{B}$ is used corresponding to which of the inductances is computed. For example, $L_{12} = \mathbf{B_{ab}} \; \mathbf{L_p} \; \mathbf{B_{cd}}^T$

$$\begin{bmatrix} V_1 \\ V_2 \\ V_3 \\ V_4 \\ V_5 \\ V_6 \\ V_7 \\ V_8 \end{bmatrix} = \begin{bmatrix} Lp_{11} & 0 & Lp_{13} & 0 & Lp_{15} & 0 & Lp_{17} & 0 \\ 0 & L_{22} & 0 & L_{24} & 0 & Lp_{26} & 0 & Lp_{28} \\ Lp_{31} & 0 & Lp_{32} & 0 & Lp_{35} & 0 & Lp_{37} & 0 \\ 0 & L_{42} & 0 & L_{44} & 0 & Lp_{46} & 0 & Lp_{48} \\ Lp_{51} & 0 & Lp_{53} & 0 & Lp_{55} & 0 & Lp_{57} & 0 \\ 0 & L_{62} & 0 & L_{64} & 0 & Lp_{66} & 0 & Lp_{68} \\ Lp_{71} & 0 & Lp_{72} & 0 & Lp_{57} & 0 & Lp_{77} & 0 \\ 0 & L_{82} & 0 & L_{84} & 0 & Lp_{86} & 0 & Lp_{88} \end{bmatrix} \begin{bmatrix} sI_1 \\ sI_2 \\ sI_3 \\ sI_4 \\ sI_5 \\ sI_6 \\ sI_7 \\ sI_8 \end{bmatrix}. \quad (5.81)$$

Compute all four inductances using the above procedure to find an analytic formula for the $\boldsymbol{L}$ matrix.

5.6 *Impedance of a loop*
As shown in Fig. 5.33, two loops are located with a center to center height difference, $h = 1$ mm. Both loops have a length $l = 5$ cm and a width $w = 3$ cm. We assume that both copper loops have a cross-section of 0.1 mm × 0.1mm. Assume that a current source of 10 mA is applied between C and D. Make a inductance-resistance PEEC Matlab program to compute the voltage between the terminals A and B as a function of frequency.

5.7 *Approximate inductance of a loop over a ground plane*
Here we make the assumption that the inductance will be reduced if we place the loop over a lossless ground plane as shown in Fig. 5.34. We assume that all the loop dimensions are the same as the ones in the previous problem. We assume that the distance to the center of the wire from the ground plane is 1 mm. Assume that the influence of the ground plane can be taken into account with a closed loop at a distance of $h = 2$ mm. Compute the new inductance for the two-loop situation. Observe that the equivalent circuit is a two-loop transformer where the secondary loop is a short circuit.

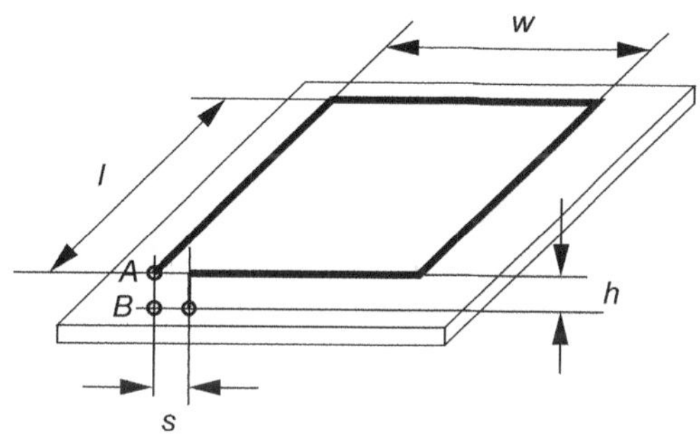

FIGURE 5.34 Single loop over a ground plane and connected to it in a point.

5.8 *Impact of lossy loop over lossy ground plane on impedance*
We use a different approximation in this example for the same loop dimensions and ground plane distances. In this approximate model, we take the losses into account. The dimensions are the same where the distance to the ground plane Fig. 5.35, $h =$ 1 mm again. For higher frequencies, the lossy ground plane impacts the inductance of the loop, in spite of the fact that it is not connected to the plane. We can make a first-order macromodel for the inductance of the loop, in the presence of the ground plane, by carving a secondary loop out of the plane below. Assume a reasonable width of two times the distance to ground for the width of the secondary ground loop.

A first-order model of both loops consists of the inductance in series with its resistance where the resistance can be computed with the usual $R = \rho\, l/A$ formula. Hence, the model consists of a transformer with a shorted secondary winding terminated by the resistance R. The inductances of the two loops are computed as is done in the previous problem geometries (Fig. 5.36).

$$\begin{bmatrix} V_1 \\ V_2 \end{bmatrix} = \begin{bmatrix} sL_{11} & sL_{12} \\ sL_{21} & sL_{22} \end{bmatrix} \begin{bmatrix} I_1 \\ I_2 \end{bmatrix}. \tag{5.82}$$

Given the equivalent circuit in Fig. 5.36, we can find an approximate expression for the situation as (5.83).

$$\frac{V_1}{sI_1} = L_{11}\left(1 - \frac{\beta^2}{1 + \frac{R}{sL_{22}}}\right), \tag{5.83}$$

where the inductive coupling factor is

$$\beta^2 = \frac{L_{12}L_{21}}{L_{11}L_{22}}.$$

Can you find a formula to estimate the frequency for which the presence of the ground plane becomes important?

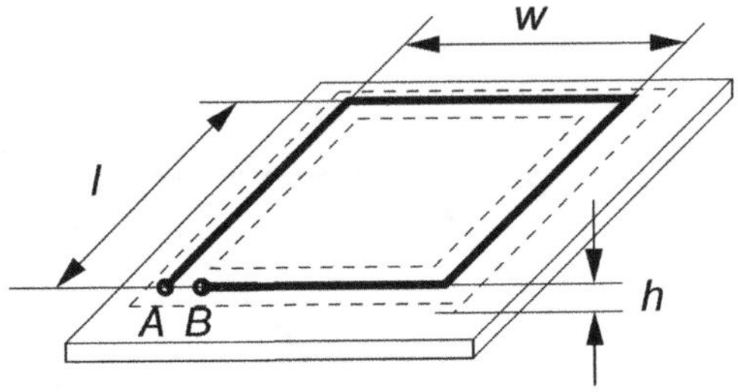

FIGURE 5.35 Lossy loop over a lossy ground plane.

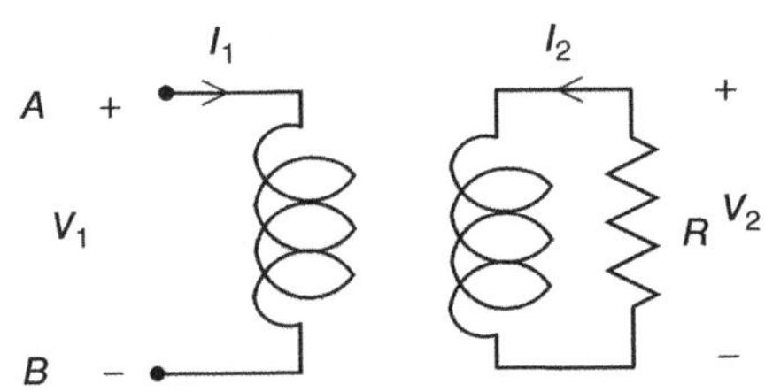

FIGURE 5.36 Equivalent circuit for single-loop example.

5.9 *Inductance of capacitor*
Capacitors have multiple uses. One of the applications is as a decoupling capacitor for a power distribution network (PDN) or for power engineering applications. For this and many other applications, we need to know not only the value of the capacitor but also inductance in a connected environment. Make an (*Lp*)PEEC model for the case shown in Fig. 5.37.

It is clear that the current in the capacitor will flow between the capacitor plates along the length. However, for our approximate inductance model, we assume that some connections exist between the two plates such that the inductance can be computed. In the most simple model, we assume that a short connection exists between the plates somewhere in the capacitor. Depending on the type of the capacitor, we can assume that the plates are shorted in different locations. We can assume that the short at the center of the capacitor plates is shown in capacitor cross section in Fig. 5.37.

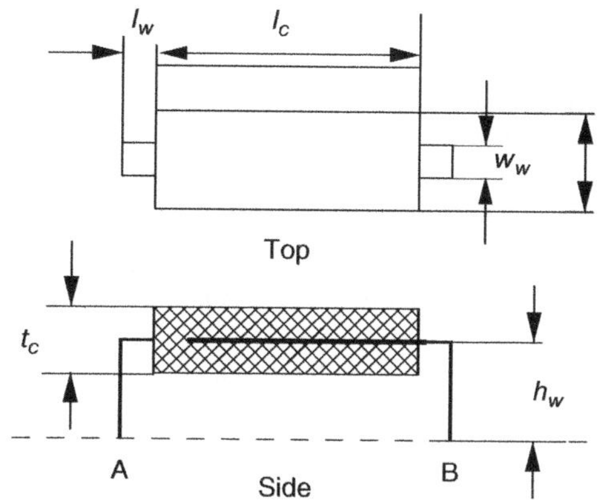

FIGURE 5.37 Capacitor model for inductance problem.

As a second case, we can study the case where the capacitive impedance is large such that crossover happens close to the side where the B contact is located. We compute the *open-loop inductance* between points A and B by simply ignoring the connection to the two points. We recognize that the open-loop inductance needs to be completed with the circuit connected to the terminals to get the actual inductance value.

Assume the following dimensions: $l_c = 5$ mm, $W_c = 2.5$ mm, $T_c = 1$ mm, and $h_W = 1.5$ mm. For the connection wires we assume that the width $w_w = 1.5$ mm, with a thickness $t_w = 0.2$ mm, and the side length $l_w = 0.7$ mm.

REFERENCES

1. F. W. Grover. *Inductance Calculations: Working Formulas and Tables*. Dover Publications, New York, 1962.
2. C. Paul. *Inductance, Loop and Partial*. John Wiley and Sons, Inc., New York, 2010.
3. A. E. Ruehli. Inductance calculations in a complex integrated circuit environment. *IBM Journal of Research and Development*, 16(5):470–481, September 1972.
4. A. E. Ruehli, A. D. Ling, and A. Ruehli Eds. *Circuit Analysis, Simulation and Design, Part 2*. Elsevier Science Publishers B.V., North-Holland, 1987.
5. P. K. Wolff and A. E. Ruehli. Inductance computations for complex three dimensional geometries. In *Proceedings of the IEEE International Symposium on Circuits and Systems*, pp. 16–19, 1981.
6. M. Kamon, M. J. Tsuk, and J. White. FastHenry: a multipole-accelerated 3-d inductance extraction program. In *Proceedings of the Design Automation Conference*, pp. 678–683, June 1993.
7. C. Holloway and E. F. Kuester. Net and partial inductance of a microstrip ground plane. *IEEE Transactions on Electromagnetic Compatibility*, 40(1):33–45, February 1998.
8. A. Devgan, H. Ji, and W. Dai. How to efficiently capture on-chip inductance effects: introducing a new circuit element K. In *Proceedings of International Conference on Computer Aided Design*, Santa Cruz, CA, pp. 150–155, November 2000.

9. K. L. Shepard and Z. Tian. Return-limited inductances: a practical approach to on-chip inductance extraction. *IEEE Transactions on Computer-Aided Design*, 19(4):425–435, April 2000.
10. S. X.-D. Tan and L. He. *Advanced Model Order Reduction Techniques in VLSI Design*. Cambridge University Press, Cambridge, 2007.
11. D. Gope, A. E. Ruehli, and V. Jandhyala. Speeding up PEEC partial inductance computations using a QR-based algorithm. *IEEE Transactions on Very Large Scale Integration (VLSI) Systems*, 15(1):60–68, January 2007.
12. A. E. Ruehli Ed. *Circuit Analysis, Simulation and Design, Part 2*. Elsevier Science Publishers B.V., North-Holland, 1987.
13. C.-K. Cheng, J. Lillis, S. Lin, and N. Chang. *Interconnect Analysis and Synthesis*. John Wiley and Sons, Inc., New York, 2000.
14. G. Zhong and C.-K. Koh. Exact closed-form formula for partial mutual inductances rectangular conductors. *IEEE Transactions on Circuits and Systems*, 50(10):1349–1353, October 2003.
15. G. Antonini and A. E. Ruehli. Fast multipole and multi-function PEEC methods. *IEEE Transactions on Mobile Computing*, 2(4):288–298, October 2003.
16. A. Muesing, J. Ekman, and J. W. Kolar. Efficient calculation of non-orthogonal partial elements for the PEEC method. *IEEE Transactions on Magnetics*, 45(3):1140–1142, March 2009.
17. G. Golup and C. Van Loan. *Matrix Computations*. The John Hopkins University Press, Baltimore, MD, 3rd edition, 1996.
18. E. B. Rosa and F. W. Grover. Formulas and tables for the calculation of mutual and self-inductance. *Bulletin of the Bureau of Standards*, 8(1):1–237, 1912.
19. G. Antonini, J. Ekman, and A. Ruehli. Toward improved time domain stability and passivity for full-wave PEEC models. In *Proceedings of the IEEE International Symposium on Electromagnetic Compatibility*, Portland, OR, August 2006.
20. E. Weber. *Electromagnetic Theory*. Dover Publications, New York, 1965.
21. C. R. Paul. *Analysis of Multiconductor Transmission Lines*. John Wiley and Sons, Inc., New York, 1992.
22. A. E. Ruehli, C. Paul, and J. Garrett. Inductance calculations using partial inductances and macromodels. In *IEEE Transactions on Electromagnetic Compatibility*, pp. 23–28, August 1995.
23. C. L. Holloway, E. F. Kuester, A. E. Ruehli, and G. Antonini. Partial and internal inductance: two of Clayton R. Paul's many passions. *IEEE Transactions on Electromagnetic Compatibility*, 55(4):600–613, April 2013.
24. M. Swaminathan and A. E. Engin. *Power Integrity Modeling and Design for Semiconductors and Systems*. Prentice Hall, Upper Saddle River, NJ, 2007.
25. E.-P. Li. *Electrical Modeling and Design for 3D System Integration*. John Wiley and Sons, Inc., New York, Hoboken, NJ, 2012.
26. J. Kim, J. Fan, A. E. Ruehli, J. Kim, and J. L. Dewniak. Inductance calculations for plane-pair area fills with vias in a power distribution network using a cavity model and partial inductances. *IEEE Transactions on Microwave Theory and Techniques*, 59(8):1909–1924, August 2011.
27. F. Zhou, A. E. Ruehli, and J. Fan. Efficient mid-frequency plane inductance computations. In *Proceedings of the IEEE International Symposium on Electromagnetic Compatibility*, Fort Lauderdale, FL, pp. 831–836, August 2010.
28. L. Li, A. E. Ruehli, and J. Fan. Accurate and efficient computation of power plane pair inductance. In *Proceedings of the IEEE International Symposium on Electromagnetic Compatibility*, Philadelphia, PA, August 2012.
29. L. Wei, L. Li, K. Shringarpure, A. E. Ruehli, E. Wheeler, J. Fan, B. Archambeault, and J. L. Drewniak. Plane-pair PEEC models for power distribution networks with sub-meshing techniques. *IEEE Transaction on Microwave Theory and Techniques*, 64(3):733–741, March 2016.

6

BUILDING PEEC MODELS

In this chapter, we consider how partial element equivalent circuit (PEEC) models are built. Details of the discretization or meshing are given in Chapter 8. Further, the fundamental concepts on how the capacitive and inductive cell partial elements are formed have already been introduced in Chapters 4 and 5.

As shown in Section 3.6, the weighted residual method (WRM) includes important approximations. Some details are discussed in Chapter 1 and in Section 3.6.

The first one is the division or meshing of conductors or other materials into finite blocks, bars, and surface elements for which we preassign fixed local current or charge distributions. An example is a bar shown in Fig. 6.1, where we assign the current to be uniformly in the x-direction. Another example is surface cells for which we assign a uniform charge density. This uniform assignment of the local current density or charge on a cell is part of what is called the basis function.

Mathematically, in general a large number of different functions are used to represent the current and voltage distributions. The choices include delta functions, piecewise constant, piecewise linear, and Rao–Wilton–Glisson (RWG) [1] set of basis and/or weighting functions. For some applications, it has been shown how triangular cells have been used for PEEC [2, 3].

The PEEC method [4] is based on the fundamental concept that basic circuit elements should be symmetrical such as conventional ones. Mathematically, this corresponds to the Galerkin WRM approach. This is accomplished by using a symmetric discrete approximation for the representation of the currents and charges for each circuit element. The basic method we use to create a finite system from the original Maxwell's equations can be viewed as WRM approximations.

We use the current density for a conductor as an example. It is shown in Fig. 6.1, where a current density is defined uniform for the conductor cross section $\mathcal{A}$. Hence, the current I is related to the current density by $J = I/\mathcal{A}$, where $\mathcal{A} = TW$ is the conductor's cross-sectional

Circuit Oriented Electromagnetic Modeling Using the PEEC Techniques, First Edition.
Albert E. Ruehli, Giulio Antonini, and Lijun Jiang.

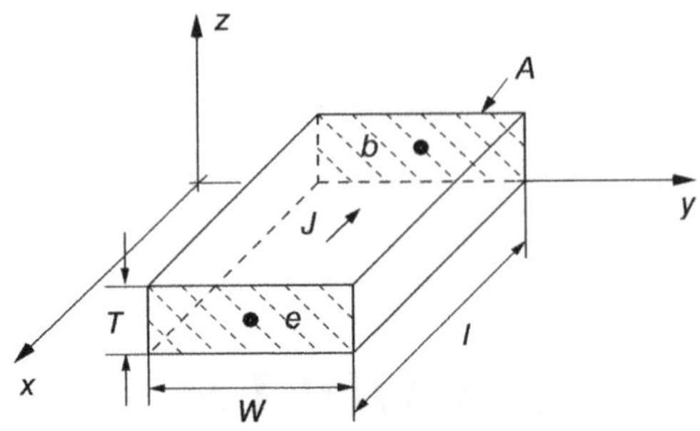

FIGURE 6.1 Example conductor bar for resistance computations.

area. In the second case, for a thin or zero thickness conductor $T = 0$, the surface current is defined as per unit width. It is $J_s = I/W$ in the unit of amperes per unit width.

In general, we include some unusual cell sizes in our PEEC models, such as the use of very large aspect ratios for the cell dimensions. Further, wherever possible we use the analytical evaluation of the partial elements. As part of this, we try to adhere to rectangular or quadrilateral cell shapes.

A specific example of the weighting process is given in (6.1) for the averaging of the fields, or

$$V_{eb} = \frac{1}{\mathcal{A}} \int_{\mathcal{A}} \int_{b}^{e} -\boldsymbol{E}(\boldsymbol{r}) \cdot \hat{\boldsymbol{x}}\, dx\, d\mathcal{A}, \tag{6.1}$$

where $\mathcal{A}$ is the cross section in the thickness T and width W direction as shown in Fig. 6.1.

We also give an example of the relation of PEEC to a Galerkin formulation. The weighting function averaging for the potential as is done in (6.10) and (6.11). The symmetric circuit elements are obtained with such weighting functions. In the mathematical domain, Galerkin proposed the idea of using the same basis and weighting function to solve both differential and integral equation problems [5] in 1915 using WRM. The PEEC method should be included in the class of the so-called discontinuous Galerkin methods [6], which also are popular today for noncircuit-oriented applications.

A key step in all WRM solution approaches is the way the resultant system of equations is solved. In PEEC, this consists of solving the modified nodal analysis (MNA) circuit matrix equations presented in Chapter 2. We could call this a new member of the WRM family.

6.1 RESISTIVE CIRCUIT ELEMENTS FOR MANHATTAN-TYPE GEOMETRIES

Resistances represent important circuit elements needed for PEEC models. So far, we did not consider resistances in our treatment of circuit elements since the topic is relatively straightforward in comparison to the capacitance and inductance models. We start with the classical form of Ohm's law (3.2c) as

$$\boldsymbol{J} = \sigma \boldsymbol{E}, \tag{6.2}$$

where $\boldsymbol{J}$ is the current density for any point in a conductor and σ is the conductivity of the material.

The total electric field $\boldsymbol{E}$ can be treated as the superposition of external fields as well as the induced scattered field, (3.54), or

$$\boldsymbol{E}(\boldsymbol{r}) = \boldsymbol{E}^{inc}(\boldsymbol{r}) - \nabla\Phi(\boldsymbol{r}) - \frac{\partial \boldsymbol{A}(\boldsymbol{r},t)}{\partial t}, \tag{6.3}$$

where the first term $\boldsymbol{E}^{inc}(\boldsymbol{r})$ on the right-hand side is an externally applied electric field at $\boldsymbol{r}$. The scalar potential Φ and vector potential $\boldsymbol{A}$ contribute to the induced scattering fields.

For most problems, applied external source fields $\boldsymbol{E}^{inc}(\boldsymbol{r})$ are zero. However, we also consider the impact of external fields for PEEC in Chapter 12. For a static case, the vector potential derivative term in (6.3) is zero. We can easily derive an equivalent circuit for the resulting equation

$$\boldsymbol{E}(\boldsymbol{r}) = -\nabla\Phi(\boldsymbol{r}). \tag{6.4}$$

In Fig. 6.1, the voltage between two points "e" for end and "b" for beginning on the conducting rectangular bar is obtained by integration of (6.4). The two end-face contacts are indicated by the crosshatched areas and the dots are used as an electrical node for the contacts. The integration of (6.4) between e and b in x yields

$$V_{be} = \Phi(b) - \Phi(e) = \int_b^e \boldsymbol{E}\cdot d\boldsymbol{\ell} = \frac{1}{\sigma}\int_b^e \boldsymbol{J}\cdot\hat{\boldsymbol{x}}\,dx. \tag{6.5}$$

As part of the circuit and WRM approximation, we place the contact to the entire end surfaces with the same potential. Hence, as shown in Fig. 6.1, the current density is uniformly defined to be $J_x = I/(TW) = I/\mathcal{A}$, where $\mathcal{A}$ is the cross-sectional area. In terms of the previous discussion on WRM, we a priori assume that the current will flow uniformly in the x-direction. In this unusual case, the *exact* current flow will be in the direction of the one assigned by the WRM model. For more complex models, current representations will be superimposed. Finally, the result for the resistance is the conventional formula

$$R_{be} = \frac{V_{be}}{I} = \frac{\ell}{\sigma TW} = \frac{\ell}{\sigma\mathcal{A}}, \tag{6.6}$$

where $\ell = |x_e - x_b|$. To be consistent with the general circuit model approach, we should also average the potential evaluation in (6.5) over the contact area on the sides, or

$$V_{be} = \frac{1}{\mathcal{A}}\int_{\mathcal{A}}\int_b^e \boldsymbol{E}(\boldsymbol{r})\cdot\hat{\boldsymbol{x}}\,dx\,d\mathcal{A} = \frac{1}{\mathcal{A}}\int_{\mathcal{V}} \boldsymbol{E}(\boldsymbol{r})\cdot\hat{\boldsymbol{x}}\,d\mathcal{V}, \tag{6.7}$$

where $\mathcal{A}$ is the cross-sectional area and $\mathcal{V}$ is the volume of the bar. The electric field $\boldsymbol{E}$ is assumed uniform over the contact. The result of (6.7) is exactly the same as (6.5). With this, we can easily compute the resistance of each rectangular bar for any discretized geometry.

The meshing or discretization for general complicated geometries is another challenging problem that is considered in Chapter 8. For more complex geometries, we have to further subdivide or mesh the conductive bars into rectangular subbars or cells. Then, we need to evaluate the resistance of each subbar in the geometry. The coupling between different bars is, in general, very limited. In fact, rectangular bar partial resistances are not coupled at all. However, we see that limited partial resistance coupling can occur for neighboring nonorthogonal elements as is evident from Section 7.2.2. Fortunately, for rectangular cells we only have to compute self-resistance terms. The resultant PEEC equivalent circuit is easy to solve using the circuit analysis technique presented in Chapter 2.

6.2 INDUCTANCE–RESISTANCE (Lp,R)PEEC MODELS

As a next step, we consider an inductance–resistance (Lp,R)PEEC model, which is a combination of the resistance model in the last section with the partial inductances in Chapter 5. Besides the fact that these models are already useful by themselves for many applications, they also help to represent a key step toward the general PEEC model in Section 6.3.

We start with the general integral equation (3.54), or

$$\boldsymbol{E}(\boldsymbol{r},t) = -\nabla\Phi(\boldsymbol{r},t) - \frac{\partial}{\partial t}\frac{\mu}{4\pi}\int_{\mathcal{V}'}\frac{\boldsymbol{J}(\boldsymbol{r}',t')}{|\boldsymbol{r}-\boldsymbol{r}'|}d\mathcal{V}'. \tag{6.8}$$

Taking (6.2) into account but not including the capacitive potential coefficient term, we obtain

$$-\nabla\Phi(\boldsymbol{r},t) = \frac{\boldsymbol{J}(\boldsymbol{r},t)}{\sigma} + \frac{\partial}{\partial t}\frac{\mu}{4\pi}\int_{\mathcal{V}'}\frac{\boldsymbol{J}(\boldsymbol{r}',t')}{|\boldsymbol{r}-\boldsymbol{r}'|}d\mathcal{V}',$$

where the vector $\boldsymbol{r}$ is placed inside or on the surface of the conductor.

For simplicity, we again start out with the single bar of Fig. 6.1 as the first geometry. As before, we integrate the electric field along the length of the current path, which is between the contacts b and e, or

$$\begin{aligned} V_{be} &= \Phi(b) - \Phi(e) = -\int_b^e \nabla\Phi(\boldsymbol{r},t)\cdot d\boldsymbol{\ell} \\ &= \frac{1}{\sigma}\int_b^e \boldsymbol{J}(\boldsymbol{r},t)\cdot d\boldsymbol{\ell} \\ &\quad + \frac{\mu}{4\pi}\frac{\partial}{\partial t}\int_{\mathcal{V}'}\int_b^e \frac{1}{|\boldsymbol{r}-\boldsymbol{r}'|}\boldsymbol{J}(\boldsymbol{r}',t')\cdot d\boldsymbol{\ell}\; d\mathcal{V}'. \end{aligned} \tag{6.9}$$

We also note that we can simplify the dot product $\boldsymbol{J}(\boldsymbol{r}',t')\cdot d\boldsymbol{\ell}$ to $J_x\, dx$ because both vectors are in the same direction as is evident from Fig. 6.1. Again, we assume that the potential on the contacts is constant. It is equivalent to take the average of the potential over the surface of contacts, which is also very important for the second term

$$\begin{aligned} V_{be} &= \frac{1}{\sigma}\frac{1}{\mathcal{A}}\int_{\mathcal{A}}\int_e^b J_x(\boldsymbol{r}',t')\, dx\, d\mathcal{A} \\ &\quad + \frac{\mu}{4\pi}\frac{\partial}{\partial t}\frac{1}{\mathcal{A}}\int_{\mathcal{A}}\int_b^e\int_{\mathcal{V}'}\frac{1}{|\boldsymbol{r}-\boldsymbol{r}'|}J_x(\boldsymbol{r}',t')\, d\mathcal{V}'\, dx\, d\mathcal{A}, \end{aligned} \tag{6.10}$$

where $\mathcal{A}$ is the surface of the contact. Finally, the uniform current density is represented as $\boldsymbol{J} = \hat{\boldsymbol{x}}\, I/\mathcal{A}$. Hence, the voltage V_{be} can be approximated as

$$V_{be} = \frac{1}{\sigma}\frac{I}{\mathcal{A}}\int_b^e dx + \frac{\mu}{4\pi}\frac{\partial}{\partial t}\frac{I}{\mathcal{A}\,\mathcal{A}'}\int_{\mathcal{V}'}\int_{\mathcal{V}}\frac{1}{|\boldsymbol{r}-\boldsymbol{r}'|}d\mathcal{V}\, d\mathcal{V}'. \tag{6.11}$$

Using what we learned about partial inductances in Chapter 5, the second part is simply contribution of the partial inductance of the bar. Hence, the circuit equations for (6.11) is rewritten as

$$V_{be} = R_1\, I + Lp_{11}\,\frac{dI}{dt}, \tag{6.12}$$

where R_1 is the partial resistance due to the loss of the conductor bar. Lp_{11} is the partial self-inductance. Their detailed definitions are

$$R_1 = \frac{1}{\sigma\mathcal{A}}\int_b^e dx \tag{6.13}$$

$$Lp_{11} = \frac{\mu}{4\pi}\frac{1}{\mathcal{A}\,\mathcal{A}'}\int_{\mathcal{V}}\int_{\mathcal{V}'}\frac{1}{|\boldsymbol{r}-\boldsymbol{r}'|}d\mathcal{V}'\, d\mathcal{V}. \tag{6.14}$$

This represents the (Lp,R)PEEC circuit model for the single bar in Fig. 6.1, which is simply Lp_{11} in series with the resistor R_1.

6.2.1 Inductance–Resistance (Lp,R)PEEC Model for Bar Conductor

Since the PEEC equivalent circuit does not form a closed loop, the inductance of the model will be an open-loop inductance according to Section 5.4. We refer the reader to Section 5.6 for an example that consists of three partial inductances. Most likely, this model is a part of a larger structure.

It is important to use this term so that we are aware of the fact that the circuit is not closed, as is also the case for a partial inductance. Fortunately, PEEC elements have electrical nodes to connect parts together such as building blocks. Some of the parts of a PEEC circuit without capacitive elements consist of partial or open-loop inductances.

The example in Fig. 6.2 is a building block that consists of a simple thin metal sheet with a finite resistivity, for example, copper. We see from Fig. 6.2 how the metal panel is subdivided into cells. For this example, we divide the current into x and y components while we assume that the thickness is small enough that we do not have to subdivide it. We need to recognize that the subdivision of the current into rectangular Manhattan components is part of the WRM approximation approach.

Hence, this reduces the unknown currents to complex $\mathbf{I}(s)$ or, in the time domain response of the currents $\mathbf{i}(t)$ in the cells. As usual, the cells carved out of the metal panel are electrically isolated from each other with infinitesimal gaps while the connections are at the nodes only. For clarity, we show in Fig. 6.2 only a few cells where the current vectors are shown. We should mention that the connections at the nodes have to obey the continuity equation that is given by Kirchhoff's current law (KCL) in Section 6.3.1.

We observe that only the cells in the same direction are inductively coupled since the dot product in (6.5) is zero for perpendicular conductors. Hence, the coupling for each element

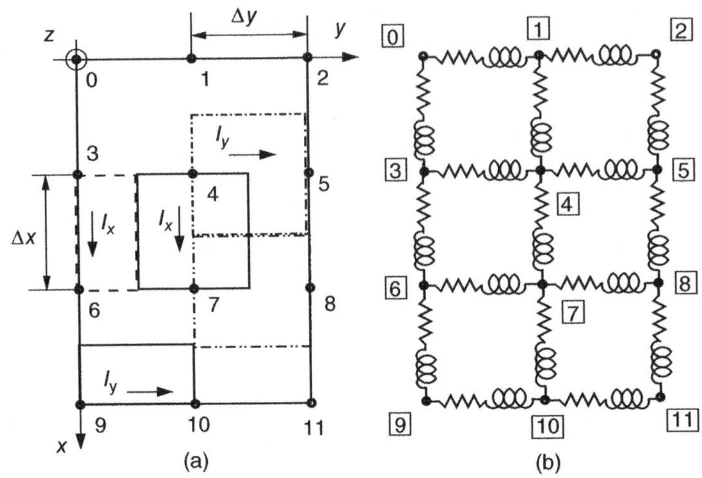

FIGURE 6.2 Subdivision of thin conducting panel into cells and PEEC circuit.

is between the parallel partial inductances only, which is similar to (6.12), or for example

$$V_k = R_k\, I_k + \sum_{\ell=1}^{N_L} Lp_{k\ell}\, \frac{dI_\ell}{dt}. \tag{6.15}$$

There are nine partial inductances in the x-direction and eight coupled inductances in the y-direction and nine coupled partial inductances in the x-direction. We note that the longitudinal cells along all sides are half-width cells.

We see that using a half cell at the edge does in general improve the solution accuracy. Also, assume that we want to connect two of the models in Fig. 6.2 to each other. Then, we simply connect the neighboring nodes together. We also observe the neighboring edge half cells merge into full cells. We consider this as the building block approach where large problems can be viewed as being built from smaller subsystems.

The second part of the process is related to the circuit equations. We note that the circuit corresponding to applying the coupling equation (6.15) to each physical cell is in Fig. 6.2a. The resultant complete (Lp,R)PEEC equivalent circuit in Fig. 6.2b needs to be solved.

Before we consider the general case, we want to observe that the (Lp,R)PEEC equivalent circuit in Fig. 6.2b does not include a ground connection. We can declare any one of the 12 nodes as the ground. Then, the circuit analysis techniques in Chapter 2 can be used to find all voltages and currents in the solved system if a current is injected at any node. However, this may not be an issue in a PEEC model that also includes capacitances with a ground node corresponding to ∞. However, care must be taken on a case-by-case situation.

6.3 GENERAL (Lp,Pp,R)PEEC MODEL DEVELOPMENT

This part of the chapter is dedicated to the more general class of PEEC models where we also include the capacitive currents in addition to the previous ones.

We subdivide these models into two different submodels. For example, in some integrated circuits (ICs), miniaturization leads to very high resistive circuits. Due to the high impedances, the currents are very small such that the inductive voltage drop $v = L\,di/dt$ is small. This case is considered first in the section on (Pp,R)PEEC equivalent circuits. Note that they also include capacitive terms from Chapter 4.

For the case where the inductances are also important, the (Lp,Pp,R)PEEC circuits are considered. This is further divided into two cases according to the working frequency: (Lp,Pp,R)PEEC and (Lp,Pp,R,τ)PEEC. Quasistatic (QS) PEEC models (Lp,Pp,R)PEEC have an upper frequency limit while the FW (Lp,Pp,R,τ)PEEC models can be extended, in principle, to any high frequency limit. First, we start with concepts that are necessary for the construction of these models.

6.3.1 Continuity Equation and KCL

The solution of a PEEC circuit also needs to satisfy the continuity equation (3.3) for the charge conservation as shown in Chapter 3.

In the time domain, it seems to be a simple approach to use the continuity equation (6.16) to replace the capacitive charge in the solution with currents in the overall solution as

$$\nabla \cdot \boldsymbol{J} + \frac{\partial q}{\partial t} = 0. \tag{6.16}$$

This result is what we call a current-only electric field integral Equation (cEFIE) formulation that is widely used for a dynamic electromagnetic solution.

However, this approach leads to an overall $1/f$ behavior that is singular for very low frequencies f. This is a problem for electronics type problems. For example, a remotely located antenna does not couple at *dc*. Then, such a solution is appropriate. From a circuit's point of view, this is similar to a high pass filter where the low frequencies are irrelevant. Hence, this approach is mostly suitable for problems that do not require a *dc* solution.

In the PEEC approach, since the capacitance and inductances are separate, a *dc* to daylight full-spectrum solution is possible by using the continuity equation (6.16). The continuity equation needs to be satisfied at each PEEC node as shown in Fig. 6.3. We can see four inductive cells joining a capacitance surface area.

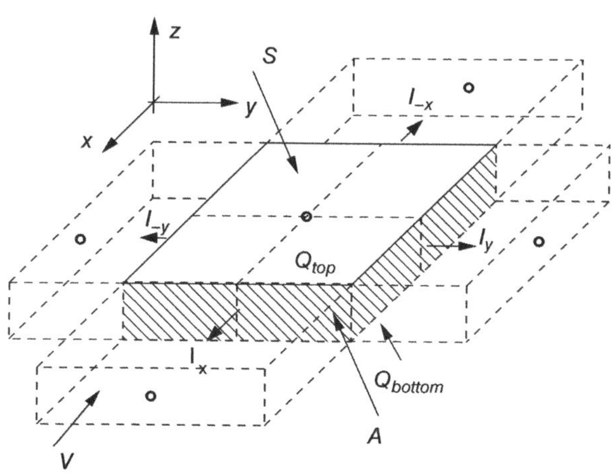

FIGURE 6.3 Example cell division with both inductive filaments and capacitive cells.

First, we apply a Gaussian surface over the area that corresponds to the node at the center of the volume covered with the capacitive cell S on both top and bottom. It is clear that, if we apply Gauss surfaces to each cell in Fig. 6.3, we will wrap up the entire volume of the center part of the conductors. Multiple applications of the continuity equation will cover the entire area.

Surface charge is present on both surfaces S of the conductive cells of the finite thickness conducting cells shown in Fig. 6.3. The top and bottom surfaces S have capacitive charge. In the following section, it is shown that the charge is on the surface only.

The Gaussian volume $\mathcal{V}$ over which we integrate the continuity equation also includes the side surface where the current flows. This results in

$$\int_{\mathcal{V}} \nabla \cdot \boldsymbol{J}\, d\mathcal{V} = \oint_{\mathcal{A}} \boldsymbol{J} \cdot \hat{\boldsymbol{n}}\, d\mathcal{A} = -\frac{\partial}{\partial t} \int_{S} q(\boldsymbol{r})\, dS, \tag{6.17}$$

where one of the Gaussian volumes $\mathcal{V}$ and the surface S includes all sides as indicated in Fig. 6.3, where the charge is on top and bottom and the current flows through the *internal* surfaces $\mathcal{A}$.

By comparing this result with Fig. 6.3, we see that the normal direction current flow occurs through the four partial inductance cross sections $\mathcal{A}$ of the conductors. Since they are interior to the structure, the charge density $q(\boldsymbol{r})$ is zero. We notice that the integrals over the side cross sections are

$$\oint_{\mathcal{A}} \boldsymbol{J} \cdot \hat{\boldsymbol{n}}\, d\mathcal{A} = I, \tag{6.18}$$

where I is the total current flowing into the four partial inductance cells.

On the other hand, the conductor top and bottom surfaces are charged with surface charge densities $q(\boldsymbol{r})$. Integrating over both top and bottom surface will yield

$$\int_{S} q(\boldsymbol{r})\, dS = Q_{top} + Q_{bottom}. \tag{6.19}$$

Finally, we add up all the contributions for all surfaces to yield

$$\sum_{k=1}^{4} I_k + \frac{d}{dt}(Q_{top} + Q_{bottom}) = 0. \tag{6.20}$$

It is clear that this is exactly the KCL equation at the node. Hence, we recognize that applying the KCL at each node is the same as applying the continuity equation. Simply stated, the solution of the circuit equations by the techniques of Chapter 2 with the above meshing will satisfy the continuity equation.

6.3.2 Relaxation Time for Charge to Surface

An important issue to validate is if the charge is indeed on the conductor surfaces even for the highest frequencies involved. The time domain equivalent is that the charges must appear on the surfaces faster than the fastest transients in the time domain responses.

We start with the Maxwell's equation (3.1c) and the displacement flux density definition $\boldsymbol{D} = \epsilon \boldsymbol{E}$.

$$\nabla \cdot \boldsymbol{D} = q = \nabla \cdot \epsilon \, \boldsymbol{E} = \epsilon \, \nabla \cdot \boldsymbol{E}, \tag{6.21}$$

where the last step is valid for piecewise continuous dielectrics. On the other hand, we can insert the conducting current $\boldsymbol{J} = \sigma \boldsymbol{E}$ into the continuity equation (6.16) to obtain

$$\nabla \cdot \sigma \boldsymbol{E} = \sigma \nabla \cdot \boldsymbol{E} = -\frac{\partial q}{\partial t}. \tag{6.22}$$

Replacing $\nabla \cdot \boldsymbol{E}$ with (6.21) yields the differential equation

$$\frac{\sigma}{\epsilon} \, q = -\frac{\partial q}{\partial t}. \tag{6.23}$$

The solution of this simple differential equation is given by

$$q = q_0 \, e^{-(\sigma/\epsilon) \, t}. \tag{6.24}$$

Hence, the time constant is $\tau = \epsilon/\sigma$. As an example, we consider the copper conductor for microelectronics. Its conductivity $\sigma = 5.78 \times 10^7$ S/m and the permittivity $\epsilon = 8.854 \times 10^{-3}$ pF/mm. Thereby, the corresponding time constant $\tau = 1.5318 \times 10^{-10}$ ns. It means that the charge density inside a conductor will disappear in the volume much faster than any transient signal. Hence, we can safely assume the charge density is zero inside conductors. The charges only reside on surfaces. Then in the interior of conductors the continuity equation (3.55) becomes

$$\nabla \cdot \boldsymbol{J}(\boldsymbol{r}, t) = 0, \tag{6.25}$$

while on the surfaces of conductors, using the surface divergence, we have

$$\nabla \cdot \boldsymbol{J}(\boldsymbol{r}, t) = -\frac{\partial q(\boldsymbol{r}, t)}{\partial t}, \tag{6.26}$$

where $\hat{\boldsymbol{n}}$ is the outward normal to the surface S.

6.3.3 Physical Aspect of the Capacitance Model

We considered most aspects of the building or PEEC models with the exception of the capacitive part. Also, stand-alone capacitance computations have been treated in Chapter 4. Here, we consider the capacitance term for PEEC models.

We again start with the total electric field $\boldsymbol{E}$ equation (3.58d)

$$\boldsymbol{E}(\boldsymbol{r}) = \boldsymbol{E}^{inc}(\boldsymbol{r}) - \nabla \Phi(\boldsymbol{r}) - \frac{\partial \boldsymbol{A}(\boldsymbol{r}, t)}{\partial t}, \tag{6.27}$$

where the capacitance part is represented by the gradient $\nabla\Phi$. The electric field generated by the scalar potential

$$\boldsymbol{E}_{\Phi}(\boldsymbol{r}) = -\nabla\Phi(\boldsymbol{r}) = -\hat{\boldsymbol{x}}\frac{\partial\Phi}{\partial x} - \hat{\boldsymbol{y}}\frac{\partial\Phi}{\partial y} - \hat{\boldsymbol{z}}\frac{\partial\Phi}{\partial z}. \tag{6.28}$$

The scalar potential Φ is related to the charge distribution by equation (3.46) as

$$\Phi(\boldsymbol{r},t) = \frac{1}{4\pi\epsilon}\int_{S'}\frac{q(\boldsymbol{r}',t')}{|\boldsymbol{r}-\boldsymbol{r}'|}dS'. \tag{6.29}$$

We can consider each direction separately by denoting $\gamma = x, y, z$ since they will turn into different capacitive elements. Hence, the terms to be added are of the form

$$E_{\Phi\gamma} = -\frac{\partial}{\partial\gamma}\left[\frac{1}{4\pi\epsilon}\int_{S'}\frac{q(\boldsymbol{r}',t')}{|\boldsymbol{r}-\boldsymbol{r}'|}dS'\right]. \tag{6.30}$$

We use the same symmetrical WRM discretization as we apply for the other circuit elements. As in Chapter 4, we are averaging potentials over the capacitive cells. Hence, we have to apply the symmetric integration to obtain voltages $v = \int E\, dl$, which is defined as

$$\frac{1}{\mathcal{A}_m}\int_{\mathcal{V}_m} f(\boldsymbol{r})\,d\mathcal{V} = \frac{1}{\mathcal{A}_m}\int_{\ell_m}\int_{\mathcal{A}_m} f(\boldsymbol{r})\,d\mathcal{A}\,d\boldsymbol{\ell}. \tag{6.31}$$

If we apply this to (6.30), we get

$$V_{\Phi\gamma} = -\frac{1}{\mathcal{A}_m}\int_{\mathcal{V}_m}\frac{\partial}{\partial\gamma}\left[\frac{1}{4\pi\epsilon}\int_{S'}\frac{q(\boldsymbol{r}',t')}{|\boldsymbol{r}-\boldsymbol{r}'|}dS'\right]d\mathcal{V}, \tag{6.32}$$

where $\mathcal{V}_m$ is a volume. The charge is confined to the surface of the conductors such that we can eliminate the conductor cross-section $\mathcal{A}_m$ in the derivation. Next, we approximate the derivative using the central finite difference scheme

$$\int_{\ell_m}\frac{\partial}{\partial\gamma}F(\gamma)d\ell \doteq \left[F\left(\gamma+\frac{\Delta\gamma}{2}\right) - F\left(\gamma-\frac{\Delta\gamma}{2}\right)\right]. \tag{6.33}$$

It means that the centers of potential cells are half a step shifted with respect to the center of the inductance cell. To clearly illustrate this, Fig. 6.4 taken from Chapter 8, Fig. 8.2 is shown here.

We see that the potential or capacitive cells are centered at the nodes, whereas the inductive cells are connected between the nodes. We also observe that some capacitive cells are corner quarter cells or side half cells in addition to the full cells.

To consistently apply the averaging in the derivation, we also average the evaluation point over the cell

$$\Phi\left(\gamma+\frac{\Delta\gamma}{2}\right) = \frac{1}{S}\int_{S}\Phi(\boldsymbol{r}_+)dS, \tag{6.34}$$

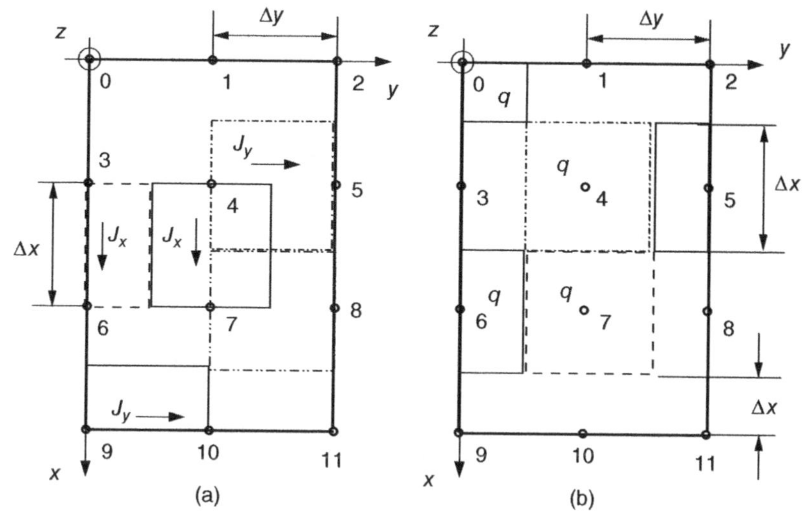

FIGURE 6.4 Inductive and capacitive cells for thin metal panel.

where $\boldsymbol{r}_+$ represents the positive half shift in position. Bringing (6.33) into (6.32), and applying the averaging process to (6.32), results in

$$\Phi = \frac{Q_+}{\epsilon S_+ S'} \int_{S_+} \int_{S'} g(\boldsymbol{r}, \boldsymbol{r}') dS' \, dS_+ - \frac{Q_-}{\epsilon S_- S'} \int_{S_-} \int_{S'} g(\boldsymbol{r}, \boldsymbol{r}') dS' \, dS_-, \tag{6.35}$$

where the charge density q over the cell is replaced by the total charge Q on that cell. g is the free space Green's function from Section 3.4.2

$$g(\boldsymbol{r}, \boldsymbol{r}') = \frac{1}{4\pi} \frac{1}{|\boldsymbol{r} - \boldsymbol{r}'|}, \tag{6.36}$$

where the time delay due to the finite value of the speed of light in the vacuum has been neglected, for example, $t = t'$ for the QS case. From this derivation, we can also define the partial potential coefficient as

$$Pp_{m\ell} = \frac{1}{\epsilon S_\ell S_m} \int_{S\ell} \int_{Sm} g(\boldsymbol{r}_m, \boldsymbol{r}_\ell) \, dS_m \, dS_\ell. \tag{6.37}$$

Potential coefficients are similar to partial inductances in terms of the fact that each cell couples to all other capacitive surface cells in the system. Hence, it corresponds to a dense coupling network and thereby a dense linear algebra matrix in the numerical model. The following section considers how the capacitance models derived in this section can be implemented into the PEEC circuits.

6.3.4 Equivalent Circuits for PEEC Capacitances

Models that include only capacitance and their computations are introduced in detail in Section 4.5.2. Here, we concentrate on the aspect that pertains to the complete PEEC

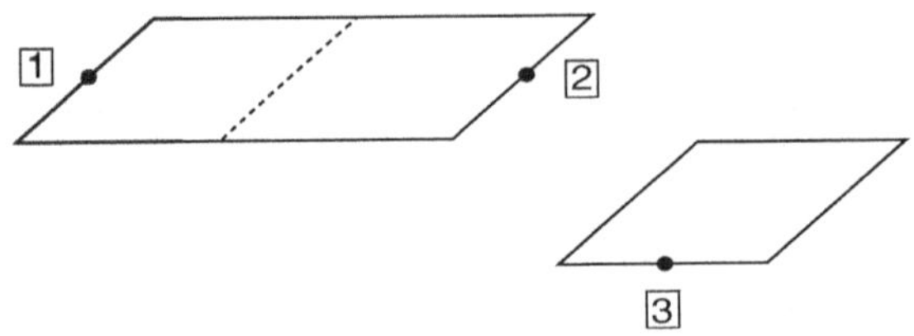

FIGURE 6.5 Three zero thickness capacitance cells example.

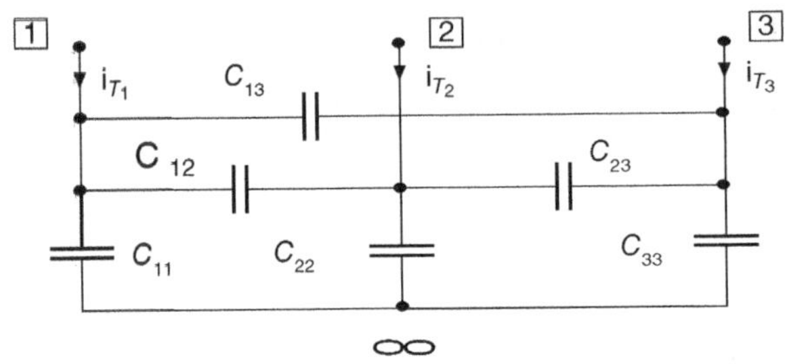

FIGURE 6.6 Example PEEC capacitance model for small three-cell example.

model construction. Equations (4.39a)–(4.39c) are used for the capacitive circuit model. Figure 4.16 is repeated for the three-cell system for convenience in Fig. 6.5. The partial coefficient of potential equations for this example are

$$\begin{aligned}\Phi_1 &= Pp_{11}Q_1 + Pp_{12}Q_2 + Pp_{13}Q_3\\ \Phi_2 &= Pp_{21}Q_1 + Pp_{22}Q_2 + Pp_{23}Q_3\\ \Phi_3 &= Pp_{31}Q_1 + Pp_{32}Q_2 + Pp_{33}Q_3.\end{aligned} \tag{6.38}$$

Chapter 4 considers how to obtain a capacitance equivalent circuit as shown in Fig. 6.6.

A more efficient model is desirable for general PEEC models, while this model is appropriate for some small problems. As was considered in Section 2.7.3, we want to connect the capacitance model at the nodes with the total current $\mathbf{i}_c$. We start with (6.38), or $\Phi = \mathbf{Pp}\ \mathbf{Q}$. To obtain an efficient formulation, we subdivide $\mathbf{Pp}$ into a diagonal part

$$\mathbf{Pd} = \begin{bmatrix} Pp_{11} & 0 & \cdots & 0\\ 0 & Pp_{22} & \cdots & 0\\ \vdots & \vdots & \vdots & \vdots\\ 0 & 0 & \cdots & Pp_{N_sN_s}\end{bmatrix} \tag{6.39}$$

and a part with the off-diagonal elements

$$\mathbf{Pp}_{-} = \begin{bmatrix} 0 & Pp_{12} & \cdots & Pp_{1N_s}\\ Pp_{21} & 0 & \cdots & Pp_{2N_s}\\ \vdots & \vdots & \vdots & \vdots\\ Pp_{N_s1} & Pp_{N_s2} & \cdots & 0\end{bmatrix}. \tag{6.40}$$

Hence, all we did so far is to split the **Pp** matrix into two parts

$$\mathbf{\Phi} = \mathbf{Pd}\,\mathbf{Q} + \mathbf{Pp}_{-}\,\mathbf{Q}, \tag{6.41}$$

where **Pd** is easily invertible since it is a diagonal matrix

$$\mathbf{Pd}^{-1} = \begin{bmatrix} \frac{1}{Pp_{11}} & 0 & \cdots & 0 \\ 0 & \frac{1}{Pp_{22}} & \cdots & 0 \\ \vdots & \vdots & \vdots & \vdots \\ 0 & 0 & \cdots & \frac{1}{Pp_{N_sN_s}} \end{bmatrix}. \tag{6.42}$$

If we left-multiply (6.41) by $\mathbf{Pd}^{-1}$, we get

$$\mathbf{Pd}^{-1}\mathbf{\Phi} = \mathbf{Pd}^{-1}\,\mathbf{Pd}\,\mathbf{Q} + \mathbf{Pd}^{-1}\mathbf{Pp}_{-}\,\mathbf{Q} = (\hat{\mathbf{I}} + \mathbf{M})\mathbf{Q}, \tag{6.43}$$

where $\hat{\mathbf{I}}$ is the identity matrix. With the definition of $\mathbf{M} = \mathbf{Pd}^{-1}\mathbf{Pp}_{-}$ as

$$\mathbf{M} = \begin{bmatrix} 0 & \frac{Pp_{12}}{Pp_{11}} & \cdots & \frac{Pp_{1N_s}}{Pp_{11}} \\ \frac{Pp_{21}}{Pp_{22}} & 0 & \cdots & \frac{Pp_{2N_s}}{Pp_{22}} \\ \vdots & \vdots & \vdots & \vdots \\ \frac{Pp_{N_s1}}{Pp_{N_sN_s}} & \frac{Pp_{N_s2}}{Pp_{N_sN_s}} & \cdots & 0 \end{bmatrix}, \tag{6.44}$$

we can simplify (6.43) by differentiating both sides with respect to time to the important form

$$\mathbf{Pd}^{-1}\frac{d\mathbf{\Phi}}{dt} = (\hat{\mathbf{I}} + \mathbf{M})\frac{d\,\mathbf{Q}}{dt}. \tag{6.45}$$

Since $d\,\mathbf{Q}/dt = \mathbf{i}_T(t)$, equation (6.45) leads to

$$\mathbf{Pd}^{-1}\frac{d\mathbf{\Phi}}{dt} - (\hat{\mathbf{I}} + \mathbf{M})\mathbf{i}_T(t) = \mathbf{0}, \tag{6.46}$$

where $\mathbf{i}_T(t)$ includes the current of all the elements connected to the capacitor nodes. We note that this equation corresponds to the equivalent circuit in Fig. 6.7 where the current controlled current sources (CCCS) are given by

$$\begin{aligned} i_{s1} &= \frac{Pp_{12}}{Pp_{11}} i_{T2} + \frac{Pp_{13}}{Pp_{11}} i_{T3} \\ i_{s2} &= \frac{Pp_{21}}{Pp_{22}} i_{T1} + \frac{Pp_{23}}{Pp_{22}} i_{T3} \\ i_{s3} &= \frac{Pp_{31}}{Pp_{33}} i_{T1} + \frac{Pp_{32}}{Pp_{33}} i_{T2}, \end{aligned} \tag{6.47}$$

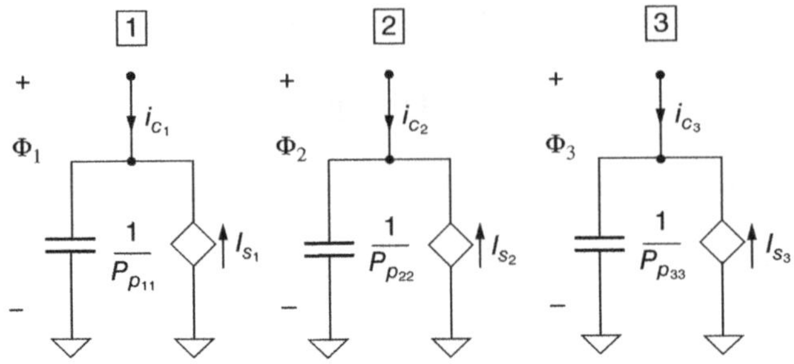

FIGURE 6.7 Current source capacitance equivalent circuit model for retarded PEEC model.

where the coefficients are represented in the **M** matrix (6.44). We should note that we can include the delay or retardation between the capacitive cells by including them in the current source models.

6.3.5 (Pp,R)PEEC Resistive Capacitive Inductor-Less Models

A (Pp,R)PEEC model without partial inductances is the only combination of partial elements we did not treat so far. Examples using these models are relevant for very high resistive circuits. Because the inductive voltage drops are relatively small compared to voltage drops over high resistive components, the inductive voltage drops are of minor importance to the overall calculation accuracy. Then, partial inductances *Lp* could be ignored in the PEEC model. Due to the continuous miniaturization of integrated chips, the internal resistive loss is increasing when the cross section of copper wires is decreasing. Hence, inductances could be ignored at low frequencies for some interconnect parasitic modeling. To construct a (Pp,R)PEEC model, the process is straightforward since all partial self- and mutual inductances can be ignored.

It is clear that we have two choices to include resistances in the MNA equations. Due to the absence of the partial inductance part, the circuit matrix can be restructured, depending on which resistor stamp is used. Conventionally, we use the Table B.3 combined with the partial inductance as outlined in Table B.7 in Appendix B to reduce the number of unknowns. For the models without *Lp*, we can also use Table B.2 stamp to further reduce unknowns.

6.3.6 Delayed (Lp,Pp,R,τ)PEEC Models

So far, Quasi Static QS models are considered in most of our derivation. Fundamentally, the QS models are only valid when the overall dimensions of the system such as the length of the signal transmission lines is small compared to the wavelength in the spectrum corresponding to the highest frequency applied by the sources. If the highest frequency corresponds to the wavelength λ that is comparable to or smaller than the overall object's size, *full-wave* (*FW*) models become necessary. *FW* means that the frequency dependency is completely considered in the modeling process for all propagation modes. The finite speed of wave propagating is considered.

Conventionally, the largest physical size of all PEEC model cells d for the subdivided system requires that for all cells $d \leq \lambda_{Max}/20$, where λ_{Max} is the working wavelength of the highest frequency component. In the time domain, the delay or retardation time is τ between any two elements, which are $\tau = |\boldsymbol{r} - \boldsymbol{r}'|/c_0$ where c_0 is the speed of light in vacuum. In the frequency domain or phasor format, the corresponding concept is the phase delay $e^{-j\beta|r-r'|}$. Here, the vectors $\boldsymbol{r}$ and $\boldsymbol{r}'$ are the points for which the delay factor is calculated.

It is important to add additional insights to this statement. Conventional TEM-mode transmission line models are physically long with a constant two-dimensional cross section. For this case, QS capacitances and inductances are appropriate 2D transmission line parameters. However, if the cross-sectional dimensions become comparable to the highest frequency wavelength, then retardation becomes relevant for the cross-section.

6.3.7 Simple Full-Wave (Lp,Pp,R,τ)PEEC Models Implementation

Using zero-volt voltage sources is a common approach to compute current unknowns in an MNA SPICE-type circuit. We use the approach from Ref. [7] to set up the PEEC equations in a conventional SPICE solver using the SPICE input language current using NMA stamps in Appendix B. We also use this approach for analytical studies for small circuits. However, for the implementation of large circuit solvers, we use the approach presented in the following section and in Section 2.7.3.

In Fig. 6.8, we show an elementary equivalent PEEC circuit for a single PEEC loop using the voltage sources. Today, we use the efficient formulation in (6.54) for an efficient solver implementation. However, for some experiments and analytical formulations, we can use the approach based on Ref. [7] where the circuit equations are formulated without using the matrix KCL given in Section 2.7.1. From the previous section, we can determine the coupling capacitive current sources in the frequency domain as

$$I_{s1} = \frac{Pp_{12}}{Pp_{11}} e^{-s\tau} I_{T2}, \quad I_{s2} = \frac{Pp_{21}}{Pp_{22}} e^{-s\tau} I_{T1}. \tag{6.48}$$

The total capacitive currents I_{T1} and I_{T2} are coupled using zero-volt and zero-resistance voltage source stamps. This is a common way to measure the current using MNA stamps in Appendix B. Hence, the delay term $e^{-s\tau}$ is included in the above potential coupling formulation. The rest of the circuit matrix is done using conventional MNA stamps in Appendix B.

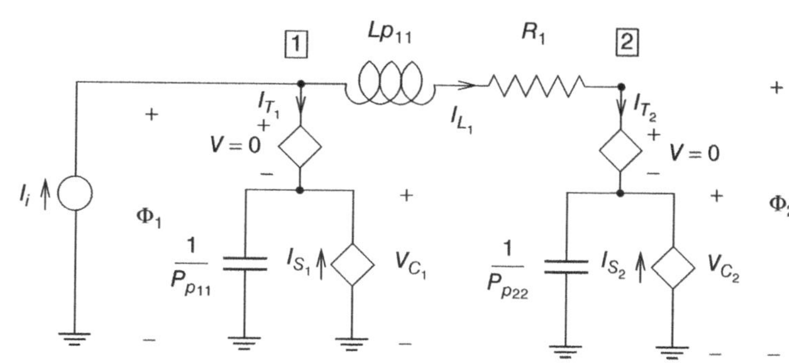

FIGURE 6.8 Small (Lp,Pp,R,τ) PEEC equivalent circuit model with current sources.

Next, we give an example for setting up the MNA equation by using conventional KCL and Kirchhoff's voltage law (KVL) equations in the frequency domain. For a FW formulation, the model has to include models for the retardation. Starting with the KCL for nodes 1 and 2, we have

$$-I_i + I_{T1} + I_{L1} = 0, \quad -I_{L1} + I_{T2} = 0. \tag{6.49}$$

The KCL at the capacitor 1 is given by

$$\frac{s}{Pp_{11}} V_{c1}(s) - \frac{Pp_{12}}{Pp_{11}} e^{-s\tau} I_{T2}(s) - I_{T1} = 0, \tag{6.50}$$

and for capacitor 2

$$\frac{s}{Pp_{22}} V_{c2}(s) - \frac{Pp_{21}}{Pp_{22}} e^{-s\tau} I_{T1}(s) - I_{T2} = 0. \tag{6.51}$$

It is seen that the zero-volt voltage source enables the separation of KCL equations by adding I_{T1} and I_{T2} to the unknowns. Note that τ is the delay between the two capacitive cells. We also have the following two KVL equations due to the added voltage sources

$$-\Phi_1 + V_{c1} = 0, \quad -\Phi_2 + V_{c2} = 0. \tag{6.52}$$

Finally, we can set up the MNA matrix from the above equations as

$$\begin{bmatrix} 1 & 0 & -1 & 0 & 0 & 0 & 0 \\ 0 & 1 & 0 & -1 & 0 & 0 & 0 \\ 0 & 0 & \frac{s}{Pp_{11}} & 0 & 0 & -1 & \frac{-Pp_{12}}{Pp_{11}} e^{-s\tau} \\ 0 & 0 & 0 & \frac{s}{Pp_{22}} & 0 & \frac{-Pp_{21}}{Pp_{22}} e^{-s\tau} & -1 \\ +1 & -1 & 0 & 0 & (-sLp_{11} - R_1) & 0 & 0 \\ 0 & 0 & 0 & 0 & 1 & 1 & 0 \\ 0 & 0 & 0 & 0 & -1 & 0 & 1 \end{bmatrix} \begin{bmatrix} \Phi_1 \\ \Phi_2 \\ V_{c1} \\ V_{c2} \\ I_{L1} \\ I_{T1} \\ I_{T2} \end{bmatrix} = \begin{bmatrix} 0 \\ 0 \\ 0 \\ 0 \\ 0 \\ I_i \\ 0 \end{bmatrix}. \tag{6.53}$$

This leads to the solution of the circuit equations in the frequency domain for one basic PEEC loop. We again note that this type of PEEC model can be implemented using the conventional SPICE input language for (QS)PEEC models.

6.4 COMPLETE PEEC MODEL WITH INPUT AND OUTPUT CONNECTIONS

In this chapter, we so far considered all aspects of PEEC models with the exception of the terminals with which the PEEC model connects to the inputs and outputs or other circuits.

We want to consider the transfer function for a full PEEC circuit. At this point, we assume that the geometry is defined and meshed and all the partial elements are computed. For most realistic problems, a small number of input sources are defined. Importantly, all the elements of the unknown vector, the potentials and inductive current can be defined as outputs. This is very convenient since this adds flexibility and we can compute quantities such as impedances.

6.4.1 Full-Wave Models

The FW circuit model has also been described in Chapter 2. We have a so-called PEEC descriptor type model of the circuits in the form of (6.54) where we also add voltage sources $\mathbf{V}_i$

$$\begin{bmatrix} s\mathbf{Pd}^{-1} & (\hat{\mathbf{I}}+\mathbf{M})\mathbf{A}_\ell \\ \mathbf{A}_\ell^T & -(s\,\mathbf{L}p+\mathbf{R}) \end{bmatrix} \begin{bmatrix} \mathbf{\Phi}_n \\ \mathbf{I}_\ell \end{bmatrix} = \begin{bmatrix} -(\hat{\mathbf{I}}+\mathbf{M})A_i\mathbf{I}_i \\ \mathbf{N}\,\mathbf{V}_i \end{bmatrix} = \mathcal{B}\mathbf{X}_i. \tag{6.54}$$

The key difference is that **Pd** is a diagonal matrix such that the inverse does not add additional costs to the solution. **M** represents the off-diagonal part of the partial coefficient of potential (6.44) matrix that does not have to be inverted.

6.4.2 Quasistatic PEEC Models

First, we take the simple case of a QS model with a capacitance model and, of course without the current source model introduced in Chapter 2. In this section, we again call the input current I_i, which is our usual notation. Figure 6.9 shows a conventional (QS)PEEC model where the capacitances can be represented by a two-terminal capacitance matrix as is done in (4.7). It is evident from this that a QS model consists of entirely conventional circuit elements, including partial mutual inductances as shown in Fig. 6.9. This capacitance model is equivalent to the full wave (FW) capacitance model (6.54) if the delays are set to zero.

$$\begin{bmatrix} s\mathbf{Pp}^{-1} & \mathbf{A}_\ell \\ \mathbf{A}_\ell^T & -(s\mathbf{Lp}+\mathbf{R}) \end{bmatrix} \begin{bmatrix} \mathbf{\Phi}_n \\ \mathbf{I}_\ell \end{bmatrix} = \begin{bmatrix} -\mathbf{A}_i\mathbf{I}_i \\ \mathbf{NV}_i \end{bmatrix} = \begin{bmatrix} -\mathbf{A}_i & \mathbf{0} \\ \mathbf{0} & \mathbf{N} \end{bmatrix} \begin{bmatrix} \mathbf{I}_i \\ \mathbf{V}_i \end{bmatrix}, \tag{6.55}$$

where we can define the input source part as

$$\begin{bmatrix} -\mathbf{A}_i & \mathbf{0} \\ \mathbf{0} & \mathbf{N} \end{bmatrix} \begin{bmatrix} \mathbf{I}_i \\ \mathbf{V}_i \end{bmatrix} = \mathbf{B}\,\mathbf{X}_i, \tag{6.56}$$

where $\mathbf{X}_i$ is the vector of inputs. Since this formulation requires the inverse $\mathbf{Cs} = \mathbf{Pp}^{-1}$, it is computationally not very efficient. Also, the model uses passive circuit elements and

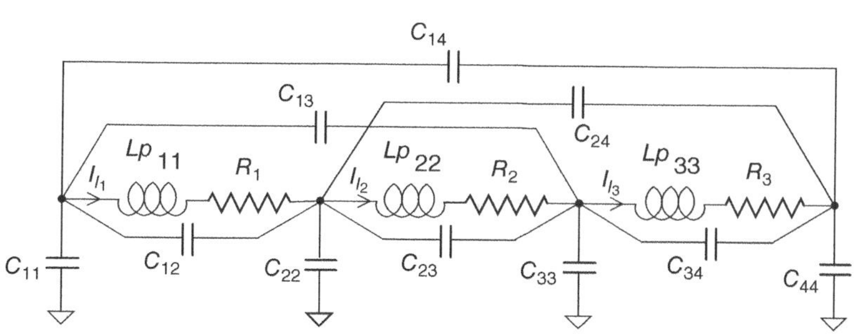

FIGURE 6.9 Quasistatic PEEC model for three fundamental loops.

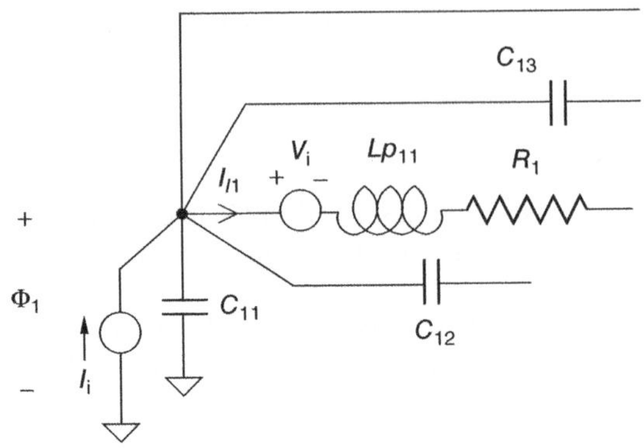

FIGURE 6.10 Example of sources associated with a PEEC model node.

only inductive couplings, while the capacitances are shown to be equivalent to capacitance model in Fig. 6.9.

We want to clearly specify the variables involved in the transfer function for the model. It is sufficient to consider a single node from Fig. 6.9 for this purpose. An input current source I_i and an input voltage source V_i is shown in Fig. 6.10. The existence of the sources at each node is specified by the selector matrix **B**.

An output vector for node 1 consists of the potential Φ_1 and/or the inductive branch current $I_{\ell 1}$ as part of the $\mathbf{I}_\ell$. The presence or the absence of sources in the circuit is represented by the matrices $\mathbf{A}_i$ and **N**. It should be evident that the input and output selector matrices are the same also for (FW)PEEC models with delays.

6.4.3 Input and Output Selectors

In a matrix form, we need to select the appropriate inputs and outputs for the problem at hand. For this, we chose the conventional form of the MNA equations that works for both the above QS and the FW equations

$$(s\,\mathcal{C} + \mathcal{G})\,\mathbf{X}(s) = \mathcal{B}\,\mathbf{X}_i(s) \tag{6.57a}$$

$$\mathbf{X}_{out}(s) = \mathcal{L}^T\mathbf{X}(s), \tag{6.57b}$$

where $\mathbf{X}_{out}$ can consist of output currents, voltages, or both and $\mathbf{X}_i$ is a vector of inputs as defined above. The matrix $\mathcal{L}$ is the output selector matrix.

The definition of the different matrices for the delay case is clear by comparing (6.54) with (6.57) where

$$\mathcal{C} = \begin{bmatrix} \mathbf{Pd}^{-1} & \mathbf{0} \\ \mathbf{0} & -\mathbf{Lp} \end{bmatrix} \tag{6.58}$$

$$\mathcal{G} = \begin{bmatrix} \mathbf{0} & (\hat{\mathbf{I}} + \mathbf{M})\mathbf{A}_\ell \\ \mathbf{A}_\ell^T & -\mathbf{R} \end{bmatrix} \tag{6.59}$$

and the right-hand side is

$$\mathcal{B}\,\mathbf{X}_i(s) = \begin{bmatrix} -(\hat{\mathbf{I}}+\mathbf{M})\mathbf{A}_i & \mathbf{0} \\ \mathbf{0} & \mathbf{N} \end{bmatrix} \begin{bmatrix} \mathbf{I}_i(s) \\ \mathbf{V}_i(s) \end{bmatrix}. \tag{6.60}$$

The solution vector in the frequency domain is

$$\mathbf{X}(s) = \begin{bmatrix} \mathbf{\Phi}_n(s) \\ \mathbf{I}_\ell(s) \end{bmatrix}, \tag{6.61}$$

that completes the description in the frequency domain.

The translation into the time domain is also rather straightforward by replacing the Laplace variable s with d/dt. Since the matrices including $\mathcal{C}$ and $\mathcal{G}$ are independent of time or s, the only difference is the system itself which is in time

$$\mathcal{C}\frac{d\mathbf{x}(t)}{dt} + \mathcal{G}\,\mathbf{x}(\mathbf{t}) = \mathcal{B}\,\mathbf{x_i}(t) \tag{6.62a}$$

$$\mathbf{x}_{out}(t) = \mathcal{L}^T\mathbf{x}(t). \tag{6.62b}$$

Note the similarity between the frequency and the time domain formulations.

As the last step in this section, we want to point out that the transfer function can be of an impedance, admittance, or a mixed type since both the inputs and outputs can be of different types. For this reason, we use the transfer function $\mathbf{H}(s)$ notation

$$\mathbf{X}_{out}(s) = \mathbf{H}(s)\,\mathbf{X}_{in}(s) = \mathcal{L}^T(s\,\mathcal{C}+\mathcal{G})^{-1}\,\mathcal{B}\,\mathbf{X}_i(s), \tag{6.63}$$

where finally the general transfer function is

$$\mathbf{H}(s) = \mathcal{L}^T(s\,\mathcal{C}+\mathcal{G})^{-1}\,\mathcal{B}. \tag{6.64}$$

Hence, the above equations represent the general system to be solved.

6.4.4 Power/Energy Type Circuit Model

For completeness, we derive a power/energy formulation for Quasi-static PEEC models. This is based on similar derivations in Ref. [8] and other circuit-theory texts. By denoting with $\mathbf{I} = \mathbf{I}_{re} + j\,\mathbf{I}_{im}$ the complex current vector, the general matrix KCL (2.37) can be written as

$$\mathbf{A}_c\mathbf{I}_c + \mathbf{A}_s\mathbf{I}_s + \mathbf{A}_\ell\mathbf{I}_\ell + \mathbf{A}_i\mathbf{I}_i = \mathbf{0} \tag{6.65}$$

for the equivalent circuit shown in Fig. 6.9. Here, $\mathbf{I}_c$ is the current for the connected capacitances, $\mathbf{I}_s$ represents the Current Controlled Current Source (CCCS) current, $\mathbf{I}_\ell$ denotes the currents flowing in the inductances while $\mathbf{I}_i$ are the currents of the external current sources. Note that this is not related to the potential external current $\mathbf{I}_{im}$ defined in this section.

The corresponding voltages for all four contributions, according to Section 2.7.2, are given by

$$\mathbf{V}_c = \mathbf{A}_c^T\, \mathbf{\Phi}_n \quad \text{or} \quad \mathbf{V}_c^T = \mathbf{\Phi}_n^T \mathbf{A}_c \tag{6.66a}$$

$$\mathbf{V}_\ell = \mathbf{A}_\ell^T\, \mathbf{\Phi}_n \quad \text{or} \quad \mathbf{V}_\ell^T = \mathbf{\Phi}_n^T \mathbf{A}_\ell \tag{6.66b}$$

$$\mathbf{V}_i = \mathbf{A}_i^T\, \mathbf{\Phi}_n \quad \text{or} \quad \mathbf{V}_i^T = \mathbf{\Phi}_n^T \mathbf{A}_i. \tag{6.66c}$$

We use the short circuit capacitances $\mathbf{Cs} = \mathbf{Pp}^{-1}$ instead of the two terminal capacitances shown in Fig. 6.9. Then, the (QS)PEEC–MNA given by (2.11) is somewhat simpler since we do not have to use the delayed capacitance model for the QS case. This leads to (6.55), or

$$\begin{bmatrix} s\,\mathbf{Pp}^{-1} & \mathbf{A}_\ell \\ \mathbf{A}_\ell^T & -(\mathbf{R} + s\mathbf{Lp}) \end{bmatrix} \cdot \begin{bmatrix} \mathbf{\Phi}_n \\ \mathbf{I}_\ell \end{bmatrix} = \begin{bmatrix} -\mathbf{A}_i \mathbf{I}_i \\ \mathbf{N}\mathbf{V}_i \end{bmatrix}. \tag{6.67}$$

Next, we form the energy type products, $\mathbf{\Phi}_n^T\, \bar{\mathbf{I}}$ where $\bar{\mathbf{I}} = \mathbf{I}_{re} - j\,\mathbf{I}_{im}$ is the conjugate complex of the current vector $\mathbf{I}$. Premultiplying (6.65) with $\mathbf{\Phi}_n^T$ yields

$$\mathbf{\Phi}_n^T \mathbf{A}_c\, \bar{\mathbf{I}}_c + \mathbf{\Phi}_n^T \mathbf{A}_s\, \bar{\mathbf{I}}_s + \mathbf{\Phi}_n^T \mathbf{A}_\ell\, \bar{\mathbf{I}}_\ell + \mathbf{\Phi}_n^T \mathbf{A}_i\, \bar{\mathbf{I}}_i = \mathbf{0}. \tag{6.68}$$

However, we noticed before in (2.38) that $\mathbf{A}_c = \hat{\mathbf{I}}$, which we use again for the identity matrix and $\mathbf{A}_s = -\hat{\mathbf{I}}$, $\mathbf{A}_i = -\hat{\mathbf{I}}$ and $\bar{\mathbf{I}}_T = \bar{\mathbf{I}}_c - \bar{\mathbf{I}}_s$. Using this fact and by comparing this result with (6.66), we can simplify the result to

$$\mathbf{\Phi}_n^T\, \bar{\mathbf{I}}_T + \mathbf{V}_\ell^T\, \bar{\mathbf{I}}_\ell = \mathbf{\Phi}_n^T\, \bar{\mathbf{I}}_i, \tag{6.69}$$

where the last term includes external currents.

Considering any single port to observe the behavior of the system since final result (6.68) includes the entire system. Hence, the right-hand side can be reduced to $\Phi_1 \bar{I}_1$. It is a standard approach to compute the input impedance Z_{in} as

$$Z_{in1} = \frac{\Phi_1\, \bar{I}_1}{I_1\, \bar{I}_1} = \frac{\Phi_1\, \bar{I}_1}{|I_1|^2}. \tag{6.70}$$

If the direction of the current is taken into account, the input impedance can be written as

$$Z_{in1} = \frac{1}{|I_{n1}|^2} [\mathbf{\Phi}_n^T\, \bar{\mathbf{I}}_T + \mathbf{\Phi}_\ell^T\, \bar{\mathbf{I}}_\ell]. \tag{6.71}$$

As a first observation, it is clear that this function has to be positive on the positive imaginary axis $s = j\omega$. The first term on the right-hand side relates to the capacitance model, whereas the last term represents the resistance and partial inductance model. The advantage of this formulation is that we can separately check the inductance/resistor and the coefficient of potential or capacitance models. It is clear that the two models have to be positive independent of each other for all frequencies.

6.4.5 Resistances, Inductance, and Capacitive Terms

So far, we did not consider the resistors in series with the inductors. We start with the inductance part of the circuit description of the

$$\mathbf{A}_\ell^T \mathbf{\Phi}_n + (s\mathbf{Lp} + \mathbf{R})\mathbf{I}_\ell = \mathbf{V}_\ell, \tag{6.72}$$

where the last step on the right-hand side is due to the matrix voltage law (6.66c). Taking the transpose of the last two parts of (6.72) and substituting it into the Lp,R part of (6.71), we get

$$Z_{in1Lp,R} = \frac{1}{|I_{n1}|^2}[\mathbf{\Phi}_\ell^T \bar{\mathbf{I}}_\ell] = \frac{1}{|I_{n1}|^2}[\mathbf{I}_\ell^T(s\mathbf{Lp}^T + \mathbf{R}^T)\mathbf{I}_\ell]. \tag{6.73}$$

Two observations further simplify this result. First, we can subdivide the equation into an inductive and a resistive part. Second, both the **Lp** and the **R** matrices are symmetrical. This leads to

$$Z_{in1,Lp,R} = s\,\frac{\mathbf{I}_\ell^T\,\mathbf{Lp}\,\mathbf{I}_\ell}{|I_{n1}|^2} + \frac{\mathbf{I}_\ell^T\,\mathbf{R}\,\mathbf{I}_\ell}{|I_{n1}|^2}. \tag{6.74}$$

It is well known that a properly computed partial inductance matrix **Lp** is symmetrical and positive definite. Also, we use the definition of positive definiteness [9].

Definition 2 ***(Positive definite matrix)*** *A symmetric $n \times n$ matrix* **M** *is called positive definite if* $\mathbf{X}^T\mathbf{M}\,\mathbf{X} > 0$ *for every nonzero vector* **X**.

It now is clear that in (6.74) $Re(Z_{inLp,R}) > 0$ in the right half of the s-plane $s = \sigma + j\omega$ where $\sigma > 0$. Again, we only consider a (QS)PEEC model.

The capacitive term can again be treated separately from the input impedance matrix (6.73) as

$$Z_{in1C} = \frac{1}{|I_{n1}|^2}[\mathbf{\Phi}_n^T \bar{\mathbf{I}}_c - \mathbf{\Phi}_n^T \bar{\mathbf{I}}_s] = \frac{1}{|I_{n1}|^2}[\mathbf{\Phi}_n^T \bar{\mathbf{I}}_T], \tag{6.75}$$

where the last part has been added for the (QS)PEEC model. We have a simpler capacitance model with one current for the connection to the nodes, which we call $\bar{\mathbf{I}}_c - \bar{\mathbf{I}}_s = \bar{\mathbf{I}}_T$. For the QS case considered, we take the first equation of (6.55) as

$$s\,\mathbf{Pp}^{-1}\,\mathbf{\Phi}_n + \mathbf{A}_\ell \mathbf{I}_\ell = -\mathbf{A}_i \mathbf{I}_i. \tag{6.76}$$

We can simplify this by using the matrix KCL (Section 2.7.1) to

$$s\,\mathbf{Pp}^{-1}\,\mathbf{\Phi}_n = \mathbf{I}_T, \tag{6.77}$$

since the capacitance model is connected directly to the nodes where $\mathbf{A}_c$ is not included since it is a unity matrix. Inserting this for $\mathbf{I}_T$ in (6.75), we get for the QS case that

$$Z_{in1C} = \frac{1}{|I_{n1}|^2}[s\,\mathbf{\Phi}_n^T\,\mathbf{Pp}^{-1}\,\bar{\mathbf{\Phi}}_n]. \tag{6.78}$$

It is clear from the positive definite Definition 2 that this part is also positive in the right half plane where $\sigma > 0$. We also know that the capacitive model can be replaced by a set of conventional coupled capacitances as is shown in Fig. 6.9. If the capacitances are computed accurately, we know that the capacitive part is passive. More details on these issues are considered in Section 4.4.1.

Hence, we can conclude the following:

Theorem 1 ***(Real positive part of port impedance)*** *The real part of the port impedance of a (QS)PEEC circuit* $Re(Z_{in1}) = Re(Z_{in1Lp,R}) + Re(Z_{in1C})$ *is positive in the right half of the* $s = \sigma + j\omega$ *plane if the matrices* **R**, **Lp**, *and* $\mathbf{Pp}^{-1}$ *are positive definite.*

Based on the above derivation (see also Refs [10] and [11]), it is shown that the transfer function $\mathbf{H}(s) = \mathcal{L}^T(s\mathcal{C} + \mathcal{G})^{-1}\mathcal{B}$ is passive. More details on passivity issues are given in Chapter 13.

6.5 TIME DOMAIN REPRESENTATION

According to (6.64), the PEEC EM model can be described by the transfer function $\mathbf{H}(s)$ where the responses are given by

$$\mathbf{X}_{out}(s) = \mathbf{H}(s) \cdot \mathbf{X}_i(s). \tag{6.79}$$

This can be transformed to the time domain to

$$\mathbf{x}_{out}(t) = \mathbf{h}(t) * \mathbf{x}_i(t), \tag{6.80}$$

where $*$ represents the convolution operator. Hence, the cascading of solutions is more time consuming in the time domain. However, techniques such as the recursive convolution approach presented in Section 2.10.2 speed up the process.

PROBLEMS

6.1 *PEEC circuit model for coupled current strips*
A rectangular strip has a width of a mm and a length of $20a$ mm. Its thickness is zero. Set $a = 2$ mm. Split the strip in the middle into two pieces that are connected at the ends. These two pieces are coupled to each other. Build a PEEC model for the strip composed of two partial inductances. Assume that the material is a Perfect Electrical Conductor (PEC) with a zero resistance. Hence, its resistance is zero. Use the closed-form partial elements given in Appendices C and D to compute the value of all components. Hint: The model has two partial inductance and four capacitive cells. Each partial inductance consists of an inductive half cell where they are sharing a node at each end.

6.2 *Assemble circuit matrix using element stamps*
Corresponding to the circuits in Fig. 6.8, assemble the circuit matrix (6.53) from the circuit equations using the MNA stamps in Appendix B.

6.3 *Compute the impedance of a loop*
A wire loop is shown in Fig. 6.11. Make a (QS)PEEC model for the loop and compute the loop impedance between A and B. Then remove branch 4 from the loop and compute the impedance of the open loop and compare the impedance results for the two cases. Also, check the impedance for using 1, 2, and 4 partial inductances per side. Use zero thickness sheet conductors of 0.5 mm width for the model using equations (C.23) and (C.25). Assume that the square loop has a side length of 30 mm. Also compute the impedance of the loop arrangements if you also include the resistance of the conductors.

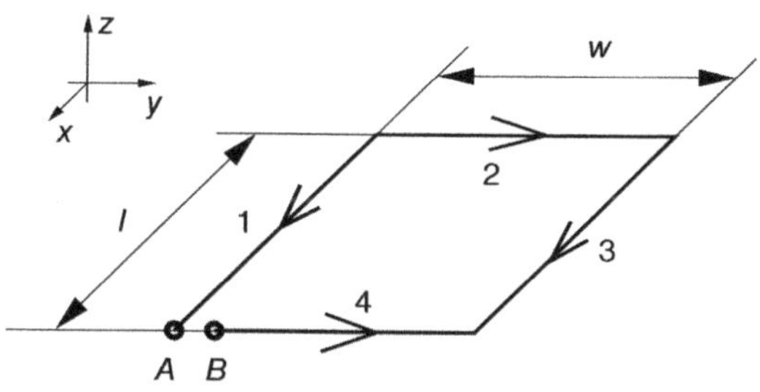

FIGURE 6.11 Single loop impedance example.

6.4 *Impedance of loop with delays*
Add delays or retardation to the loop impedance in Problem 6.3. This is accomplished by adding delays or retardation to the coupled partial inductances and the partial potential coefficients. The simplest way to do this is to assume that the delay can be computed from node to node for the partial potential coefficients. For the partial inductances, we calculate the delay from center to center of the cells and for the partial coefficients of potential we use the node-to-node delay. Make sure that you extend the highest frequency such that the added delay changes the result from the one obtained from Problem 6.3. Remember that for the highest frequency the cell size should be such that the wavelength of the frequency is such that the smallest wavelength is $\leq\lambda/20$.

6.5 *PEEC circuit model for the external conductor*
Assume that the current filament has a rectangular cross section with a width equal to a mm and the height equal to $a/2$ mm. The length of the filament is $10a$ mm and the current on it is uniformly along the longitudinal direction. Build the PEEC circuit model of this current filament by ignoring all couplings to other filaments possibly positioned nearby. Assume the material is copper with the conductivity $5.96 \times 10^7\ S/m$ and $a = 1$ mm. Use the closed form given in Appendices C and D to compute the value of all components.

REFERENCES

1. S. M. Rao, D. R. Wilton, and A. W. Glisson. Electromagnetic scattering by surfaces of arbitrary shape. *IEEE Transactions on Antennas and Propagation*, 30:409–418, May 1982.
2. Y. Zhu and A. Cangellaris. *Multigrid Finite Element Methods for Electromagnetic Field Modeling*. John Wiley and Sons, Inc., New York, 8th edition, 2006.

3. A. Rong and A. C. Cangellaris. Comprehensive broadband electromagnetic modeling of on-chip interconnects with a surface discretization-based generalized PEEC method. *IEEE Transactions on Advanced Packaging*, 28(3):434–444, August 2005.

4. A. E. Ruehli. Electrical analysis of interconnections in a solid state circuit environment. In *ISSCD, IEEE International Solid-State Circuits Conference*, Volume XV, pp. 64–65, February 1972.

5. B. G. Galerkina. On the life and scientific career of academician B. G. Galerkin. *Izvestiya Akademii nauk SSSR, Otdelenie tekhnicheskikh nauk*, 8:1159–1164, 1951.

6. J. S. Hesthaven and T. Warburton. *Nodal Discontinuous Galerkin Methods: Algorithms, Analysis, and Applications, Springer Texts in Applied Mathematics 54*. Springer-Verlag, New York, 2008.

7. W. Pinello, A. C. Cangellaris, and A. Ruehli. Transient electromagnetic analysis and model complexity reduction using the partial element equivalent circuit formulation. In *Massachusetts Institute of Technology*, Volume 1, pp. 63–66, June 1997.

8. D. Hazony. *Elements of Network Synthesis*. Reinhold Publishing Corporation, New York, 1963.

9. G. Golup and C. Van Loan. *Matrix Computations*. The John Hopkins University Press, Baltimore, MD, 3rd edition, 1996.

10. N. Balabanian, T. A. Bickart, and S. Seshu. *Electrical Network Theory*. John Wiley and Sons, Inc., New York, 1969.

11. M. Celik, L. Pileggi, and A. Odabasioglu. *IC Interconnect Analysis*. Kluwer Academic Publishers, Boston/Dordrecht/London, 2002.

7

NONORTHOGONAL PEEC MODELS

The previous chapter is a prelude to this chapter. The fundamental techniques for constructing the nonorthogonal partial element equivalent circuit (PEEC) models are similar to the orthogonal ones. However, the compute time required for nonorthogonal structures can be significantly larger. As we learn in this chapter, the computations of the partial inductances **Lp** and coefficients of potential **Pp** matrices are much more challenging as well as more time consuming. Further, all nonorthogonal partial inductances are coupled, unlike orthogonal cases where only partial inductances in the same direction are coupled by partial mutual inductances. The circuit matrix will be much more dense. It is clear that a good compromise is the use of orthogonal cells in combination with nonorthogonal cells. Hence, orthogonal cells are used as much as possible. This leads to the situation where orthogonal partial elements are coupled to nonorthogonal partial elements.

Early on in the history of PEEC models, simplified nonorthogonal geometries were approximated by rectangular bars with arbitrary orientations [1–4]. Other approaches have also been proposed for nonorthogonal PEEC models using triangular cells for the problems at hand [5–7]. The quadrilateral or hexahedral PEEC formulation has been shown to work well, for example, Refs [8–10]. The quadrilateral and hexahedral shapes can be used to represent different conductor as well as dielectric cells. Further, quadrilateral surface cells have been used successfully for electromagnetic (EM) antenna type modeling for other integral equation approach [11, 12]. Importantly, large aspect ratio cells can be used for quadrilateral and hexahedral cells. An example of two quadrilateral inductive cells is shown in Fig. 7.1. Two inductance half cells are used in this example.

In this text, we do not consider triangular cell shapes for several reasons. Perhaps the most important one is because of the features of heterogeneous electromagnetic circuit (EM/Ckt) problem, which we want to solve. Using triangular cells may lead to additional unknowns since the aspect ratio of the cells has to be relatively small. This limited aspect ratio can lead to a large number of unknowns for the problems of interest. Each triangular cell has three current directions. Hence, if triangular cells would also be employed, the

Circuit Oriented Electromagnetic Modeling Using the PEEC Techniques, First Edition.
Albert E. Ruehli, Giulio Antonini, and Lijun Jiang.

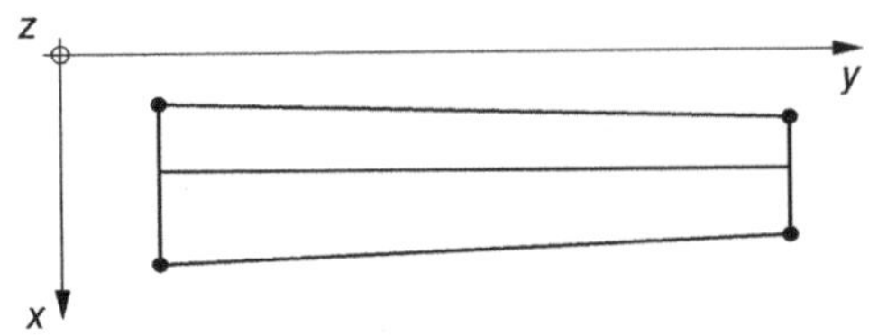

FIGURE 7.1 Example of two long quadrilateral inductive cells.

computational complexity of evaluating integrals increases to three fundamental cell types have to include rectangles, quadrilaterals, and triangles.

It is intriguing to observe that the topology of the orthogonal and quadrilateral PEEC equivalent circuit models is exactly the same since the additional partial mutual inductances are not shown in the schematics. However, the additional inductive couplings must include nonorthogonal partial mutual inductances.

The modeling of nonorthogonal shapes is becoming more important for the large class of higher frequency problems of interest. Many examples are in the signal and power integrity problem area such as the modeling of high voltage towers [13] as well as some printed circuit boards formulations that use nonorthogonal cells [8, 14–16]. Comparisons with measurements and solutions obtained with other techniques have been made in Refs [17, 18]. The validation of such results should be an ongoing process.

7.1 REPRESENTATION OF NONORTHOGONAL SHAPES

It is clear that one of the more complicated parts for the nonorthogonal shape modeling is the coordinate system that represents the geometry. This is an important part of this chapter.

A few concepts are fundamental for the treatment of nonorthogonal shapes. First, the *global* coordinate system determines the location of all objects independent of their local orientation. This global coordinate system is a conventional, orthogonal one where an object k is assumed to be located at a point (x_k, y_k, z_k). How can we then represent an object with an arbitrary orientation? In general, this *local* object can be an orthogonal, quadrilateral, and/or hexahedral conductor or dielectric block or cell. Hence, we need to define the center point of the kth object (x_k, y_k, z_k) and its orientation with respect to the general coordinates.

In the global orthogonal coordinates, a vector from the origin to the point is given by $\boldsymbol{r} = x_k\hat{\boldsymbol{x}} + x_y\hat{\boldsymbol{y}} + x_z\hat{\boldsymbol{z}}$ where the global unit vectors $\hat{\boldsymbol{x}}$, $\hat{\boldsymbol{y}}$ and $\hat{\boldsymbol{z}}$ are position independent. To emphasize the vector $\boldsymbol{r}$ represents always the global coordinate system. However, a typical case has many local coordinates for the many nonorthogonals subsystems that are represented with different orientations in the global system.

It is important to understand that each separate local coordinate system is locally coupled to the global coordinates. Hence, the local coordinates allow the arbitrary orientation of the local objects. We call the local coordinates (a, b, c). We use a, b, c to distinguish them from circuit variables such as voltage v. It is clear that we need to relate the local coordinates to the global ones.

We start out with the description of the geometrical objects that can be orthogonal and/or nonorthogonal. The more difficult issue is the representation of nonorthogonal geometries. The *global* coordinate system represents all parts of the geometry. Ultimately, all bodies need to be described in global coordinates such that the interactions can be represented.

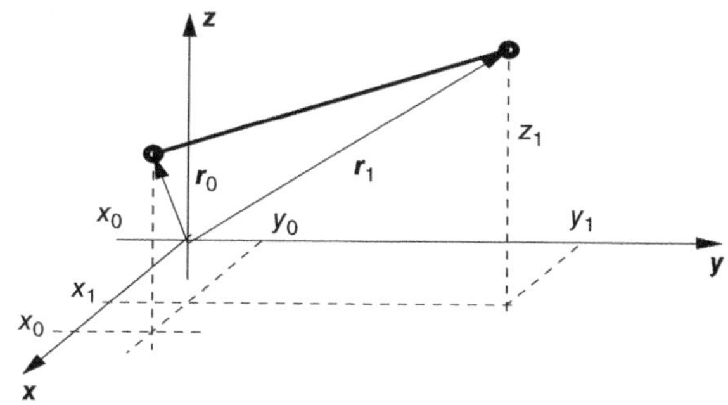

FIGURE 7.2 Line in global 3D coordinate system.

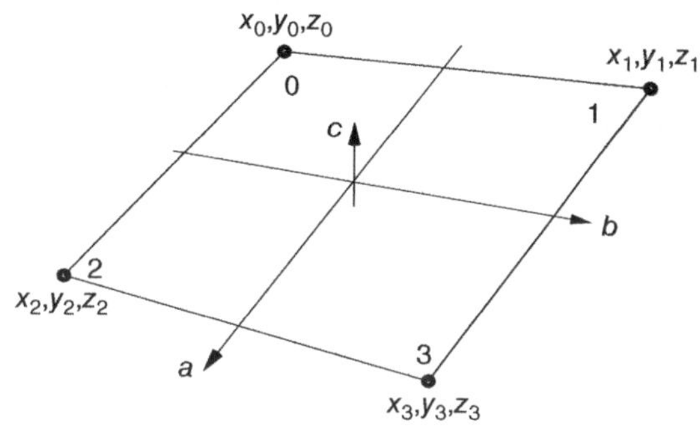

FIGURE 7.3 Basic quadrilateral element with local and global coordinates.

A simple nonorthogonal example is the case where a straight wire connects between x_0, y_0, z_0 and x_1, y_1, z_1, which is shown in Fig. 7.2. For example, we would like to represent the global x coordinate in terms of the local coordinate variable a. Then, we can represent all points in x between the two end points x_0 and x_1 as

$$x(a) = \frac{1-a}{2}\,x_0 + \frac{1+a}{2}\,x_1 \tag{7.1}$$

by choosing a point a where $-1 \leq a \leq 1$. It is evident that if $a = -1$, then $x(-1) = x_0$ and if $a = 1$ then $x(1) = x_1$.

The local unit vectors are defined as $\hat{\boldsymbol{a}}$, $\hat{\boldsymbol{b}}$, and $\hat{\boldsymbol{c}}$. Details about nonorthogonal coordinate systems can be found in several texts, for example, Ref. [19]. Figure 7.3 shows an example of a zero thickness quadrilateral element with its local coordinate system. For example, we need to be able to specify the location of points on the surface of the quadrilateral. In this text, we assume that all local cell surfaces for all objects such as quadrilateral as well as hexahedrons are planar. From Fig. 7.3, it is clear that the quadrilateral is uniquely defined by the coordinates of four plane corner points. This introduces a linear dependence of some of the coordinates. The global coordinate center point is at the origin of the local coordinates shown in Fig. 7.3.

Next, we present a simple approach for a unique definition of the corners of the local coordinates in the global system. This is accomplished in a clever way by the local variables a, b where $a \in [-1, +1]$ and $b \in [-1, +1]$ as will be given below. Based on Fig. 7.3, the corner location corresponds to $a = -1$ or $a = 1$ and $b = -1$ or $b = 1$. Hence, it is convenient

TABLE 7.1 Local corner specification for quadrilateral surface.

Variable a	Variable b	Logic a	Logic b	Corner assigned
−1	−1	0	0	0
−1	1	0	1	1
1	−1	1	0	2
1	1	1	1	3

to use *logical* variable to identify the corners. We assign a logical zero $a = 0$ to the variable $a = -1$ and a logical $a = 1$ to the variable $a = 1$, and so on. Hence, Table 7.1 can be constructed for the corner representation.

From this, it is easy to uniquely assign logical codes to the corners in Fig. 7.3.

With this definition of the corners, we can uniquely relate the location of a point on the quadrilateral surface to a point in the global rectangular coordinate system x, y, z. This is accomplished by the following conventional mappings:

$$N_0(a, b) = 1/4(1 - a)(1 - b) \tag{7.2a}$$

$$N_1(a, b) = 1/4(1 - a)(1 + b) \tag{7.2b}$$

$$N_2(a, b) = 1/4(1 + a)(1 - b) \tag{7.2c}$$

$$N_3(a, b) = 1/4(1 + a)(1 + b) \tag{7.2d}$$

and

$$x = \sum_{k=0}^{3} N_k(a, b)\, x_k \tag{7.3a}$$

$$y = \sum_{k=0}^{3} N_k(a, b)\, y_k \tag{7.3b}$$

$$z = \sum_{k=0}^{3} N_k(a, b)\, z_k, \tag{7.3c}$$

where (a, b) is located on the quadrilateral surface and (x, y, z) is the corresponding point in the global coordinates.

7.1.1 Hexahedral Bodies

It is clear that for many partial inductance and resistance cells and dielectric bodies, etc., we need finite thickness cells. For this case, we need three coordinates (a, b, c) to describe a point, where $c \in [-1, +1]$ is added to the quadrilateral sheet representation as shown in Fig. 7.4. Clearly, the model is an extension of the quadrilateral case. We only have to add the equations for the three local directions.

First, the corner assignment for eight corners is shown in Table 7.2.

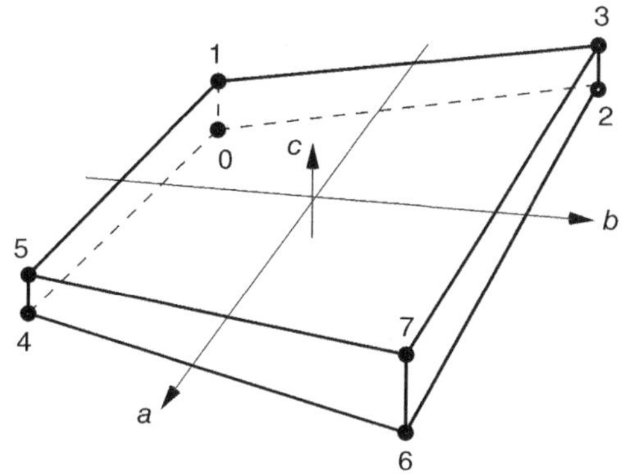

FIGURE 7.4 Basic hexahedral element or object with local coordinates.

TABLE 7.2 Local corner specification for hexahedral body.

Logic *a*	Logic *b*	Logic *c*	Corner assigned
0	0	0	0
0	0	1	1
0	1	0	2
0	1	1	3
1	0	0	4
1	0	1	5
1	1	0	6
1	1	1	7

The coefficients for (7.5) are given by

$$N_0(a,b,c) = 1/8(1-a)(1-b)(1-c) \tag{7.4a}$$

$$N_1(a,b,c) = 1/8(1-a)(1-b)(1+c) \tag{7.4b}$$

$$N_2(a,b,c) = 1/8(1-a)(1+b)(1-c) \tag{7.4c}$$

$$N_3(a,b,c) = 1/8(1-a)(1+b)(1+c) \tag{7.4d}$$

$$N_4(a,b,c) = 1/8(1+a)(1-b)(1-c) \tag{7.4e}$$

$$N_5(a,b,c) = 1/8(1+a)(1-b)(1+c) \tag{7.4f}$$

$$N_6(a,b,c) = 1/8(1+a)(1+b)(1-c) \tag{7.4g}$$

$$N_7(a,b,c) = 1/8(1+a)(1+b)(1+c), \tag{7.4h}$$

where all the local coordinate variables are between $[-1, +1]$. Because node indices are unique, again we can obtain the mappings to the global coordinates as

$$x = \sum_{k=0}^{7} N_k(a, b, c)\, x_k$$
$$y = \sum_{k=0}^{7} N_k(a, b, c)\, y_k \tag{7.5}$$
$$z = \sum_{k=0}^{7} N_k(a, b, c)\, z_k.$$

At this point, we have a unique mapping between a local point located anywhere in the hexahedral shape to the global coordinates x, y, z. The combined coordinate systems represent a convenient way to represent nonorthogonal coordinates for the shapes at hand.

7.1.2 Derivatives of the Local Coordinates

Hopefully, it is clear now that local coordinates pertain to a specific body. The local variables are valid only on the surfaces or inside the volume of the body. The main purpose of the local coordinates is to facilitate the evaluation of PEEC circuit elements the nonorthogonal partial inductances, potential coefficient, and the resistances. For these computations, we not only need relations between the global coordinate vector $\boldsymbol{r}$ and its local coordinates but also local derivatives for the local coordinates and other information.

For example, for a quadrilateral shape in Fig. 7.3, we can simply find the general derivatives

$$\frac{\partial N_0}{\partial a} = -\frac{1}{4}(1-b), \quad \frac{\partial N_0}{\partial b} = -\frac{1}{4}(1-a) \tag{7.6a}$$

$$\frac{\partial N_1}{\partial a} = -\frac{1}{4}(1+b), \quad \frac{\partial N_1}{\partial b} = +\frac{1}{4}(1-a) \tag{7.6b}$$

$$\frac{\partial N_2}{\partial a} = +\frac{1}{4}(1-b), \quad \frac{\partial N_2}{\partial b} = -\frac{1}{4}(1+a) \tag{7.6c}$$

$$\frac{\partial N_3}{\partial a} = +\frac{1}{4}(1+b), \quad \frac{\partial N_3}{\partial b} = +\frac{1}{4}(1+a). \tag{7.6d}$$

Based on (7.3), we can obtain the desired derivatives using (7.6a)–(7.6d)

$$\frac{\partial x}{\partial a} = \frac{\partial N_0}{\partial a}x_0 + \frac{\partial N_1}{\partial a}x_1 + \frac{\partial N_2}{\partial a}x_2 + \frac{\partial N_3}{\partial a}x_3 \tag{7.7a}$$

$$\frac{\partial y}{\partial a} = \frac{\partial N_0}{\partial a}y_0 + \frac{\partial N_1}{\partial a}y_1 + \frac{\partial N_2}{\partial a}y_2 + \frac{\partial N_3}{\partial a}y_3 \tag{7.7b}$$

$$\frac{\partial z}{\partial a} = \frac{\partial N_0}{\partial a}z_0 + \frac{\partial N_1}{\partial a}z_1 + \frac{\partial N_2}{\partial a}z_2 + \frac{\partial N_3}{\partial a}z_3, \tag{7.7c}$$

where the other two sets of derivatives with respect to b and c are specified by replacing a by b or c. We finally have all the components for the evaluation of the important tangential vector with respect to the local coordinates that are

$$\begin{aligned}\frac{\partial \boldsymbol{r}}{\partial a} &= \frac{\partial x}{\partial a}\hat{\boldsymbol{x}} + \frac{\partial y}{\partial a}\hat{\boldsymbol{y}} + \frac{\partial z}{\partial a}\hat{\boldsymbol{z}} \\ \frac{\partial \boldsymbol{r}}{\partial b} &= \frac{\partial x}{\partial b}\hat{\boldsymbol{x}} + \frac{\partial y}{\partial b}\hat{\boldsymbol{y}} + \frac{\partial z}{\partial b}\hat{\boldsymbol{z}} \\ \frac{\partial \boldsymbol{r}}{\partial c} &= \frac{\partial x}{\partial c}\hat{\boldsymbol{x}} + \frac{\partial y}{\partial c}\hat{\boldsymbol{y}} + \frac{\partial z}{\partial c}\hat{\boldsymbol{z}}.\end{aligned} \tag{7.8}$$

The importance of these derivatives is evident in the following section.

7.2 SPECIFICATION OF NONORTHOGONAL PARTIAL ELEMENTS

The derivations of the vectors in the previous section provide the foundation for the formulation of the nonorthogonal partial elements. For the partial coefficients of potential **Pp**, zero thickness rectangular or quadrilateral shapes are used since the charge is a surface quantity. For resistances **R** and partial inductances **Lp**, rectangular or hexahedral boxes are used. As a practical solution for uniform thin partial inductances or resistance, we can utilize quadrilateral shapes with a uniform thickness without having to resort to more complex hexahedral representations. It is clear that we want to use nonorthogonal coordinates only where it is necessary. For example, for the coupling between a rectangular and a nonorthogonal shape, we represent each shape with the appropriate coordinates. Of course, we can adhere to the global rectangular coordinates for the rectangular Manhattan object.

We use a zero thickness or very thin quadrilateral conductive sheet shown in Fig. 7.5 to motivate the derivation of the next step. An inductive/resistance half cell between nodes 0 and 2 is illustrated on the quadrilateral shape with the current density vectors. Further, the crosshatched area attached to node 1 is an example of a capacitive quarter cell. We recognize that in both cases, full cells can be formed by attaching other sheets to the same nodes forming larger bodies. As an example, two inductive half cells from two different quads can be connected between nodes 0 and 2 and four capacitive quarter cells may be connected to node 1 to form full capacitive cell.

The direction of the current flow in Fig. 7.5 from node 0 to node 2 is given by a vector that changes direction with position on the half cell. We can compute the tangential direction vectors $\hat{\boldsymbol{a}}, \hat{\boldsymbol{b}}, \hat{\boldsymbol{c}}$ from the tangential vector by first forming the magnitude of the vectors

$$h_a = \left|\frac{\partial \boldsymbol{r}}{\partial a}\right|, \; h_b = \left|\frac{\partial \boldsymbol{r}}{\partial b}\right|, \; h_c = \left|\frac{\partial \boldsymbol{r}}{\partial c}\right|. \tag{7.9}$$

The unit vectors are found by dividing the vectors by the magnitude, or

$$\hat{\boldsymbol{a}} = \frac{1}{h_a}\frac{\partial \boldsymbol{r}}{\partial a}, \qquad \hat{\boldsymbol{b}} = \frac{1}{h_b}\frac{\partial \boldsymbol{r}}{\partial b}, \qquad \hat{\boldsymbol{c}} = \frac{1}{h_c}\frac{\partial \boldsymbol{r}}{\partial c}. \tag{7.10}$$

These unit vectors give the direction of the current anywhere on the cells.

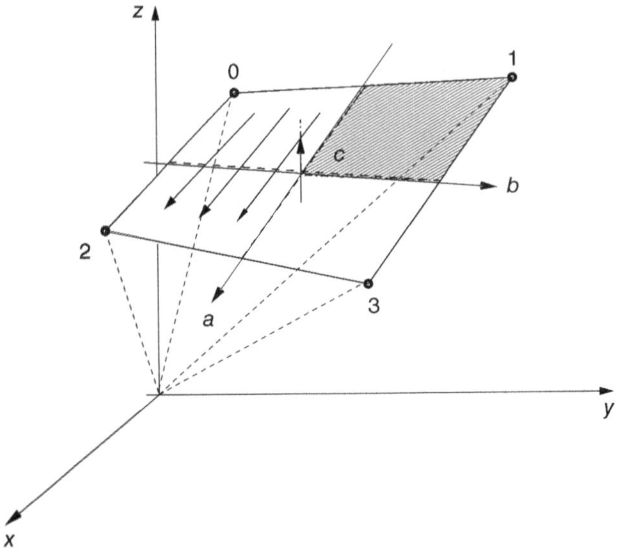

FIGURE 7.5 Quadrilateral patch with four inductive half cells.

For completeness, we also want to introduce that the square of the magnitude is also called the *metric coefficient* [19]. For example,

$$h_a^2 = \frac{\partial \mathbf{r}}{\partial a} \cdot \frac{\partial \mathbf{r}}{\partial a} = \left(\frac{\partial x}{\partial a}\right)^2 + \left(\frac{\partial y}{\partial a}\right)^2 + \left(\frac{\partial z}{\partial a}\right)^2. \tag{7.11}$$

7.2.1 Discretization of Conductor and Dielectric Geometries

Chapter 3 shows the integral equation representing the solution to be discretized, for example, (3.54) for orthogonal structures whose solution is given in Chapter 6.

The basic hexahedral element or object for the nonorthogonal element is shown in Fig. 7.4 and an example for the connection between quadrilateral and orthogonal surface elements is shown in Fig. 7.6. As we do for rectangular coordinates, we set up the inductive and capacitive meshes. For example, we see how the quarter cells connected to the β node form the capacitance plate. However, it is clear that the total cell potential coefficients are also computed by combining the quarter cell values. The inductive cells are formed by half cells or by combining two half cells into a full cell. To represent the current flow in the orthogonal cells, we use the conventional current density or weighting function $J = I/(WT)$, where I and J are the current and current density while W and T are the cell width and thickness, respectively.

For the nonorthogonal case, we assume that the conductor thickness is also subdivided into cells. As an example, the thickness divisions take the skin effect into account as considered in Chapter 9. This leads to the hexahedral elements in Fig. 7.4 for the nonorthogonal case. It directly leads to a nonorthogonal version of the volume filament (VFI) skin-effect model presented in Section 9.3 especially for conductors with a constant thickness.

The generalization of the current distribution weighting for nonorthogonal hexahedral cell shapes is given by

$$J_a = \frac{h_a}{\left|\frac{\partial r}{\partial a} \cdot \left(\frac{\partial r}{\partial b} \times \frac{\partial r}{\partial c}\right)\right|} I_a = w_a I_a, \tag{7.12}$$

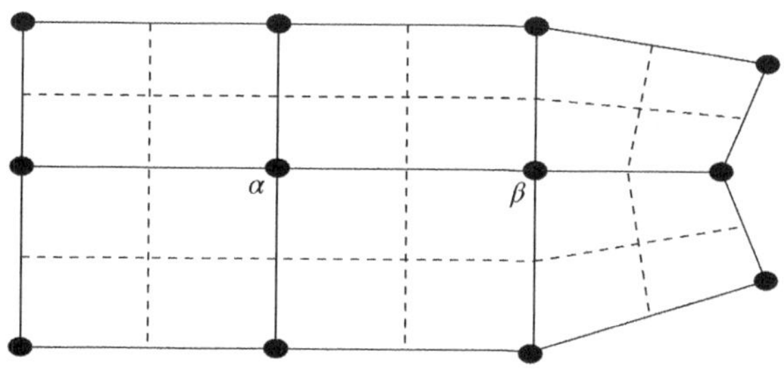

FIGURE 7.6 Geometry with rectangular and quadrilateral elements.

where J_b and J_c can easily be found by permuting the indices. We call the quotient in (7.12) the weight w_a so that it is simplified to $J_a = w_a I_a$. We shall note that all the above quantities are a function of the local position coordinates (a, b, c).

Next, we use an integral or inner product operator that essentially implements $V = \int \boldsymbol{E} \cdot \boldsymbol{d\ell}$ for the nonorthogonal case, or

$$V_a = \int_a \int_b \int_c w_a \, \hat{\boldsymbol{a}} \cdot \boldsymbol{E}(a, b, c) \left| \frac{\partial \boldsymbol{r}}{\partial a} \cdot \left(\frac{\partial \boldsymbol{r}}{\partial b} \times \frac{\partial \boldsymbol{r}}{\partial c} \right) \right| da \, db \, dc. \tag{7.13}$$

Where $\boldsymbol{E}(a, b, c)$ is the $\boldsymbol{E}$ field term to be integrated. In fact, we need to apply the inner product to each term of (3.54) to transform each term to a voltage drop across a PEEC circuit element in the Kirchhoff's voltage law (KVL) equation.

The first element on the right-hand side terms (3.54), equivalent to the derivation for rectangular coordination in section 6.1, leads to the series resistance term in the form

$$R_a = \frac{1}{\sigma} \int_a \int_b \int_c \frac{h_a^{\,2}}{\left| \frac{\partial r}{\partial a} \cdot \left(\frac{\partial r}{\partial b} \times \frac{\partial r}{\partial c} \right) \right|} \, da \, db \, dc. \tag{7.14}$$

To give a specific example of a resistance case, if we consider the cell between corners 0 and 2 in Fig. 7.5. The resistance connected to corners will be

$$R(0, 2) = \frac{1}{\sigma} \int_{a=-1}^{1} \int_{b=-1}^{0} \int_{c=-1}^{1} \frac{h_a^{\,2}}{\left| \frac{\partial r}{\partial a} \cdot \left(\frac{\partial r}{\partial b} \times \frac{\partial r}{\partial c} \right) \right|} \, da \, db \, dc, \tag{7.15}$$

where we assign a uniform thickness to the layer in the c direction.

The second right-hand side term of (3.54) yields the partial inductance corresponding to (5.17). The orthogonal Manhattan case is repeated here for convenience:

$$Lp_{km} = \frac{1}{\mathcal{A}_k \, \mathcal{A}_m} \frac{\mu_0}{4\pi} \int_{\mathcal{A}_k} \int_{\mathcal{A}_m} \int_{\ell_k} \int_{\ell_m} \frac{\hat{\boldsymbol{t}}_k \cdot \hat{\boldsymbol{t}}_m}{R_{k,m}} \, d\ell_m \, d\ell_k \, d\mathcal{A}_m \, d\mathcal{A}_k. \tag{7.16}$$

For the nonorthogonal case, after applying (7.13) becomes the generalization of the partial inductance concept for nonorthogonal problems, or

$$Lp_{aa'} = \mu_0 \int_a \int_b \int_c \int_{a'} \int_{b'} \int_{c'} (\hat{\boldsymbol{a}} \cdot \hat{\boldsymbol{a}}') \, h_{a'} \; g(\boldsymbol{r}, \boldsymbol{r}') \, da' \, db' \, dc' \, h_a \, da \, db \, dc, \tag{7.17}$$

where the Green's function $g(\boldsymbol{r},\boldsymbol{r}')$ still is evaluated with the same global coordinates $\boldsymbol{r}(a,b,c)$ and $\boldsymbol{r}'(a',b',c')$. This clearly makes the analytical evaluation of the integral more difficult in the nonorthogonal coordinates. We also want to give a specific example for the evaluation of the integral for the quad sheet in Fig. 7.5. In this case, also assume a small thickness in the c direction where we compute the partial mutual inductance between the half cells with corners (0,2) and (1,3), which is

$$Lp(0,2;1,3) = \mu_0 \int_{a=-1}^{1} \int_{b=-1}^{0} \int_{c=-1}^{1} \int_{a'=-1}^{1} \int_{b'=0}^{1} \int_{c'=-1}^{1}$$
$$[(\hat{\boldsymbol{a}} \cdot \hat{\boldsymbol{a}}')\, h_{a'}\ \ g(\boldsymbol{r},\boldsymbol{r}')] da'\ db'\ dc'\ h_a\ da\ db\ dc. \tag{7.18}$$

The dot-product $(\hat{\boldsymbol{a}} \cdot \hat{\boldsymbol{a}}')$ in (7.18) can be evaluated in several different ways depending on the details of the integration techniques used in Appendix C. Also, as an approximation we may evaluate the unit vectors at the center of each of the half cells.

The charge density is of a similar form as (7.12). Of course, the charge does not have a directional dependence. The volume charge density q_v is given by

$$q_v(a,b,c) = \frac{1}{\left| \frac{\partial \boldsymbol{r}}{\partial a} \cdot \left(\frac{\partial \boldsymbol{r}}{\partial b} \times \frac{\partial \boldsymbol{r}}{\partial c} \right) \right|} Q, \tag{7.19}$$

where Q is the total charge in the volume cell. For conventional conductors, the charge is located on the surface cells only. The gradient of the potential in the $\hat{\boldsymbol{a}}$ direction is evaluated to be

$$\nabla\Phi|_a = \frac{1}{h_a} \frac{\partial \Phi}{\partial a}. \tag{7.20}$$

To maintain the symmetry for the circuit elements or coefficients, we approximate the derivative using an integrated average over the two corner cells corresponding to the derivative where the spacing is the projection in the $\hat{\boldsymbol{a}}$ direction for the center-to-center distance d_a. This approximation implies that (7.20) can be simplified to

$$\nabla\Phi|_a \approx \frac{1}{h_a\, d_a} \int_{S_\ell} \Phi(a,b,c)\, d\, S_\ell, \tag{7.21}$$

where S_ℓ is a surface in the appropriate surface direction.

Finally, by substituting for the potential Φ as in (3.58b) into (7.21) and by canceling the appropriate terms, we get for the normalized coefficients of potential

$$Pp_{aa'} = \frac{1}{\epsilon_0} \int_a \int_b \int_{a'} \int_{b'} g(\boldsymbol{r},\boldsymbol{r}')\ da'\ db'\ da\ db, \tag{7.22}$$

where the global coordinates $\boldsymbol{r}$ and $\boldsymbol{r}'$ are evaluated on the cell related to the cell of interest.

We also want to give an example for the evaluation of a Pp. In Fig. 7.5, we chose the capacitive quarter cells, one is the crosshatched attached to corner 1 and to the quarter cell attached at corner 2.

$$Pp(1,2) = \frac{1}{\epsilon_0} \int_{a=-1}^{0} \int_{b=0}^{1} \int_{a'=0}^{1} \int_{b'=-1}^{0} g(\boldsymbol{r},\boldsymbol{r}')\, da'\, db'\, da\, db. \tag{7.23}$$

The PEEC circuit for a quadrilateral element is shown in Fig. 7.7. It consists of basic KVL loops. Specifically, a KVL loop involves two nodes with a partial inductance and a resistance in series that is closed by the capacitances to infinity. As is evident from Fig. 7.5, the part of the model represented by Fig. 7.7 includes the inductive half cells and the capacitive quarter cells.

Of course, we can simplify neighboring quads with some partial elements that are connected to the same nodes in parallel. Specifically, the two inductive half cells connected to α and β in Fig. 7.6 could be combined. Again, the PEEC circuit topology for the orthogonal and the nonorthogonal cases are the same, with the exception of the circuit element values and the additional partial mutuals. Importantly, the same modified nodal analysis (MNA) circuit solver can be used for all cases.

Using the excess capacitance model in Section 10.4.6, we can also treat dielectric hexagonal blocks. The fundamental formulation is same for both situations. Similar to the orthogonal case, the dielectrics are represented with the same circuit elements. Using the nonorthogonal excess capacitance of a dielectric cell that is equivalent to (10.68) as

$$C_e = \epsilon_0(\epsilon_r - 1) \Big/ \int_a \int_b \int_c \frac{h_a^{\,2}}{\left| \frac{\partial \boldsymbol{r}}{\partial a} \cdot \left(\frac{\partial \boldsymbol{r}}{\partial b} \times \frac{\partial \boldsymbol{r}}{\partial c} \right) \right|}\, da\, db\, dc, \tag{7.24}$$

where ϵ_r is the dielectric constant of the dielectric cell. The equivalent circuit shown in Fig. 7.8 for the dielectric excess capacitance model is copied for convenience from

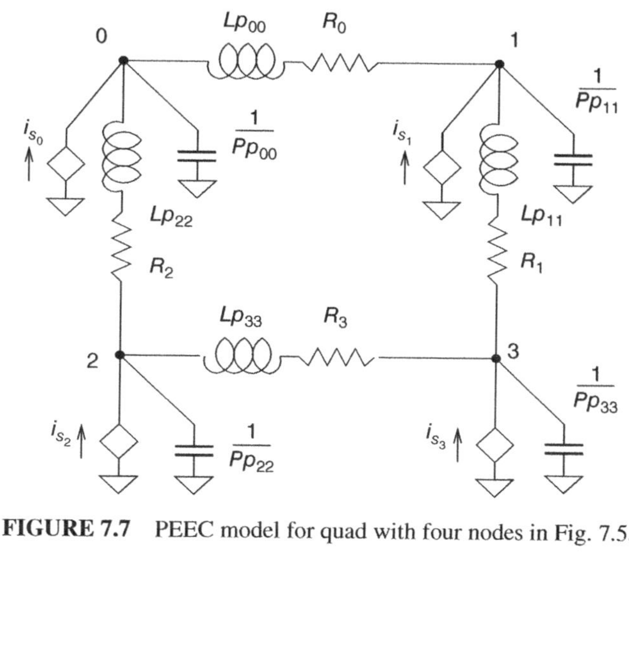

FIGURE 7.7 PEEC model for quad with four nodes in Fig. 7.5.

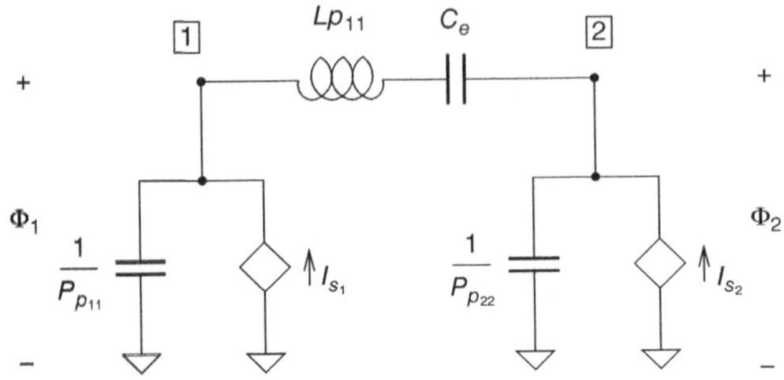

FIGURE 7.8 Model for volume capacitance from Section 10.4.6.

Fig. 10.18. Importantly, we simply can take the nonorthogonal partial inductance in (7.17) and the series resistance in (7.14). The details of the derivation are given in Section 10.4.6.

7.2.2 Continuity Equation and KCL for Nonorthogonal Geometries

The continuity equation for the rectangular PEEC models is considered in Section 6.3.1. The continuity equation must also be satisfied for the nonorthogonal case at the cell level for the currents and charges. Its differential form is given by (3.3) or $\nabla \cdot \boldsymbol{J} + \frac{\partial q}{\partial t} = 0$ where again $\boldsymbol{J}$ is the current density and q is the surface charge density. The continuity equation needs to be applied at the location of each node corresponding to Figs. 7.3–7.5.

Unlike in Fig. 6.3, for the model we only show a quarter of the area in Fig. 7.9 for which the continuity equation is applied. Here, we show only one quarter of the centre cell unlike in Fig. 6.3, which shows the cell for rectangular coordinates. Since only one quarter of the elements surrounding the node is shown, we assume that the surface element in Fig. 7.9 may be connected to other similar surfaces along the *a–c* and the *b–c* surfaces. Hence, the volume for which the continuity equation is applied consists of the corners that are involved in the geometry surrounding the node(s). It is sufficient to consider only the corner elements by ignoring the internal surfaces shown in Fig. 7.9 for simplicity.

Integrating the continuity equation over the corner yields

$$\int_{\mathcal{V}} \nabla \cdot \boldsymbol{J} d\mathcal{V} = -\frac{\partial}{\partial t} \int_{\mathcal{V}} q \, d\mathcal{V} = \int_{S} \boldsymbol{J} \cdot \hat{\boldsymbol{n}} \, dS, \tag{7.25}$$

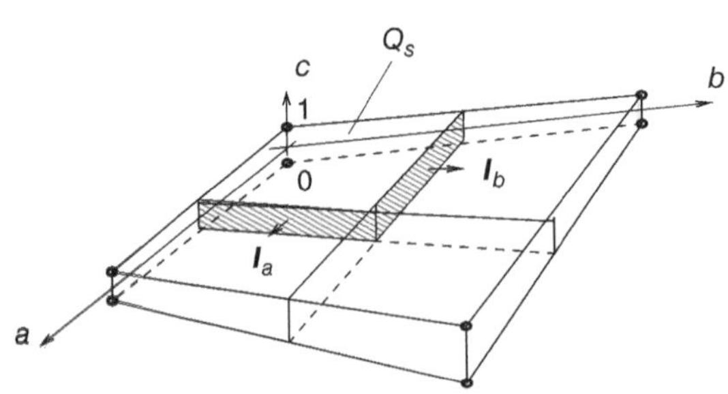

FIGURE 7.9 Corner of cell for continuity equation.

where the divergence theorem (3.33) [20] is used in the last step and the vector $\hat{\boldsymbol{n}}$ is normal to the surface S. The volume integral part pertains to the charge on the top and bottom $a-b$ surface corners connected to nodes 1 and 0, which are charged as indicated with Q_s in Fig. 7.9. If the nodes 0 and 1 are shorted, then the charge density consists of two δ-functions at the surfaces of the conductors with the surface charge $q(a,b,c)$ and the contributions at $c=\pm 1$ are

$$Q_{\pm} = \int_{a=-1}^{0}\int_{b=-1}^{0} q_{\pm}(a,b)\left|\frac{\partial \boldsymbol{r}}{\partial a}\times\frac{\partial \boldsymbol{r}}{\partial b}\right| da\, db \tag{7.26}$$

and where the surface charge can be found from (7.19) as $q_{\pm}(a,b) = Q_{\pm}\Big/\left|\frac{\partial r}{\partial a}\times\frac{\partial r}{\partial b}\right|$. Substituting this into (7.26) yields the charges $Q_{\pm}$ on the corner surfaces.

Similarly, the currents associated with the corner nodes 0,1 are flowing through the cross-sectional areas indicated by I_a in the a direction and I_b in the b direction. The current through the section of conductor cross section for I_a is given by

$$I_a = \int_{b=-1}^{0}\int_{c=-1}^{0} I_a\, w_a\, \hat{\boldsymbol{a}}\cdot\left|\frac{\partial \boldsymbol{r}}{\partial b}\times\frac{\partial \boldsymbol{r}}{\partial c}\right| db\, dc \tag{7.27}$$

Where w_a is given by (7.12).
Again, the same relationship holds for the equation for the b coordinate. Adding all terms pertaining to the continuity equation, we get for one corner only

$$\frac{dQ_+}{dt}+\frac{dQ_-}{dt}+I_a+I_b=0, \tag{7.28}$$

which can be recognized as the Kirchhoff's current law (KCL). It is evident that the continuity equation is satisfied since the admittance part of the MNA [21] formulation method is based on summing up all the current contributions at a node. Since the partial inductances extend from one node to a neighbor node, we need to subdivide the partial inductances into two halves where the part at the node is instantaneous and where the second part which belongs to a neighboring cell must include retarded partial mutual inductances.

7.3 EVALUATION OF PARTIAL ELEMENTS FOR NONORTHOGONAL PEEC CIRCUITS

For nonorthogonal geometries, the partial element evaluations are more challenging. In all cases, the singular Green's function is more difficult for the partial self-terms. Of course, the singularity can be eliminated if an analytic integration is possible.

The nearest neighbor results in coupling elements that are largest in magnitude. Therefore, these coefficients usually require more care. The magnitude sensitivity decreases with the distance between the cells for the partial elements. However, this is not the case for the radiation or phase sensitivity that is almost the same for all distances between the cells.

The partial element evaluation represents an important part of the overall solution. For this reason – and for the efficient evaluation of all elements – the so-called *multifunction* approach was devised, which is outlined in Appendix E. Issues of importance are both the accuracy and speed for the partial element evaluation. A general evaluation of partial elements for nonorthogonal structures consists of the combination of analytic and numerical techniques. Some of the formulations are presented in Appendices C and D.

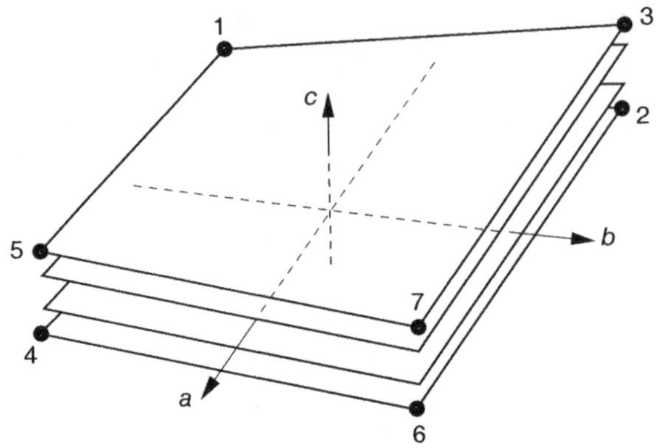

FIGURE 7.10 Example for numerical solution of c direction layered representation.

We consider an example for partial element evaluation using a mixed numerical–analytic combination. The example involves a quadrilateral cell problem where the thickness c is represented by a set of layers or sheets as shown in Fig. 7.10. In this example, the thickness of the 3D hexahedral structure is assumed to be thin compared to the other cell dimensions. Hence, the numerical integration is applied in the thin c direction as shown with the sheets in Fig. 7.10.

If we use local coordinates, the kth sheet is specified by $a \in [-1, +1]$, $b \in [-1, +1]$, and $c \in [c_{k-1}, c_k]$. The values of the $\{c_k\}$ are the values obtained from the Gaussian quadrature rule for the Lvl levels. Hence, each inductive cell is represented by four sheets as is shown in Fig. 7.10. There are thus a total of four $(Lvl) - 1$ sheets for each inductive cell shown in Fig. 7.10. The value of Lvl determines the accuracy for a given aspect ratio of the cell and the accuracy required.

For the inductance computation, the partial inductance of each sheet cell is required with respect to all the other sheet cells. Importantly, the integral (7.17) is reduced to the evaluation Nc fourfold integrals of the following form

$$Lp_{aa'} = \mu_0 \int_a \int_b \int_{a'} \int_{b'} (\hat{\boldsymbol{a}} \cdot \hat{\boldsymbol{a}}')\, h_{a'}\;\; g[\boldsymbol{r}(a, b, c), \boldsymbol{r}(a', b', c')]\, h_a\, da'\, db'\, da\, db, \quad (7.29)$$

where Nc is the product of the number of layers represented by each of the two cells. The partial inductance problem is reduced to a fourfold integral with the surface subdivisions that is easier to evaluate even if numerical integration is used.

As an example, we can use rectangular or triangular subcells to compute the self- or mutual value for two cells. The important issue is that at the PEEC level, we still want to use rectangular, quadrilateral, or hexahedral cells. Assuming that we subdivide a local coordinate sheet $Cell = a \in [-1, +1], b \in [-1, +1]$ into subcells, then we have subdivisions that are chosen to maintain a reasonable aspect ratio as well as the required accuracy. Fortunately, analytical formulas also exist for the evaluation of quasistatic triangle surface integrals such that the singularity can be avoided [24]. The integral over the smaller of the two cells can be performed numerically.

As another example, the integration in (7.17) with respect to a, b and c can be performed by using a nine-point sixth-order accurate product Gaussian quadrature rule. Once we have

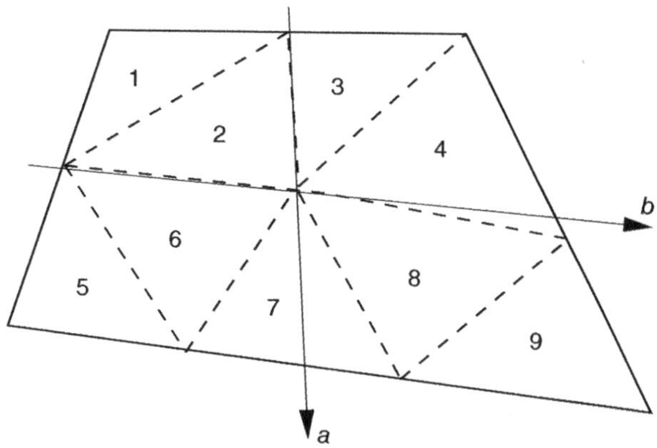

FIGURE 7.11 Example for subdividing quadrilateral sheet with triangles or rectangles for evaluation of integral for quadrilateral external shapes.

obtained an approximation S_1 to the integral (7.17) using this method, we divide each cell into twice as many c layers as before, by doubling Lvl, and repeat the procedure, obtaining another approximation, S_2. Then, we can use a second- order Richardson extrapolation to obtain our final estimate $S = (4S_2 - S_1)/3$ [22, 25].

Analytical formulas are available for rectangular shapes. Therefore, it is always desirable to use as many rectangular cells as possible. But it is also possible and actually convenient to use Gaussian quadrature for rectangles when accuracy is not critical. For example, for the far distance interaction for the static field, the Gaussian quadrature rule can achieve convenience and satisfactory accuracy without excessive compute time.

If triangles are used, most calculations are performed in the numerical regime. Gaussian quadrature rules as shown in Appendix E will be a convenient facility to handle the integrals.

The overall integral can lead to a time-consuming evaluation for too many subdivisions. Hence, we would like the number of subareas as small as possible. The compute time and accuracy are the key issues why this is such an important issue. Hence, the evaluation of the integrals is predominantly performed analytically, given the techniques in Appendices C and D. In the example in Fig. 7.11, results in a partial self (Lp_{11} or Pp_{11}) double integral, which leads to the following two integrations over the subareas

$$\int_K \int_M = \sum_{k=1}^{Nk} \sum_{m>k}^{Nm} \int_k \int_m . \tag{7.30}$$

Assuming that we want to compute a self Lp_{11} or Pp_{11} for the example in Fig. 7.11, we set $Nk = Nm = 9$. We want to point out that the practical evaluations of (7.30) can be simplified due to the symmetry of the subintegrals and other issues if some shapes are the same. If we have to compute the partial mutual Lp_{km} or Pp_{km}, we can take advantage of the symmetry of the coefficients.

We note that the subdivision of the cells into subcells allows the additional refinement of the phase by the assignment of the local retardation in the time and frequency domains. Only a small amount of additional compute time results since the subcells are located on the same cell without introducing additional element in the MNA unknown vector. This serves

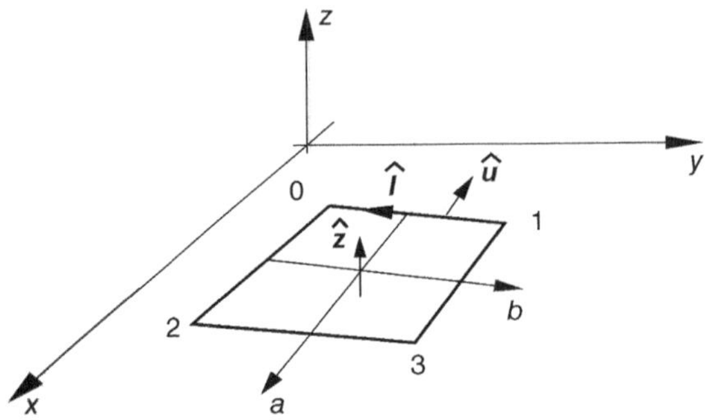

FIGURE 7.12 Orthogonal quadrilateral for partial self-potential term.

as phase refinement that is used in the passivity enhancement techniques as presented in Section 13.5.2.

The techniques presented in this chapter results in a refined solution for nonorthogonal shapes with large aspect ratio cells.

7.3.1 Analytic Solution for Quadrilateral Cells in a Plane

In this section, we present the analytical evaluation of the zero thickness integrals for quadrilateral cells located on a plane. This derivation would be too lengthy for the Appendices C and D. In addition, there are other results that are useful such as the integral over filament with the kernel R. The integrals are complicated in spite of the fact that they do not have singularities in the integrand. Remember that the original integrals have $1/R$ singularities for the self-terms. In this section, we are using global coordinates for the evaluation of the integrals since an analytic result can be obtained in global coordinates for this case.

The original work was presented in Ref. [23], whereas the new solution was led by D. Romano in a recent work. The evaluation of the partial potential coefficient using local coordinates is given by (7.24). A self-term for a quadrilateral cell is shown in Fig. 7.12 and the coupling situation for two quadrilateral cells is given in Fig. 7.13.

We observe that the key difference between the partial coefficients of potential and the partial inductances is a different multiplication factor as well as the dot-product between the current directions. We have to approximate the dot-product for partial inductances by using average current directions. However, the accuracy is reduced if the current direction changes considerably over at least one of the cells. This may require further subdivisions of the cells.

The geometry of the problem of interest is shown in Fig. 7.13 for the general case for two quadrilaterals, whereas Fig. 7.12 represents the special case for the self-term. In this computation, we have to assume that both surfaces are located in the $z = 0$ plane in the general coordinate system (x, y, z). The two zero thickness quadrilateral cells 1 and 2 are specified using the local corner coordinates 0–3, with the associated global coordinates x, y, z with the surfaces S_1 for quadrilateral 1 and S_2 for 2 corresponding to the primed coordinates. The integral to be evaluated for the zero thickness Pp or Lp is

$$I_T = \int_{S_2}\int_{S_1} \frac{1}{|\boldsymbol{r} - \boldsymbol{r}'|}\, dS\, dS' = \int_{S_2}\int_{S_1} \frac{1}{R}\, dS\, dS' \tag{7.31a}$$

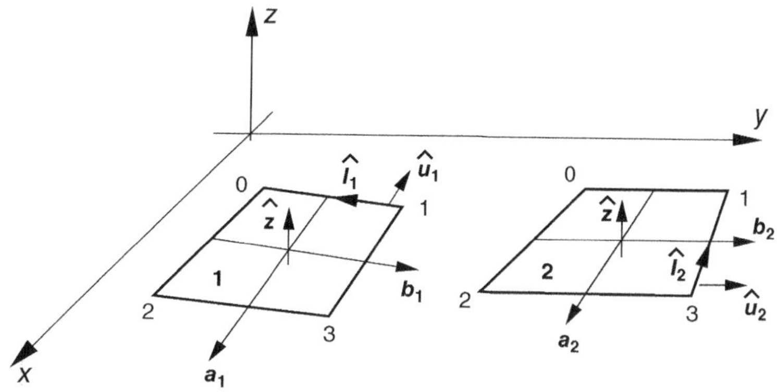

FIGURE 7.13 Orthogonal coplanar quadrilateral for partial mutual term.

with

$$R = \sqrt{(x - x')^2 + (y - y')^2}, \tag{7.31b}$$

where $\boldsymbol{r}$ and $\boldsymbol{r}'$ denote the source and observation points on S_1 and S_2. For the two generic quadrilaterals S_1 and S_2, the following identity holds [24]:

$$R^{-1} = -\nabla_{S_1} \cdot \nabla_{S_2} R, \tag{7.32}$$

where ∇_{S_1} and ∇_{S_2} are the surface operators, and where the second one operates on the primed coordinates, respectively. By means of the surface gradient and the surface divergence theorem [19], (7.31a) is transformed to

$$I_T = -\oint_{\partial S_2}\oint_{\partial S_1} R\,(\hat{\boldsymbol{u}} \cdot \hat{\boldsymbol{u}}')\; dl\; dl', \tag{7.33}$$

where ∂S_1 and ∂S_2 represent the circumferences of the quadrilateral shapes. We should repeat that they are in the $z = 0$ plane. Also, $\hat{\boldsymbol{u}}$ and $\hat{\boldsymbol{u}}'$ are the outward vectors normal to ∂S_1 and ∂S_2 as shown in Figs 7.12 and 7.13. They are related to the unit vector tangential to the quadrilateral sides $\hat{\boldsymbol{l}}$ and $\hat{\boldsymbol{l}}'$ and they are normal to their surfaces unit vector $\hat{z}$, by means of

$$\hat{\boldsymbol{u}} = \hat{\boldsymbol{l}} \times \hat{z} \tag{7.34a}$$

$$\hat{\boldsymbol{u}}' = \hat{\boldsymbol{l}}' \times \hat{z}'. \tag{7.34b}$$

Application of (7.33) to S_1 and S_2 gives a 16-term sum of line integrals

$$I_T = -\sum_{i=1}^{4}\sum_{j=1}^{4} \int_{l_i'}\int_{l_j} R\,(\hat{\boldsymbol{u}}_j \cdot \hat{\boldsymbol{u}}_i')\; dl_j\; dl_i'. \tag{7.35}$$

Since the sides of the quadrilaterals are straight lines $\hat{\boldsymbol{u}} \cdot \hat{\boldsymbol{u}}' = k$ for each pair of sides l and l', then each term of (7.35) becomes

$$I_l = k \int_{l'} \int_l R \; dl \; dl' = k \, I_p. \tag{7.36}$$

The analytical integration of integrals I_p depends on the configuration of two sides l and l'.

If the two sides are perpendicular where $k = \hat{\boldsymbol{u}} \cdot \hat{\boldsymbol{u}}' = 0$, then the computation of I_p is zero. Hence, removing the evaluation saves compute time. Of course, the x and y axes are exchanged if the integration is set up in the wrong direction. The analytical evaluation of I_p is considered in the following section for the general case as well as the special case when two sides coincide is considered.

7.3.2 General Case for Evaluation of Integral I_p

It is clear from the previous section that to evaluate I_T, we have to solve the integrals I_p between two segments l and l' as shown in Fig. 7.14. Hence, the integrals involve two lines l with the end points $P_1(x_1, y_1)$, $P_2(x_2, y_2)$ and l' with the end points $P_3(x_3, y_3)$ and $P_4(x_4, y_4)$, which can substitute from (7.35) in the global reference system (x, y, z) for l as

$$\begin{aligned} y &= mx + q \\ m &= \frac{y_2 - y_1}{x_2 - x_1}, \quad q = y_1 - mx_1, \quad dl = \sqrt{1 + m^2} dx, \end{aligned} \tag{7.37}$$

and for l' as

$$\begin{aligned} y' &= m'x' + q' \\ m' &= \frac{y_4 - y_3}{x_4 - x_3}, \quad q' = y_3 - m'x_3, \quad dl' = \sqrt{1 + m'^2} dx'. \end{aligned} \tag{7.38}$$

Substituting (7.31b), (7.37), and (7.38) in (7.36) one obtains

$$I_p = \int_{l'} \int_l \sqrt{(x - x')^2 + (y - y')^2} \; dl \; dl'. \tag{7.39}$$

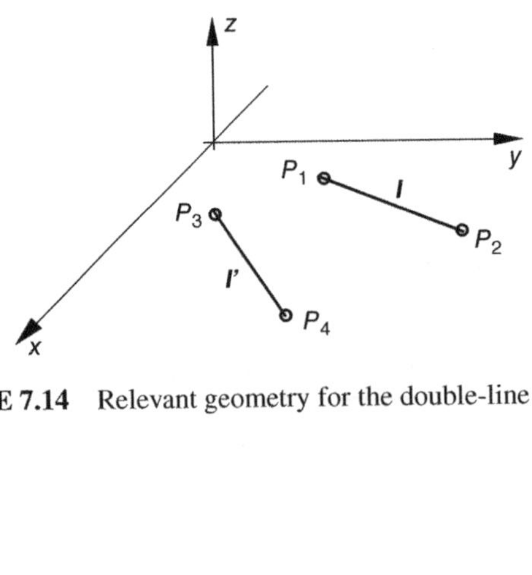

FIGURE 7.14 Relevant geometry for the double-line integrals.

In (7.39), the internal integral is solved by using the formula (7.56a)

$$I = \int_l \sqrt{(x-x')^2 + (y-y')^2}\, dl = \sqrt{1+m^2} \int_{x_1}^{x_2} \sqrt{cx^2+bx+c}\; dx$$

$$= \sqrt{1+m^2} \left[\frac{(2cx_2+b)}{4c} \sqrt{cx_2^2+bx_2+a} \right.$$

$$+ \frac{\Delta}{8c\sqrt{c}} \ln \left(\frac{b+2cx_2}{2\sqrt{c}} + \sqrt{cx_2^2+bx_2+a} \right)$$

$$\left. - \frac{(2cx_1+b)}{4c} \sqrt{cx_1^2+bx_1+a} - \frac{\Delta}{8c\sqrt{c}} \ln \left(\frac{b+2cx_1}{2\sqrt{c}} + \sqrt{cx_1^2+bx_1+a} \right) \right], \quad (7.40)$$

where it is defined that

$$\Delta = 4ac - b^2$$
$$a = x'^2 + q^2 + y'^2 - 2qy'$$
$$b = 2(mq - x' - my')$$
$$c = 1 + m^2. \quad (7.41)$$

Substituting (7.40) in (7.39) one has

$$I_p = \int_{l'} I \quad dl' = \sqrt{1+m^2} \left[(I_1 + I_2) - (I_3 + I_4) \right]. \quad (7.42)$$

The integrals I_i for $i = 1, \ldots, 4$ on the right-hand side of (7.42) refer to the four terms shown in (7.40). These four integrals are evaluated next.

For I_1 which is

$$I_1 = \int_{l'} \frac{(2cx_2+b)}{4c} \sqrt{cx_2^2+bx_2+a}\; dl' \quad (7.43)$$

by substituting in the first and the last equations of (7.37)

$$I_1 = \int_{l'} \frac{(2cx_2+b)}{4c} \sqrt{cx_2^2+bx_2+a}\; dl'$$

$$= \sqrt{1+m'^2} \int_{x_3}^{x_4} \frac{(2cx_2+b)}{4c} \sqrt{cx_2^2+bx_2+a}\; dx'$$

$$= \sqrt{1+m'^2} \left[K_1 \int_{x_3}^{x_4} \sqrt{Cx'^2+Bx'+A}\; dx' - K_2 \int_{x_3}^{x_4} x' \sqrt{Cx'^2+Bx'+A}\; dx' \right], \quad (7.44)$$

where

$$\begin{aligned}
A &= (1+m^2)x_2^2 - 2m(q'-q)x_2 + (q'-q)^2 \\
B &= 2[m'(q'-q) - (1+mm')x_2] \\
C &= 1 + m'^2 \\
K_1 &= \frac{x_2}{2} - \frac{m(q'-q)}{2(1+m^2)} \\
K_2 &= \frac{1+mm'}{2(1+m^2)}.
\end{aligned} \tag{7.45}$$

The integrals in (7.44) are solved by using the formulas (7.56 a) and (7.56 b). Following the same steps, an expression for I_3 is obtained, which is the same as I_1 in (7.44). The only difference is in the expressions for the coefficients A, B, and K_1, which are equal to the ones in (7.45) if x_2 is substituted by x_1.

For I_2

$$I_2 = \int_{l'} \frac{\Delta}{8c\sqrt{c}} \ln\left(\frac{b+2cx_2}{2\sqrt{c}} + \sqrt{cx_2^2 + bx_2 + a}\right) dl' \tag{7.46}$$

by substituting in the first and the last equations of (7.37)

$$\begin{aligned}
I_2 &= \int_{l'} \frac{\Delta}{8c\sqrt{c}} \ln\left(\frac{b+2cx_2}{2\sqrt{c}} + \sqrt{cx_2^2 + bx_2 + a}\right) dl' \\
&= \sqrt{1+m'^2} \int_{x_3}^{x_4} \frac{\Delta}{8c\sqrt{c}} \ln\left(\frac{b+2cx_2}{2\sqrt{c}} + \sqrt{cx_2^2 + bx_2 + a}\right) dx' \\
&= \sqrt{1+m'^2} \left[D \int_{x_3}^{x_4} x'^2 \ln\left(Gx' + H + \sqrt{Cx'^2 + Bx' + A}\right) dx' \right. \\
&\quad + E \int_{x_3}^{x_4} x' \ln\left(Gx' + H + \sqrt{Cx'^2 + Bx' + A}\right) dx' \\
&\quad \left. + F \int_{x_3}^{x_4} \ln\left(Gx' + H + \sqrt{Cx'^2 + Bx' + A}\right) dx' \right]
\end{aligned} \tag{7.47}$$

where

$$\begin{aligned}
G &= -\frac{mm'+1}{\sqrt{1+m^2}} \\
H &= \frac{(1+m^2)x_2 - m(q'-q)}{\sqrt{1+m^2}}
\end{aligned}$$

$$D = \frac{(m' - m)^2}{2(1 + m^2)^{3/2}}$$

$$E = \frac{(m' - m)(q' - q)}{(1 + m^2)^{3/2}}$$

$$F = \frac{(q' - q)^2}{2(1 + m^2)^{3/2}}. \tag{7.48}$$

In the above expressions, the coefficients A, B, and C are given in (7.45). In order to further simplify the analytical solutions of the integrals of (7.47), it is possible to set $w' = Gx' + H$ that can be written as

$$\int_{x_3}^{x_4} \ln\left(Gx' + H + \sqrt{Cx'^2 + Bx' + A}\right) dx'$$
$$= \frac{1}{G} \int_{inf}^{sup} \ln\left(w' + \sqrt{Sw'^2 + Tw' + U}\right) dw' = \frac{1}{G} I_F(w', S, T, U)\Big|_{inf}^{sup} \tag{7.49a}$$

$$\int_{x_3}^{x_4} x' \ln\left(Gx' + H + \sqrt{Cx'^2 + Bx' + A}\right) dx'$$
$$= \frac{1}{G^2}\left[\int_{inf}^{sup} w' \ln\left(w' + \sqrt{Sw'^2 + Tw' + U}\right) dw' - HI_F(w', S, T, U)\Big|_{inf}^{sup}\right]$$
$$= \frac{1}{G^2}\left[I_E(w', S, T, U)\Big|_{inf}^{sup} - HI_F(w', S, T, U)\Big|_{inf}^{sup}\right] \tag{7.49b}$$

$$\int_{x_3}^{x_4} x'^2 \ln\left(Gx' + H + \sqrt{Cx'^2 + Bx' + A}\right) dx'$$
$$= \frac{1}{G^3}\left[\int_{inf}^{sup} w'^2 \ln\left(w' + \sqrt{Sw'^2 + Tw' + U}\right) dw'\right.$$
$$\left. + H^2 I_F(w', S, T, U)\Big|_{inf}^{sup} - 2HI_E(w', S, T, U)\Big|_{inf}^{sup}\right]$$
$$= \frac{1}{G^3}\left[I_D(w', S, T, U)\Big|_{inf}^{sup} + H^2 I_F(w', S, T, U)\Big|_{inf}^{sup} - 2HI_E(w', S, T, U)\Big|_{inf}^{sup}\right] \tag{7.49c}$$

with

$$inf = Gx_3 + H$$

$$sup = Gx_4 + H$$

$$S = C/G^2$$

$$T = B/G - 2CH/G^2$$
$$U = A - BH/G + CH^2/G^2. \quad (7.50)$$

We should exclude from the solution for $G = 0$, which results for the case when the two segments are perpendicular. For this case, the evaluation of I_p is eliminated as mentioned in Section 7.3.2. The analytical solution for the integrals I_D, I_E, and I_F is given in (7.56) and (7.58).

Following the same steps, the expression for I_4 is obtained: it is completely equal to that for I_2 in (7.47). The only difference is in the expressions of the coefficients A, B, and H that are equal to those in (7.45) and (7.48) with x_2 substituted by x.

7.3.3 Evaluation of Integral I_p When Two Sides l Coincide

For two sides l coincident, equation (7.38) can be written as

$$l' : y' = mx' + q$$
$$m = \frac{y_2 - y_1}{x_2 - x_1}, \quad q = y_1 - mx_1, \quad dl' = \sqrt{1 + m^2} \ \ dx'. \quad (7.51)$$

Hence, the coefficients a, b, and c in equation (7.41) can be written as

$$a = (1 + m)^2 x'^2$$
$$b = -2(1 + m^2)x'$$
$$c = 1 + m^2. \quad (7.52)$$

In this case, $\Delta = 4ac - b^2 = 0 \, \forall x'$. This implies that equation (7.40) turns into

$$I = \int_l \sqrt{(x - x')^2 + (y - y')^2 dl} = \sqrt{1 + m^2} \int_{x_1}^{x_2} \sqrt{cx^2 + bx + c dx} \quad (7.53)$$
$$= \sqrt{1 + m^2} \left[\frac{(2cx_2 + b)}{4c} \sqrt{cx_2^2 + bx_2 + a} - \frac{(2cx_1 + b)}{4c} \sqrt{cx_1^2 + bx_1 + a} \right].$$

Hence, equation (7.42) turns into

$$I_p = \int_{l'} I \ dl' = \sqrt{1 + m^2} \left[I_1 - I_3 \right]. \quad (7.54)$$

The expression of I_1 is the same as that of equation (7.43) with

$$A = (1 + m^2)x_2^2$$
$$B = -2(1 + m^2)x_2$$

$$C = 1 + m^2$$
$$K_1 = \frac{x_2}{2}$$
$$K_2 = \frac{1}{2}. \tag{7.55}$$

The solution of the integrals required is given next. If we define $\boldsymbol{g(x) = cx^2 + bx + a}$, then the first and the second integrals are with the solution given as

$$\int \sqrt{\boldsymbol{g(x)}}\, \boldsymbol{dx} = \frac{b + 2cx}{4c}\sqrt{g(x)} + \frac{4ac - b^2}{8c\sqrt{c}} \ln\left(\frac{b + 2cx}{2\sqrt{c}} + \sqrt{g(x)}\right) \tag{7.56a}$$

$$\int \boldsymbol{x}\sqrt{\boldsymbol{g(x)}}\, \boldsymbol{dx} = \left[\frac{cx^2 + bx + a}{3} - \frac{bx}{4} - \frac{b^2}{8c}\right] \frac{\sqrt{g(x)}}{c}$$
$$- \frac{(4ac - b^2)b}{16c^2\sqrt{c}} \ln\left(\frac{b + 2cx}{2\sqrt{c}} + \sqrt{g(x)}\right) \tag{7.56b}$$

$$\int \ln(\boldsymbol{x} + \sqrt{\boldsymbol{g(x)}})\, \boldsymbol{dx} = [\ln(x + \sqrt{g(x)}) - 1]x$$
$$+ \frac{b}{2f}\ln(g(x) - x^2) + \frac{\sqrt{-p}}{f}\tan^{-1}\left(\frac{b + fx}{\sqrt{-p}}\right) - \frac{b}{f\sqrt{c}}\ln(b + 2cx + 2\sqrt{cg(x)})$$
$$+ \frac{h}{r}\ln\left(\frac{b + fx + \sqrt{p}}{s(x) - \sqrt{8g(x)(b^2 + b\sqrt{p} - v/2)}}\right)$$
$$+ \frac{e}{q}\ln\left(\frac{-b - fx + \sqrt{p}}{t(x) + \sqrt{8g(x)(b^2 - b\sqrt{p} - v/2)}}\right) \tag{7.57a}$$

$$\int \boldsymbol{x}\ln(\boldsymbol{x} + \sqrt{\boldsymbol{g(x)}})\, \boldsymbol{dx} = \frac{1}{2}\left[\ln(x + \sqrt{g(x)}) - \frac{1}{2}\right]x^2$$
$$- \frac{b\sqrt{g(x)}}{2cf} - \frac{b^2 - v/2}{2f^2}\ln(g(x) - x^2) - \frac{b\sqrt{-p}}{f^2}\tan^{-1}\left(\frac{b + fx}{\sqrt{-p}}\right) + \frac{bx}{2f}$$
$$+ \frac{(3cb^2 - b^2 - vc)}{2f^2c^{3/2}}\ln(b + 2cx + 2\sqrt{cg(x)})$$
$$+ \frac{h'}{r'}\ln\left(\frac{s(x) - \sqrt{8g(x)(b^2 + b\sqrt{p} - v/2)}}{b + fx + \sqrt{p}}\right)$$

$$+\frac{e'}{q'}\ln\left(\frac{t(x)+\sqrt{8g(x)(b^2-b\sqrt{p}-v/2)}}{-b-fx+\sqrt{p}}\right) \tag{7.57b}$$

$$\int x^2\ln(x+\sqrt{g(x)})\,dx=\frac{1}{3}\left[\ln(x+\sqrt{g(x)})-\frac{1}{3}\right]x^3$$

$$-\frac{u(x)\sqrt{g(x)}}{6f^2c^2}+\frac{2b^3-(3/2)vb}{3f^3}\ln(g(x)-x^2)$$

$$-\frac{4b^4-5vb^2+v^2}{3f^3\sqrt{-p}}\tan^{-1}\left(\frac{b+fx}{\sqrt{-p}}\right)$$

$$+\frac{bw}{6f^3c^{5/2}}\ln(b+2cx+2\sqrt{cg(x)})+\frac{bx^2}{6f}+\frac{2x(v/2-b^2)}{3f^2}$$

$$+\frac{h''}{r''}\ln\left(\frac{b+fx+\sqrt{p}}{s(x)-\sqrt{8g(x)(b^2+b\sqrt{p}-v/2)}}\right)$$

$$+\frac{e''}{q''}\ln\left(\frac{-b-fx+\sqrt{p}}{t(x)+\sqrt{8g(x)(b^2-b\sqrt{p}-v/2)}}\right), \tag{7.57c}$$

where the definitions used are

$$f=2(c-1)$$

$$v=2fa$$

$$p=b^2-v$$

$$h=b^3+b^2\sqrt{p}-bv-\sqrt{p}\,v/2$$

$$e=-b^3+b^2\sqrt{p}+bv-\sqrt{p}\,v/2$$

$$r=f\sqrt{2p}\sqrt{b^2+b\sqrt{p}-v/2}$$

$$q=f\sqrt{2p}\sqrt{b^2-b\sqrt{p}-v/2}$$

$$s(x)=b^2+2bx+b\sqrt{p}-v+2cx\sqrt{p}$$

$$t(x)=-b^2-2bx+b\sqrt{p}+v+2cx\sqrt{p}$$

$$h'=\sqrt{2}[b^4+b^3\sqrt{p}-(5/4)vb^2-(3/4)vb\sqrt{p}+v^2/4]$$

$$
\begin{aligned}
e' &= \sqrt{2}[-b^4 + b^3\sqrt{p} + (5/4)vb^2 - (3/4)vb\sqrt{p} - v^2/4] \\
r' &= 2f^2\sqrt{p}\sqrt{b^2 + b\sqrt{p} - v/2} \\
q' &= 2f^2\sqrt{p}\sqrt{b^2 - b\sqrt{p} - v/2} \\
u(x) &= b^2(3 - 7c) + fcbx + 2vc \\
w &= b^2(-15c^2 + 10c - 3) + 12ca(3c^2 - 4c + 1) \\
h'' &= 2\sqrt{2}[b^5 + b^4\sqrt{p} - (3/2)vb^3 - vb^2\sqrt{p} + bv^2/2 + \sqrt{p}\,v^2/8] \\
e'' &= 2\sqrt{2}[-b^5 + b^4\sqrt{p} + (3/2)vb^3 - vb^2\sqrt{p} - bv^2/2 + \sqrt{p}\,v^2/8] \\
r'' &= 3f^3\sqrt{p}\sqrt{b^2 + b\sqrt{p} - v/2} \\
q'' &= 3f^3\sqrt{p}\sqrt{b^2 - b\sqrt{p} - v/2}.
\end{aligned} \tag{7.58}
$$

It is clear that the analytic solutions are not trivial to implement. However, it was found that they are much faster for closely located conductors. Also, the singularity for self-terms is automatically taken care of. The above equation is best used for very close cells while approximate solutions are more efficient for cells which are further apart.

PROBLEMS

7.1 *Conversion of coordinates*
Coordinate conversion is required for nonorthogonal, global coordinates to a quadrilateral element shown in Fig. 7.3. Show that points (x, y, z) on the quadrilateral element can be converted into a point in the local a, b by the formulation in Section 7.1 to a point on the surfaces of the quadrilateral shape. We note that the quadrilateral surface is bounded by the cornerpoints points 0 to 3. Show that the points (a, b) in the $-1, +1$ range correspond to global coordinate points (x, y, z) by using equation (7.3).

7.2 *Continuity equation*
Explain the function of the continuity equation in relation to the equivalent circuit in Fig. 7.7. Show that equation (7.27) is based on the continuity equation for the nonorthogonal geometry shown.

7.3 *Resistance for nonorthogonal geometries*
Prove equation (7.14) for the resistance of the nonorthogonal geometries. Also, we don't consider possible resistive couplings due to the nonorthogonal overlap of resistive cells. Find a formula for the resistive coupling between overlapping cells, which is an extension of (7.14). An example for resistive coupling in Fig. 7.3 for two overlapping inductance half cells which are connected between nodes 0 and 2 and the other one is connected between nodes 0 and 1. Of course the quadrilateral is not rectangular.

7.4 *PEEC inductance circuit model for coupled current strips*
This problem is very similar to the first problem in Chapter 6. However, the conductor shape is trapezoidal rather than rectangular. The thickness $T = 1$ mm and the corner point is $X_c = y_c = 30$ mm and the width $w = 10$ mm for the trapezoidal strip in Fig. 7.15. The same way as in the first problem in Chapter 6, split the strip in the middle into two pieces that are connected at the ends. These two half-cell pieces are coupled to each other. Build an inductance–resistance (Lp, R) PEEC model for the strip two partial. Make an R, L equivalent circuit for the problem. Is the model an *open*-loop model?

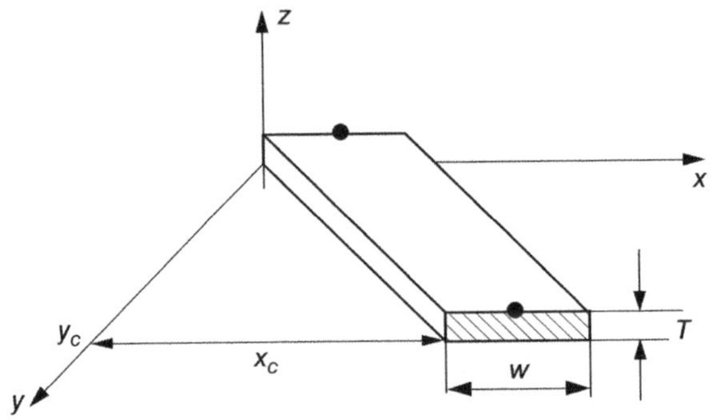

FIGURE 7.15 Orthogonal quadrilateral for partial self-potential term.

REFERENCES

1. P. K. Wolff and A. E. Ruehli. Inductance computations for complex three dimensional geometries. In *Proceedings of the IEEE International Symposium on Circuits and Systems*, pp. 16–19, 1981.
2. R. F. Milsom, K. J. Scott, G. Clark, J. C. McEntegart, S. Ahmed, and F. N. Soper. FACET – a CAE system for RF analogue simulation including layout. In *Proceedings of the Design Automation Conference*, Volume 26, pp. 622–625, 1989.
3. R.-B. Wu, C.-N. Kuo, and K. K. Chang. Inductance and resistance computations for three-dimensional multiconductor interconnection structures. *IEEE Transactions on Microwave Theory and Techniques*, MTT-40(2):263–270, February 1992.
4. M. Kamon, M. J. Tsuk, and J. White. FastHenry: a multipole-accelerated 3-d inductance extraction program. In *Proceedings of the Design Automation Conference*, pp. 678–683, June 1993.
5. Y. Wang, V. Jandjayla, and C.-J. R. Shi. Coupled electromagnetic-circuit simulation of arbitrary-shaped conducting structures. In *Digest of Electrical Performance of Electronic Packaging*, Volume 10, Boston, MA, pp. 233–236, October 2001.
6. A. Rong and A. C. Cangellaris. Generalized PEEC models for three-dimensional interconnect structures and integrated passives of arbitrary shapes. In *Digest of Electrical Performance of Electronic Packaging*, Volume 10, Boston, MA, pp. 225–228, October 2001.
7. M. Tröscher and H. Katzier. Efficient PEEC modeling of multi-layer boards and multi-chip modules. In *Proceedings International Zurich Symposium on EMC*, Volume 14, Zürich, Switzerland, pp. 269–273, February 2001.
8. A. E. Ruehli, G. Antonini, and A. Orlandi. Extension of the partial element equivalent circuit method to non-rectangular geometries. In *Proceedings of the IEEE International Symposium on Electromagnetic Compatibility*, Seattle, WA, pp. 728–733, August 1999.

9. G. Antonini, A. Ruehli, and J. Esch. Non orthogonal PEEC formulation for time and frequency domain modeling. In *Proceedings of the IEEE International Symposium on Electromagnetic Compatibility*, Minneapolis, MN, August 2002.
10. A. E. Ruehli, G. Antonini, J. Esch, J. Ekman, A. Mayo, and A. Orlandi. Non-orthogonal PEEC formulation for time and frequency domain EM and circuit modeling. In *Proceedings of the IEEE International Symposium on Electromagnetic Compatibility*, Volume 45, pp. 167–176, May 2003.
11. B. M. Kolundzija and B. D. Popovic. Entire-domain Galerkin method for analysis of metallic antennas and scatterers. *IEE Proceedings H Microwaves, Antennas and Propagation*, 140(1):1–10, January 1993.
12. H. Singer, H.-D. Brüns, and G. Bürger. State of the art in the moment method. In *Proceedings of the IEEE International Symposium on Electromagnetic Compatibility*, Santa Clara, CA, pp. 122–227, August 1996.
13. G. Antonini, S. Cristina, and A. Orlandi. PEEC modeling of high voltage tower under direct and nearby lightning strike. In *Proceedings of the 10th International Symposium on High Voltage Engineering ISH*, Volume 10, Montreal, CA, August 1997.
14. K.-B. Wu, G.-H. Shiue, W.-D. Guo, C.-M. Lin, and R.-B. Wu. Delaunay-Voronoi modeling of power-ground planes with source port corrections. *IEEE Transactions on Advanced Packaging*, 31(2):303–310, May 2008.
15. H. Shi, J. Fan, and J. Drewniak. Modeling multilayered PCB power-bus designs using an MPIE based circuit extraction technique. In *Proceedings of the IEEE International Symposium on Electromagnetic Compatibility*, Denver, CO, pp. 647–651, August 1998.
16. J. F. Fan, H. Shi, A. Orlandi, J. L. Knighten, and J. L. Drewniak. Modeling DC power-bus structures with vertical discontinuities using a circuit extraction approach based on a mixed-potential integral equation formulation. *IEEE Transactions on Advanced Packaging*, 24:143–157, May 2001.
17. A. E. Ruehli, G. Papadopoulos, E. Michielssen, A. Cangellaris, and K. Aygun. Transient analysis for printed circuit board problem using two different solvers. In *Proceedings of the IEEE International Symposium on Electromagnetic Compatibility*, Montreal, CA, pp. 1327–1332, August 2001.
18. B. Archambeault and A. E. Ruehli. Analysis of power/ground-plane EMI decoupling performance using the partial-element equivalent circuit technique. *IEEE Transactions on Electromagnetic Compatibility*, 43(4):437–445, November 2001.
19. C.-T. Tai. *Generalized Vector and Dyadic Analysis*. IEEE Press, New York, 1992.
20. S. Ramo, J. R. Whinnery, and T. Van Duzer. *Fields and Waves in Communication Electronics*. John Wiley and Sons, Inc., New York, 1994.
21. C. Ho, A. Ruehli, and P. Brennan. The modified nodal approach to network analysis. *IEEE Transactions on Circuits and Systems*, 22(6):504–509, June 1975.
22. A. E. Ruehli. Equivalent circuit models for three dimensional multiconductor systems. *IEEE Transactions on Microwave Theory and Techniques*, MTT-22(3):216–221, March 1974.
23. G. Antonini, A. Orlandi, and A. Ruehli. Analytical integration of quasi-static potential integrals on non-orthogonal coplanar quadrilaterals for the PEEC method. *IEEE Transactions on Electromagnetic Compatibility*, 44(2):399–403, May 2002.
24. P. Arcioni, M. Bressan, and L. Perregrini. On the evaluation of the double surface integrals arising in the application of the boundary integral method to 3-d problems. *IEEE Transactions on Microwave Theory and Techniques*, 45(3):436–439, March 1997.
25. W. Gander, M. Gander, and M. Kwok. *Scientific Computing, An introduction using Maple and MATLAB*, Springer-Verlag, Berlin and New York, 2014.

8

GEOMETRICAL DESCRIPTION AND MESHING

In this chapter, we present an overview of issues for the meshing for partial element equivalent circuit (PEEC) models. We specifically consider applications in the electromagnetic modeling (EMM), electromagnetic compatibility (EMC) areas such as interconnect modeling for power and signal integrity (SI) as well as power engineering area models. Hence, these structures include different complex mixed circuit and electromagnetic (EM) aspects. The EM structures are in general very heterogeneous and they may be connected to electronic circuits that are described in the circuit domain.

Early on in the history of PEEC models, simplified nonorthogonal geometries were modeled in terms of rectangular bars [1–4]. While the use of rectangular bars is very efficient, it also limits the geometries that can be represented with these shapes. Hence, the consistent inclusion of nonorthogonal shapes with the rectangular ones is important and evolving. In general, we need to consider a multitude of shapes for the applications of interest. Rectangular blocks are used for some parts such as printed circuit boards (PCBs) and transmission line conductors. Different round conductors are used for connection wire models of different shapes. Very thin conductors are modeled with lines or quadrilateral cells or other approximations, and volumes are represented with hexahedral models and if possible rectangular block shapes.

The solution using the Rao–Wilton–Glisson (RWG) [5] triangles is suitable for special applications. These techniques use a large number of small computationally efficient triangular cell basis functions [6] such as they are used in some integral equation modeling methods [7]. For these techniques, a larger number of smaller cells is used and the computation per cell is minimized while the number of unknowns is increased. It is clear that the use of triangular-shaped cells is a good approach for a relatively homogeneously structured high-frequency problems such as airplanes [6]. Triangles are also used in some PEEC codes, for example, Refs [9–11].

Circuit Oriented Electromagnetic Modeling Using the PEEC Techniques, First Edition.
Albert E. Ruehli, Giulio Antonini, and Lijun Jiang.

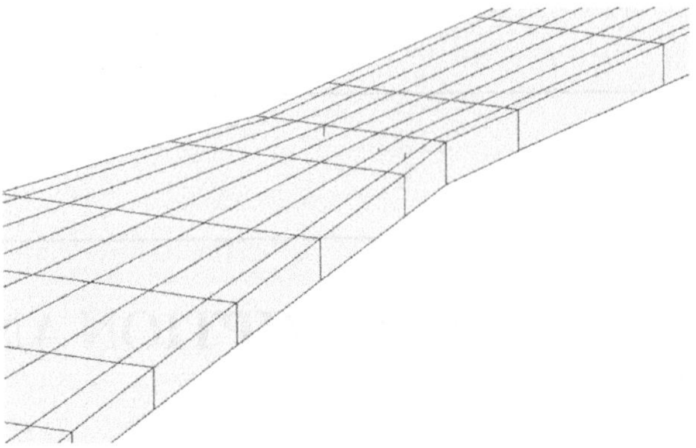

FIGURE 8.1 A wire section that is meshed with long, shallow angle cells.

In the version of PEEC considered in this text, we avoid the use of triangular shapes. Three of the advantages are better meshing of large aspect ratio cells, the reduced number of basis functions required, and the ease of representing shallow angle structures as shown in Fig. 8.1. Another issue is the elimination of an entire set of coupling integrals and all the triangle-to-rectangle coupled elements that are needed for mixed rectangular and triangular elements. Of course, this is due to the partial or totally analytic partial elements that we prefer to use for PEEC models.

8.1 GENERAL ASPECTS OF PEEC MODEL MESHING REQUIREMENTS

Today, a wide range of different meshing techniques are employed in the general area of integral equation solution techniques. An interesting set of applications consists of the modeling of antenna type structures for a multitude of shapes [12]. These applications often involve computations for a single frequency or for a limited range of frequencies. Also, these types of geometries have fewer closely spaced overlapping areas. We consider closely spaced surfaces or bodies that are in general more difficult to mesh in Section 8.4.2.

It is important to understand details of the meshing requirements for the different PEEC classes of problem. Since, overall efficiency can be gained by meshing, it is desirable to tailor the meshing part to the solution.

A simplifying issue is that we mostly address signal integrity (SI), power integrity (PI), and noise integrity (NI) problems. Usually, the accuracy of a SPICE circuit solver is in the range of 10^{-2} to 10^{-3}. This is sufficient for this class of problems. For this reason, we tailor our PEEC EM solutions in this accuracy range. One of these accuracy choices was that we considered that all cells had planar surfaces [13].

One of the advantages of this approach is that we can employ analytic integrations for the partial elements, which yields more flexibility. For SI/PI/NI problems, the difference in the size of geometrical objects involved can be very large. This requires additional care in choosing approximate meshes. The use of different size meshes for conductors and dielectrics needs more intelligence built into the mesher.

As shown in Section 4.4.1, the choice of the cell segmentation has a strong influence on the accuracy for closely spaced conductors as well as other objects. Hence, the meshing approach must include special techniques for closely spaced conductors or dielectrics.

A technique called projection meshing must be employed. The inaccuracies introduced by close surfaces that are not meshed with projection methods were first presented in Ref. [14], and is also considered in Section 4.4.1 as well as in Refs [15–17] and [18]. Details are again considered in Section 8.4.

Nonorthogonal shapes are represented by quadrilateral surface and hexahedral three-dimensional elements. We found this to be very effective for the approximation of the currents and surface charges. Quadrilateral surface cells have been used successfully for EM modeling using integral equation solution approach, for example, Refs [19, 20].

A smart mesher can save a large number of unknowns in part due to the large cell aspect ratios that can be tolerated in PEEC due to the analytic partial elements used. This result is accomplished by the additional work, which allows much larger aspect ratios for the cells generated by the mesher. As mentioned above, triangles [5] are not used due to their sensitivity with respect to the large aspect ratios. As is shown in Fig. 8.19 we can always subdivide triangles into quadrilateral shapes.

The parallel development of the mesher and the PEEC solver is very desirable since the mesher needs to have information about the capabilities and limitations of the solver. This implies that even at the mesher level, we have to understand more about the problems and the solution details. Some important aspects are the inclusion of both wide-band skin-effect and dielectric loss models. Another important issue is minimizing the local complexity of the meshing while preserving sufficient overall accuracy. This is somewhat different from the approach in finite element solvers where the accuracy is improved iteratively during the solution.

There are two fundamental issues that determine the size of the cells used to represent the geometry. For high-frequency or fast time domain solutions, a maximum dimension of a cell is given by λ/F where λ is the shortest wavelength in the spectrum and $F = 10, \ldots, 30$, where $\lambda/20$ is leading generally to a good solution, at least for high-frequency problems. Hence, this may limit the maximum dimension of all cells.

Second, especially for heterogeneous geometries, the sufficiently accurate representation of a current path may require even smaller cells compared to the wavelength. In the absence of other close conductors, the number of cells on conductors will be determined by the current flow or the largest frequency involved. The current flow is determined by the connections at the nodes and the required accuracy. A good aspect of the approach is that results with a reasonable accuracy can be obtained with careful meshing. It is obvious that for these cases, the use of large nonuniform cell size differences can reduce the number of cells or unknowns required in a problem.

8.2 OUTLINE OF SOME MESHING TECHNIQUES AVAILABLE TODAY

Meshing has become a separate area of research. Many different texts are available dedicated specifically to the general subject, for example, Refs [13, 21, 22]. The theory and practice of meshing are sophisticated areas. A multitude of different techniques exist today. It is useful to consider the fundamental framework for meshing, which is considered in Ref. [22], since it is relevant for our applications. It divides the techniques into three basic groups, which are outlined next.

Algebraic methods: This class of methods can be very widely interpreted. It essentially consists of meshes that are obtained with algebraic mappings. A large class of these methods are related to the nonorthogonal coordinates considered in Chapter 7. This approach is very flexible since the approach includes many different heuristic mapping methods.

PDE-based methods: In this approach, simplified partial differential equations are solved. For example, solution to Laplace's equation is used for meshing details inside the region of interest. Often, we see approaches that may be in part algebraic and also based on partial differential equation (PDE) solutions.

Multiblock methods: Multiblock methods break up the overall region or problem into subproblems that are solved somewhat separately. Of course, they have to match at the boundaries. These methods are of importance for our class of problems. They often consist of different elements that are best considered as separately meshed areas. Ultimately, the matching of the different parts has to be consistent. Some of these issues are considered in this chapter.

8.2.1 Meshing Example for Rectangular Block

First, we give a fundamental example of a PEEC mesh. It is best to start from the most simple case, which is a thin, rectangular sheet as we considered in the previous two chapters. Of course, we used simple PEEC meshing examples previously starting with Chapter 2. Unfortunately, most real-life examples are more complicated than what we have considered so far.

We use the simple example in Fig. 8.2 for a thin sheet of metal. Also, we assume that it exists in free space without the presence of other conductor or dielectric objects in its vicinity.

Some of the inductive/resistive cells are shown in Fig. 8.2a, whereas some of the capacitance cells are shown in Fig. 8.2b. We use half cells at the edge for both the inductive and the capacitive cells. The advantage of this is that this object can be connected at the nodes. Then two touching half cells can connect into full cells. Hence, we have a flexible approach where structures can be built from blocks that connect together in a systematic way. Again, for clarity, not all the inductive cells connected to a node are shown. For example, in Fig. 8.2a, four partial inductances are connected to node 4. As is apparent from Fig. 8.2b, we also connect a capacitive surface cell to the node on both the upper and the lower surfaces.

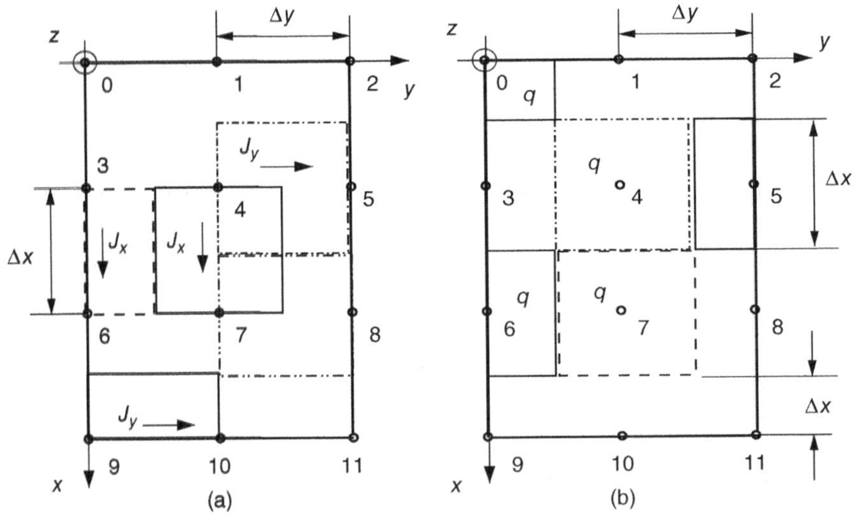

FIGURE 8.2 Inductive and capacitive mesh for sheet or block.

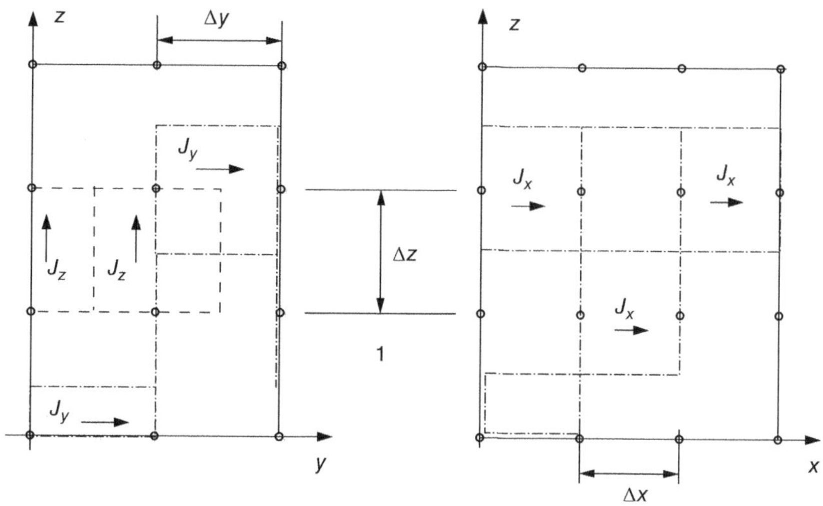

FIGURE 8.3 Inductance volume cells for rectangular conductor.

Internal nodes of a thick conductor block or brick will not have capacitive cells associated with them. In Section 6.3.2, it was shown that the internal charge is zero. However, in the volume filament (VFI) model considered in Chapter 9, the internal nodes represent a key contribution to the skin-effect model.

Figure 8.3 together with Fig. 8.2 can be considered the complete meshing for a rectangular block where all six faces have surface cells such as the ones shown in Fig. 8.2b. Hence, the cube example has four nodes in the x-direction, three nodes in the y-direction, and four nodes in the z-direction, and the block or brick has a total of 48 nodes.

Finally, we give a 3D meshing example to illustrate the half and quarter cells at the corner in Fig. 8.4. The purpose of this example is to illustrate some of the cells involved at the corner of a conductor or dielectric block.

8.2.2 Multiblock Meshing Methods

We start out with the concept of *multiblock meshing* [22] outlined above. This is clearly a top-down hierarchical approach. We may have important details that we can give to the mesher, for the specific problem about the subblocks at hand. This may help to break up the problem into multiple domains that are meshed separately in detail. The connection between the sections is accomplished by connecting appropriate nodes as is illustrated in the previous section.

The meshing of subproblems is generally much less complex than the meshing of the total problem. Hence, wherever possible, we subdivide our problems into sections. An obvious example for using is a connector where different parts of the system are designed separately before being connected together. However, an overall model may need to include such parts as a connector. This may be the starting point for partitioning into different mesh subsections. The reduction of the overall complexity by this approach may be substantial.

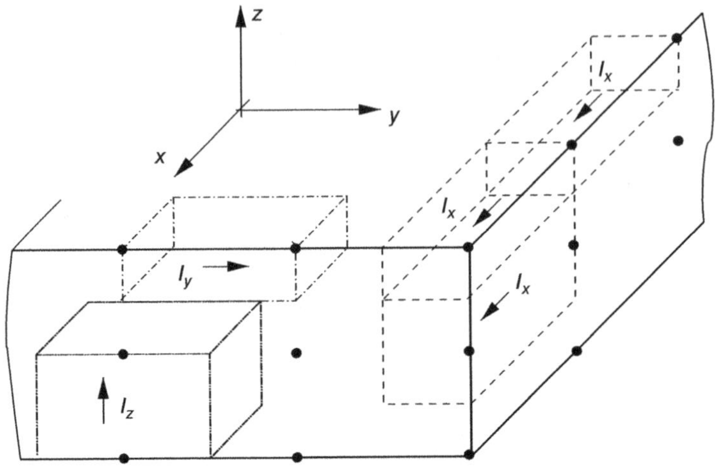

FIGURE 8.4 3D meshing of a corner for finite thickness plane.

8.2.3 Meshing of Nonorthogonal Subproblems

The meshing of subregions is, in many cases, constrained by the boundary regions. It is clear that each subregion needs to exhibit the appropriate boundary nodes to share with other subsections by coinciding nodes. First, we have to find the appropriate constraints for the location of all boundary nodes.

Again, each subcell at the interface is designed to be flat. We give in this section an example for a quad that is located in the $z = 0$ plane. The global corners are given in Table 8.1.

To submesh the quad-block into suitable subquads and cells for the PEEC model, we simply apply (7.5) for the mapping of the x and y locations. Hence, the nodes of the sub-circuit are given by choosing a and b as $-1, -0.5, 0, 0.5, 1$. The result for the submesh is shown in Fig. 8.5 which was created with a small Matlab program. To obtain this model a single plane piece shown in Fig. 7.5, section 7.2, is used by laying out with subdivisions. Also shown in Fig. 8.5, is an example of local coordinate labels to the internal nodes are further used in sections 8.2.5 and 8.3.

8.2.4 Adjustment of Block Boundary Nodes

In order to connect two different blocks together, the nodes of all of the block boundaries need to be adjusted or its number increased or reduced. It is clear that if we have a mosaic of different blocks, we have to match all the nodes to the adjoining boundary

TABLE 8.1 Global corner specification for quadrilateral block surface example.

Corner number	x-Location	y-Location	z-Location
0	1	1	0
1	1.5	3	0
2	2.5	0.8	0
3	2	3.2	0

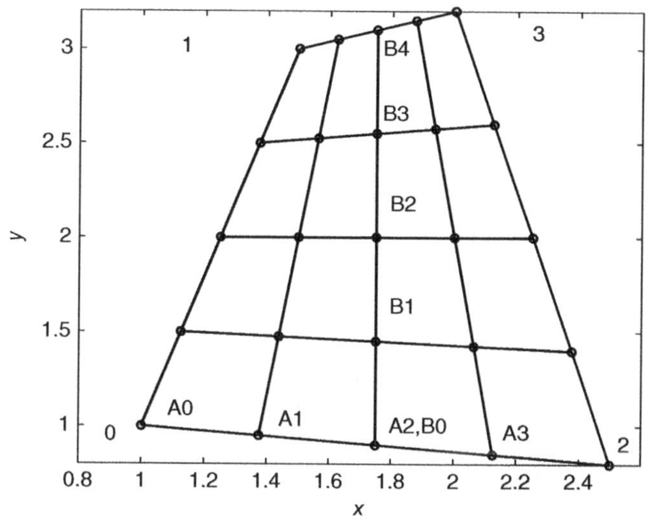

FIGURE 8.5 Submeshing for a flat nonorthogonal quadrilateral block patch.

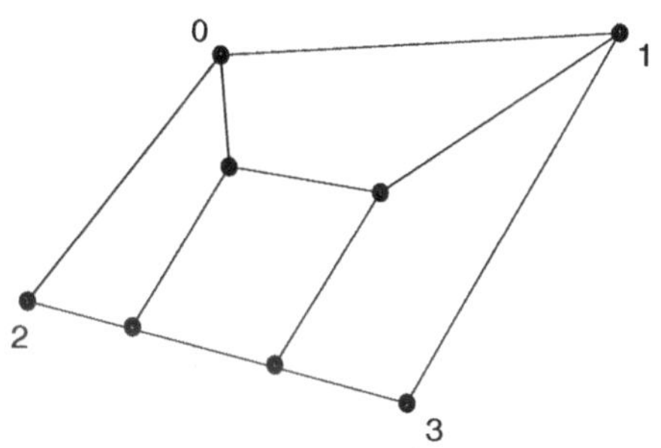

FIGURE 8.6 An example for the introduction of additional boundary nodes.

nodes. Fortunately, techniques exist for the change in boundary node locations. Examples are available, for example, Ref. [23]. In Fig. 8.6, we give an example where two additional nodes are introduced between the global nodes 2 and 3. Of course, the subquads generated by this process need the labeling of the new quads and nodes in the global coordinates.

It is obvious that additional nodes are added at the side of the boundary where they are necessary.

8.2.5 Contacts Between the EM and Circuit Parts

Contacts must be made between the different blocks for a multi-block environment. The entire problem, including the EM parts is solved as a single problem in the time or frequency domain solution.

It is evident from the previous section that the connection to a geometrical element can be established at each of the electrical nodes of a body. Each node is labeled by a specific location in terms of the local (A, B, C) coordinates shown in Fig. 8.5 for the block. An

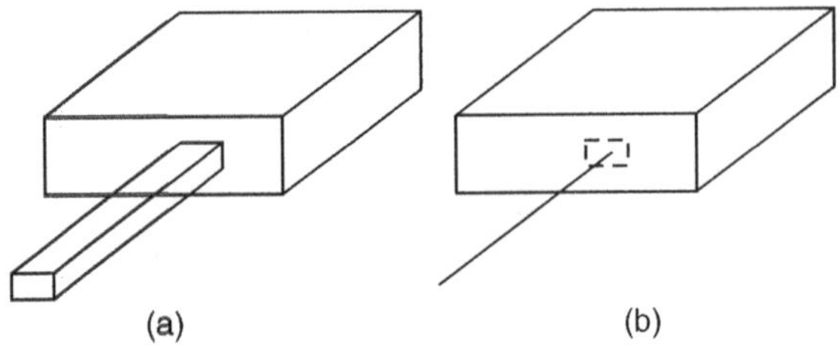

FIGURE 8.7 Contact and macromodel for contact.

example is $A = 0$, $B = 4$, and $C = 2$. These coordinates are unlike the usual local coordinates. Rather, they represent integer coordinates that enumerate the nodes along each side of a body, which may consist of multiple cells. As an example, $A = 0$ means the first node along the A side. The (A, B, C) notation is used instead of (X, Y, Z) since the overall body may not be Manhattan brick shaped. An example of the joining of nodes is shown in Fig. 8.10.

It is still not trivial to correctly connect the electrical circuit to the electromagnetic model. Inaccuracies may occur near the connections. For example, a wire contact attached to a conducting body as shown in Fig. 8.7a may create a large number of cells and unknowns. However, this can be avoided with clever approximations. If we attempt to replace the contact with a single wire macromodel, as shown in the example in Fig 8.7b, then the contact area must be chosen to be large enough as is shown by the dashed area in Fig. 8.7b.

Essentially, the contact modeling is not trivial. Often an insufficient number or too many cells are used in the models of the details of the connections. The size of the contact is not defined by the conductor and the size and the resistance will increase with cell refinements. To avoid this, we have to use a cell with a fixed size with the wire cross section shown by dashed lines in Fig. 8.7b. Hence, the connection size could be locally defined independent of the number of surrounding cells used for the mesh.

This is an example that shows that care must be taken when more efficient, simplified models are used. A similar simplifying technique can be used for a wire which is a via connection through a plane which does not make a connection to this plane as is considered in Fig. 8.17. In fact, for a very simple model, a wire can connect through a ground plane where we completely ignore the presence of a hole in the plane through which the wire is connected.

8.2.6 Nonorthogonal Coordinate System for Geometries

The basis of the meshing consists of the description for the conductor, dielectric as well as other objects such as sources, loads, and potentially connected integrated and other circuits. We suggest that the reader reviews Section 7.1, as an introduction to the relation of the local coordinates (a, b, c) to the global system $\boldsymbol{r}$.

Three-dimensional geometries with an arbitrary orientation represent an important case. Fortunately, this is just an extension of the case in (7.1) represented in (7.2). We consider the case in Fig. 8.8 that consist of a zero thickness quadrilateral shape. In fact, the more general quadrilateral shape in Fig. 8.9 also includes a potentially variable thickness of the shape by adding a c coordinate where $-1 \leq c \leq 1$ much like the a and b variables.

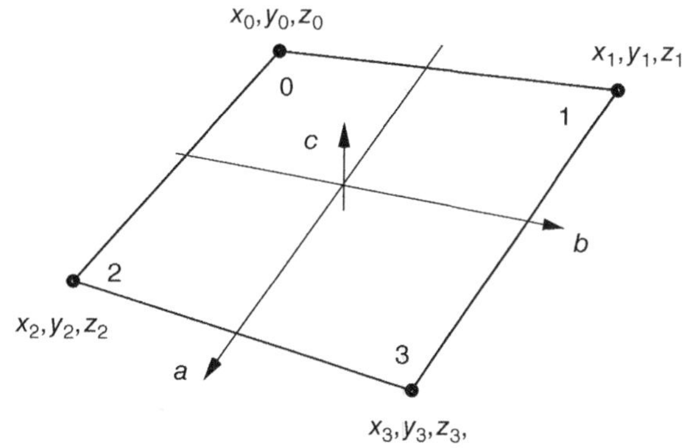

FIGURE 8.8 Basic quadrilateral object with local coordinates.

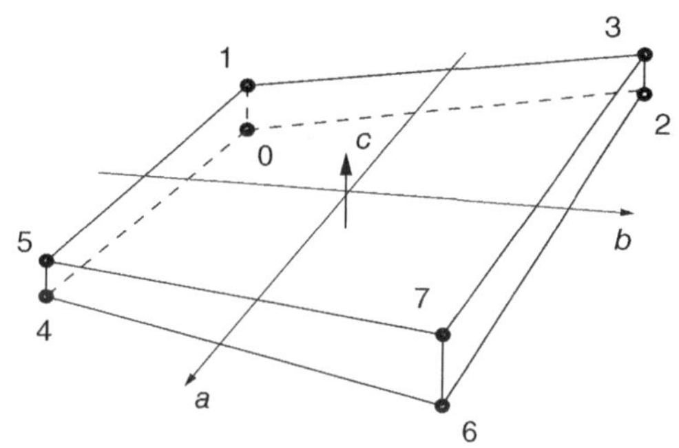

FIGURE 8.9 Basic hexahedral element or object with local coordinates.

The equations for this case are

$$\begin{aligned} \boldsymbol{r}(a,b,c) = {} & \frac{(1-a)(1-b)(1-c)}{8}\boldsymbol{r}_0 + \frac{(1-a)(1-b)(1+c)}{8}\boldsymbol{r}_1 \\ & + \frac{(1-a)(1+b)(1-c)}{8}\boldsymbol{r}_2 + \frac{(1-a)(1+b)(1+c)}{8}\boldsymbol{r}_3 \\ & + \frac{(1+a)(1-b)(1-c)}{8}\boldsymbol{r}_4 + \frac{(1+a)(1-b)(1+c)}{8}\boldsymbol{r}_5 \\ & + \frac{(1+a)(1+b)(1-c)}{8}\boldsymbol{r}_6 + \frac{(1+a)(1+b)(1+c)}{8}\boldsymbol{r}_7. \end{aligned} \tag{8.1}$$

The coordinate system used for a hexahedral element is shown in Fig. 8.9. Again, the local coordinates identify the location of any point belonging to the hexahedron in terms of variables a, b, c where $s \in [-1, +1]$, where s represents a, b, or c. The purpose of all this is to uniquely map a coordinate (a, b, c) into a point in the global coordinates $\boldsymbol{r}$. We accomplish this by interpolating a point vector from eight point vectors $\boldsymbol{r}_k$ where $k = 0, \ldots, 7$ with the coordinates x_k, y_k, z_k. It will be clear that the corners of the hexahedron are reached when $a, b, c = \pm 1$.

We use the same node assignment as presented in Table 7.1. This creates an assignment of the orientation of the quadrilateral of hexahedral surfaces in the mesh as shown in Fig. 8.9 for the hexahedron using the binary code for the symbols a, b, c, where the -1 coordinates map into logical zeros, and $+1$ coordinates map into logical ones. The order of the logical variables is abc. Hence, for example, the binary code $abc = 011$ corresponds to the corner $a = 0$, $b = 1$, and $c = 1$ and its decimal equivalent is 3 for corner 3 as can be verified in Fig. 8.9. This makes the assignment unique and easy to remember. As indicated above, all local coordinates have to relate back to the global x, y, z coordinates. Therefore, a unique representation is needed for the mapping from a local point a, b, c on an object to the global point $\boldsymbol{r}$. Mapping a point in the above hexahedron from a local coordinate point a, b, c into a global coordinate point x, y, z is described by (7.3) as

$$x = \sum_{k=0}^{7} N_k(a, b, c) x_k \tag{8.2a}$$

$$y = \sum_{k=0}^{7} N_k(a, b, c) y_k \tag{8.2b}$$

$$z = \sum_{k=0}^{7} N_k(a, b, c) z_k. \tag{8.2c}$$

The coefficients in (8.2) are given by (7.4) where $s \in [-1, +1]$ and again $s = a, b, c$. The close relation to the binary variables is evident.

With this, we are in a position to also express the tangential vectors with respect to the local coordinates as

$$\frac{\partial \boldsymbol{r}}{\partial s} = \frac{\partial x}{\partial s}\hat{\boldsymbol{x}} + \frac{\partial y}{\partial s}\hat{\boldsymbol{y}} + \frac{\partial z}{\partial s}\hat{\boldsymbol{z}}, \tag{8.3}$$

where the derivatives are found from (8.2a) as presented in detail in Section 7.1.2. Finally, the magnitude of the tangential vector $h_s = \left|\frac{\partial \boldsymbol{r}}{\partial s}\right|$ where the position-dependent unit vectors can be determined from $\hat{\boldsymbol{s}} = \left(\frac{\partial \boldsymbol{r}}{\partial s}\right) \Big/ h_s$ where again $s = a, b, c$. With this, we are prepared to formulate the geometrical aspects for the nonorthogonal PEEC circuit elements.

8.3 SPICE TYPE GEOMETRY DESCRIPTION

In the PEEC circuit solution, terminal or nodal variables are associated with each of the resultant circuit elements. They will be collected in the modified nodal analysis (MNA) circuit solver part in the vector of unknowns. The solution vector variables are quantities such as the potentials $\boldsymbol{\Phi}$, and other conventional circuit variables such as $\boldsymbol{i}, \boldsymbol{v}, \boldsymbol{q}$ where $\boldsymbol{i}$ is the current, $\boldsymbol{v}$ is the voltage, and $\boldsymbol{q}$ is the charge, if necessary. If no other circuit elements are included, the solution vector $(\boldsymbol{\Phi}, \boldsymbol{i})^T$ is used as the only unknown since this vector directly yields the most useful EM circuit output variables.

The general PEEC model can involve conventional lumped circuit elements as well as physical shapes as shown, for example, in Fig. 8.5. Hence, the question is if we could describe the geometrical objects using the SPICE circuit language such that the connection between the SPICE circuit elements and the geometrical objects can directly be established.

As is evident from the above representation of quadrilateral surfaces or the hexahedral blocks can be used as the geometrical units. In most cases, a *multiblock*-type mesh description of the geometry is suitable. Also, we subdivide mixed circuits and EM problems including the connections between the problem domains.

In the circuit domain, the SPICE language is used for the circuit description as a standard. Unfortunately, the details of the language are not well standardized while the SPICE language is used in numerous circuit solvers using subcircuits to describe blocks of circuits.

In a *multiblock* geometrical meshing environment, we can also create meshed parts that correspond to the subcircuits in the circuit domain. Hence, finite conductor and dielectric blocks can be implemented for the general combined orthogonal and nonorthogonal three-dimensional geometries. Two basic statements are used in a form that is compatible with hexahedral structures for the nonorthogonal PEEC approach [17].

An interesting simplification for the description file for the geometry can be achieved for problems of interest. Conventionally, the shapes of the partial elements are described by the geometry files, a shape at the time. This means that each cell for the partial elements is described by a single statement. However, we can also describe geometrical objects that include multiple partial elements or cells. Specifically, the geometrical section shown in Fig. 8.9 can be a unit that later will be modeled in terms of the partial elements. An example of the labeling of the subdivisions is given in Fig. 8.5 and it is used in this section.

An important aspect of the joined geometry-circuit description is the identification of the nodes that are well defined for the PEEC cells and the nodes can be uniquely labeled. Hence, SPICE circuit elements can be connected to the nodes.

The labeling of nodes is rather simple. The nodes of a block can uniquely be identified by the block number and the node identified by the three dimensions a, b, or c, which may correspond to x, y, z for orthogonal shapes. Hence, for a block, the number of subdivisions can be specified by the number of subdivisions in each direction for the block as **ADIV**, **BDIV** and **CDIV**. Hence, the cell subdivision nodes can uniquely be specified starting at the corner $a = -1, b = -1, c = -1$, and so on. The label for the node at the corner would be $(A0, B0, C0)$. Hence, the label for the next nodes along the A edge could be $(A1, B0, C0)$, ...

Finally, the description of a block conductor with a conductivity $\rho = 1.724e - 8$ [kΩ mm] or also [Ω m].

```
.pr1 ADIV=2 BDIV=3 CDIV=1 RhoV=1.724e-8Ohm*m
.Bh1 pr1 0mm 0mm 0mm
         0mm 1mm 0mm
         2mm 0mm 0mm
         2mm 3mm 0mm
         0mm 0mm 7mm
         0mm 1mm 0mm
         2mm 0mm 7mm
         2mm 3mm 7mm
```

The .pr1 information specifies the properties of the body and the .Bh1 uses the .pr1 statement and the coordinate points for the eight vertices of a hexahedral body, where the x, y, z of the global coordinates are given.

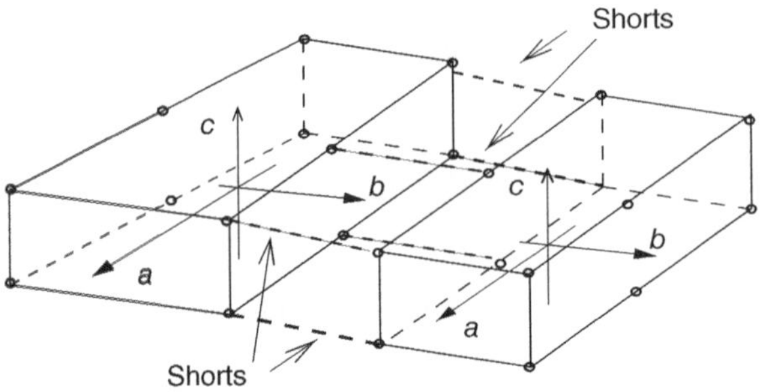

FIGURE 8.10 Example of shorting between two hexahedral bodies.

8.3.1 Shorting of Adjoining Bodies

Most *multiblock* elements are connected to other ones to construct large geometries. The connection nodes to the body are uniquely defined in the above example where a connection is possible at each of the nodes defined above. It is clear that nodes of the neighboring conductors should be at matching physical location within a tolerance as is considered in Section 8.4.4. Electrically, the bodies are connected by joining nodes as shown in Fig. 8.10. A SPICE-type statement is used to indicate the shorting process in the input file. An example of a shorting SPICE-type statement is

```
.short  Bh1(A0,B1,C1) Bh2(A0,B0,C1)
```

for one connection in the back of the body. This will hide some inside surfaces once the two bodies in Fig. 8.10 are merged. These surfaces will be declared to be charge free.

8.4 DETAILED PROPERTIES OF MESHING ALGORITHMS

The size of the PEEC surface or volume mesh may be limited by other constraints. For example, a given cell C_m, which may represent a partial inductance or a potential coefficient, which has a *diameter* D_m defined by $D_m = max(|\boldsymbol{r}_1|, |\boldsymbol{r}_2|)$, where $\boldsymbol{r}_1, \boldsymbol{r}_2$ are the diagonals of a cell. The size of the mesh cells is clearly determined by two different issues: (a) the fact that the *diameter* should not exceed the $D_m < \lambda/20$ for high-frequency impedance problems; and (b) by the sufficient details of the representation of the current or charge distributions.

The current distribution issue is sometimes called the *Laplace factor* since it may locally be the local solution of a PDE [22], which could be the Laplace equation. The solution to Laplace equation can be useful in the determination of locally dense conductor situations. Meshes such as the one in Fig. 8.1 result where the current flow is taken into account by the mesh.

The appropriate mesh size is a combination between the two factors considered in the previous two paragraphs. Of course, from an efficiency point of view, we would like to make the cells as large as possible. However, the size of the mesh cell is limited by the smaller of the two criteria – possibly in all three spatial directions.

For PEEC, we also assume that some of the mesh cells may have a very large aspect ratio R_a in the different spatial directions (a, b, c) or (x, y, z) for rectangular cells. For example, for a zero thickness rectangular cell, we define the aspect ratios as $R_{a,min} \leq |x_\ell / y_\ell| \geq R_{a,max}$. The exact definition of large depends on the problem details. As two examples, a large aspect ratio can be $R_{a,min} \leq 0.1$ and $R_{a.max} \geq 10$ or $R_{a,min} \leq 0.01$ and $R_{a.max} \geq 100$. Similarly, for nonorthogonal shapes, we can use the distance, for example, in the a direction by taking the distance corresponding to the points $a = -1$ and $a = +1$, which maps to the global (x, y, z) coordinates map to using (8.1) $a_\ell = |\boldsymbol{r}(1, b, c) - \boldsymbol{r}(-1, b, c)|$. Then, the distance ratio of interest may be $|a_\ell / b_\ell|$ with the length b_ℓ defined in the same way. Again, the relevant aspect ratios that need to be computed depend on the problem details.

The ratio R_a may also be limited by the quality of the implementation of the partial element integrals considered in the Appendices C–E. Partial elements need to be optimized such that they remain accurate even for very large aspect ratio cells while not resulting in large compute times. This is discussed in Section 5.5.2 for inductances.

The meshing needs to be surface and volume *compatible* since we want to use the meshing for combined volume (V)PEEC and surface (S)PEEC and combined (V,S)PEEC [24] models. Fortunately, the theory for quadrilateral and hexahedral meshing is well developed today, for example, Refs [25–28]. Hence, we can benefit directly from this large body of work.

8.4.1 Nonuniform Meshing Algorithm for Efficient PEEC Models

Nonuniform mesh layers are used for efficient skin-effect models as is shown in Chapter 9. Examples of skin-effect models with variable thickness layers are given in Ref. [29]. It was also shown that efficient capacitance models can be constructed using nonuniform mesh sizes, for example, Ref. [30]. Further, in for example Ref. [31], better results were obtained if the cell sizes were decreased toward the ends of a transmission line structure. However, we need to observe that nonuniform cells may be constrained by other meshing requirements such as projection requirements. These issues are considered below.

We consider the layer or mesh thickness of the nonuniform skin-effect models. This is used in several models presented in Chapter 9. We assume that the problem starts with a conducting layer as shown in Fig. 8.11.

In general, we assume that the conductor has a finite thickness and we are meshing it symmetrically toward the center. This assumes that we care about the current flow on both sides of the conducting sheet. We can adjust the meshing to other situations. For example, we may have transmission lines only on one side of the conducting layer or ground plane. In this case, we need the mesh refinement only on one side of the conductor.

Next, we consider how the layer thickness in Fig. 8.11 is subdivided. For this reason, we consider the thickness direction in more detail as shown in Fig. 8.12.

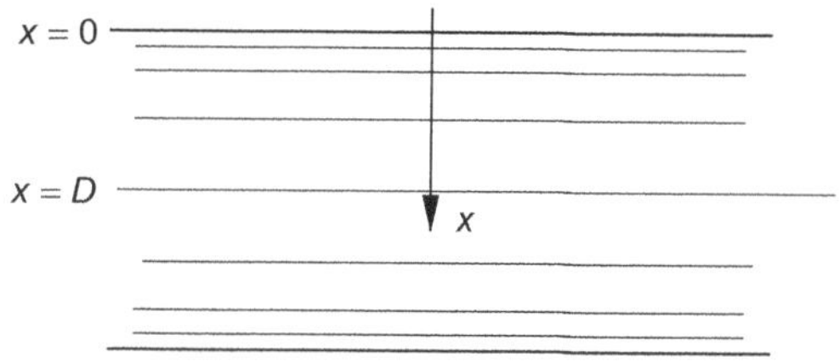

FIGURE 8.11 Example of conductive sheet of thickness 2D with skin-effect layers.

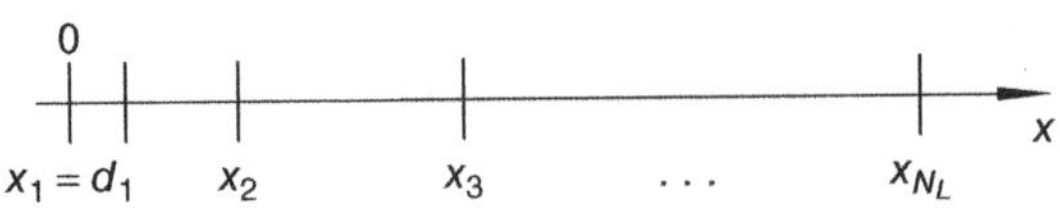

FIGURE 8.12 Example mesh for increasing layer thickness.

Without loss of generality, we assign x to the thickness variable direction. Usually, the thin first layer is chosen to be smaller than the skin-depth δ at the highest frequency [29]. This layer has the thickness d_1 in Fig. 8.12.

As is presented in Chapter 9, the current density decays away from the surface. A compromise between the minimum number of layers and the model accuracy determines the increase β, which is the factor by which the thickness of the layer is increased in comparison to a previous layer.

In (8.4), we compute the location of the layers, starting at the surface of the conductor toward the center.

$$\begin{aligned}
x_1 &= d_1 \\
x_2 &= x_1 + \beta\, d_1 \\
x_3 &= x_2 + \beta^2\, d_1 \\
&\vdots \\
x_i &= x_{i-1} + \beta^{i-1}\, d_1 \\
&\vdots \\
x_N &= x_{N-1} + \beta^{(N-1)}\, d_1.
\end{aligned} \tag{8.4}$$

The goal is to subdivide the conductor with a reasonable number of cells or layers while keeping the increase in the layer thickness $1.1 \leq \beta \leq 3$ such that it provides a reasonable accuracy. Also, the number of layers should be as small as possible, such as $N_L \leq 12$. We should note that the exponential increase in the layers can rapidly result in large overall conductor thickness. Else, the factor β for the center thickness of the conductor can be increased.

From (8.4), we can add the total thickness of all layers as

$$D_L = (1 + \beta + \beta^2 + \cdots + \beta^k + \cdots + \beta^{(N-1)})\, d_1. \tag{8.5}$$

Hence, this is

$$D_L = d_1 \sum_{k=0}^{N_L-1} \beta^k = d_1 \frac{1 - \beta^{N_L}}{1 - \beta}, \tag{8.6}$$

where the last expression is the usual sum for a geometric series.

Using (8.6), we can find a compromise between the number of layers N_L and the thickness of the thin layer d_1 as well as the increase in the layer thickness β. Several examples of layered models are given in the following chapter.

The consequence of nonuniform meshing can be the creation of cells that are too large due to the fast increase in the layer thickness. This may result in a maximum cell size condition $d_k < \lambda/20$. The parameters of the algorithm such as β can be adjusted in the limit value of the maximum cell size.

Also, for very thick conductors, we may want to model the center of a conductor with uniform cells once a certain layer thickness has been reached.

8.4.2 α Cell Projection Algorithm

An important issue has led to the α cell projection algorithm. It was recognized early on that the cells between closely overlapping conductors must be lined up or projected on each other. It was shown in Ref. [14] that poor results are obtained if nonprojected cells are used. The same issue is also substantiated in Section 4.4.

We consider the realistic example where several layers are located in close proximity to a narrow conductor. The boundaries for all closely located cells need to be lined up. This is even more difficult to accomplish for a conductor or dielectric with other closely located bodies on both sides. The cells on both sides need to be closely projected on each other. It becomes even more complicated for multiple close layers where the cell size should be different on both sides of layers. For this case, it is desirable to attempt a slow increasing or decreasing in the projection between the cell boundaries. We call the technique used for the cell size change α-projection algorithm. The motivation is given by a permissible angle α by which the cell size projection changes.

We illustrate the α projection issue in Fig. 8.13, where it is required that the cell width at the bottom surface should be larger than the small width W of the top conductor. This is accomplished by increasing the cell width for each layer in Fig. 8.13 to W_L. For the layer that is at distance d from the narrow top conductor, the cell width is increased to

$$W_L = W + d \tan(\alpha). \tag{8.7}$$

It is clear that fewer unknowns result if larger cells can be used. Experimentally, the best results have been obtained where the angle α is somewhat increased with distance d in the example in Fig. 8.13, rather than being the same for all layers.

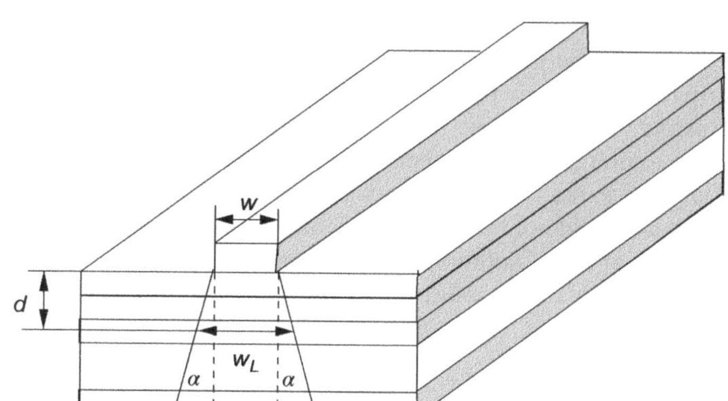

FIGURE 8.13 Example of α-projection for layers under narrow conductor.

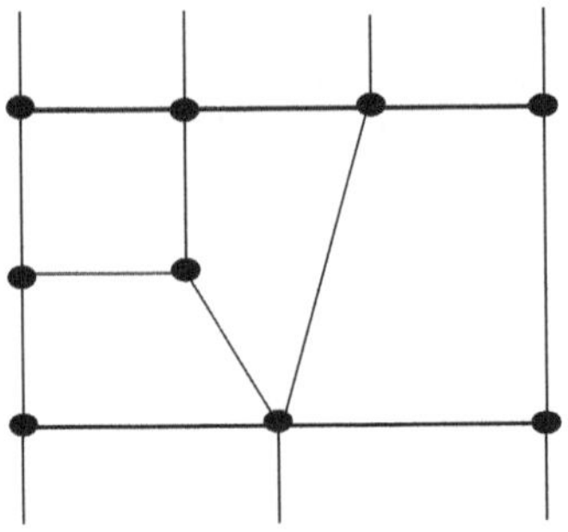

FIGURE 8.14 Reduction of four nodes (top) to three nodes (bottom) surfaces.

Another issue related to the example in Fig. 8.13 is that conductors also need to be subdivided along the cell width W. This represents the fact that the current also distributes toward the edge of the conductor. To accomplish this, further width subdivisions may be needed for the conductor with the overall width W. This is part of the VFI skin-effect model in Section 9.3. Hence, it would be very costly to also project the small width of the cells to the layers.

Provided that we want to use nonorthogonal cells, we can use the node matching technique in Fig. 8.14 to eliminate as many small cells as possible. Of course, we want to use quadrilateral or hexahedral cell for the transition. The assumption for this approach is that the thickness of the layers below is not small compared to the spacing of the subdivision of the conductor of total width W. It is evident that the spacing of the nodes cannot be changed much since it is important to maintain a moderate side-to-side projection ratio.

8.4.3 Smoothing and Tolerancing

Other well-known processing steps performed on existing or new mesh structures are *smoothing* and *tolerancing*. The ultimate goal of this is again to reduce the number of unknowns with little loss of accuracy. In these approaches, the meshes are changed and the geometrical objects may be slightly approximated to simplify the meshing. In some large aspect ratio cases, the goal is to use as many rectangular or cubical cells as possible since they are computationally more efficient. In some cases, this results in a considerable reduction of the number of unknowns, which can also help to simplify the joining of the different multiblock parts.

A nontrivial aspect is to determine the error that results from the slight change in the geometry due to the tolerance in the geometry. It is clear that the geometrical changes must be very small compared to the shortest wavelength corresponding to the highest frequency. Also, the current flow should not be significantly altered. The low-frequency current flow is especially important in PEEC solutions since the same solver can be employed for very low frequencies that include the *dc* solution.

8.4.4 Node Relaxation

Several techniques can be used to simplify the meshing problem. We can associate a tolerance sphere for each node, which is called the *node relaxation* approach. In this approach, nodes within the particular tolerance spheres will be connected together.

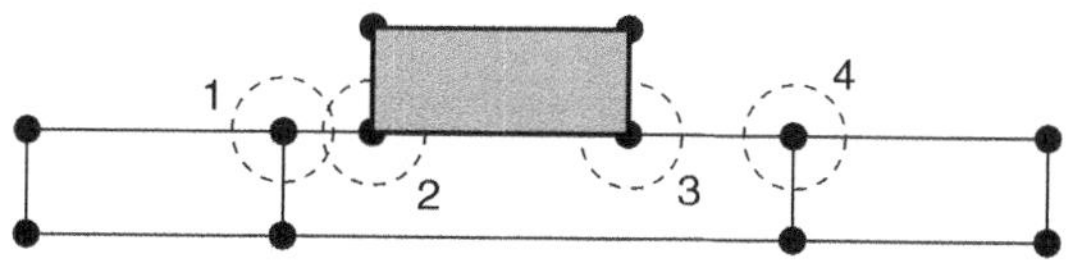

FIGURE 8.15 Example for node joining with tolerance circle test.

This is illustrated in Fig. 8.15, where the two left nodes 1 and 2 are shorted together since the tolerance spheres are overlapping. On the right side of Fig. 8.15, we show that nodes 3 and 4 are too far apart to be connected since the tolerance spheres do not overlap. The equivalent circuit for the shorted nodes can be simplified by joining partial elements. In fact, this simplification is also used in the joining of multiblocks in Section 8.3.1.

Next, we illustrate some applications of *node relaxation*-based *tolerances*. The node shift or relaxation leads to larger approximations for coarse meshes and objects since the tolerance sphere should be proportional to the mesh size.

A first application example relates to electrical connections and measuring probes. Some of these probes have given geometrical distances between the contacts. Electrical probes or contacts can be used to observe port voltages or currents. Also, electrical current or voltage type sources are used to apply currents or voltages to the physical structures or to connect general SPICE-type equivalent circuits to the electromagnetic geometries. Using tolerance circles or spheres will make it easier to place the contacts to a more uniform mesh to the PEEC mesh.

We give a second example for the use of approximate models. We often need solution for package models to study the overall long-time behavior. However, for these long-time solution the waveform details are less important For this case, it is desirable to decrease the meshing accuracy by switching to a coarser mesh. For a more coarse mesh, we found that it is easier to use the node positioning relaxation algorithm to let the contacts, etc., snap to locations in the new mesh. Usually, the simplified mesh leads to considerably reduced compute times.

For the third application, we return to the α-projection technique given in the previous section in Fig. 8.13. However, if the layers are approximated by rectangular type cells, the application of node relaxation can be used if the angle α is small. In Fig. 8.16, we give an example where the tolerance spheres are overlapping such that the nodes are connected together.

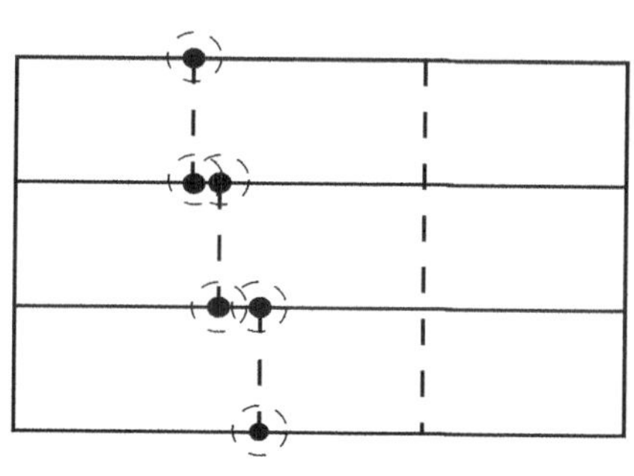

FIGURE 8.16 Node relaxation with rectangular cells.

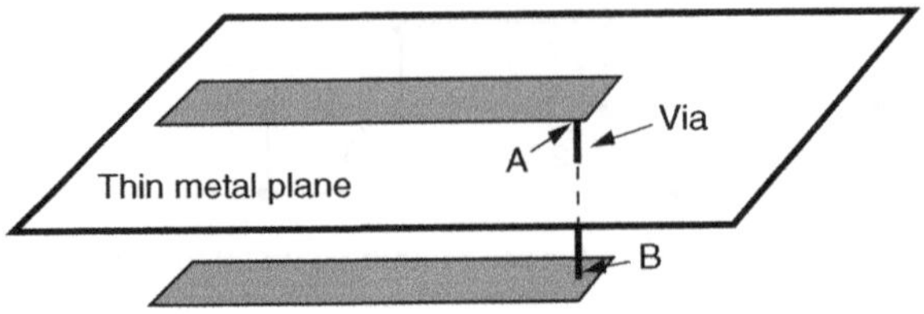

FIGURE 8.17 Via connection without meshed hole in plane.

8.5 AUTOMATIC GENERATION OF GEOMETRICAL OBJECTS

The mesh generation in this section consists of several examples. Hopefully, it provides the fundamental concepts for the generation of geometrical objects.

We start out with a simple example. Today's PCBs have many layers. This implies that multiple via connections and decoupling capacitors are involved. Simplifications can be built into PEEC model meshing. In this example, a via connection between the points A and B as shown in Fig. 8.17 has not to be connected to the center plane at the via connection location.

For low frequencies, the impact on the unconnected plane can be ignored. This is considered in Section 8.2.5. On the other hand, the connection between the upper and lower conductors can be established in the circuit domain by replacing the connection with a short circuit or a small macromodel such as a resistor in series with a partial inductance connected between node points A and B. It is clear that the meshing of plane in the middle is simplified considerably by approach.

8.5.1 Automatic Meshing Techniques for Thin and Other Objects

Physically thin conductor can be represented by a zero thickness meshing approximation. The key issue clearly is the joining of the thin planes as shown in Fig. 8.18. Further approximations can be made later by adding some thickness in the PEEC plane models. For the meshing, the planes are replaced by zero thickness planes located at the center of the actual conductor thickness as shown in Fig. 8.18. The planes need to be connected at the intersection where the nodes from both planes are connected. Importantly, the dominant physics and circuit properties are preserved by this simplification.

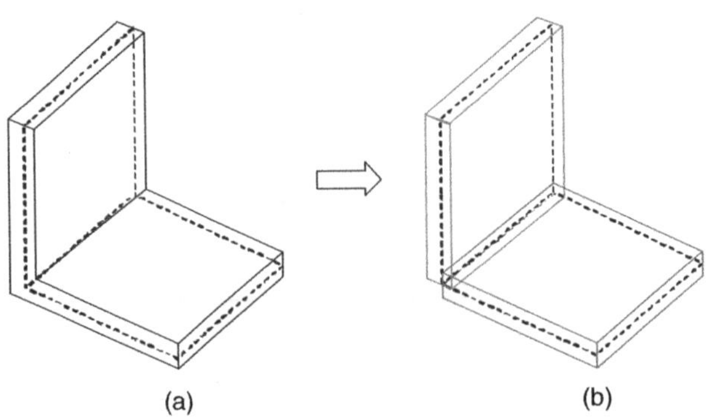

FIGURE 8.18 Zero thickness approximation of an L-shaped 3D object.

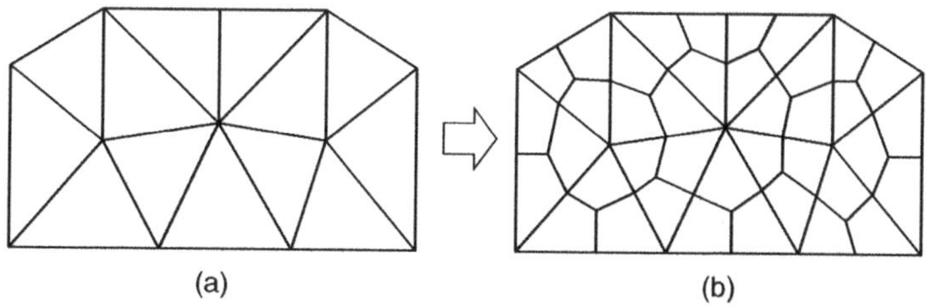

FIGURE 8.19 Triangle to quadrilateral subdivision.

Today, different techniques exist for the meshing of complex geometries. Among the techniques for mesh generation, a distinction has been made between *Indirect Methods* and *Direct Methods*.

The first indirect approach generates the quadrilateral surfaces by starting from an initial triangular mesh (e.g., Delaunay [22]). Quadrilateral shapes are then created inside the triangles. This approach is simple, robust, and well established. The easiest way to generate quadrilateral shapes is to subdivide each triangle into three quadrilaterals with the center points of the three sides as vertices and a barycenter of neighboring triangles, as shown in Fig. 8.19.

The second class, called *direct methods*, creates quadrilaterals directly without using an initial geometrical decomposition. This increases the control over the shape of the individual quadrilaterals. It is always desirable to create rectangular type shapes where possible. The obvious reason for this is the parameters associated to rectangles can be computed efficiently and accurately. Also, rectangular shapes allow very long cells, which is useful for efficiency improvements.

An example of a meshed thin metal structure is given in Fig. 8.20.

The quality Q of a mesh can be defined as

$$Q = \frac{\sum_{i=1}^{n_q} A_i \sum_{j=1}^{4} \max\{0, 1 - |\frac{2}{\pi}\theta_{i,j} - 1|\}}{\sum_{i=1}^{n_q} A_i}, \tag{8.8}$$

where $\theta_{i,j}$ is the jth angle of the ith quadrilateral, A_i is its area, and n_q is the total number of quadrilaterals.

The direct methods provide meshes of higher overall quality than indirect methods. The issues of interest are execution time, quality of the mesh, and total number of quadrilaterals created. Some of the direct methods of interest are the *looping algorithm*, the domain decomposition through geometric optimization, and *bubble mesh*. We only briefly consider the principles of the looping algorithm.

8.5.2 Looping Algorithm Example

To illustrate some of the techniques used, we give a short introduction into the so-called looping algorithm. Sluiter and Hansen [32], as well as Talbert and Parkinson [33], have developed a recursive algorithm for the decomposition of a planar domain into quadrilateral shapes.

It starts out with a planar domain with a fixed contour, which consists of piecewise linear segments as shown in Fig. 8.18. The required mesh density is prespecified according

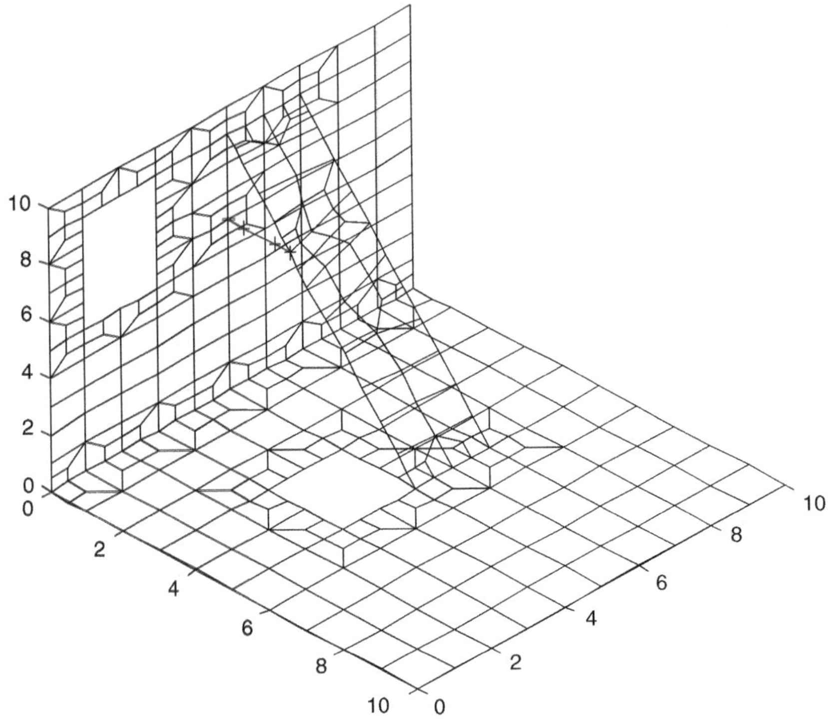

FIGURE 8.20 Example of small thickness meshed structure EMC problem.

to some local requirements. Hence, it starts out with the closed polygonal made of n nodes and n edges.

At each step, the algorithm subdivides the oval polygon through a segment (known as *split line*) of length L_{split}, where the ends connect to two of the n nodes as shown in Fig. 8.21. Other quantities needed in the algorithm are the length of the polynomial diagonal D_{pol} as well as the two areas of the subdomains A_1 and A_2 and the four angles 1–4 shown in Fig. 8.21.

Chae and Jeong proposed in Ref. [34] to associate each possible *split line* with a value that takes all the abovementioned parameters. The best *split line* is identified as minimizing

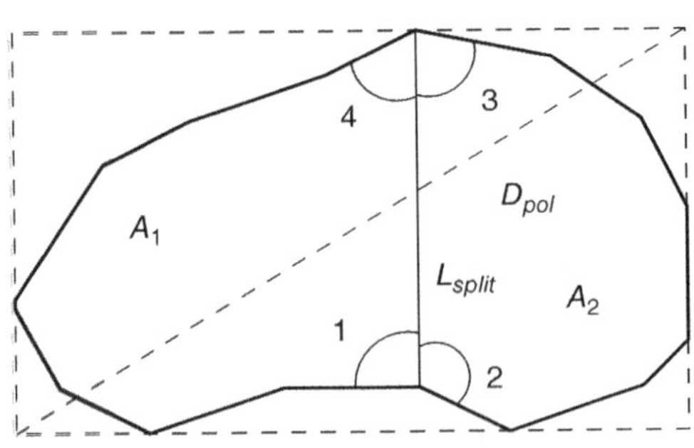

FIGURE 8.21 Examples of possible *split line*.

a function that is empirically obtained as combination of all the previous defined parameters. Recursively subdividing the original domain in subdomains leads to obtain polygons with four, six, or eight sides. In our case, quadrilaterals are directly obtained.

8.6 MESHING OF SOME THREE DIMENSIONAL PRE-DETERMINED SHAPES

A technique that has proven to be successful in commercial EM solvers is the generation of common physical shapes. Some fundamentals for such an approach given here are based on the work in Refs [23, 35]. These authors call a generator of such a shape a *generatrix*. For compatibility reasons, we also will adhere to their other assigned names with slight additions. They are as follows:

- *BodRev* This represents bodies of revolution. This represents a class of special shapes or generatrices like a circle that can be created by a rotation.
- *BodTrans* Such a body is created by the translation of another body or generatrix.
- *BodTwoGen* This body is generated by the use of two generatrices. An example is a tube created by meshing two circles.

Many complex structures can be modeled by quadrilaterals and seven hexahedral shapes can be made by combining generatrices and the use of two generatrices (*BodTwoGen*). Differently shaped bodies can be obtained from two generatrices. The two BodRev shapes do not exactly have to be the same.

As we show, models of circles, arcs, ellipses, squares, rectangles, and spirals can be created. Thus, simple but fairly general algorithms for geometrical modeling can be obtained. For example, a circular waveguide and coaxial lines can be modeled with *BodRev* shapes. Structures can efficiently be modeled by combining a spiral quadric generatrix and a body of two generatrices *BodTwoGen* [23].

8.6.1 Generation Techniques and Meshing of Special Shapes Like Circles

We briefly consider the generation of the meshing of special shapes such as circles. This is the generatrix according to the work in Ref. [35]. A function for circular shapes is based on the familiar equations for circles and ellipses.

$$\left(\frac{x}{\alpha}\right)^{2/t} + \left(\frac{y}{\beta}\right)^{2/t} = 1, \quad \alpha, \beta > 0, \quad t \geq 0, \tag{8.9}$$

where 2α and 2β represent length dimensions of the main axes along x- and y-coordinates, and the parameter t determines the general shape of this function. For example, a circle or ellipse is obtained for $t = 1$, a rectangle is obtained in principle for $t = 0$, and a rhomboid is obtained for $t = 2$. In order that arcs and spirals can also be defined, (8.9) is modified into the spiral quadratic function, described by the following parametric equations

$$x = q\,\alpha(1 + c\varphi)\cos\varphi, \quad y = q\,\beta(1 + \gamma\,\varphi)\sin\varphi \tag{8.10a}$$

$$(\cos\varphi)^{2/t} + (\sin\varphi)^{2/t} = q^{2/t}, \quad \varphi_1 \leq \varphi \leq \varphi_2, \tag{8.10b}$$

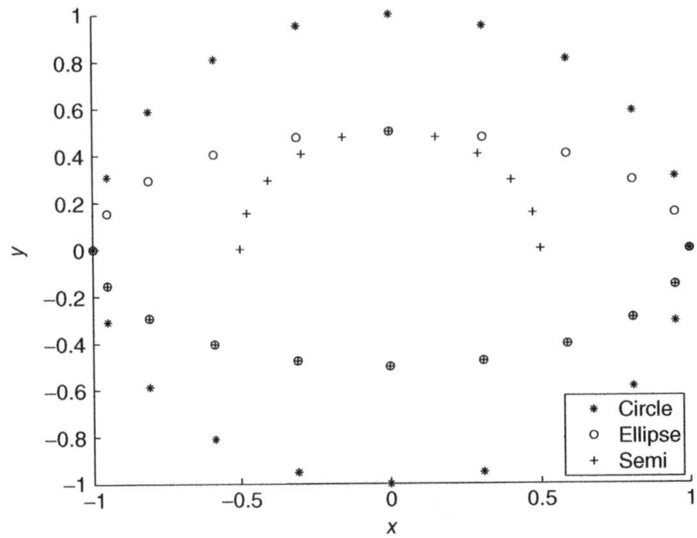

FIGURE 8.22 Example node point arrangements created by generatrix equation.

where q is a scale factor.

The parameter φ is an angle measured from the x-coordinate axis and takes values from φ_1 to φ_2. If $\varphi_2 - \varphi_1 < 360°$ various types of arcs are obtained. If γ is different from zero and $\varphi_2 - \varphi_1 > 360°$, different spiral functions are obtained.

A generatrix will be used as a basis which gives the quadrilaterals to place the PEEC models. Positions of the nodes are defined by angles φ:

$$\varphi_i = \varphi_1 + (i-1)\frac{\varphi_2 - \varphi_1}{n}, \quad i = 1, \dots, n, \tag{8.11}$$

where n is the number of nodes.

We chose $t = 1$ for all the experiments presented below. Three examples are shown in Fig. 8.22 for the application of (8.11). The parameters in the *Circle* case are $\alpha = 1$, $\beta = 1$, $\gamma = 0$, and for $N = 20$ nodal points. For the *Ellipse*, the parameters are $\alpha = 1$, $\beta = 0.5$, $\gamma = 0$, and $N = 20$ nodal points. Finally, for the *Semicircle* $\alpha = 0.5$, $\beta = 0.5$, $\gamma = 0$, and $N = 10$ nodal points only.

In the Fig. 8.23 example, we set the spiral parameter to $\gamma = 0.001$, $\alpha = 1$, $\beta = 1$, and $N = 20$ nodal points. The spiral coils are located in the plane $x = 0$.

The rotation (C.2), (C.3), (C.4), and translation (C.5) formulas in Appendix C are used to move a body into the appropriate position in the global coordinate systems x, y, z.

8.6.2 Bodies Generated by Using Generatrices

Many bodies encountered in different geometries can be generated using multiple generatrices. It is clear from the previous section that we can create different node sets for the interconnections. Generatrices can be translated, rotated, and scaled into appropriate meshing structure orientations. An example of a zero thickness structure generated by two generatrices is shown in Fig. 8.24.

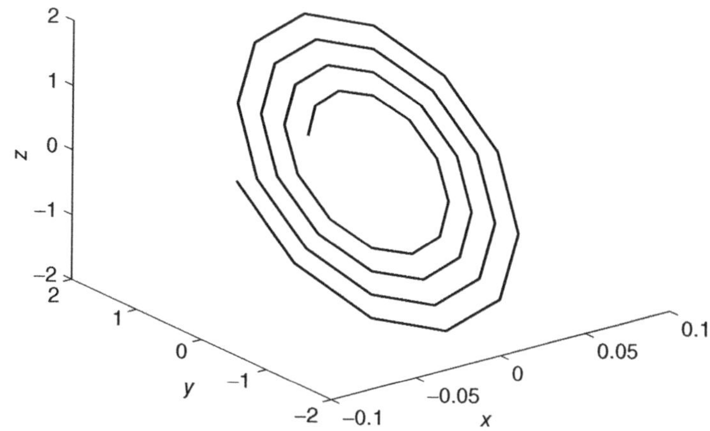

FIGURE 8.23 Spiral of node points generated with generatrix equation.

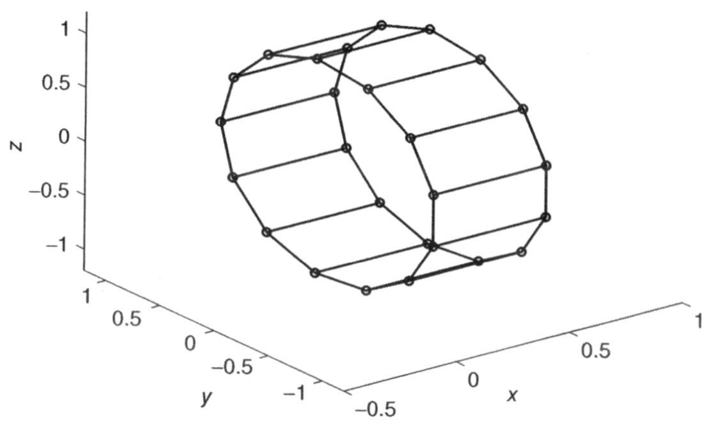

FIGURE 8.24 Small section of mesh formed with two generatrices.

8.7 APPROXIMATIONS WITH SIMPLIFIED MESHES

Approximate solutions are sufficient in some cases for meshes. Again, conductor surfaces are approximated by zero thickness sheets in cases where the skin effect plays only a minor role and if the thickness is constant. These approximations can result in a significant reduction of the compute time while the fundamental result of interest can be obtained. A large class of such basic problems exist today, of which some are low-frequency problems or quasistatic ones.

Other approximations of the geometry can be made such as the representation of round wires or other round parts of the geometry by rectangular approximations. The result sometimes is a model that only includes rectangular shapes. Then, rectangular approximations can be used for the partial element computations.

Figure 8.25 shows a flowchart producing the PEEC-oriented zero thickness meshes. With the input geometries and corresponding electrical properties, the whole problem is

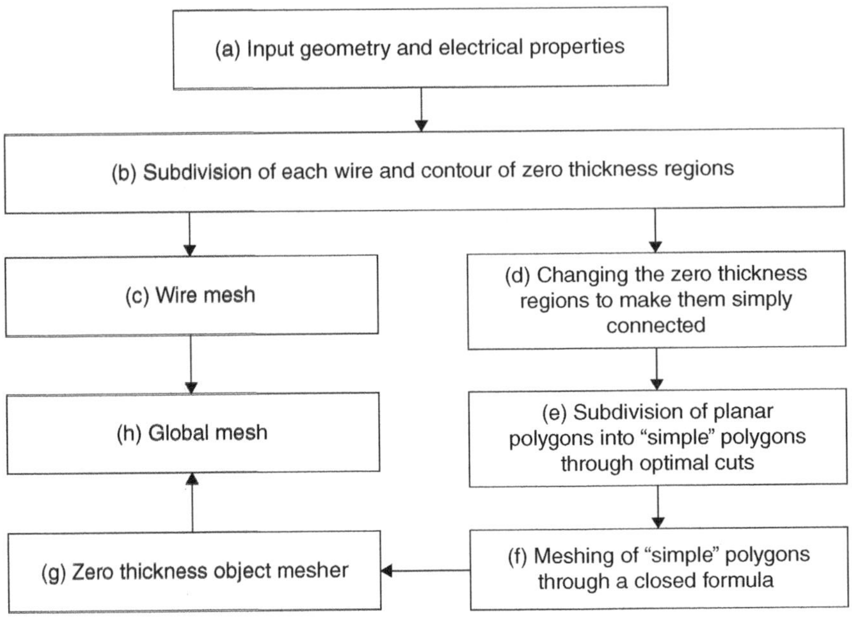

FIGURE 8.25 Outline of steps in a PEEC-oriented mesher for zero thickness structures.

divided into small regions for the next-step, zero-thickness approximation. Then wires are partitioned, simplified using rectangular shapes, and then meshed first. Other shapes are simplified into zero thickness structures that are properly connected to maintain the current continuity. The simplified structures are subdivided into many planar polygons. Next, the planar polygons are subdivided into simple polygons such as rectangles that are created with optimal cuts. Simple polygons can easily be meshed and partial elements are computed using formulas given in Appendices C, D and E. For nonrectangular shapes, a mesher for flat zero thickness elements can be used. This leads to point where the matrix elements are computed and are stamped into the MNA matrix for the final solution.

8.8 MESH GENERATION CODES

The different mesher requirements for a PEEC solver are evident at this point. A key factor is that partial inductances and partial potential coefficients are meshed separately. In many cases, the partial inductance part needs a volume mesh for skin-effect models, whereas the partial potential coefficients require a surface mesh. In this, PEEC is different from other conventional EM solvers. Therefore, special mesh reorganization will be needed. Fortunately, conventional mesh generators can be used for experimental purposes. Especially for educational purposes, such meshes are adequate. We suggest two different sources that can be used for this purpose.

The first tool called *meshgrid* can generate 2D and 3D meshes in Matlab: *http://www.mathworks.com/help/matlab/ref/meshgrid.html*. For example, the 2D running command in meshgrid in Matlab is

$$[X, Y] = meshgrid(xgv, ygv).$$

This mesher replicates the mesh vectors *xgv* and *ygv* to produce a full mesh. The outputs *X* and *Y* are coordinates for the mesh points. By reshuffling these points, regular 2D PEEC meshes can be obtained. Both input mesh vectors and output coordinate arrays can be three dimensional. This is especially important since PEEC needs volumetric meshes for the VFI skin-effect currents in 3D conductors. However, *meshgrid* does not directly provide filament-type meshes, which makes it less convenient to use for 3D volumetric problems.

The second useful experimental mesh tool is called gmesh, which is available under *http://gmsh.info/*. It is an open-source three-dimensional finite element mesh generator with a built-in computer aided design (CAD) engine and postprocessor. It is distributed under the terms of the GNU General Public License (GPL).

Gmsh can partition the geometry into lines, triangles, quadrangles, tetrahedra, prisms, hexahedra, and pyramids. It works in a bottom-up fashion. First, lines are discretized. Then, the meshes of lines are used to mesh surfaces. Finally, the surface meshes are used to mesh volumes. Details about *Gmsh* and tutorials can be found from the *Gmsh*'s website.

Again, these tools are mostly useful for educational purposes.

PROBLEMS

8.1 *Meshing in vicinity of contacts*
The meshing for contacts in a plane is tested in this problem. We want to compute the open loop inductance between two contacts located on the same plane. A thin plate with the meshing similar to the one shown in Fig. 8.2 is assumed which is 4 cm long and 2 cm wide. Two 5 mm by 5 mm contacts are placed at 1 cm and 3 cm along the length in the middle, 1 cm from each side. Compute the open-loop inductance between the two contacts for different mesh densities for the plate. How dense does the meshing have to be to get a good answer? Asking in a different way, we would like to find the smallest number of node for which we get a good value of the inductance. This problem relates to the contact issue presented in Fig. 8.7.

8.2 *Point located on a quadrilateral element*
Construct a quadrilateral element as shown in Fig. 8.8, which is not parallel to the global (x, y, z) coordinates. Note that the element must have a planar surface. How do you choose the four corner points such that the surface is planar? Use the equations from Section 7.1 to compute the matrix of global coordinates (x, y, z) that correspond to the uniform steps in $-1 \geq a \geq 1$ and $-1 \geq b \geq 1$, where each of the two local variables (a, b) are incremented by steps size of 0.1.

8.3 *Using vertical cell projection*
The strip conductor of a strip line has the width of 0.8 mm and a length equal to 20 mm. At a vertical conductor-to-conductor distance of 0.2 mm, a rectangular ground strip below it has the width equal to 4 mm with the length also equal to 20 mm. The zero thickness conductors are lined up along the length and the top conductor is centered in the middle over the ground strip. Compute the loop inductance between the conductors at one end while they are shorted at the other end. The projection theory in section 8.4.2 that we at least have to subdivide the bottom ground conductor such that the bottom conductor width is also subdivided into several strips such that an 0.8 mm strip inductive cell is cut out of the bottom conductor for accuracy reasons. Note for a simple model all the conductor strips are connected together at the end with the exception of the input port for the strip conductor.

8.4 *Using node relaxation*
Assume that we want to compute the partial self-inductance of a conductor with almost parallel sides along the length as is shown in the example in Fig. 7.1. We want to approximate such a conductor with a few rectangular partial inductance cells. We assume that the conductor is 10 mm long and its width changes from 1 to 1.3 mm. Use several sections of rectangular conductor shapes using the tolerance node relaxation in Fig. 8.15. Compute the total partial inductance of the entire strip from end to end. Test the result with several different subsections. This relates to the work in Ref. [36] where rectangular cells are used for nonorthogonal shapes.

8.5 *Using mesh generatrix techniques*
Use the formulation in Section 8.6.1 to duplicate the node arrangements in Fig. 8.22.

REFERENCES

1. P. K. Wolff and A. E. Ruehli. Inductance computations for complex three dimensional geometries. In *Proceedings of the IEEE International Symposium on Circuits and Systems*, pp. 16–19, 1981.
2. R. F. Milsom, K. J. Scott, G. Clark, J. C. McEntegart, S. Ahmed, and F. N. Soper. FACET – a CAE system for RF analogue simulation including layout. In *Proceedings of the Design Automation Conference*, Volume 26, pp. 622–625, 1989.
3. R.-B. Wu, C.-N. Kuo, and K. K. Chang. Inductance and resistance computations for three-dimensional multiconductor interconnection structures. *IEEE Transactions on Microwave Theory and Techniques*, MTT-40(2):263–270, February 1992.
4. M. Kamon, M. J. Tsuk, and J. White. FastHenry: a multipole-accelerated 3-d inductance extraction program. In *Proceedings of the Design Automation Conference*, pp. 678–683, June 1993.
5. A. W. Glisson and D. R. Wilson. Simple and efficient numerical methods for problems of electromagnetic radiation and scattering from surfaces. *IEEE Transactions on Antennas and Propagation*, 28:593–603, September 1980.
6. S. M. Rao, D. R. Wilton, and A. W. Glisson. Electromagnetic scattering by surfaces of arbitrary shape. *IEEE Transactions on Antennas and Propagation*, 30:409–418, May 1982.
7. J. M. Jin. *The Finite Element Method in Electromagnetics*. John Wiley and Sons, Inc., New York, 2nd edition, 2002.
8. A. Rong and A. C. Cangellaris. Comprehensive broadband electromagnetic modeling of on-chip interconnects with a surface discretization-based generalized PEEC method. *IEEE Transactions on Advanced Packaging*, 28(3):434–444, August 2005.
9. Y. Wang, V. Jandjayla, and C.-J. R. Shi. Coupled electromagnetic-circuit simulation of arbitrary-shaped conducting structures. In *Digest of Electrical Performance of Electronic Packaging*, Volume 10, Boston, MA, pp. 233–236, October 2001.
10. A. Rong and A. C. Cangellaris. Generalized PEEC models for three-dimensional interconnect structures and integrated passives of arbitrary shapes. In *Digest of Electrical Performance of Electronic Packaging*, Volume 10, Boston, MA, pp. 225–228, October 2001.
11. M. Tröscher and H. Katzier. Efficient PEEC modeling of multi-layer boards and multi-chip modules. In *Proceedings International Zurich Symposium on EMC*, Volume 14, Zürich, Switzerland, pp. 269–273, February 2001.
12. C. A. Balanis. *Antenna Theory: Analysis and Design*. John Wiley and Sons, Inc., New York, 2005.
13. S. Steinberg and P. Knupp. *Fundamentals of Grid Generation*. CRC Press, Boca Raton, FL, Ann Arbor, MI, London, Tokyo, 1994.

14. A. E. Ruehli, P. A. Brennan, and H. W. Young. Recent progress in capacitance computation methods. In *Proceedings of the IEEE International Symposium on Circuits and Systems*, Phoenix, AZ, pp. 135–138, 1975.

15. A. E. Ruehli, G. Antonini, and A. Orlandi. Extension of the partial element equivalent circuit method to non-rectangular geometries. In *Proceedings of the IEEE International Symposium on Electromagnetic Compatibility*, Seattle, WA, pp. 728–733, August 1999.

16. G. Antonini, A. Ruehli, and J. Esch. Non orthogonal PEEC formulation for time and frequency domain modeling. In *Proceedings of the IEEE International Symposium on Electromagnetic Compatibility*, Minneapolis, MN, August 2002.

17. A. E. Ruehli and G. Antonini. On modeling accuracy of EMI problems using PEEC. In *Proceedings of the IEEE International Symposium on Electromagnetic Compatibility*, Boston, MA, August 2003.

18. B. Rubin. Physics-based gridding for electrical package analysis codes. *IEEE Transactions on Advanced Packaging*, 46(4):828–838, November 2010.

19. B. M. Kolundzija and B. D. Popovic. Entire-domain Galerkin method for analysis of metallic antennas and scatterers. *IEE Proceedings H Microwaves, Antennas and Propagation*, 140(1):1–10, January 1993.

20. H. Singer, H.-D. Brüns, and G. Bürger. State of the art in the moment method. In *Proceedings of the IEEE International Symposium on Electromagnetic Compatibility*, Santa Clara, CA, pp. 122–227, August 1996.

21. H. Edelsbrunner. *Geometry and Topology for Mesh Generation*. Cambridge University Press, Cambridge, 2001.

22. P. J. Frey and P.-L. George. *Mesh Generation, Applications to Finite Elements*. John Wiley & Sons, Inc., Hoboken, NJ, 2nd edition, 2008.

23. B. M. Kolundzija. Automatic mesh generation using single- and double-node segmentation techniques. *IEEE Antennas and Propagation Magazine*, 40(2):30–38, August 1998.

24. A. E. Ruehli, D. Gope, and V. Jandhyala. Mixed volume and surface PEEC circuit and electromagnetic solver. In *Proceedings International Zurich Symposium on EMC*, Volume 16, Zürich, Switzerland, pp. 263–267, February 2005.

25. T. D. Blacker and M. B. Stephenson. Paving: a new approach to automated quadrilateral mesh generation. *International Journal for Numerical Methods in Engineering*, 32(4):811–847, September 1991.

26. M. A. Price, C. G. Armstrong, and M. A. Sabin. Hexahedral mesh generation by medial surface subdivision: Part 1. Solids with convex edges. *International Journal for Numerical Methods in Engineering*, 38(19):3335–3359, October 1995.

27. V. Akcelik, B. Jaramaz, and O. Ghattas. Nearly orthogonal two-dimensional grid generation with aspect ratio control. *Journal of Computational Physics*, 171(2):805–821, August 2001.

28. S. Owen. A survey of unstructured mesh generation technology. In *Proceedings of 7th International Meshing Roundtable*, pp. 239–267, October 1998.

29. L. Daniel, A. Sangiovanni-Vincentelli, and J. White. Proximity templates for modeling of skin and proximity effects on packages and high-frequency interconnect. In *Proceedings of International Conference on Computer Aided Design*, pp. 326–333, June 2002.

30. A. E. Ruehli and P. A. Brennan. Efficient capacitance calculations for three-dimensional multiconductor systems. *IEEE Transactions on Microwave Theory and Techniques*, 21(2):76–82, February 1973.

31. A. Woo, T. Yioultsis, and A. C. Cangellaris. *BestFit:* a SPICE-compatible model for efficient, passive, broadband transmission-line analysis for dispersive interconnects. In *Digest of Electrical Performance of Electronic Packaging*, Volume 12, Princeton, NJ, pp. 367–370, October 2003.

32. M. L. Sluiter and D. L. Hansen. A general purpose automatic mesh generator for shell and solid finite elements. *ASME Computers in Engineering*, 3:29–34, 1993.

33. J. A. Talbert and A. R. Parkinson. Development of an automatic, two-dimensional finite element mesh generator using quadrilateral elements and Bezier curve boundary definition. *International Journal for Numerical Methods in Engineering*, 29(7):1551–1567, May 1991.

34. J.-H. Jeong and S.-W. Chae. Unstructured surface meshing using operators. In *Proceedings of the 6th International Meshing Roundtable*, Sandia National Laboratories, pp. 281–291, 1997.

35. A. Djordjević and B. Kolundzija. Spiral super-quadric generatrix and bodies of two generatrices in automated parameterization of 3-d geometries. *IEEE Transactions on Microwave Theory and Techniques*, 45(6):864–866, June 1997.

36. Y. Hackl, P. Scholz, W. Ackermann, and T. Weiland. Multifunction approach and specialized numerical integration algorithms for fast inductance evaluations in nonorthogonal PEEC systems. *IEEE Transactions on Electromagnetic Compatibility*, 57(5):1155–1163, October 2015.

9

SKIN EFFECT MODELING

The skin effect results in nonuniform current flow in conductors for higher frequencies or fast transients in the time domain. Essentially, the current density is reduced in the inner parts of conductors. In fact, the modeling of skin-effect has a long history, for example, [1]. Skin-effect is an important area of early research for large transmission lines (TLs), transformers, and power stations. Today, power engineering applications are even more important.

The solution of most electromagnetic (EM) problems include the modeling of the broadband skin-effect loss for the conducting planes and 3D conductors, for example, Refs [2, 3], or [4]. Three-dimensional problems are present in connectors, on printed circuit boards (PCBs) and other common geometries. Frequency domain applications include radio frequency (RF)/microwave on-chip circuit issues such as the modeling of coils with the correct Q-factor, which is sensitive to losses. Skin-effect losses are also important for modeling of the antenna efficiency and gains in an integrated circuit (IC) environment. Clearly, the inclusion of the skin effect in solving large problems is important. The assumed current flow in a model can be in one direction (1D), two directions (2D), or all three directions (3D).

Reliable solutions exist today for the one direction (1D) current flow in TLs [5]. However, the 2D or 3D current flow as is the case of many partial element equivalent circuit (PEEC) models are much more challenging. The solution of these large EM problems include the modeling of the broadband skin-effect loss for conducting planes and 3D conductors.

It is clear today that the modeling of problems with a true three-dimensional current flow that is frequency or time dependent is very challenging. It is also very desirable that the model for the frequency- or time-dependent 3D current flow includes a dc solution for the circuit class of problems we consider. We observe from our research work that in general, the low-frequency inductive behavior is more difficult to model while the prediction

Circuit Oriented Electromagnetic Modeling Using the PEEC Techniques, First Edition.
Albert E. Ruehli, Giulio Antonini, and Lijun Jiang.

of resistive dependence seems to be less challenging. Further, the computation with such models in an EM solver can be difficult and costly in both compute time and memory requirements.

It is well known that for single frequency or narrow band high frequencies, the currents are confined close to the surface, which makes the prediction of the current flow easier for problems that are in this class. We call the solution approach for this class the EXP solutions since they are based on the exponential 1D skin-effect formulas.

9.1 TRANSMISSION LINE BASED MODELS

In the last decade, skin-effect loss models for TLs both in the time and frequency domains have been a key research issue. The current flow in a 2D cross section [5–9], is 1D along the length. Such models are much easier to construct for several reasons. The transverse electromagnetic (TEM) mode allows the decoupling of the inductance and capacitance solution, which is computed based on 2D solvers. We should note that this separation is also the case for PEEC models as is evident from the previous chapters. Usually, the inductance/skin-effect loss problem is solved separately from the rest of the TL solution. The two-dimensional cross-sectional dependence of the current distribution has to be taken into account in wide-frequency-band models. However, these models are still based on a 1D current flow direction, assuming infinitely long parallel conductors or – practically – very long conductors. The contribution of the end effects will be ignored. Also, a realistic TL problem usually involves only a small number of different cross sections for which we can precompute the skin-effect model data. This also has been done for on-chip skin-effect models [10, 11]. The simplifying aspect is that one-direction current flow for TLs is dependent only on the distribution of the current in the 2D cross-sectional dimension. PEEC models can be used to model the 2D cross-sectional structures, but conventional TL models [5] are more efficient.

9.1.1 Anomalous Skin-Effect Loss and Surface Roughness

For completeness, we want to consider two relatively new issues for our applications of skin-effect loss models for the problems of interest. Mainly, this is a result of the continuous miniaturization and higher speed of semiconductor circuits. Conductors with a thickness of much smaller that 1 μm are common today in IC chips. This leads to new issues.

For such on-chip problems, as the wire dimensions are reaching dimensions below 0.2 μm, the so-called anomalous skin effect is increasing the surface resistivity for a layer that is about 0.04 μm for copper [12]. The resistivity becomes higher than what we specify for thicker conductors. Typical values for the conductivity are given in the Tables in Appendix A. Due to the small thickness of some of the interconnects on a chip, the 0.04 μm layers represent an appreciable part of the thickness and it will increase the resistance of the interconnect.

This issue has also been considered in a paper where examples are given for the resistivity degradation toward the conductor edge [13]. It is clear that the implementation of the anomalous skin effect for very small conductors requires volume filament (VFI) or global surface impedance (GSI) conductor models. These volume-based models are discussed in this chapter.

The second issue that can increase the resistance of conductors is surface conductor roughness. In this situation, the boundary of the conductors is not smooth and both the resistance and the inductance are impacted. It is especially important for PCBs where conductor roughness is used to enhance the adhesiveness of the layers. Of course, this makes the surface roughness problem worse.

Statistical methods have been used frequently for surface scattering problems, for example, Ref. [14]. These methods also have been used for surface roughness applications [15–18]. These computations can become very time consuming for reasonably large problems and we need to resort to approaches such as Refs [19–21], where the roughness is taken into account with a modification of the material properties. However, we are not considering these issues here.

9.1.2 Current Flow Direction and Coordinate Dependence

As mentioned in the introduction to this chapter, the current flow direction is a fundamental aspect of any skin-effect model. For example, a one-directional current flowing in the x-direction may depend on two coordinates, for example, $J_x(y, z)$. We next give a more extensive list of possibilities.

- The current flow can be restricted in one dimension J_x and the variation can depend on one space dimension. Example: $J_x(y)$. This may be the case for a very thin, narrow metal strip. For rectangular cross sections, we may have $J_x(y, z)$ for which we have a 1D current flow that depends on two variables y, z.
- Even for a thin plane, the current flow can be in two directions, J_x and J_y. Then the space dependence of the current is two dimensional, (x, y).
- Finally, the general case is where we have a current flow in all three directions J_x, J_y, and J_z, which can be (x, y, z) dependent.

The key issue to understand the difference between the dependence of the current on the coordinates (x, y, z). Another issue is the number of directions the current flow is restricted to by the formulation.

9.2 ONE DIMENSIONAL CURRENT FLOW TECHNIQUES

9.2.1 Analytical 1D Current Flow Models

The analytical 1D skin-effect model for a conducting half-space [22, 23] is the most simple case. Such an interface between a free space and a conductor is shown in Fig. 9.1. For this case, the skin depth is inversely proportional to the square root of frequency given by (9.1).

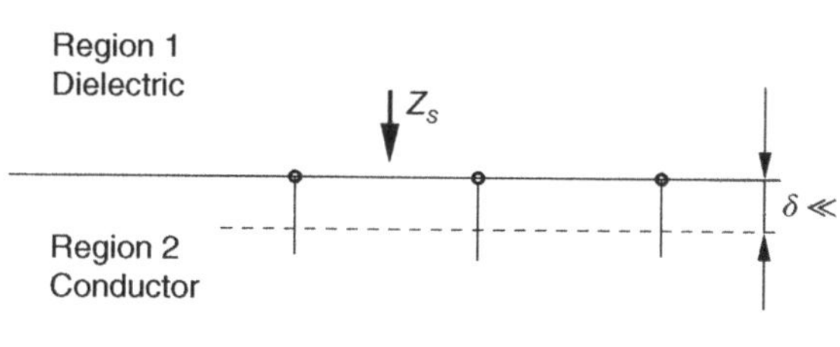

FIGURE 9.1 Conductor interface for 1D skin-effect model.

Also, the high-frequency resistance increases with frequency as $\sqrt{f}$ due to the reduced effective cross section for the current flow. We refer to one of the EM texts [22, 23] on the derivation of the skin-effect equations for a flat or a small curvature one-dimensional surface, which can include some circular cross sections. In these simple cases, the solution can be obtained analytically.

We call the class of skin-effect models, which are based on the conventional exponential decay of the 1D current density in conductors, EXP type model. Several authors have developed different frequency domain 1D models, probably in part unaware of each other's research. Some of these works are [11, 24–26]. The first model presented in this chapter for thin conductors is related to this type of model. We found that applying an EXP-based model to nonthin conductors was difficult [27]. We have to recognize that the EXP model is locally oriented and that, for thicker conductors, the representation inside the conductors requires additional currents, as we discuss below. However, we show in Section 9.2.3 that applying 1D type models for very thin conductor problems can yield efficient solutions.

9.2.2 Narrow Band High-Frequency Skin-Effect Models

Narrow frequency band or single high-frequency skin-effect models are the most simple to construct. We can use a 1D EXP model for current penetration or skin depth δ given by Ramo et al. [22]

$$\delta = \frac{1}{\sqrt{\pi f \mu \sigma}}, \tag{9.1}$$

where f is the frequency, σ is the conductivity of the conductor and μ is the permeability, which is usually μ_0. It is evident from this that at high frequencies, the skin depth restricts the current to a well-defined path close to the surface. We assume that a conductor or ground plane is, in this case, thicker than the skin depth δ. Hence, we can assume that the two sides of the plane are isolated from each other. As shown in Fig. 9.2, we assume that we cut out a surface cell of a size Δx and Δy where at a given angular frequency ω, the equivalent current penetration thickness is δ. This results in an equivalent impedance of the cell of $Z_s = R_s + j\omega L_s$, with

$$R_s = \frac{\Delta x}{\sigma \delta \Delta y}, \quad L_s = \frac{\Delta x}{\omega \sigma \delta \Delta y}. \tag{9.2}$$

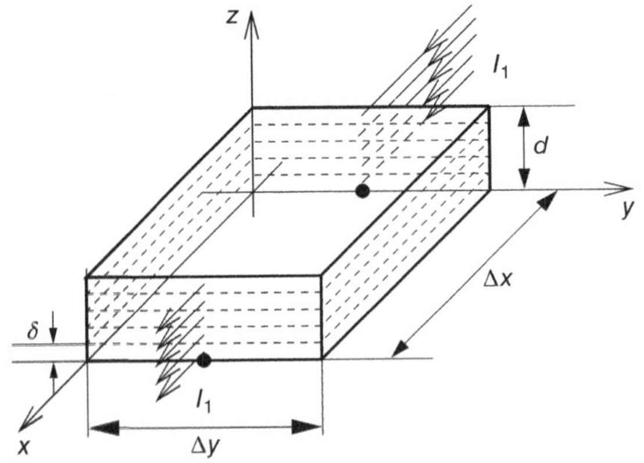

FIGURE 9.2 Section of ground plane with layers for current flow.

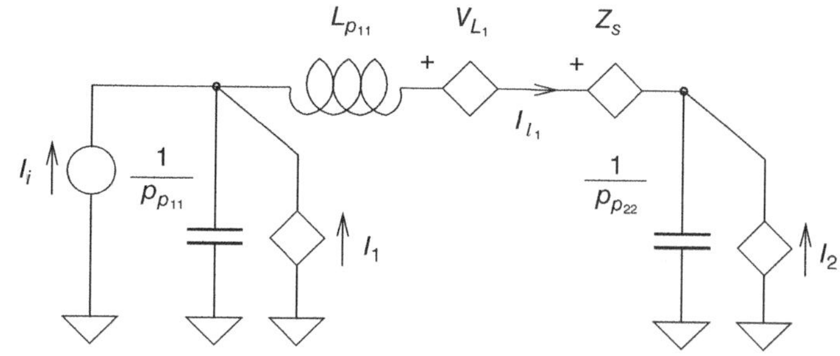

FIGURE 9.3 PEEC loop with additional skin-effect impedance source.

Further, due to the small diffusion distance δ into the conductor, we also can assume that the lateral loss coupling between two neighboring conductor surface cells is small. As a consequence, all the skin-effect loss couplings to other surface cells can be ignored.

To couple the two models, we simply place the above impedance $Z_s = R_s + j\omega L_s$ into the PEEC circuit in Fig. 9.3, where the current in the model is $I = I_{\ell 1}$.

9.2.3 Approximate GSI Thin Conductor Skin-Effect Model

In the following two sections, we consider the class of 1D current flow general surface impedance GSI models. In these models, the mutual inductive coupling between the inside of a conductor and the outside of the surface are approximated. They relate the inductance part to the internal–external inductance concepts that have been used for many years [28, 29]. Also, these issues have been reviewed in a paper dedicated to Clayton Paul [30]. Figure 9.3 shows how the model is embedded in the PEEC model. The internal inductance and loss model is based on a differential equation (DE) formulation. This model is connected in series and is shown as the impedance Z_s in Fig. 9.3.

In this section, we treat the case where the conductors are thin, whereas thick conductors are considered in the following section. The external inductive model consists of a zero thickness partial inductance Lp_{11}, which is connected in series to the internal DE-based model Z_s representing the impedance in Fig. 9.3 for the conductor surfaces [31]. This avoids the inductive couplings from the inside of the VFI skin-effect model. At least for thin conductors, the approach can be less costly than the VFI model, since many couplings are ignored.

The solution of the current diffusion problem inside the conductor is fundamentally different from the external PEEC propagation model. For the thin conductor problem, the skin-effect model must represent the EXP current penetration on opposite surfaces. Hence, the mesh cells on both side surfaces need to be lined up and can be coupled. This corresponds to top and bottom cells in Fig. 9.2. Due to the assumed small conductor thickness, the coupling to other surface cells including the neighboring cells is ignored. This leads to a very sparse internal impedance coupling matrix. For broadband applications, the skin-depth changes widely over the frequencies of interest. Hence, the impact of the skin effect on the impedance is large for thin conductor areas.

The next task is to provide an EM and finally a circuit model for the diffusion of current inside the thin conductor. Especially for the time domain, we want to keep the order or the number of poles or circuit elements of this skin-effect model as low as possible. The fundamental idea is based on the derivation of an appropriate equivalent circuit for the EXP function based on [1]. This skin-effect model was designed for one-dimensional current dependence for a single surface. However, for PEEC we want a more general model that includes both sides of the thin conductor. Fortunately, the approach can be applied in the context of partial inductances [32], while the original model [1] was designed for a 2D TL model with a 1D current flow.

We approximate the model by breaking the thickness into layers as shown in Fig. 9.2. As we consider below, we do not use uniform layers. This leads to an equivalent circuit model in the z-direction as is indicated by the dashed lines. This is the dominant skin-effect direction for the thin layer structures. The Maxwell equations for the interior of the conductor are

$$\nabla \times \boldsymbol{E} = -\mu \frac{\partial \boldsymbol{H}}{\partial t} \tag{9.3}$$

and

$$\nabla \times \boldsymbol{H} = \boldsymbol{J} = \sigma \boldsymbol{E}. \tag{9.4}$$

We observe that for the conductor, a simplified solution is that the following fields are nonzero: $\boldsymbol{E} = \hat{\boldsymbol{x}} E_x$ and $\boldsymbol{H} = \hat{\boldsymbol{y}} H_y$. Therefore, this reduces (9.3) to

$$\frac{\partial E_x}{\partial z} = -\mu \frac{\partial H_y}{\partial t}. \tag{9.5}$$

Further, (9.4) will also reduce to a simple form

$$\frac{\partial H_y}{\partial z} = -\sigma E_x, \tag{9.6}$$

since in the derivation we discretize the plane thickness in the z-direction only as shown in Fig. 9.2. Of course, we are also using the usual subdivision of the conductor into PEEC cells for the currents in the x- and y-directions shown in Fig. 6.2.

We start with $z_1 = 0$ and we label the thickness of the layers as $k = 1, \ldots, N_k$ where N_k is the number of layers. Hence, the top layer ends at $z_{N_k} = d$. Also, the thickness of the layers is given by $\Delta z = z_{k+1} - z_k$.

To derive an equivalent circuit, we start with (9.5) and we multiply it by the cell length Δx. The voltage drop along the layer k in the x-direction is given by $\Delta x E_{x,k} = V_{x,k}$, which leads to

$$V_{x,k+1} - V_{x,k} = -\mu \Delta x \Delta z \frac{\partial H_{y,k}}{\partial t}. \tag{9.7}$$

To convert this into an equivalent circuit, we define the differential inductance of layer k in the conventional way as

$$L_k = \frac{\Phi_k}{I_{x,k}} = \frac{B_{y,k} \Delta x \Delta z_k}{I_{x,k}}, \tag{9.8}$$

which results in $B_{y,k} = \mu H_{y,k}$ and $H_{y,k} = I_{x,k}/\Delta y$ to

$$L_k = \frac{\mu \Delta x \Delta z_k}{\Delta y}. \tag{9.9}$$

With this, we can simplify (9.7) to

$$V_{x,k+1} - V_{x,k} = -L_k \frac{\partial I_{x,k}}{\partial t}, \tag{9.10}$$

which is the circuit equation for an inductance element.

Similarly, starting from (9.6), we again replace the magnetic field with $H_{y,k} = I_{y,k}/\Delta y$ and $E_x = V_x/\Delta x$ to obtain

$$\frac{1}{\Delta y}\frac{\partial I_{x,k}}{\partial z} = -\sigma \frac{V_{x,k}}{\Delta x}. \tag{9.11}$$

This leads to the local resistance of the block as

$$\frac{1}{R_k} = \frac{\sigma \Delta Z_k \Delta y}{\Delta x}. \tag{9.12}$$

If we numerically approximate the derivative $\partial I_{x,k}/\partial z = (I_{x,k} - I_{x,k+1})/\Delta z$, we can finally find that

$$I_{x,k} - I_{x,k+1} = \frac{V_{x,k}}{R_k}. \tag{9.13}$$

It is a small step to show that (9.7) and (9.13) correspond to the equivalent circuit in Fig. 9.4. Note that we assigned half the differential inductance to each vertical branch to get a symmetric circuit. On purpose, we make sure that the surface impedance between the terminals A and C is

$$\lim_{\omega \to \infty} \text{Im}(Z) = 0, \tag{9.14}$$

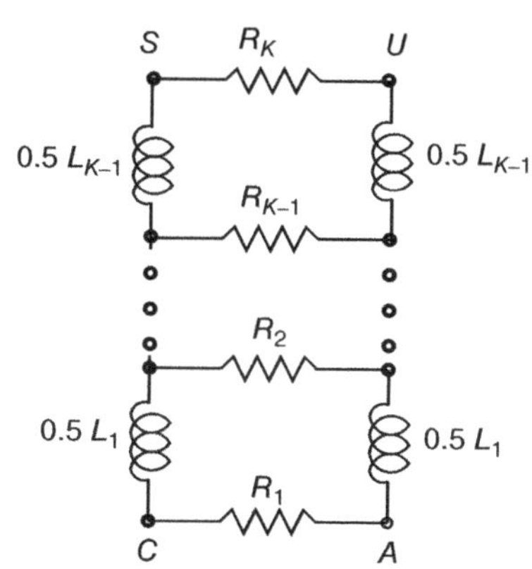

FIGURE 9.4 Two-dimensional skin-effect circuit for the layered model in Fig. 9.2.

such that the circuit will become resistive for very high frequencies. This eliminates unwanted resonances at very high frequencies where the model becomes invalid due to the discretization. This relates to the *passivication* work in Chapter 13. This circuit skin-effect part of the model is automatically Hilbert consistent since we represent it with the passive nondelay equivalent circuit in Fig. 9.4 [33].

9.2.4 Physics-Based Macromodel

A *macromodel* is a model that is much smaller with less unknowns than the original full model. However, it has the same main properties as that of the original model for the application at hand. Hence, features that are not important are lost.

A *reduced order model* (*ROM*), which is a macromodel, can be obtained by many different approaches such as Vector Fitting [34]. A technique for obtaining a *ROM* is called a model order reduction (*MOR*) method since it reduces the number of states. In general, this is obtained by reducing the number of unknown currents and voltages in the system.

A special feature of such a model is if it is mathematically oriented or if it is *physics based*. The advantage of the physics-based model is that the equivalent circuit corresponds to a physical layout. For many applications, it is more intuitive.

Here, we use a direct *physics-based MOR model*, which we call *PB-MOR*, which is accomplished directly based on a physical simplification of a model.

In the 1D model in Fig. 9.4, we have reduced the thickness of the cells toward the current carrying surfaces. This results in a drastic reduction in the number of layers required and the number of eigenvalues or poles. We call the model obtained with the nonuniform meshing in the z-direction physics-based global surface impedance (PB-GSI) and also the thin-GSI model since it applies only to thin conductors. We note that in Ref. [35], the cells are also reduced in thickness toward the surfaces for a VFI model. This results in a less expensive model. In fact, it is essential for this model.

As a test example for the PM-GSI model, we start out with a 1D model based on the derivation in the previous section with the equivalent circuit in Fig. 9.4. The thickness of the first layer is chosen to be $d_1 = \delta/2$, where δ is the skin depth for the highest frequency in the spectrum. The cell thickness is increased by a factor β starting from the surface layer with the dominant current flow. So, for L layers, the thickness d of the conductor and the thickness multiplying factor for the thickness are related by the finite geometric series with the sum

$$d = d_1 \frac{1 - \beta^{N_L}}{1 - \beta}, \tag{9.15}$$

where d_1 is again the thickness of the first layer and N_L is the total number of layers. A derivation of (9.15) is given in section (8.4.1).

We would like to limit the number of layers as much as possible to keep the number of poles low in the MOR representation. To test our 1D model, we compare it with the 1D-EXP model in Fig. 9.5 for the inductance and in Fig. 9.6 for the resistance. The results show very good agreement for a cell, which is 30 μm thick with a seven-layer PM-GSI model. We are well aware that at low frequencies, a simple 1D-EXP model does not give a correct result as is apparent from both figures. However, the equivalent circuit yields the correct low-frequency answer. As considered above, the reduction in inductance for high frequencies in the PM-GSI model is due to the high frequency limit introduced by the number of sections used in the model.

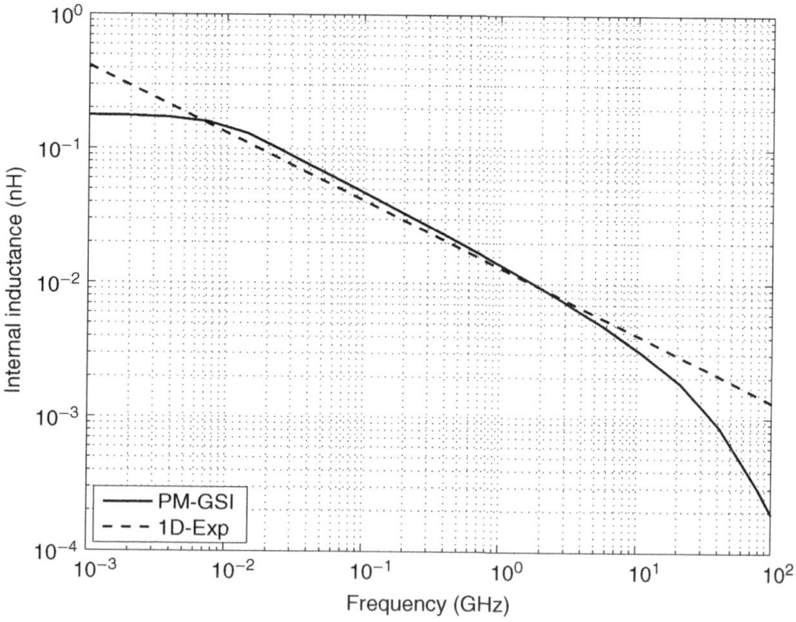

FIGURE 9.5 Comparison between PM-GSI and 1D-EXP model internal inductance.

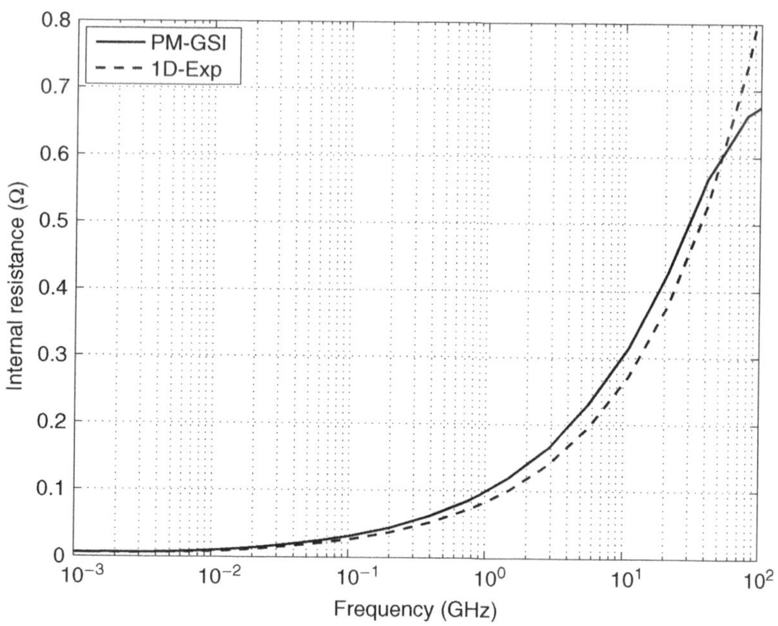

FIGURE 9.6 Comparison between PM-GSI and 1D-EXP model internal resistance.

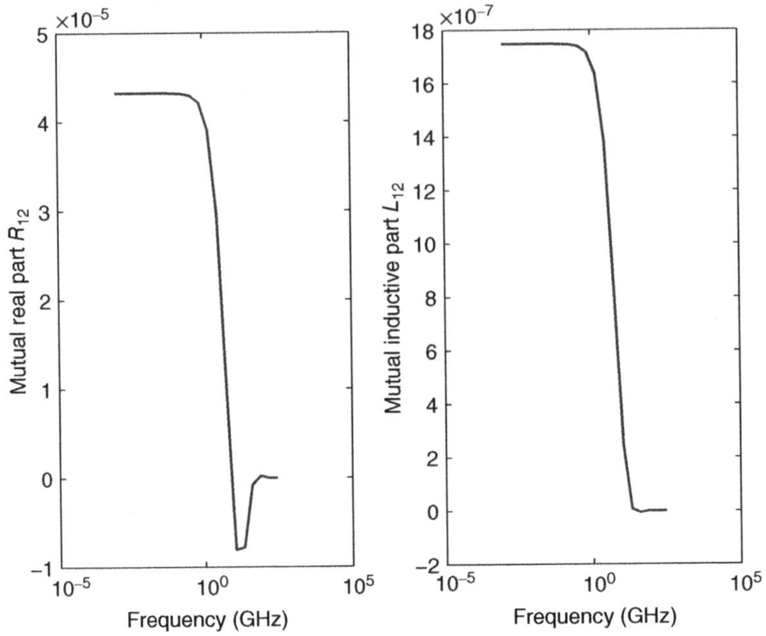

FIGURE 9.7 Resistance and inductance for Z_{12} for 2 μm layer.

For electrically thin conductors, it is essential to include the top-to-bottom conductor coupling since the internal coupling between the top and the bottom layers is very strong. However, the coupling to the neighboring cells is relatively weak. We added an additional port to the model shown in Fig. 9.4 to compute the transimpedance between the ports. For the C to A port, we have Z_{11} and for the S to U, we have Z_{12} as shown in Fig. 9.4. We give an example for the evaluation of the real and inductive parts of Z_{12}. In the example, with a layer thickness of 2 μm, we have four increasing layers and four symmetrically decreasing layers toward the other surface. In the results in Fig. 9.7, the imaginary part is divided by ω to get the inductive coupling. Hence, we have the desired two-port model for the cells on the opposite side of the conductors.

9.2.5 Frequency Domain Solver for Physics-Based Macromodel

The evaluation of impedances of the ladder circuit shown in Fig. 9.4 in the frequency domain can be solved efficiently due to the simple analytical solution of such ladder circuits that are available [36]. Hence, this can speed up the computation of the impedances that can be directly stamped into the modified nodal analysis (MNA) matrix in the frequency domain. The MNA matrix for the PEEC circuit with the impedance is similar to (2.45)

$$\begin{bmatrix} s\mathbf{Pd}^{-1} & (\hat{\mathbf{I}}+\mathbf{M})\mathbf{A}_\ell \\ \mathbf{A}_\ell^T & -[s\mathbf{Lp}+\mathbf{Z}_s(s)] \end{bmatrix} \begin{bmatrix} \mathbf{\Phi}_n \\ \boldsymbol{I}_\ell \end{bmatrix} = \begin{bmatrix} -(\hat{\mathbf{I}}+\mathbf{M})\mathbf{A}_i\mathbf{I}_i \\ \mathbf{0} \end{bmatrix}, \tag{9.16}$$

where $\mathbf{Lp}$ is the zero thickness surface partial inductance matrix, accounting for the magnetic field coupling occurring among surface currents, $\mathbf{Z}_s$ is the surface impedance accounting for the skin effect, **Pd** and **M** are coefficients of potential matrices describing the electric field coupling among surface charges, $\mathbf{A}_\ell$ is the connectivity matrix, and $\boldsymbol{I}_i(s)$ represents potential current sources at the nodes.

9.2.6 Approximate Thin Wire Skin-Effect Loss Model

The model in the previous section is efficient for thin sheet-type geometries. Another similar class of problems are thin wires as they often occur in IC interconnect wires and wire bonds. An example of a wire segment is shown in Fig. 9.8. Starting out with (9.3) and (9.4), a cylindrical skin-effect model can be derived provided that the proximity effect is small enough for multiple parallel wires. This is the case if the center-to-center distance between the parallel wires is >2*d*.

The current is strictly in the z-direction with a circular symmetric current density $\boldsymbol{J} = J_z\hat{z}$ and the length of the tube is $\Delta z = z_e - z_b$ is sufficiently long for the approximation. Then, the internal field is centered around the dashed line in Fig. 9.8. Hence, the magnetic field in the ϕ direction in a cylindrical coordinate system is given by $\boldsymbol{H} = H_\phi\hat{\boldsymbol{\phi}}$. These assumptions considerably simplify the vector operations.

If we take (9.4) and integrate it over the cross section, we get

$$\int_S (\nabla \times \boldsymbol{H}) \cdot \hat{\boldsymbol{n}}\, ds = \oint_\ell \boldsymbol{H} \cdot d\ell = \int_S \boldsymbol{J} \cdot \hat{\boldsymbol{n}}\, ds, \tag{9.17}$$

where the second step is based on Stokes' theorem (3.34). We can get

$$I_z(r,t) = 2\pi r H_\phi(r,t) \tag{9.18}$$

for $0 \le r \le a$ in our specific case.

Next, we take Maxwell's equation (3.1a) or (9.3) where we used $\boldsymbol{J} = \sigma\boldsymbol{E}$ to get

$$\nabla \times \boldsymbol{J} = -\mu\sigma\frac{\partial \boldsymbol{H}}{\partial t}. \tag{9.19}$$

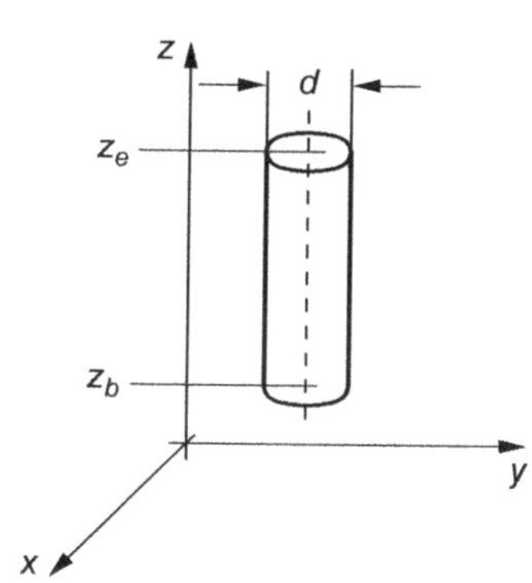

FIGURE 9.8 Section of round wire for internal skin-effect models.

We have to take the cross-product in cylindrical coordinate for the conditions where $\boldsymbol{J} = J_z\hat{\boldsymbol{z}}$. Also, the magnetic field intensity has only the component $\boldsymbol{H} = H_\phi\hat{\boldsymbol{\phi}}$. This leads to

$$\frac{\partial J_z(\rho, t)}{\partial \rho} = \mu\sigma \frac{\partial H_\phi(\rho, t)}{\partial t}. \tag{9.20}$$

From this and (9.18), we finally get a first-order DE

$$\frac{\partial J_z(\rho, t)}{\partial \rho} = \frac{\mu\sigma}{2\pi\rho} \frac{\partial I_z(\rho, t)}{\partial t}. \tag{9.21}$$

The goal is a circuit model for the internal part of the cylindrical conductor, which we subdivide into the $\Delta\rho$ tube segments as is shown in Fig. 9.9. The current in each tube is of the form

$$J_z(k, t) = \frac{I_{k-1}(t) - I_k(t)}{\mathcal{A}_k} \tag{9.22}$$

for tube k and $\mathcal{A}_k$ is its cross-sectional area and $I_k(t)$ is the current in the tube segment. All currents will be in the z-direction.

With this, we can approximate the derivative on the left-hand side of (9.21) such that the equation can be written for the diffusion case shown in Fig. 9.9 for the first two tubes as

$$\frac{I_0 - I_1}{\sigma\mathcal{A}_1} - \frac{I_1 - I_2}{\sigma\mathcal{A}_2} = \frac{\mu(\rho_{AV1} - \rho_{AV2})}{2\pi\rho_1} \frac{\partial I_1(\rho, t)}{\partial t}, \tag{9.23}$$

where the average radius of the tubes is $\rho_{AV1} = (\rho_0 + \rho_1)/2$, and so on. Since all currents represent the total z-directed current in each tube, we can rewrite (9.23) by multiplying by $\Delta z = |z_e - z_s|$ in terms of circuit elements where

$$L_1 = \frac{\mu|z_e - z_s|(\rho_{AV1} - \rho_{AV2})}{2\pi\rho_1} \tag{9.24}$$

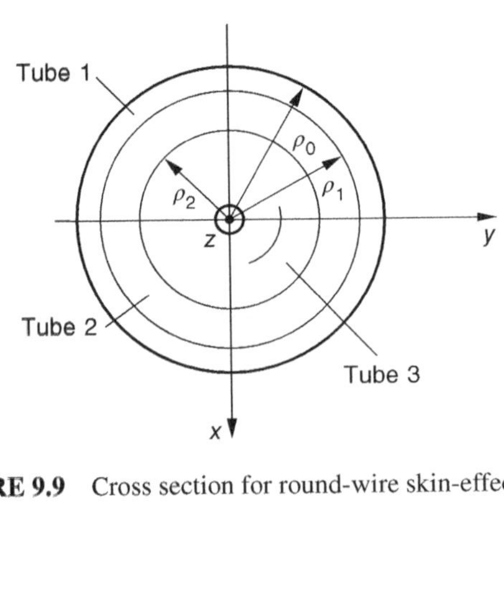

FIGURE 9.9 Cross section for round-wire skin-effect model.

as the differential diffusion inductance. Also, on the left-hand side of the equation, we recognize the resistors of the form $R_1 = |z_e - z_s|/(\sigma \mathcal{A}_1)$. With this, we can recognize the circuit equation

$$R_1 I_1 + R_2 I_1 + L_1 \frac{\partial I_1}{\partial t} - R_2 I_2 = 0 \tag{9.25}$$

corresponds to an equivalent circuit. We should note in Fig. 9.9 that the current in tube 2 is $I_1 - I_2$. All other loops are of the same form since we are progressing toward the center of the round conductor. This leads to the equivalent circuit in Fig. 9.10. Note that the partial inductance of the outer zero thickness shell (9.28) has been added to the model. We emphasize that since this model is a self-term, it can be applied to any orientation for a cylindrical conductor in the global coordinate system.

The overall solution approach is based on the approximate internal–external inductance approximation for inductance [30]. A formula for the partial inductance of the external tube is given in Appendix C. The partial inductance integral to be solved for a section of wire is (5.17) where

$$Lp_{m\ell} = \frac{1}{\mathcal{A}_m \, \mathcal{A}_\ell} \frac{\mu}{4\pi} \int_{\mathcal{V}_m} \int_{\mathcal{V}_\ell} \frac{\hat{\boldsymbol{t}}_m \cdot \hat{\boldsymbol{t}}_\ell}{R_{m,\ell}} \, d\mathcal{V}_m \, d\mathcal{V}_\ell, \tag{9.26}$$

where $\hat{\boldsymbol{t}} = \hat{z}$ are the unit vectors tangential to the currents and where $R_{m,\ell}$ is the distance metric between the source and observation points. For a round bar, the partial inductance is approximated as [28, 29]

$$Lp_{11} \simeq \frac{\mu_0}{2\pi} \ell \left[\ln \left(\frac{\ell}{a} + \sqrt{\left(\frac{\ell}{a}\right)^2 + 1} \right) - \sqrt{1 + \left(\frac{a}{\ell}\right)^2} + \frac{a}{\ell} \right], \tag{9.27}$$

where a is the wire radius and ℓ is the length. We originally assumed that this is an approximation to the low-frequency inductance for a cylindrical conductor. However, the value we obtained from it seems to be closer to that of a tube, at least for the layer aspect ratios we are considering. This approximate equation seems to be an average value for the partial inductance.

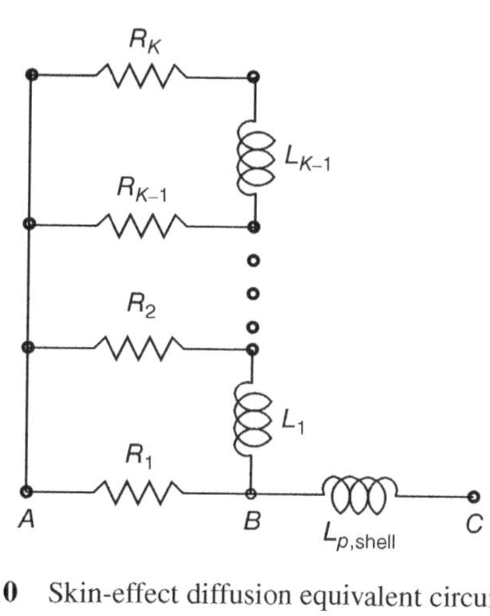

FIGURE 9.10 Skin-effect diffusion equivalent circuit for cylinder.

For this model, an important partial inductance model represents the high-frequency limit. This infinite frequency limit is the partial inductance of a zero thickness cylindrical tube with a radius a. The partial inductance (9.26) for the cylinder reduces to

$$Lp_{11} = \frac{\mu}{4\pi^2} \int_{\phi=0}^{\pi} \int_{z=0}^{\ell} \int_{z'=0}^{\ell} \frac{dz\, dz'\, d\phi}{\sqrt{4a^2 \sin^2(\phi/2) + (z-z')^2}} \tag{9.28}$$

by recognizing that the symmetry can be used to reduce the fourfold integral to a threefold integral. We were able to analytically solve the integrals for the case of interest where the length ℓ of the wire is longer than the diameter $d = 2\,a$. The result for the tube conductor or high-frequency limit is given by (C.17) as

$$\begin{aligned} Lp_{11} = \frac{\mu\ell}{4} \Bigg[& \left(\frac{k^2}{480} + \frac{k^4}{1280} + \frac{1}{3600} \right) \pi^3 \\ & + \left(\frac{1}{18} - \frac{k^2}{24} \right) \pi \\ & + (-2\,\log(\ell) + 6\,\log(2) + 2 + 2\,\log(a) - 4\,\log(k\pi)) \frac{1}{\pi} + \frac{8\,a}{\ell\,\pi^2} \Bigg], \end{aligned} \tag{9.29}$$

where $k = d\,/\,\ell$.

Evaluating (9.29) resulted in 0.54995 nH which is the high-frequency limit for $a = 0.05$ mm and $\ell = 1$ mm while the approximate formula (9.27) led to 0.54775 nH. Hence, the approximate formula does not represent a low-frequency result for the full cylinder, since its value is supposed to be larger than the high-frequency result. Figures 9.11

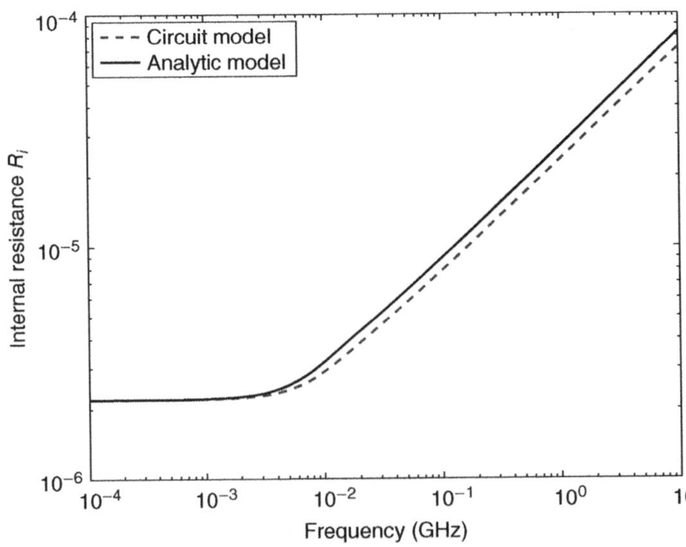

FIGURE 9.11 Internal skin-effect resistance in kiloohms for cylinder.

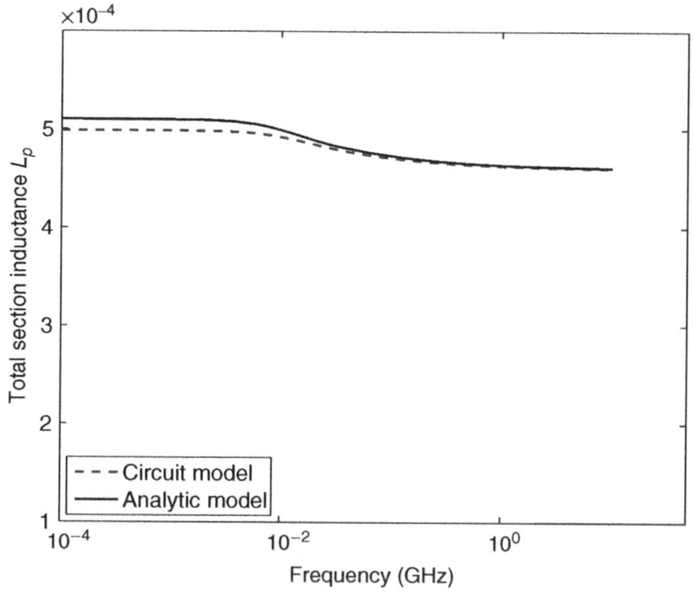

FIGURE 9.12 Resultant inner differential and external partial inductance in microhenry.

and 9.12 show a comparison for the internal differential inductance using the analytical model (9.29) and the equivalent circuit model in Fig. 9.10.

We note that this model is very similar to the small thickness GSI model. Hence, it is applicable for a practical solver.

We should note that the work in Ref. [37] also considers some mutual couplings and the radiation resistance for the cylinder. The wire model from this chapter is used in section 13.5.6 to model the high frequency loss for wire structures.

9.3 3D VOLUME FILAMENT (VFI) SKIN-EFFECT MODEL

The VFI model requires in general the 3D meshing of the conductor by the methods in Chapter 8. This model has been part of the PEEC method from the beginning [38]. Of course, the model applies also to the nonorthogonal geometries [39] which are considered in Chapter 7. Hence, most techniques in this chapter apply to nonorthogonal shapes.

VFI skin-effect models can be used for many applications. An example is the on-chip wiring [40] where the skin effect is not well pronounced. In this case, very few subdivisions of the wire are needed. This keeps the number of couplings low and the overall efficiency high.

For other applications where the skin depth δ is much smaller than the cross-sectional dimension and a straightforward implementation of the approach is not efficient. Very small filament conductors are required on the surface to represent the skin effect for this case. This has led to more sophisticated implementation of the VFI, which we consider after introducing more details.

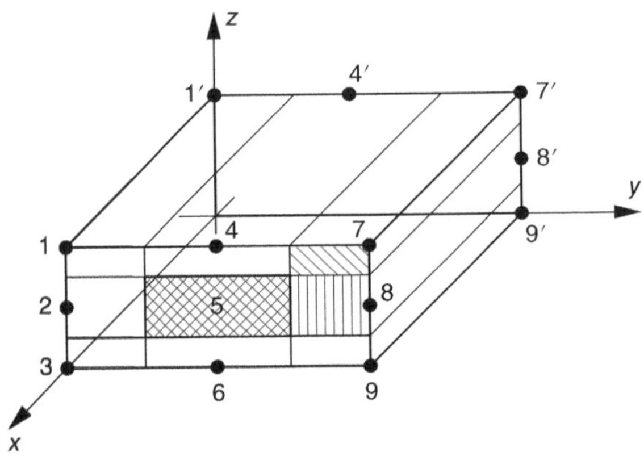

FIGURE 9.13 Volume filament subdivisions for a single conductor bar with 1D current flow.

In this section, we treat wide-frequency-band skin-effect models since they are necessary for many applications. In general, several approximations can be made to speed up the solution. Here, we consider approximation based approaches while we will not consider matrix based speed-up techniques, for example, Ref. [41]. The following section is based on these approaches.

9.3.1 Approximate 3D VFI Model with 1D Current Flow

We start this section with the model in Fig. 9.13. Such a model may be applied for an approximate PEEC model where it is clear that the current flow is in the x-direction. We should note that the models considered in this section are based on the 1D assumptions about the current flow. The cross section is subdivided with a conventional PEEC mesh.

In the model in Fig. 9.13, each subbar, which we call a filament, can carry a different current. All surface cells, which excludes node 5, connect to a capacitive part for a complete PEEC model. However, since there are no cross connections, errors may be introduced in the current flow representation especially for nonstraight wires. We should note that the connective surfaces at the ends will not carry capacitive cells since the next section will join at the interface. Of course, for a complete model, capacitive surface cells are connected only to the external surface nodes.

9.3.2 Shorts at the Intersections

As another approximation, we can short all filament conductors at each joining section where we connect the ends of all the filament conductors together as shown in the example in Fig. 9.14. This approach has been pursued in Refs [42, 43]. It is clear that this approach is most suitable for high-aspect-ratio conductors where the conductor cross section is small compared to the section length.

From the equivalent circuit in Fig. 9.14, we can see that the dc or low-frequency current for each of the filament conductors I_k is determined by

$$I_k = \frac{V_{11'}}{R_k} = \frac{V_{11'}\,\sigma \mathcal{A}_k}{\ell}, \tag{9.30}$$

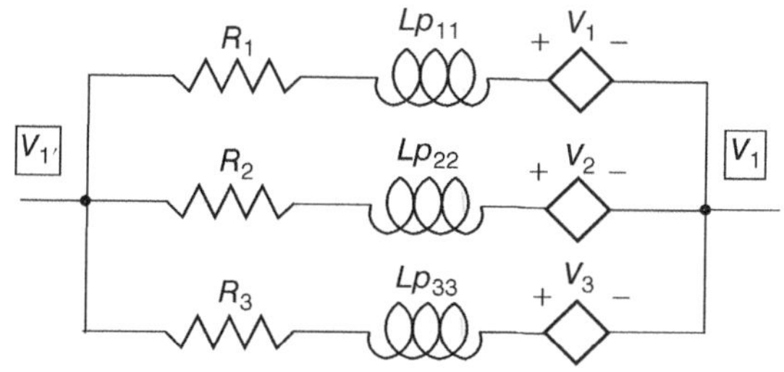

FIGURE 9.14 Equivalent circuit for 1D current VFI section with shorts.

where for this case R_k is the filament resistance and $\mathcal{A}_k$ is its cross-sectional area. Then, from this the total current will be

$$I_{tot} = \sum_{k=1}^{N} I_k = V_{11'} \frac{\sigma}{\ell} \sum_{k=1}^{N} \mathcal{A}_k, \tag{9.31}$$

since the length ℓ of all filament conductors is same. This shows that the current distributes itself uniformly where each filament carries a current proportional to its cross-sectional area. The inductive impedance at high frequencies ωL dominates over the resistance and the current flow is determined by the inductances. We observe that the coupled voltages due to the partial mutual inductances are larger toward the center of the conductor. They are represented by the voltage sources V_k in Fig. 9.14. We fundamentally can understand that the high-frequency current will be concentrated toward the periphery of the conductor due to the stronger mutual coupling for the internal conductor filaments. This is a physical interpretation of the skin effect.

9.3.3 Proximity Effect

We use a two-conductor example in Fig. 9.15 to consider the proximity effect. It consists of how the presence of the second conductor will impact the current flow or the partial inductance of the first conductor.

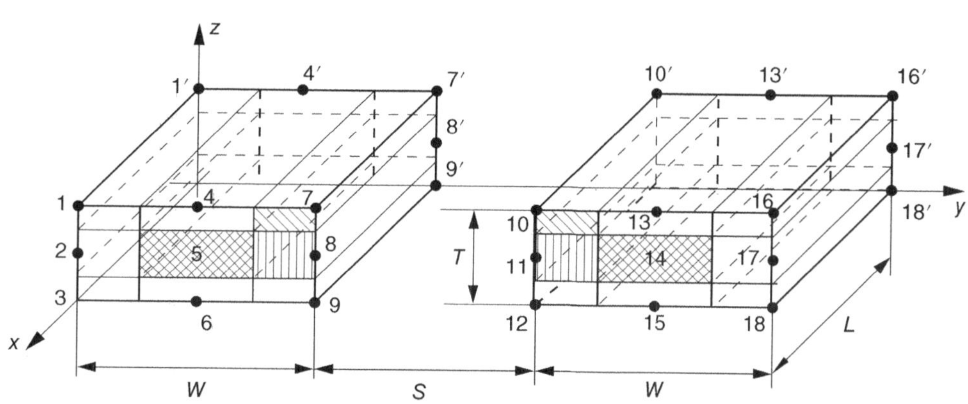

FIGURE 9.15 Two volume filament (VFI) model bars separated by distance s.

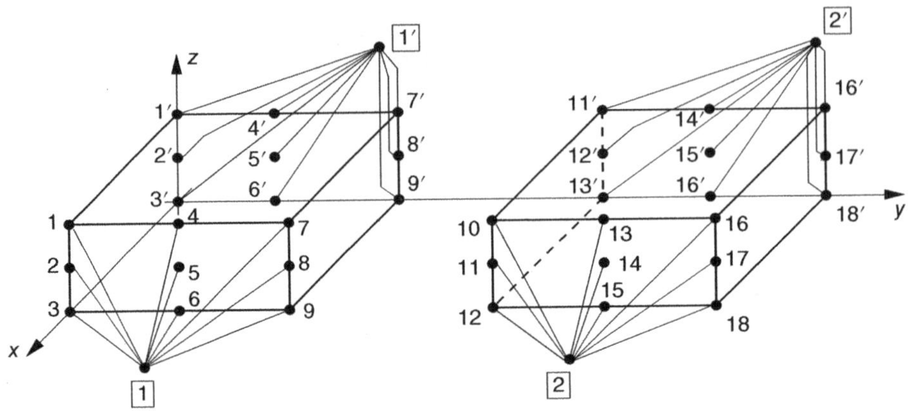

FIGURE 9.16 Shorts connections between faces for the two-conductor example.

To study the proximity effect, we again use the same VFI 1D current flow model for both conductors where the ends are shorted as shown in Fig. 9.16. This shorted model was also used for including the skin effect in thin conductors in Ref. [44].

9.3.4 Circuit Equations for Proximity Effect Study

The equivalent circuit for this 1D-VFI model is shown in Fig. 9.17. All the filament conductors are in parallel. We choose both conductors to be lined up and to be of the same length ℓ, width W, and thickness T while the edge-to-edge separation is S. For each conductor, we divide the width into N_W filament conductors and the thickness into N_T filaments while the current density is uniform in all filament conductors. Hence, the resistance of each filament is given by $R_k = \frac{\ell}{\sigma WT} N_W N_T$. The partial self- and mutual inductances are computed for each filament conductor from Appendix C by using formula (C.35) and (C.39). The voltage for

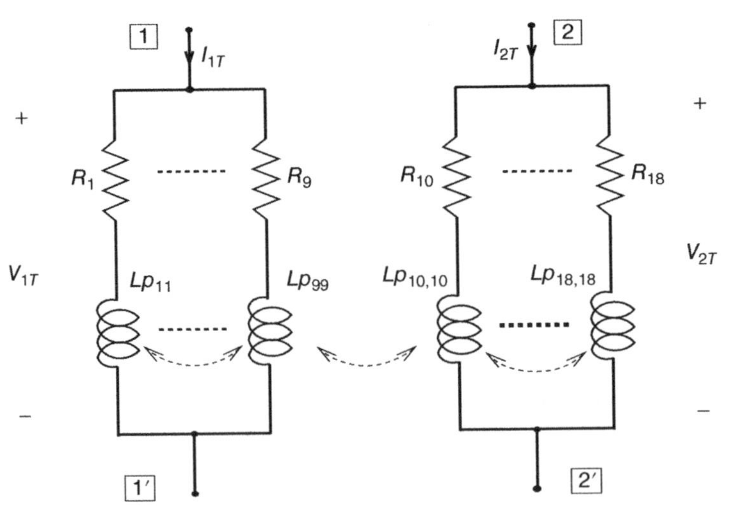

FIGURE 9.17 VFI circuit model for the two-conductor problem in Fig. 9.16.

each filament bar in Fig. 9.17 is given by

$$V_k = R_k I_k + sLp_{kk} I_k + s \sum_{k \neq m} Lp_{km} I_m. \tag{9.32}$$

Corresponding to Fig. 9.17, we can write a matrix system for all the filament bars as (9.32)

$$\begin{bmatrix} V_1 \\ V_2 \end{bmatrix} = \begin{bmatrix} R_1 + s\mathbf{Lp}_{11} & s\mathbf{Lp}_{12} \\ s\mathbf{Lp}_{21} & \mathbf{R}_2 + \mathbf{Lp}_{22} \end{bmatrix} \begin{bmatrix} \mathbf{I}_1 \\ \mathbf{I}_2 \end{bmatrix}, \tag{9.33}$$

where

$$\mathbf{V}_1 = \begin{bmatrix} V_1 \\ V_2 \\ \vdots \\ V_N \end{bmatrix}, \quad V_2 = \begin{bmatrix} V_{N+1} \\ V_{N+2} \\ \vdots \\ V_{2N} \end{bmatrix} \tag{9.34}$$

with

$$\mathbf{R}_1 = \mathrm{diag}(R_1, R_2, \ \dots \ , R_N), \ \mathbf{R}_2 = \mathrm{diag}(R_{N+1}, R_{N+2}, \ \dots \ , R_{2N}). \tag{9.35}$$

Next, we would like to compute the behavior of the terminal model corresponding to the external variables, which is of the form

$$\begin{bmatrix} \mathbf{V}_{1T} \\ \mathbf{V}_{2T} \end{bmatrix} = \begin{bmatrix} \mathbf{Z}_{11} & \mathbf{Z}_{12} \\ \mathbf{Z}_{21} & \mathbf{Z}_{22} \end{bmatrix} \begin{bmatrix} \mathbf{I}_{1T} \\ \mathbf{I}_{2T} \end{bmatrix}. \tag{9.36}$$

It is clear that due to the short circuit connections that $V_1 = V_2 = \cdots = V_N = V_{1T}$ and $V_{N+1} = V_{N+2} = V_{2N} = V_{2T}$. We start by setting the voltages along the filaments to

$$\mathbf{V}_{1T} = \begin{bmatrix} V_{1T} \\ V_{1T} \\ \vdots \\ V_{1T} \end{bmatrix}, \quad \mathbf{V}_{2T} = \begin{bmatrix} V_{2T} \\ V_{2T} \\ \vdots \\ V_{2T} \end{bmatrix}. \tag{9.37}$$

If we solve, or invert the system (9.33), we get

$$\begin{bmatrix} \mathbf{I}_{1T} \\ \mathbf{I}_{2T} \end{bmatrix} = \begin{bmatrix} \mathbf{Z}_{11} & \mathbf{Z}_{12} \\ \mathbf{Z}_{21} & \mathbf{Z}_{22} \end{bmatrix}^{-1} \begin{bmatrix} \mathbf{V}_{1T} \\ \mathbf{V}_{2T} \end{bmatrix} = \begin{bmatrix} \mathbf{Y}_{11} & \mathbf{Y}_{12} \\ \mathbf{Y}_{21} & \mathbf{Y}_{22} \end{bmatrix} \begin{bmatrix} \mathbf{V}_{1T} \\ \mathbf{V}_{2T} \end{bmatrix}. \tag{9.38}$$

Finally, we can obtain the 2×2 impedance matrix of interest

$$\begin{bmatrix} V_{1T} \\ V_{2T} \end{bmatrix} = \begin{bmatrix} Z_{11} & Z_{12} \\ Z_{21} & Z_{22} \end{bmatrix} \begin{bmatrix} I_{1T} \\ I_{2T} \end{bmatrix} \tag{9.39}$$

where I_{1T} is the sum of currents of the vector $\boldsymbol{I}_{1T}$, which contains all the filament currents on conductor 1, and I_{2T} is the sum of currents of the vector $\boldsymbol{I}_{2T}$, which yields all the filament currents on conductor 2. Hence, the impedances can easily be obtained by summing up all the currents.

$$Z_{11} = \frac{V_{1T}}{\sum_{k=1}^{N} I_k}, \quad Z_{21} = \frac{V_{1T}}{\sum_{k=N+1}^{2N} I_k}. \tag{9.40}$$

We finally can rewrite this in a form where each of the four impedances Z_{ij} consists of a frequency-dependent resistance and inductance, or

$$\begin{bmatrix} V_{1T} \\ V_{2T} \end{bmatrix} = \begin{bmatrix} R_{11}(\omega) + j\omega L_{11}(\omega) & R_{12}(\omega) + j\omega L_{12}(\omega) \\ R_{21}(\omega) + j\omega L_{21}(\omega) & R_{22}(\omega) + j\omega L_{22}(\omega) \end{bmatrix} \begin{bmatrix} I_{1T} \\ I_{2T} \end{bmatrix}. \tag{9.41}$$

Next, we give an example for the frequency-dependent parameters for two rectangular conductors in Fig. 9.15 for the above derivation. They consist of two bars with $\ell = 1$ cm, $W = T = 50$ μm, and spacing $S = 55$ to 100 μm. The subdivision of the conductors with nine segments along W and T results in a total of 81 filament conductors for each of the two conductors. Hence, the matrix size in (9.32) is 162 × 162. The choice of the number of subdivisions should be such that the cross-sectional dimensions of each subbar is less than a skin depth at the highest frequency of interest. The results given are for a frequency range of 100 kHz to 100 MHz. For nine divisions, the thickness and width of each subbar will be one skin depth at 141 MHz.

Figure 9.18 shows the total real part for the bar impedance V_{1T}/I_{1T}, in isolation and in the presence of the second conductor, for different spacings. Observe that the resistance or real part begins to deviate from its dc value around 10 MHz and varies as 10 dB/decade above that as expected. The bar $W = T = 50$ μm is 2 skin depths at 6.987 MHz. Also shown is the result assuming that the current is uniformly distributed over the cross section. Observe that the bars are very close and the proximity effect causes the resistance to be significantly higher than the isolated bar when skin effect is well pronounced. The second conductor is shorted at its end. This is somewhat different from a true 3D VFI model.

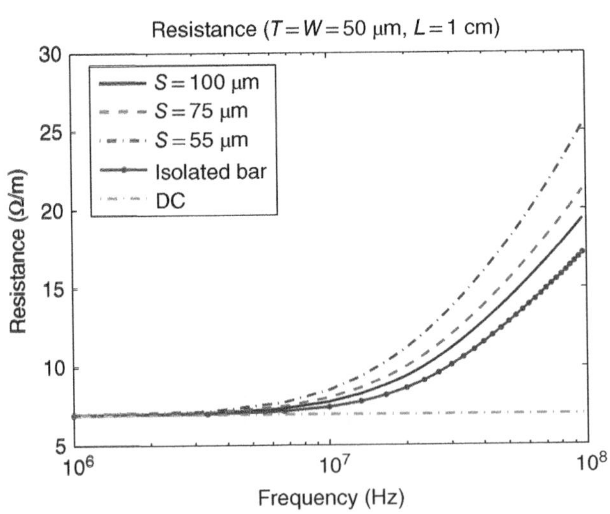

FIGURE 9.18 Resistance of bar versus frequency.

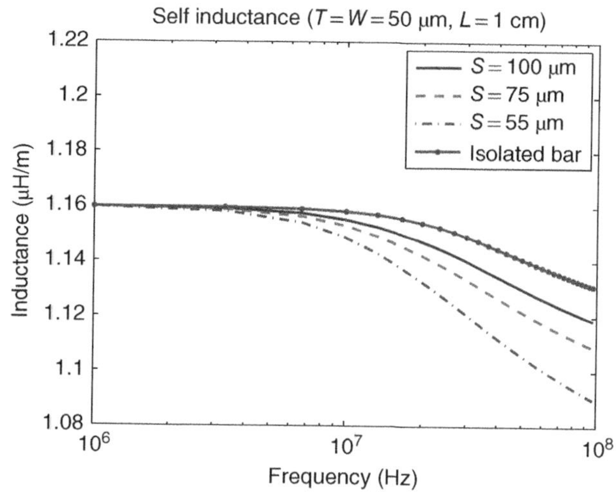

FIGURE 9.19 Inductance of bar versus frequency.

L_{11}, in isolation and in the presence of the other bar, for different spacings. The proximity effect is clearly visible in the reduction of the self-inductance as the bars get closer for frequencies larger than 10 MHz in Fig. 9.19.

In Fig. 9.20, we show the total mutual inductance between the conductors. Observe that this total mutual inductance changes very little with frequency and is also relatively independent of the presence of the other bar. Hence, a single voltage source is sufficient to take the bar-to-bar coupling into account where the coupling model is a constant mutual inductance. This is shown in the macromodel in Fig. 9.19, which consists of a model for each bar in isolation together with the coupling model for the bars.

The above VFI model represents a convenient approach for long conductors. Hence, it is only recommended for long quasi-2D models. Other interesting skin-effect issues occur where the current flow at low frequencies is much different from the high-frequency

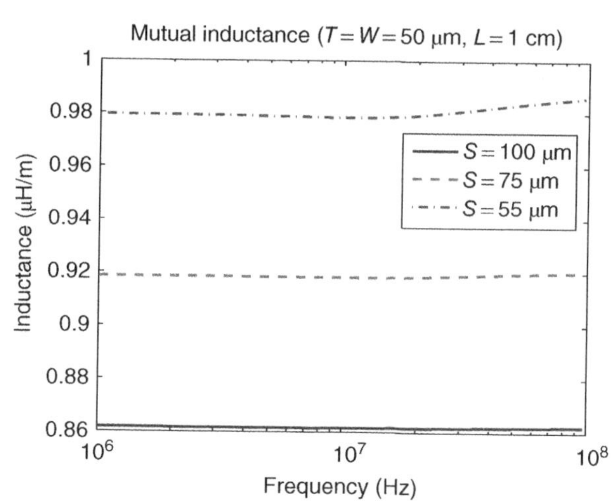

FIGURE 9.20 Mutual inductance of bar versus frequency.

path. It was shown in Ref. [45] that large inductance variations can occur for nonstraight conductors. In fact, in this example low-frequency inductance was shown to be almost 10 times larger than the high-frequency inductance.

9.3.5 Full 3D Current Flow Skin-Effect Models

Nonuniform meshing is a key factor in the efficiency of the VFI model. The large impact is evident from Sections 9.2.3 and 9.2.6. For true 3D models, it is important to use the nonuniform meshes in all three space dimensions such that all outside layers are thin. In fact, the same rules as in Fig. 9.2 can be applied for the choice of the outside layer thickness being a fraction of the skin depth δ at the highest frequency of interest. The layer thickness should be increased by a factor of 1.1 to 3 depending on the specifics as suggested in Section 8.4.1. With such an increase, it is clear that only a few layers are needed with such a substantial thickness increase for most geometries. This results in large cells in the interior of the conductor. Without any speedup technique, only medium-sized problems can be solved on a computer with moderate memory.

Other improvements may apply for the modeling of a special class of problems. For example, it was suggested in Ref. [11], that for some geometries, a set of template current cell can be precomputed so that some of the computational work is done before the actual computation of the current.

9.3.6 Equivalent Circuit for 3D VFI Model

We give an example of the meshing and equivalent circuit for a 3D VFI model for a metal body with a finite thickness. To give details on the meshing, we show the conventional horizontal x–y PEEC cells for a particular level in Fig. 9.21. For most conventional cases, the cells at all levels must be vertically aligned as is shown in Fig. 9.2. The vertical layer-to-layer connection and a top view of this connection is shown in Fig. 9.22. Hence, all three directions of currents internal to the conductors are properly represented in this conventional PEEC mesh. The PEEC equivalent circuit corresponding to a conductor node is shown in Fig. 9.23. Different approximations have been made to include the partial inductance couplings [46–48].

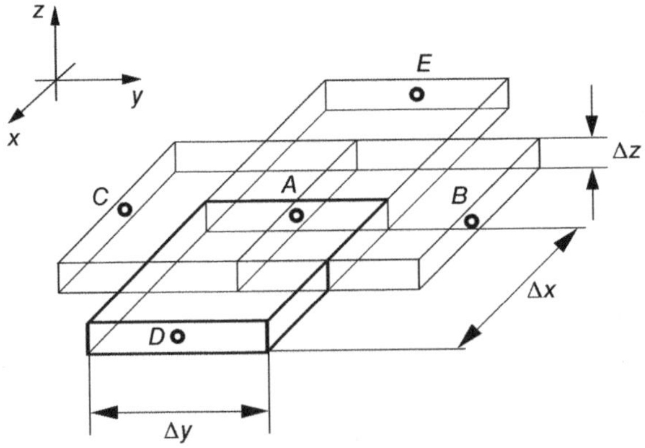

FIGURE 9.21 Horizontal partitioning of one layer of a metal sheet.

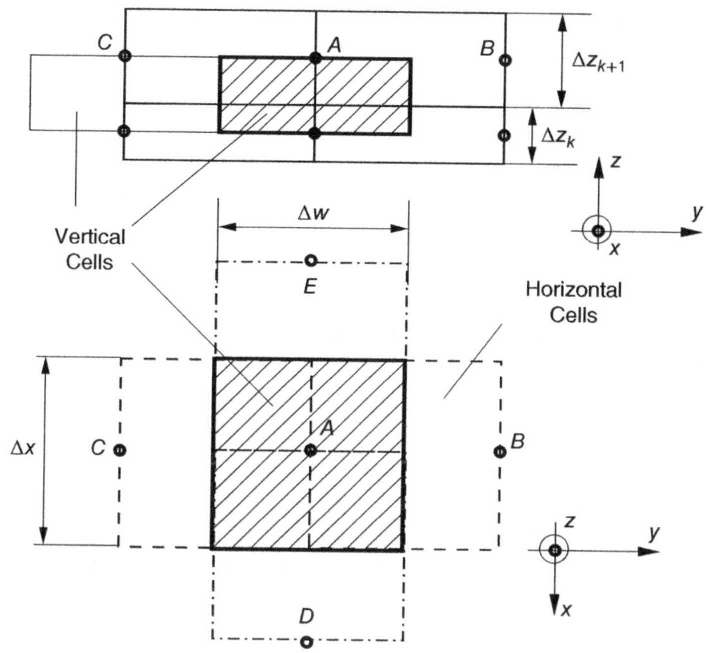

FIGURE 9.22 Example of vertical cell to connect the skin-effect layers.

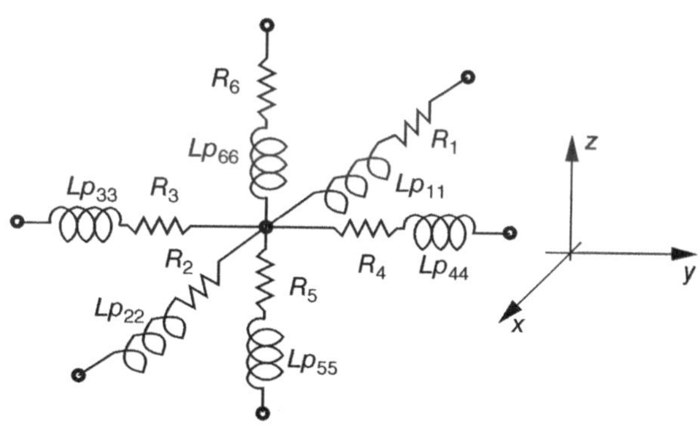

FIGURE 9.23 VFI conductor internal 3D Skin-effect PEEC circuit.

In Ref. [27], we attempted to construct a full three-dimensional model using the internal–external inductance approximations for general 3D conductors. This model is an extension of the 1D GSI in Section 9.2.3. The conductor-to-conductor coupling is used in an internal 3D DE model. Hence, the fundamental approach is not much different from the thin conductor model in Section 9.2.3. The challenging aspect of this problem is that we need a true three-dimensional current flow representation. An example of the work is shown in Figs. 9.30 and 9.31. It is evident from these results that the accuracy of these results is marginal. We need to observe that the internal–external surface model used in this approach is an approximation. From this work, we conclude that it is less suitable for thick conductors, whereas this model works well for the thin GSI model.

9.3.7 Surface Equivalence Theorem-Based Skin-Effect Model

Surface integral equation-based skin-effect models are another important class of solution approaches. Surface-based models are presented in Refs [23, 49] and for 2D models in Ref. [8]. In these approaches, unknowns are not placed inside the conductors as is done in volume models as is shown in Section 11.4. We recognized before that a local approximation with a conventional surface impedance model 9.2.1 leads to the same very high frequency-limited solution only for a very thin layer that corresponds to a strong skin effect.

Theoretically, surface-type integral equation method can result in good models for skin-effect losses for piecewise homogeneous conductors. However, problems may occur in a numerical implementation due to the large difference in resistance between the conductor and the surrounding dielectric, which makes the resultant linear system highly unbalanced. Also, the general solution is based on four integral equations, two in Region 1 and two in Region 2 in Fig. 9.1 as is shown in Section 11.4.

In this section, we consider a simplified surface approach called generalized impedance boundary condition (GIBC) approach given in Refs [50, 51]. This new approach is specifically tailored to problems with lossy materials. The GIBC solution considered here is presented with the currents-only electrical field integral equation (EFIE) formulation presented in (3.69) in Chapter 3, which does not give a dc solution.

In this approach, it is necessary to have a closed surface S that separates space into the external region we call 1 and interior region 2, using the equivalence principle, the current-only electric field integral equation (cEFIE) for conductors with a closed surface can be written as

$$
\begin{aligned}
&j\omega\mu_1\, \hat{t}\cdot \int_S \left(\bar{\mathbf{I}}+\frac{\nabla\nabla}{\beta_1^2}\right) g_1(\mathbf{r},\mathbf{r}') \cdot \boldsymbol{J}_{se}(\mathbf{r}')\, dS' \\
&-\hat{t}\cdot \int_S \nabla\, g_1(\mathbf{r},\mathbf{r}') \times \boldsymbol{J}_{sm}(\mathbf{r}')\, dS' = -\hat{t}\cdot \boldsymbol{E}^{inc}(\mathbf{r}),
\end{aligned}
\tag{9.42}
$$

where $\boldsymbol{r} \in S^-$. S^- is the inside of the closed boundary surface. The vector $\hat{t}$ is tangential to the boundary surface S and g_1 is the scalar free space Green's function based in region 1, which is the external region. Also, β_1 is the free space wave number in region 1 while $\boldsymbol{J}_{se}$ is the equivalent electric current on the surface S, and $\boldsymbol{J}_{sm}$ is the equivalent magnetic current on the surface S.

An analog magnetic surface integral equation magnetic field integral equation (MFIE) is also used. A PEEC-type MFIE formulation is presented in Section 11.4.1. The MFIE equation for the magnetic field for the skin-effect model is

$$
\begin{aligned}
&-\hat{t}\cdot \int_S \nabla\, g_1(\mathbf{r},\mathbf{r}') \times \boldsymbol{J}_{se}(\mathbf{r}')\, dS' \\
&+j\omega\mu_1\, \hat{t}\cdot \int_S \left(\bar{\mathbf{I}}+\frac{\nabla\nabla}{\beta_1^2}\right) g_1(\mathbf{r},\mathbf{r}') \cdot \boldsymbol{J}_{sm}(\mathbf{r}')\, dS' = -\hat{t}\cdot \boldsymbol{H}^{inc}(\mathbf{r}).
\end{aligned}
\tag{9.43}
$$

For convenience, we name the so-called operator

$$\mathcal{L}_{vE}(\mathbf{r},\mathbf{r}')\,\boldsymbol{J}(\mathbf{r}') = j\omega\mu_v \int_S \left(\bar{\mathbf{I}} + \frac{\nabla\nabla}{\beta_v^2}\right) g_v(\mathbf{r},\mathbf{r}')\cdot \boldsymbol{J}(\mathbf{r}')\, dS' \tag{9.44}$$

and also

$$\mathcal{K}_{vE}(\mathbf{r},\mathbf{r}')\boldsymbol{J}(\mathbf{r}') = -\int_S \nabla g_v(\mathbf{r},\mathbf{r}') \times \boldsymbol{J}_{se}(\mathbf{r}')\, dS', \tag{9.45}$$

where v is the index of the region. For this case, $v = 1$ or $v = 2$. The above cEFIE formulation can be written as

$$\hat{\boldsymbol{t}}\cdot\mathcal{L}_{1E}(\mathbf{r},\mathbf{r}')\,\boldsymbol{J}_{se}(\mathbf{r}') + \hat{\boldsymbol{t}}\cdot\mathcal{K}_{1E}(\mathbf{r},\mathbf{r}')\boldsymbol{J}_{sm}(\mathbf{r}') = -\hat{\boldsymbol{t}}\cdot\boldsymbol{E}^{inc}(\mathbf{r}) \tag{9.46}$$

and the cMFIE formulation is simplified to

$$\hat{\boldsymbol{t}}\cdot\mathcal{K}_{1E}(\mathbf{r},\mathbf{r}')\,\boldsymbol{J}_{se}(\mathbf{r}') + \hat{\boldsymbol{t}}\cdot\mathcal{L}_{1E}(\mathbf{r},\mathbf{r}')\boldsymbol{J}_{sm}(\mathbf{r}') = -\hat{\boldsymbol{t}}\cdot\boldsymbol{H}^{inc}(\mathbf{r}), \tag{9.47}$$

where $\boldsymbol{r} \in S^-$ which is for the exterior region since the Green's function is based in region 1.

For the interior region 2, a similar formulation can be achieved since there is no incident field inside. The cEFIE for the internal region is

$$\begin{aligned} j\omega\mu_2\,\hat{\boldsymbol{t}}\cdot &\int_S \left(\bar{\mathbf{I}} + \frac{\nabla\nabla}{\beta_2^2}\right) g_2(\mathbf{r},\mathbf{r}')\cdot \boldsymbol{J}_{se}(\mathbf{r}')\, dS' \\ &-\hat{\boldsymbol{t}}\cdot \int_S \nabla\, g_2(\mathbf{r},\mathbf{r}') \times \boldsymbol{J}_{sm}(\mathbf{r}')\, dS' = 0, \end{aligned} \tag{9.48}$$

where $\boldsymbol{r} \in S^+$, which is right outside the boundary surface S or conductor. Further, the cMFIE for the interior region is

$$\begin{aligned} -\hat{\boldsymbol{t}}\cdot &\int_S \nabla\, g_2(\mathbf{r},\mathbf{r}') \times \boldsymbol{J}_{se}(\mathbf{r}')\, dS' \\ +\, j\omega\mu_2\,\hat{\boldsymbol{t}}\cdot &\int_S \left(\bar{\mathbf{I}} + \frac{\nabla\nabla}{\beta_2^2}\right) g_2(\mathbf{r},\mathbf{r}')\cdot \boldsymbol{J}_{sm}(\mathbf{r}')\, dS' = 0. \end{aligned} \tag{9.49}$$

By using operator representations, the inside surface integral equations can be simplified to

$$\hat{\boldsymbol{t}}\cdot\mathcal{L}_{2E}(\mathbf{r},\mathbf{r}')\,\boldsymbol{J}_{se}(\mathbf{r}') + \hat{\boldsymbol{t}}\cdot\mathcal{K}_{2E}(\mathbf{r},\mathbf{r}')\boldsymbol{J}_{sm}(\mathbf{r}') = 0 \tag{9.50}$$

and

$$\hat{\boldsymbol{t}}\cdot\mathcal{K}_{2E}(\mathbf{r},\mathbf{r}')\,\boldsymbol{J}_{se}(\mathbf{r}') + \hat{\boldsymbol{t}}\cdot\mathcal{L}_{2E}(\mathbf{r},\mathbf{r}')\boldsymbol{J}_{sm}(\mathbf{r}') = 0, \tag{9.51}$$

where $\boldsymbol{r} \in S^+$ on the inside surface.

By setting the field point $\boldsymbol{r}$ on the boundary surface S, and combining the exterior region cEFIE (9.46) with the internal cMFIE (9.51), a cEFIE for only the equivalent current $\boldsymbol{J}_{se}$ can be obtained. The interesting result is that we obtain an equivalent scalar impedance Z_s for the equivalent electric current as

$$\left[\hat{\boldsymbol{t}} \cdot \mathcal{L}_{1E}(\mathbf{r},\mathbf{r}') - \hat{\boldsymbol{t}} \cdot \mathcal{K}_{1E}(\hat{\boldsymbol{t}} \cdot \mathcal{L}_{2E})^{-1}\, \hat{\boldsymbol{t}} \cdot \mathcal{K}_{2E}\right] \boldsymbol{J}_{se}(\mathbf{r}') = -\hat{\boldsymbol{t}} \cdot \boldsymbol{E}^{inc}(\mathbf{r}) \tag{9.52}$$

or

$$[\,\hat{\boldsymbol{t}} \cdot \mathcal{L}_{1E}(\mathbf{r},\mathbf{r}') - Z_s(\mathbf{r},\mathbf{r}')]\, \boldsymbol{J}_{se}(r') = -\hat{\boldsymbol{t}} \cdot \boldsymbol{E}^{inc}(\mathbf{r}) \tag{9.53}$$

which is expanded

$$j\omega\mu_1\, \hat{\boldsymbol{t}} \cdot \int_S \left(\bar{\mathbf{I}} + \frac{\nabla\nabla}{\beta_1^2}\right) g_1(\mathbf{r},\mathbf{r}') \cdot \boldsymbol{J}_{se}(\mathbf{r}')\, dS' - Z_s(\mathbf{r},\mathbf{r}')\boldsymbol{J}_{se}(\mathbf{r}') = -\hat{\boldsymbol{t}} \cdot \boldsymbol{E}^{inc}(\mathbf{r}), \tag{9.54}$$

where Z_s is the GIBC that can be implemented in a PEEC model as shown in Fig. 9.3,

$$\begin{aligned} Z_s(\mathbf{r},\mathbf{r}')\boldsymbol{J}_{se}(\mathbf{r}') &= \hat{\boldsymbol{t}} \cdot \mathcal{K}_{1E}(\hat{\boldsymbol{t}} \cdot \mathcal{L}_{2E})^{-1}\hat{\boldsymbol{t}} \cdot \mathcal{K}_{2E}(\mathbf{r},\mathbf{r}')\boldsymbol{J}_{se}(\mathbf{r}') \\ &= \hat{\boldsymbol{t}} \cdot \int_S dS'''\, \nabla g_1(\mathbf{r},\mathbf{r}''') \\ &\quad \times \left[\hat{\boldsymbol{t}} \cdot j\omega\mu_2 \int_S \left(\bar{\mathbf{I}} + \frac{\nabla'''\nabla'''}{\beta_2^2}\right) g_2(\mathbf{r}''',\mathbf{r}'')\cdot\right]^{-1} dS'' \\ &\quad \hat{\boldsymbol{t}} \cdot \int_S \nabla''\, g_2(\mathbf{r}'',\mathbf{r}') \times \boldsymbol{J}_{se}(\mathbf{r}')\, dS'. \end{aligned} \tag{9.55}$$

It is noted that the operator inverse $(\hat{\boldsymbol{t}} \cdot \mathcal{L}_{2E})^{-1}$ is evaluated through the multifrontal method [52]. The multifrontal method is a direct method for solving the linear equation inversion. It organizes operations during the factorization of sparse matrices and uses the low rank condition. A tree is employed to handle the dependency between partial factorizations.

The GIBC method employs the conventional delta gap feed model for a finite size cross section where the feed current is only applied to the surface. This is different from VFI, which feeds the current throughout the entire conductor cross section. Hence, the surface feeding scheme used in GIBC could generate an error in the result if an insufficient number of cells are used. However, we found that the results for a large number of cells agreed very well with the VFI model. In the following section, in Figs. 9.32 and 9.33, results are presented where we compare the accuracy of the GIBC model to the VFI model.

9.4 COMPARISONS OF DIFFERENT SKIN-EFFECT MODELS

The purpose of this section is to illustrate the features of the different models presented in this chapter. Importantly, the chosen problems are such that the current path changes with frequency or time, depending on the type of solver used. We again want to emphasize that for many applications broadband models, which are valid for a wide frequency range, are of importance. Of course, this is also true for time domain applications. It is clearly difficult to obtain such broadband skin-effect results for a large class of 3D structures using approximations. Fortunately, this was pointed out to the first author of the book by Pinello before the end of the last century [53]. Yet, this issue is generally not well understood. We assume for the experiments that all conductors are copper embedded in air. The external air environment in all our solutions is represented using an EFIE model described. However, we exclude the dielectric part of the PEEC model to make certain that the skin-effect results are not masked by resonances due to the external part of the geometry.

The PEEC VFI model was used in all comparisons, which includes full 3D current inside the conductors. The GSI model was implemented and applied to thin as well as thick conductors. Finally, a GIBC surface formulation solver [50] was used in the comparison. All computations are done on a single processor with 8 GB of random access memory. So, this limited some of the very high frequency accuracy. For very high frequency results, a larger number of small cells and computer memory are required.

The examples given are interesting wide frequency range problems where the 3D current path is not predetermined. Specifically, the horseshoe (HShoe) problem consists of one layer shown in Fig. 9.24, but it has wide flat corners that lead to a current redistribution with frequency. The geometrical data for this example is shown in Table 9.1. The additional data required is $w = 10\,\mu\text{m}$ and the gap width is $w_c = 0.2\,\mu\text{m}$. We used two different models for the contacts area at the inside of the gap. In one case, we use a 0.4 μm section at the contact surface with a conductivity that is 10^4 times larger than copper. This leads to contacts at the entire cross-sectional surface. Alternatively, point contacts are used at the front of the gap as shown in Fig. 9.24.

The second problem in Fig. 9.25 consists of an L-shaped conductor with a close ground plane. At high frequencies, the current will take the path that minimizes the inductance right

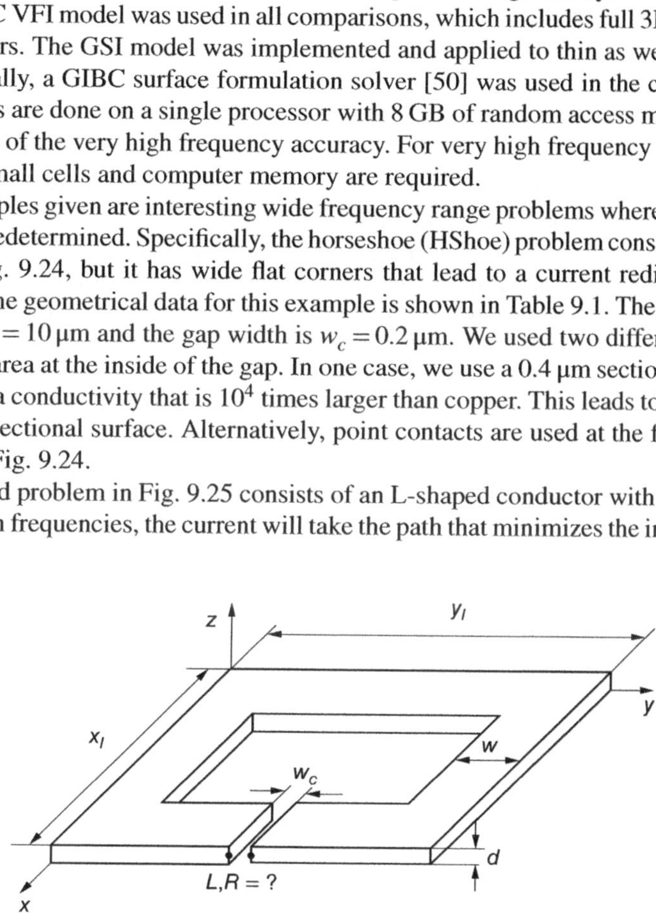

FIGURE 9.24 Geometry of U-shaped test problem called *HShoe*.

TABLE 9.1 Dimensions in micrometers for the example problems.

Problem name	x_g	x_w	x_ℓ	y_g	y_w	y_ℓ	d	z_w	z_g
HShoe1			35			50	1		
HShoe6			35			50	6		
LShape5	400	80	320	400	80	320	5	50	5
LShape18	400	80	320	400	80	320	18	50	10

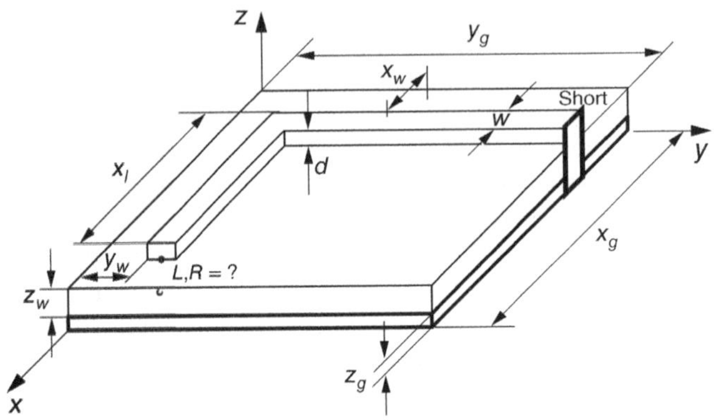

FIGURE 9.25 Shorted L-shaped conductor over ground-plane model.

under the L-shaped wire. However, at low frequencies, the current takes the path of least resistance that is across the plane. The lengths of the L conductors are relatively small for this redistribution to be very large. Hence, the change in inductance is relatively modest. Much larger changes in L and R have been obtained for longer lines [45].

The dimensions for the L-shaped problem Fig. 9.25 are also given in Table 9.1 for both the thin conductor LShape5 and the thick conductor LShape18. We chose a very wide conductor of 80 µm since it leads to a more interesting current behavior. The short-to-ground at the end of the top L-shaped conductor is accomplished with a strip that extends over the entire width from the L-shaped conductor to the ground plane as is shown in Fig. 9.25. The strip thickness is the same as that of the LShape conductor.

9.4.1 Thin Conductor Results

In this section, we utilize the thin-GSI model in Section 9.2.3. The first problem is the 1 µm thick and 10 µm wide HShoe1. We compare thin-GSI with the 3D-VFI solver results. The results are given in Figs. 9.26 and 9.27 for the inductance and resistance. Further, for the thin L-shaped conductor problem, both the ground plane and the L-shaped conductor are chosen to be 5 µm thick. We again compare the 3D-VFI solution with the thin-GSI solution. We observe that the agreement between the solutions in Figs. 9.28 and 9.29 is good. The difference in the high-frequency resistance is due to the limited number of cells used in the 3D-VFI solution.

9.4.2 Thick Conductor Results

Our comparison with thicker conductor problems is first for the thick GSI solution in Section 9.3. The result we give is for the HShoe problem where the conductor thickness is 6 µm with a width of 10 µm. The results shown in Fig. 9.30 for the resistance and Fig. 9.31 for the inductance were obtained with a simplified approximate form of the approach in Ref. [54]. It is evident that the approach does not yield a satisfactory solution.

For the second thick conductor example, we compare GIBC solution with the 3D-VFI results for the HShoe problem with the thick $d = 6$ µm version in Fig. 9.32 for the resistance and Fig. 9.33 for the inductance. The agreement is good with the exception of the very high frequency response, which is due to the limited number of cells used for the 3D-VFI model.

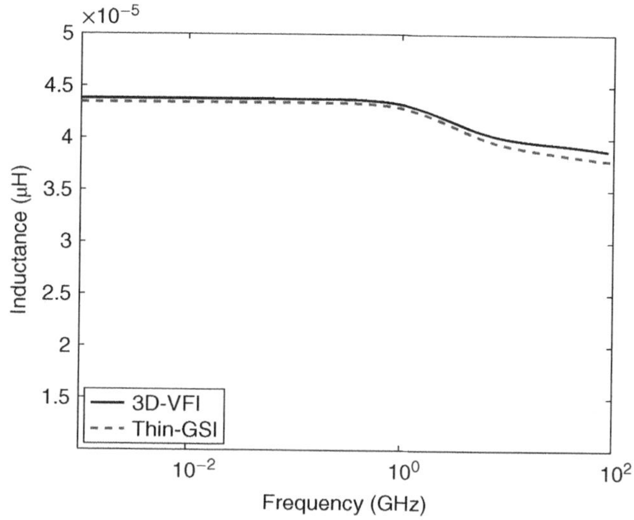

FIGURE 9.26 Inductance for HShoe with 1 μm conductor with thin GSI solver.

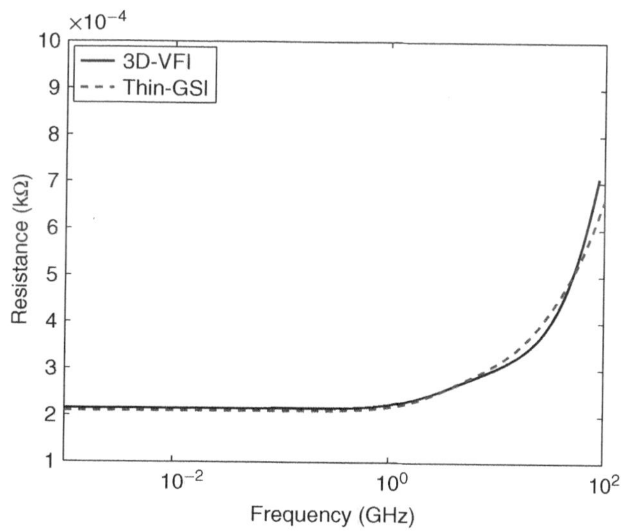

FIGURE 9.27 Resistance for HShoe with 1 μm conductor with thin GSI solver.

9.4.3 Comparison of Example Results

It is difficult to make direct comparisons between models obtained with different solvers that are programmed and running in different environments. However, we can evaluate models on how well they work in general. The problem size can vary from very simple experimental models to a very large number of elements. Further, as pointed out in the introduction to the chapters, the ability to model details of the current direction is very important for 3D models.

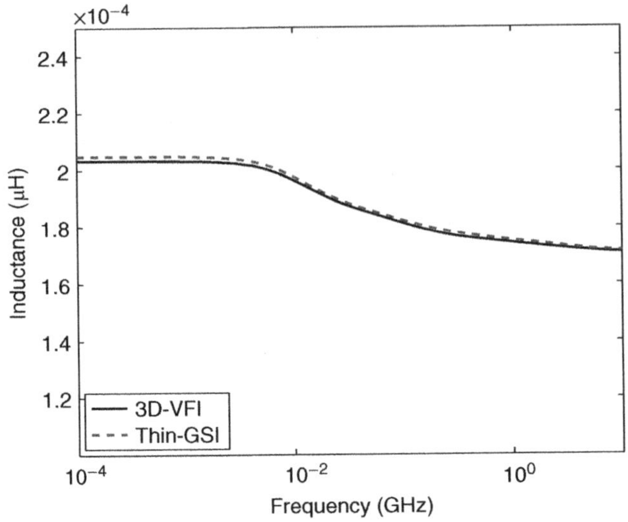

FIGURE 9.28 Inductance for thin LShape conductor over ground plane.

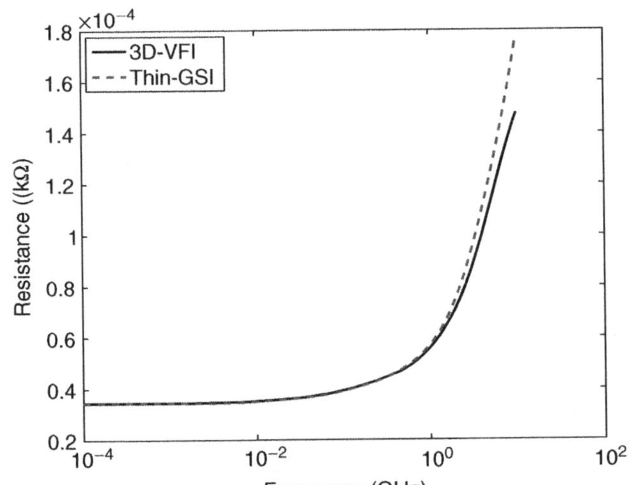

FIGURE 9.29 Resistance for thin LShape conductor over ground plane.

Problems consisting of thin planes in Section 9.2.1 and/or wires that can speed up the solutions are shown in Section 9.2.3. However, the same external–internal inductance GSI model does not work well for non thin conductors as shown in Fig. 9.30. Fundamentally, the widely used internal–external inductance model [28, 29, 30] is too approximate for thick conductor problems. This is not surprising since it is an electrical current-only approximation to the surface IE boundary conditions.

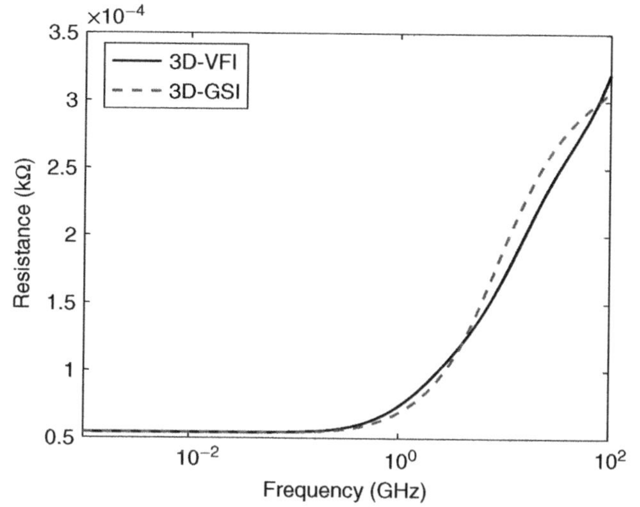

FIGURE 9.30 Resistance for HShoe, $d = 6$ μm.

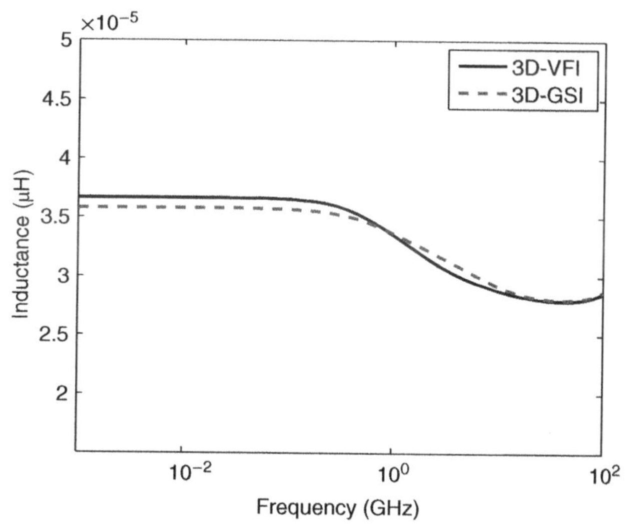

FIGURE 9.31 Inductance for Hshoe, $d = 6$ μm.

Finally, the surface equivalence GIBC solution that uses appropriate boundary conditions compares well with the VFI model in Section 9.3.7. However, the GIBC model requires a sufficiently large number of surface cells to form a good closed surface for a reasonably accurate result. This aspect is different for the VFI model that works well with a reduced accuracy even for a moderate number of cells or filaments.

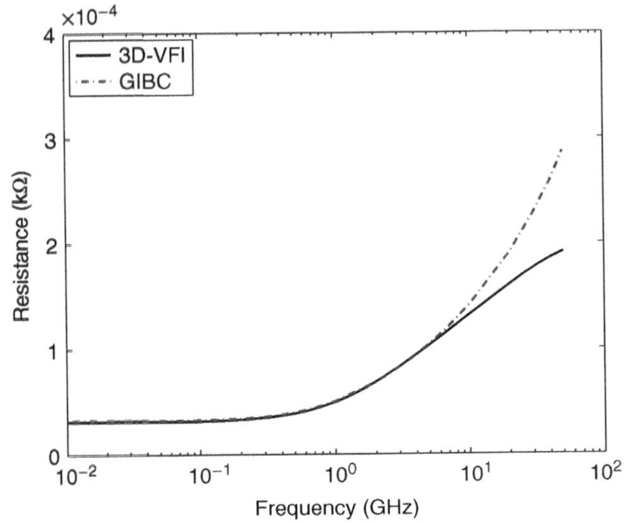

FIGURE 9.32 Resistance for HShoe with $d = 6$ μm conductor with GIBC solver.

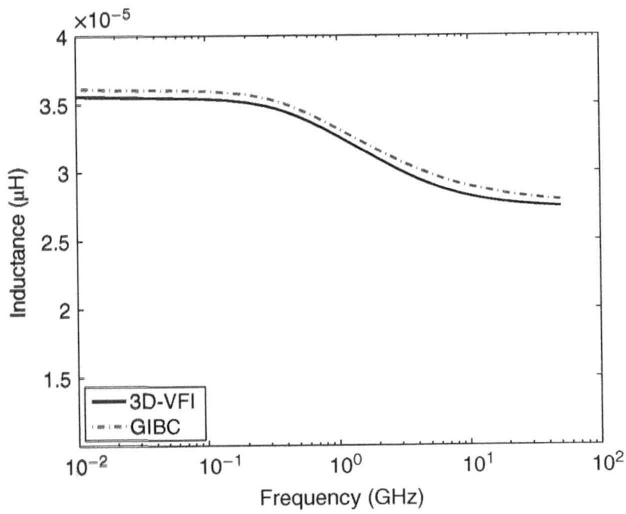

FIGURE 9.33 Inductance for HShoe with $d = 6$ μm conductor with GIBC solver.

PROBLEMS

9.1 *Volume filament approach*

Using the VFI approach, compute the impedance matrix of two parallel conductors (copper $\sigma = 5.8 \times 10^7$ S/m), illustrated in Fig. 9.34. Use the subdivisions as shown in Fig. 9.15 and the shorted ends as in Fig. 9.16.

Consider the dimensions $x_{s1} = x_{s2} = 0, x_{e1} = x_{e2} = 2$ cm, $y_{s1} = 0, y_{e1} = 0.1$ cm, $y_{s2} = y_{e1} + 0.2$ cm, $y_{e2} = y_{s2} + 0.1$ cm, $z_{s1} = z_{s2} = 0, z_{e1} = z_{e2} = 50$ μm. The result will be a frequency-dependent 2×2 impedance matrix.

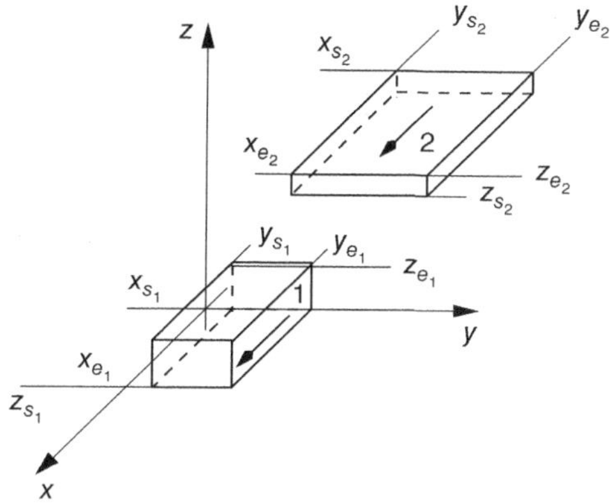

FIGURE 9.34 Two rectangular conductors.

9.2 *Construct a Single-conductor VFI skin-effect model*
Construct a single-conductor skin-effect model similar to the previous problem. Compute the impedance from end to end of the rectangular conductor (copper $\sigma = 5.8 \times 10^7$ S/m), illustrated in Fig. 9.35. Consider $x_s = 0, x_e = 5$ mm, $y_s = 0, y_e = 20$ μm, $z_s = 0$ μm, $z_e = 10$ μm. Compute the length impedance in the x-direction of the conductor at 1 MHz, 1 GHz, and 10 GHz. Use the model in Fig. 9.13 with subdivisions along the length x.

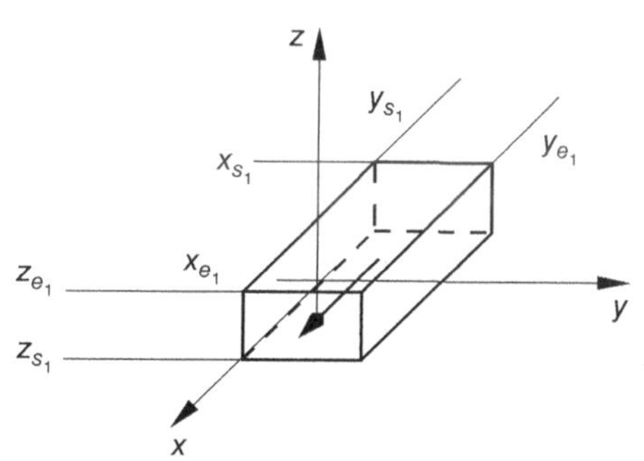

FIGURE 9.35 A rectangular bar with thickness.

9.3 *Skin effect using 1D model in Section 9.2.2*
Compare the impedance of the previous problem with a skin effect with a model that is constructed with the high-frequency 1D skin-effect model in Section 9.2.2. Hint: Assume that all four side surfaces are represented with such a 1D impedance model. Increase the cross section to $y_e = 3$ mm and $z_e = 1$ mm. Compare the result between the VFI model and the 1D surface impedance model for this larger cross section.

9.4 *Skin-effect circuit model for tube*
Build a five-layer tube equivalent circuit skin-effect model for a conductor as shown in Fig. 9.8 with $d = 0.1$ mm and $\ell = z_e - z_b = 0.5$ mm. With one segment along the length use (9.29) for the computation of the impedance along the length as a function of frequency. Compare the result with the conventional result obtained with the Bessel function result, for example, Ref. [22, p. 184].

9.5 *Generalized impedance boundary condition*
Derive the generalized impedance boundary condition (GIBC) given in (9.55).

REFERENCES

1. H. A. Wheeler. Formulas for the skin effect. *Proceedings of the IRE*, 30(9):412–424, September 1942.
2. A. E. Ruehli and A. C. Cangellaris. Progress in the methodologies for the electrical modeling of interconnect and electronic packages. *Proceedings of the IEEE*, 89(5):740–771, May 2001.
3. M. Swaminathan and A. E. Engin. *Power Integrity Modeling and Design for Semiconductors and Systems*. Prentice Hall, Upper Saddle River, NJ, 2007.
4. E.-P. Li, C. Wei, A. C. Cangellaris, E. X. Liu, M. D'Amore, J. Kim, and T. Sudo. Review of EMC analysis technologies for packaging, PCB and novel interconnects. *IEEE Transactions on Electromagnetic Compatibility*, 52(8):248–265, May 2010.
5. C. R. Paul. *Analysis of Multiconductor Transmission Lines*. IEEE Press, John Wiley Interscience, Hoboken, NJ, 2nd edition, 2008.
6. C. S. Yen, Z. Fazarinc, and R. L. Wheeler. Time-domain skin-effect model for transient analysis of lossy transmission lines. *Proceedings of the IEEE*, 70(7):750–757, July 1982.
7. T. R. Arabi, A. T. Murphy, T. K. Sarkar, R. F. Harrington, and A. R. Djordjevic. On the modeling of conductor and substrate losses in multiconductor, multidielectric transmission line systems. *IEEE Transactions on Microwave Theory and Techniques*, 39(7):1090–1097, July 1991.
8. M. J. Tsuk and J. A. Kong. A hybrid method for the calculation of the resistance and inductance of transmission lines with arbitrary cross sections. *IEEE Transactions on Microwave Theory and Techniques*, 39(8):1338–1347, August 1991.
9. F. Bouzidi, H. Aubert, D. Bajon, and H. Baudrand. Equivalent network representation of boundary conditions involving generalized trial quantities–application to lossy transmission lines with finite metallization thickness. *IEEE Transactions on Microwave Theory and Techniques*, MTT-45(6):869–876, June 1997.
10. A. Sangiovanni-Vincentelli, L. Daniel, and J. White. Using conduction modes as global basis functions for efficient electromagnetic analysis of on-chip interconnect. In *Proceedings of the Design Automation Conference*, pp. 563–566, June 2001.
11. L. Daniel, A. Sangiovanni-Vincentelli, and J. White. Proximity templates for modeling of skin and proximity effects on packages and high-frequency interconnect. In *Proceedings of International Conference on Computer Aided Design*, pp. 326–333, June 2002.
12. R. Matick. *Transmission Lines for Digital and Communication Networks*. McGraw-Hill Book Company, New York, 1969.
13. A. Tsuhiya and H. Onodera. Gradient resistivity method for numerical evaluation of anomalous skin effect. In *IEEE Workshop on Signal Propagation on Interconnects*, Naples, Italy, pp. 139–142, May 2011.
14. L. Tsang, J. A. Kong, K.-H. Ding, and C. O. Ao. *Scattering of Electromagnetic Waves*. John Wiley and Sons, Inc., New York, 2001.
15. Z. Zhu and J. White. FastSies: a fast stochastic integral equation solver for modeling the rough surface effect. In *Proceedings of International Conference on Computer Aided Design*, Santa Clara, CA, pp. 675–682, November 2005.
16. H. Braunisch, X. Gu, A. Camacho-Bragado, and L. Tsang. Off-chip rough-metal-surface propagation loss model and correlation with measurements. In *Proceedings of IEEE Electronic Components and Technology Conference (ECTC)*, Reno, NV, pp. 785–791, May 2007.

17. T. Moselhy and L. Daniel. Stochastic high order basis functions for volume integral equation with surface roughness. In *Digest of Electrical Performance o Electronic Packaging*, Atlanta, GA, pp. 73–76, October 2007.

18. L. Tsang, H. Braunisch, R. Ding, and X. Gu. Random rough surface effects on wave propagation in interconnects. *IEEE Transactions on Advanced Packaging*, 33(4):839–856, November 2010.

19. E. Hammerstad and O. Jensen. Accurate models for microstrip computer-aided design. In *1980 IEEE MTT-S International Microwave symposium Digest*, pp. 407–409, May 1980.

20. A. C. Scogna and W. Schauer. Stripline simulation model with tapered cross section and conductor surface profile. In *Proceedings of the IEEE International Symposium on Electromagnetic Compatibility*, Honolulu, HI, July 2007.

21. Y. Shlepnev. Roughness characterization for interconnect analysis. In *Proceedings of the IEEE International Symposium on Electromagnetic Compatibility*, Long Beach, CA, August 2011.

22. S. Ramo, J. R. Whinnery, and T. Van Duzer. *Fields and Waves in Communication Electronics*. John Wiley and Sons, Inc., New York, 1994.

23. C. A. Balanis. *Advanced Engineering Electromagnetics*. John Wiley and Sons, Inc., New York, 1989.

24. D. De Zutter and L. Knockaert. Skin effect modeling based on differential surface admittance operator. *IEEE Transactions on Microwave Theory and Techniques*, 53(8):2526–2538, August 2005.

25. G. Antonini and A. E. Ruehli. Combined loss mechanism and stability model for the partial element equivalent circuit technique. In *Proceedings of Applied Computational Electromagnetics Conference*, Verona, Italy, March 2007.

26. M. Al-Qedra, J. Aronsson, and V. Okhmatovski. A novel skin-effect based surface impedance formulation for broadband modeling of 3-d interconnects with electric field integral equation. *IEEE Transactions on Microwave Theory and Techniques*, 58(12):3872–3881, December 2010.

27. A. E. Ruehli, G. Antonini, and L. Jiang. Skin effect loss models for time and frequency domain PEEC solver. *Proceedings of the IEEE*, 101(2):451–472, February 2013.

28. F. W. Grover. *Inductance Calculations: Working Formulas and Tables*. Dover Publications, New York, 1962.

29. C. Paul. *Inductance, Loop and Partial*. John Wiley and Sons, Inc., New York, 2010.

30. C. L. Holloway, E. F. Kuester, A. E. Ruehli, and G. Antonini. Partial and internal inductance: two of Clayton R. Paul's many passions. *IEEE Transactions on Electromagnetic Compatibility*, 55(4):600–613, April 2013.

31. K. M. Coperich, A. Ruehli, and A. C. Cangellaris. Enhanced skin effect for partial element equivalent circuit PEEC models. In *Digest of Electrical Performance of Electronic Packaging*, Volume 8, San Diego, CA, pp. 189–192, October 1999.

32. K. Coperich, A. E. Ruehli, and A. C. Cangellaris. Enhanced skin effect for partial-element equivalent circuit PEEC models. *IEEE Transactions on Microwave Theory and Techniques*, 48(9):1435–1442, September 2000.

33. M. Celik, L. Pileggi, and A. Odabasioglu. *IC Interconnect Analysis*. Kluwer Academic Publishers, Boston/Dordrecht/London, 2002.

34. B. Gustavsen and A. Semlyen. Rational approximation of frequency domain responses by Vector Fitting. *IEEE Transactions on Power Apparatus and Systems*, 14(3):1052–1061, July 1999.

35. J. White, L. Daniel, and A. Sangiovanni-Vincentelli. Interconnect electromagnetic modeling using conduction models as global basis function. In *Digest of Electrical Performance of Electronic Packaging*, Volume 9, Scottsdale, AZ, pp. 203–206, October 2000.

36. S. C. Dutta Roy. Terminal impedances and transfer functions of ladder networks. *Proceedings of the Institution of Electrical Engineers*, 111(10):1653–1658, October 1964.

37. A. E. Ruehli, G. Antonini, and L. Jiang. Skin-effect model for round wires in PEEC. In *IEEE EMC Europe, International Symposia on EMC*, Rome, Italy, September 2012.

38. A. E. Ruehli. Equivalent circuit models for three dimensional multiconductor systems. *IEEE Transactions on Microwave Theory and Techniques*, MTT-22(3):216–221, March 1974.

39. A. E. Ruehli, G. Antonini, J. Esch, J. Ekman, A. Mayo, and A. Orlandi. Non-orthogonal PEEC formulation for time and frequency domain EM and circuit modeling. In *Proceedings of the IEEE International Symposium on Electromagnetic Compatibility*, Volume 45, pp. 167–176, May 2003.

40. P. J. Restle, A. E. Ruehli, S. G. Walker, and G. Papadopoulos. Full-wave PEEC time-domain method for the modeling of on-chip interconnects. *IEEE Transactions on Computer-Aided Design of Integrated Circuits and Systems*, 20(7):877–886, July 2001.

41. D. Gope, A. E. Ruehli, and V. Jandhyala. Speeding up PEEC partial inductance computations using a QR-based algorithm. *IEEE Transactions on Very Large Scale Integration (VLSI) Systems*, 15(1):60–68, January 2007.

42. M. Kamon, M. J. Tsuk, and J. White. FastHenry: a multipole-accelerated 3-D inductance extraction program. In *Proceedings of the Design Automation Conference*, pp. 678–683, June 1993.

43. L. Yang and Z. Wang. Fast extraction for 3-D inductance and resistance in interconnects. In *Proceedings of International Conference on Communications, Circuits and Systems*, pp. 1334–1338, August 2002.

44. M. Kamon, M. J. Tsuk, and J. K. White. FastHenry: a multipole-accelerated 3-D inductance extraction program. *IEEE Transactions on Microwave Theory and Techniques*, 42(9):1750–1758, September 1994.

45. A. E. Ruehli, N. Kulaza, and J. Pivnichny. Inductance of nonstraight conductors close to a ground plane. *IEEE Transactions on Microwave Theory and Techniques*, 23(8):706–708, August 1975.

46. D. Gope and V. Jandhyala. Efficient solution of EFIE via low-rank compression of multilevel predetermined interactions. *IEEE Transactions on Antennas and Propagation*, 53(10):3324–3333, October 2005.

47. D. Gope, A. E. Ruehli, and V. Jandhyala. Speeding up PEEC partial inductance computation using a QR-based algorithm. *IEEE Transactions on Very Large Scale Integration (VLSI) Systems*, 15(1):60–68, January 2007.

48. F. Zhou, A. E. Ruehli, and J. Fan. Efficient mid-frequency plane inductance computations. In *Proceedings of the IEEE International Symposium on Electromagnetic Compatibility*, Fort Lauderdale, FL, pp. 831–836, August 2010.

49. W. C. Chew. *Waves and Fields in Inhomogeneous Media*. IEEE Press, New York, 1995.

50. Z. G. Qian and W. C. Chew. Generalized impedance boundary condition. In *Proceedings of IEEE Antennas and Propagation Society International Symposium*, Honolulu, HI, June 2007.

51. Z.-G. Qian and W. C. Chew. Fast full-wave surface integral equation solver for multiscale structure modeling. *IEEE Transactions on Antennas and Propagation*, 57(11):268–272, November 2009.

52. J. W. H. Liu. The multifrontal method for sparse matrix solution: theory and practice. *SIAM Review*, 34(1):82–109, March 1992.

53. W. Pinello. Personal communication with A. Ruehli.

54. G. Antonini and A. E. Ruehli. Skin-effect model for time and frequency domain PEEC solver. In *Proceedings of the IEEE International Symposium on Electromagnetic Compatibility*, Long Beach, CA, August 2011.

10

PEEC MODELS FOR DIELECTRICS

Many practical problems include dielectric insulating materials, substrates, dielectric coatings, and other applications of dielectric materials. Hence, it is important to also include dielectrics in partial element equivalent circuit (PEEC) models. A fundamental, physics-based representation of the dielectric issues is presented in [1]. We concentrate on inclusion of several different dielectric circuit-oriented models in PEEC to cover a wide range of applications.

Specifically, three different issues are considered in this chapter. First, we treat the electrical description of dielectric materials in the time and frequency domains. Second, we consider different approaches for the description of the properties of dielectrics such that they can be included in the solution. Third, we present approaches for the inclusion of dielectrics in a PEEC model.

10.1 ELECTRICAL MODELS FOR DIELECTRIC MATERIALS

10.1.1 Frequency and Time Domain Models for Dielectric Materials

Dielectrics can be lossless or lossy, and the dielectric constant ε_r can be frequency or time dependent or independent. Hence, models must be available and suitable for many different cases. Today, the inclusion of frequency or time dependence in dielectrics is important as the ever-increasing frequencies present challenges.

Clearly, the most simple dielectric model in PEEC is if a uniform relative dielectric constant $\varepsilon_r > 1$ can be used. Hence, the assumption is that the entire dielectric space consists of the same material. A fair number of problems can be represented with such a model, which is straightforward to implement. We replace ε_0 by the new dielectric constant $\varepsilon_0\varepsilon_r$ for all partial potential coefficients in the frequency domain. For the lossless dielectric case,

Circuit Oriented Electromagnetic Modeling Using the PEEC Techniques, First Edition.
Albert E. Ruehli, Giulio Antonini, and Lijun Jiang.

the same approach can be applied in the time domain. This approach is considered in more detail in Section 10.4.1.

Fortunately, most of the dielectric models can be represented with an equivalent circuit description such that they can naturally fit into the PEEC circuit environment.

In the frequency domain, a general formulation for the relative permittivity of lossy material is specified in the form

$$\varepsilon(\omega) = \varepsilon'(\omega) - j\,\varepsilon''(\omega) = \varepsilon'(\omega)\,[1 - j\tan\delta(\omega)], \tag{10.1}$$

where ε' is the real part of the dielectric constant, ε'' represents the loss part, and $\omega = 2\pi f$ is the angular frequency.

Perhaps another more clear notation used for (10.1) is

$$\varepsilon(\omega) = \varepsilon_{re}(\omega) - j\,\varepsilon_{im}(\omega) = \varepsilon_{re}(\omega)\,[1 - j\tan\delta(\omega)], \tag{10.2}$$

where the notation used for the real part is $\varepsilon' = \varepsilon_{re}$ and for the imaginary part $\varepsilon'' = \varepsilon_{im}$, Ref. [2]. The loss tangent is defined by $\varepsilon_{im}(\omega) = \varepsilon_{re}(\omega)\ \tan\ \delta(\omega)$. For convenience, we use both notations in this chapter.

For conventional dielectrics, a slow decrease with frequency in the important region is observed for ε' for realistic lossy dielectrics. The loss tangent is less predictable in its variation with frequency. If it is not zero for $\omega \to 0$, then the material also has a dc resistance. Furthermore, we want to point out that a loss tangent $\tan\delta$ which is constant with frequency is inconsistent with the behavior of ε' as discussed in the following section. It is important to recognize that we do not have complete data over the entire frequency range $0 \leq f \leq \infty$ from physical measurement. Unfortunately, this results in problems finding dielectric models for the new, faster technologies and denser layout IC designs, for example, Refs [3, 4].

A question that is beyond the scope of this book is the availability of accurate broadband data for the complex permittivity $\varepsilon(\omega) = \varepsilon_{re}(\omega) - j\varepsilon_{im}(\omega)$. However, very accurate broadband data for dielectrics above the 100 GHz range is harder to obtain. Very often, a loss tangent is specified for material properties. As is evident from (10.2)

$$\tan\ \delta(\omega) = \frac{-\mathrm{Im}[\varepsilon(\omega)]}{\mathrm{Re}[\varepsilon(\omega)]} = \frac{\varepsilon_{im}(\omega)}{\varepsilon_{re}(\omega)}. \tag{10.3}$$

From (10.10), it is also evident that the loss tangent $\tan\ \delta(\omega)$ is an odd function of frequency, as is clearly pointed out in Ref. [5]. Importantly, Hilbert consistent, causal models of dispersive and lossy dielectrics cannot have a constant relative permittivity ε_r. Hilbert consistency or the equivalent Kramers–Krönig relation is considered in Section 10.1.3.

10.1.2 Models for Lossy Dielectric Materials

The models for dielectrics must be suitable to represent different material behavior such that they can be embedded in a PEEC solution as efficiently as possible. As was pointed out in, for example, Refs [6, 7], significant mismatches have been observed for model with lossy dielectrics between measurements and results for 2D quasistatic (QS) models used for 3D interconnect problems. This is especially true if they are used with transmission line tranverse electromagnetic mode (TEM) mode type models. A better approximation is required for quasistatic (QS)PEEC models with losses included. The full-wave (FW) PEEC models are even more challenging from an implementation point of view.

10.1.3 Permittivity Properties of Dielectrics

The properties of dielectrics are mostly given in the frequency domain rather than the time domain. In general, we need to know both ε' and ε'' to properly use the material. The interesting aspect is that once either the real or imaginary part of the dielectric model is known for the entire frequency range $0 \leq f \leq \infty$, then the other part is uniquely determined by the so-called Hilbert transform. Experimentally, we lack the complete knowledge of the data for the entire frequency range. From a theoretical point of view, if the material satisfies the Hilbert relations between the real and the imaginary parts [8], then we say that it is *Hilbert consistent*, which is clearly an important feature for obtaining passive models considered in Chapter 13.

We use the Hilbert transform between the real and imaginary parts $\varepsilon'(\omega)$ and $\varepsilon''(\omega)$. For dielectrics, it is also called the Kramers–Krönig relations [10] that are

$$\varepsilon'(\omega) = 1 + \frac{1}{\pi} P \int_{-\infty}^{+\infty} \frac{\varepsilon''(x)}{x-\omega}\, dx \tag{10.4a}$$

$$\varepsilon''(\omega) = -\frac{1}{\pi} P \int_{-\infty}^{+\infty} \frac{[\varepsilon'(x)-1]}{x-\omega}\, dx, \tag{10.4b}$$

where P denotes the principal value. Fortunately, this implies that the singular points are such that for the numerical integration of the integrals, we can ignore the behavior in the immediate small neighborhoods on both sides near the singularities. An equivalent form that is easier to implement is given by

$$\varepsilon'\omega) = 1 + \frac{2}{\pi} P \int_{0}^{+\infty} \frac{\varepsilon''(x)}{x-\omega}\, dx \tag{10.5a}$$

$$\varepsilon''(\omega) = -\frac{2}{\pi} P \int_{0}^{+\infty} \frac{[\varepsilon'(x)-1]}{x-\omega}\, dx, \tag{10.5b}$$

which uses positive frequencies only.

Hilbert consistency between the real and imaginary parts of the relative dielectric constant is essential for the construction of passive models, for example, Ref. [5, 8].

10.1.4 Electrical Permittivity Model for Time Domain

We start from (3.2b) in Chapter 3, which is $\boldsymbol{D}(\omega) = \varepsilon(\omega)\boldsymbol{E}(\omega)$. We also need to define the electric susceptibility χ, which is given as

$$\varepsilon(\omega) = \varepsilon_0(1 + \chi(\omega)). \tag{10.6}$$

With this, the susceptibility can be written as

$$\boldsymbol{D}(\boldsymbol{r},t) = \varepsilon_0 \boldsymbol{E}(\boldsymbol{r},t) + \varepsilon_0 \int_0^t \chi(\tau)\boldsymbol{E}(\boldsymbol{r},t-\tau)\, d\tau \tag{10.7}$$

using the usual time domain convolution. To obtain $\chi(t)$, we have to use the inverse Fourier transform of the frequency dependent dielectric constant such that we obtain

$$\chi(\tau) = \frac{1}{2\pi} \int_{-\infty}^{+\infty} \frac{\varepsilon(\omega)}{\varepsilon_0 \quad - 1} e^{-j\omega\tau} d\omega. \tag{10.8}$$

Of course, this corresponds to the time-to-frequency forward Fourier transform

$$\varepsilon(\omega) = \varepsilon_0 + \varepsilon_0 \int_0^{\infty} \chi(\tau) e^{j\omega\tau}\, d\tau. \tag{10.9}$$

If the quantities $\boldsymbol{D}(\boldsymbol{r},t)$ and $\boldsymbol{E}(\boldsymbol{r},t)$ are positive real functions [11, 12], we have for real frequencies ω that

$$\varepsilon'(-\omega) = \varepsilon'(\omega), \quad \varepsilon''(-\omega) = -\varepsilon''(\omega), \tag{10.10}$$

where ε' is an even function. Causality of the electromagnetic (EM) solution is in question if the relationships (10.10) are violated [5, 8].

10.1.5 Causal Models for Dispersive and Lossy Dielectrics

We pointed out in Section 10.1.3 that Hilbert consistent models are required to construct valid EM solutions for lossy dielectrics. Theoretically, we assume that all the relevant quantities are known for the entire frequency range $0 \leq f \leq \infty$. However, we are lucky if the electrical permittivity $\varepsilon(\omega)$ is known for a sufficient number of frequency decades. Furthermore, it is well known that many EM-solver techniques do not provide a dc solution.

A class of circuit-oriented models is represented by a ratio of polynomials in s where all coefficients are real. A general band-limited susceptibility function $\chi(s)$ is of the form

$$\chi(s) = \frac{B_m(s)}{A_n(s)}, \qquad n > m. \tag{10.11}$$

In these polynomials, the order of the denominator must be larger than the numerator such that the model leads to $\chi(S) = 0$ at the limit $s \to \infty$. Thus, the general form of the electrical permittivity $\varepsilon(s)$ is

$$\varepsilon(s) = \varepsilon_0\, \varepsilon_\infty + \varepsilon_0 \chi(s), \tag{10.12}$$

where ε_∞ describes the permittivity at $\lim s \to \infty$.

The most common and simple dispersive dielectric model is the single-pole Debye model:

$$\varepsilon(s) = \varepsilon_0\, \varepsilon_\infty + \varepsilon_0 \frac{\varepsilon_S - \varepsilon_\infty}{1 + s\tau}, \tag{10.13}$$

where τ is the relaxation constant. A Lorentzian model is defined as

$$\varepsilon(s) = \varepsilon_0\, \varepsilon_\infty + \varepsilon_0 \frac{(\varepsilon_S - \varepsilon_\infty)\omega_0^2}{\omega_0^2 + 2s\delta + s^2}, \tag{10.14}$$

where δ is the relaxation constant and ω_0 is the resonance angular frequency. The Debye model has its main applications in the subterahertz range, whereas the Lorentz model yields a better approximation for higher resonance models. We also observe that a model with a single real or complex pole pair is a very narrow-band model valid only for a limited frequency range.

In general, broadband models are obtained by addition of several models of the form of (10.13) or (10.14). However, we have to point out the computational cost which each additional term implies. These models are of the general form

$$\varepsilon(s) = \varepsilon_0 \left[\varepsilon_\infty + \sum_{m=1}^{N_D} \frac{(\varepsilon_{DS,m} - \varepsilon_{D\infty,m})}{1 + s\tau_m} + \sum_{n=1}^{N_L} \frac{(\varepsilon_{LS,n} - \varepsilon_{L\infty,n})\omega_{0,n}^2}{s^2 + 2s\delta_n + \omega_{0,n}^2} \right], \tag{10.15}$$

where parameters ε_∞, $\varepsilon_{DS,m}$, $\varepsilon_{D\infty,m}$, $(m = 1, \ldots, N_D)$, $\varepsilon_{LS,n}$ and $\varepsilon_{L\infty,n}$, $(n = 1, \ldots, N_L)$ are all positive and satisfy the following conditions:

$$\varepsilon_S = \varepsilon_0 \left(\sum_{m=1}^{N_D} \varepsilon_{DS,m} + \sum_{n=1}^{N_L} \varepsilon_{LS,n} \right) \tag{10.16a}$$

$$\varepsilon_\infty = \varepsilon_0 \left(\sum_{m=1}^{N_D} \varepsilon_{D\infty,m} + \sum_{n=1}^{N_L} \varepsilon_{L\infty,n} \right), \tag{10.16b}$$

where ε_S and ε_∞ represent the permittivity at *dc* and at infinite frequency. In these equations, N_D represents the number of Debye terms according to (10.13) and N_L are the number of Lorentz terms according to (10.14).

Fortunately, a loss model can be designed for high-speed electrical interconnect applications using typical dielectric materials such as FR-4. As an example, a result is given in Fig. 10.1, where we show the magnitude of permittivity and loss tangent for DriClad. The results, which are valid for about four decades, are obtained with a fourth-order Debye model, where $\varepsilon_S = 4.1$, $\varepsilon_\infty = 3.79$, and the other parameters chosen as shown in Section 10.3. The Debye model parameters are carefully chosen terms ($N_D = 1, \ldots, 4$) to keep the computational cost at a minimum.

For completeness, we also include two additional models. For example, for polymer and composite materials, the Cole–Cole dispersion law is also widely used [13, 14], which is

$$\varepsilon(s) = \varepsilon_0\varepsilon_\infty + \varepsilon_0 \frac{\varepsilon_S - \varepsilon_\infty}{\left[1 - (s\tau)^{1-\alpha}\right]}, \qquad 0 < \alpha < 1. \tag{10.17}$$

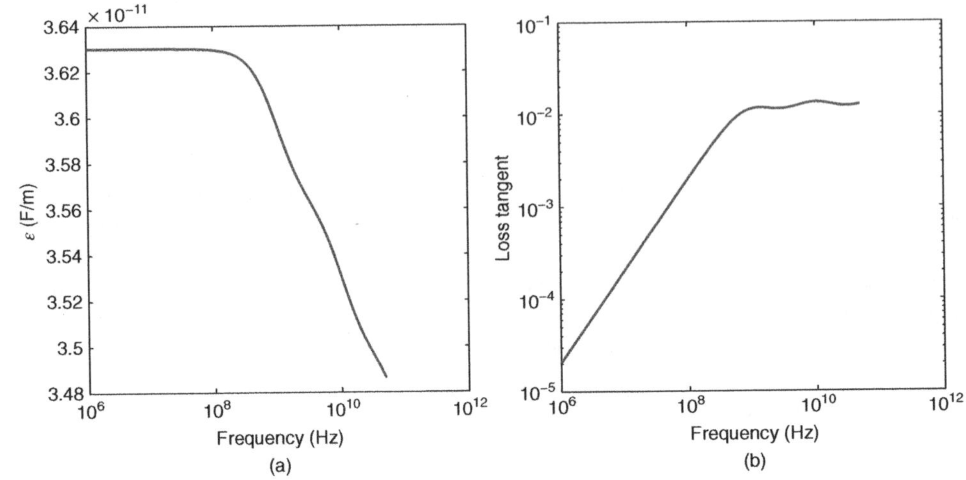

FIGURE 10.1 Fourth-order Debye model for DriClad. (a) Magnitude of permittivity. (b) Loss tangent.

This model can also be approximated by fitting a sum of third and fourth terms in (10.15). We should note that (10.17) is a special case of the more general dispersion law described by the Havriliak–Negami function [15, 16]

$$\varepsilon(s) = \varepsilon_0 \varepsilon_\infty + \varepsilon_0 \frac{\varepsilon_S - \varepsilon_\infty}{\left[1 - (s\tau)^{1-\alpha}\right]^{\beta}} . \tag{10.18}$$

In general, these formulas can also be used in multiterm form (10.15).

10.2 CIRCUIT ORIENTED MODELS FOR DISPERSIVE DIELECTRICS

Circuit models for dielectrics are of importance for several reasons. We considered techniques in Chapter 2 on the inclusion of frequency-dependent circuits in the PEEC–modified nodal analysis (MNA) time solution. The best choice for the implementation of the dispersive model depends on the complexity of the model. The recursive convolution approach in Section 2.10.2 is preferable for the inclusion of very complicated circuit models with a large number of circuit elements and internal nodes.

An equivalent circuit-based approach is very suitable for a PEEC model with a smaller number of nodes since it allows the use of the conventional solver techniques presented in Chapter 2. Another important factor is always for a model to be suitable for both the time and the frequency domain. This simplifies the solution in both domains.

In the following sections, equivalent circuits are presented for the Debye model, the Lorentz model, and other general combined models.

10.2.1 Simple Debye Medium Circuit Model for Dielectric Block

We first consider the one-pole Debye model (10.13) for lossy dielectrics. The equivalent circuit is used from the dielectric PEEC model in Section 10.4.5. This one-pole model

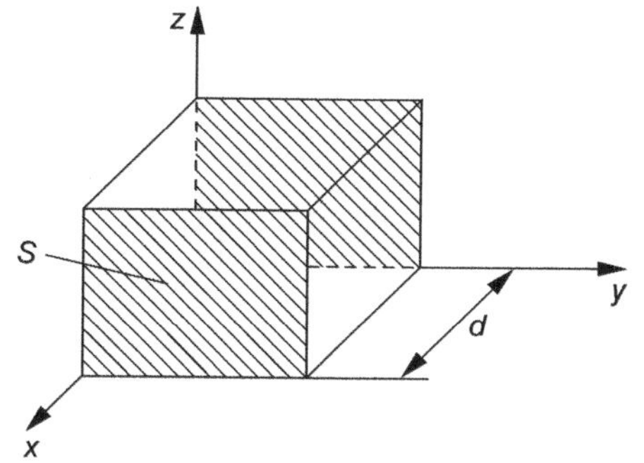

FIGURE 10.2 Rectangular block of dielectric material.

is the most simple for dielectrics with loss resulting in the least number of unknowns or compute time. However, a single-pole model can only accommodate a limited frequency range. Figure 10.2 shows the geometry for the dielectric bar cell for which we apply the model. We assume that the crosshatched areas are the contact surfaces.

The admittance of the block of dielectric material in Fig. 10.2 for a Debye model is

$$Y_c(s) = sS/d\varepsilon_0 \left[\varepsilon_\infty + \frac{(\varepsilon_S - \varepsilon_\infty)}{1 + s\,\tau}\right], \tag{10.19}$$

where the model parameters to be specified are ε_∞, ε_S and τ.

It is easy to identify the equivalent circuit parameters. The capacitive admittance can be rewritten as

$$Y_c(s) = sC_{D,\infty} + Y_{RC}(s), \tag{10.20}$$

where the capacitance at $f \rightarrow \infty$, $C_{D,\infty}$ is given by

$$C_{D,\infty} = \varepsilon_0\,\varepsilon_\infty\,\frac{S}{d}. \tag{10.21}$$

The RC series parameters are

$$C_D = \varepsilon_0\,(\varepsilon_S - \varepsilon_\infty)\,\frac{S}{d} \tag{10.22}$$

and

$$R_D = \frac{\tau\,d}{S\,\varepsilon_0\,(\varepsilon_S - \varepsilon_\infty)}. \tag{10.23}$$

The corresponding equivalent circuit is shown in Fig. 10.3.

Again, this simple model can only represent a limited variation of the permittivity with frequency. The implementation of the circuit model in the MNA solution works equally well for both the frequency and the time domains since all the elements in the equivalent circuit are frequency independent.

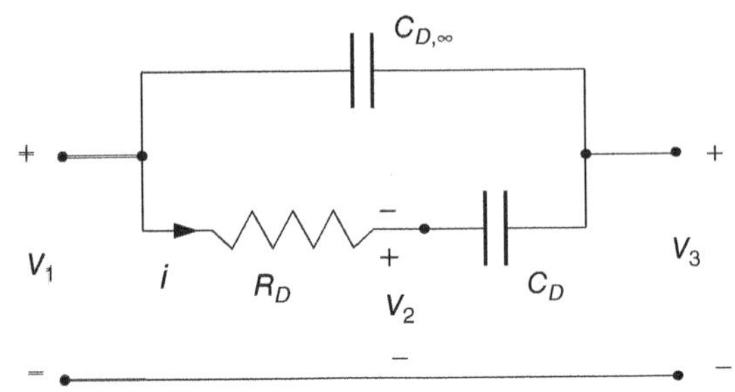

FIGURE 10.3 Equivalent circuit for a one-pole Debye medium.

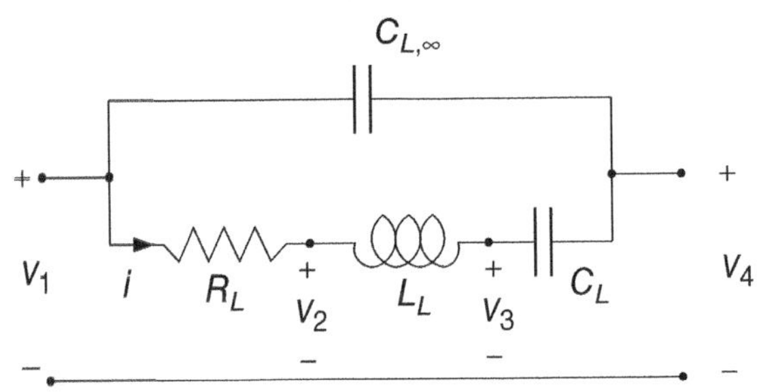

FIGURE 10.4 Complex pole equivalent circuit for a Lorentz medium.

10.2.2 Simple Capacitance Model for Lorentz Media

The Lorentz media equations add somewhat more flexibility over the Debye model, which is necessary for some materials since it involves possibly complex pole behavior. Applying the same reasoning as in the Debye model case, the equivalent circuit for the model is given in Fig. 10.4.

The admittance for the model branch is given by

$$Y_c(s) = sS/d\varepsilon_0 \left[\varepsilon_\infty + \frac{(\varepsilon_S - \varepsilon_\infty)\,\omega_0^2}{s^2 + 2\,s\,\delta + \omega_0^2} \right], \tag{10.24}$$

where the shape of the dielectric block is the same as for the Debye model in Fig. 10.2. The first term in (10.24) is again the admittance of the excess capacitance at infinite frequency and the second one is treated as an admittance $Y_{RLC}(s)$ for the series connection in the equivalent circuit in Fig. 10.4 such that we have

$$Y_c(s) = sC_{L,\infty} + Y_{RLC}(s), \tag{10.25}$$

where $C_{L,\infty} = \varepsilon_0 \varepsilon_\infty S/d$. It is a small task in circuit analysis to find the values of the circuit elements. Some simple manipulations allow us to synthesize $Y_{RLC}(s)$ as a series *RLC* equivalent circuit with the following values of circuit parameters:

$$\begin{cases} L_L = 1/(\varepsilon_0 S/d(\varepsilon_S - \varepsilon_\infty)\omega_0^2) \\ R_L = 2\delta/(\varepsilon_0 S/d(\varepsilon_S - \varepsilon_\infty)\omega_0^2) \\ C_L = \varepsilon_0 S/d(\varepsilon_S - \varepsilon_\infty) \end{cases} \tag{10.26}$$

in the equivalent circuit in Fig. 10.4.

We notice that at very high frequencies, the admittance of the capacitance that dominates is $Y_c(s) = sC_{L,\infty}$. For very low frequencies, the admittance of the circuit is $Y_c(s) \cong s\varepsilon_0\varepsilon_S S/d$.

We can verify that the equivalent circuit is also consistent for the nondispersive case. This can be obtained for the low-frequency range assuming that $\delta = 0$. Then in Fig. 10.4, the series *RLC* circuit reduces to the capacitor C_L. This shows that the global circuit is equivalent to the static excess capacitance $C_{LS} = C_L + C_{L\infty} = \varepsilon_0\varepsilon_S S/d$.

To enhance the accuracy of a model, we have to add the losses to a model by including a loss with the appropriate ε and a tan δ. We note that a frequency-independent loss factor is not physically consistent since it does not represent the high frequencies correctly [5].

10.3 MULTI-POLE DEBYE MODEL

The models presented in the previous two sections allow only a limited frequency variation because of the small number of poles used. However, the frequency dependence can be improved by adding more sections to the model. The Debye model with multiple poles is widely used, and the complexity is limited only by the need to keep the number of poles as small as possible. Additional poles added to a model cover a broader frequency range. Specifically, we consider a multipole Debye model, where potentially only one pole is used for each frequency decade to maintain the computational efficiency.

It is possible to design a physical consistent Debye model starting from the limited set of data, ε_S, tan δ, and τ_1, where tan δ and τ_1 now have the meaning of loss tangent and time constant of the first relaxation process. In the following, we refer to Ref. [17] where a methodology is presented to obtain analytically the Debye model parameters for substrates with a constant loss tangent over a specified bandwidth. Hence, a multipole Debye model leads to

$$\varepsilon(s) = \varepsilon_0\left[\varepsilon_\infty + \sum_{i=1}^{N_D}\frac{(\varepsilon_{S,i} - \varepsilon_{\infty,i})}{1 + s\tau_i}\right]$$

$$= \varepsilon_0\left[\varepsilon_\infty + \sum_{i=1}^{N_D}\frac{\Delta\varepsilon_i}{1 + s\tau_i}\right], \tag{10.27}$$

where the number of poles is determined by N_D.

Figure 10.5 shows the circuit representation of the Debye model where $C_{D\,\infty}$, determined by ε_∞, impacts the constant highest frequency behavior of the dielectric.

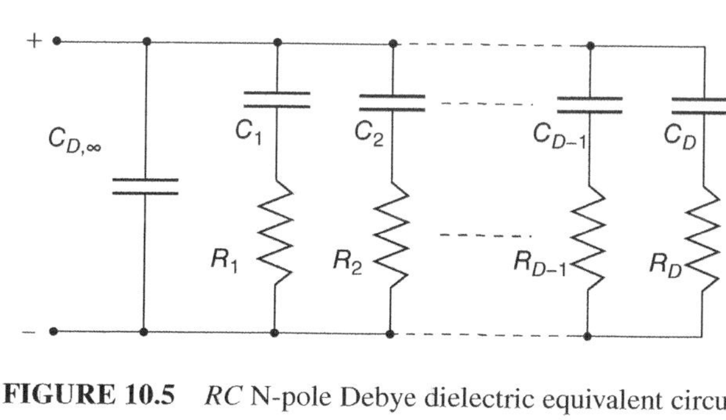

FIGURE 10.5 *RC* N-pole Debye dielectric equivalent circuit.

In order to obtain a sufficiently smooth loss tangent over several decades, multiple *RC* branches have to be included. According to Ref. [17], the values of the circuit elements in *RC* branches in the equivalent network can be evaluated as

$$C_i = C = 2k \tan\delta \, C_\infty \tag{10.28a}$$

$$R_i = \frac{1}{\omega_i C}, \qquad \omega_{i+1} = 10\,\omega_i \quad \mapsto \tau_{i+1} = 0.1\,\tau_i \tag{10.28b}$$

for $i = 1, \ldots, N_D$; k is a scale factor set to 1 for $N_D = 1$ and to 0.9 for $N_D > 1$. The capacitances C_i and C_∞ must meet the low frequency condition

$$C_{dc} = C_\infty + N_D\, C. \tag{10.29}$$

Actually, C_{dc} corresponds to ε_S and the knowledge of $\varepsilon_S \rightarrow C_{dc}$, tan δ, and the number of poles N_D allow us to obtain ε_∞ that determines C_∞ from

$$\varepsilon_\infty = \frac{\varepsilon_S}{1 + 2\,k \tan\delta\, N_D}. \tag{10.30}$$

Once ε_S and ε_∞ are known, all the other parameters can be evaluated from the analogy between the Debye model (10.27) and the *RC* circuit

$$\Delta\varepsilon_i = \varepsilon_{S,i} - \varepsilon_{\infty,i} \;\rightarrow\; \varepsilon_{\infty,i} = \varepsilon_{S,i} - \Delta\varepsilon_i, \quad i = 1, \ldots, N_D. \tag{10.31}$$

We should note that the multipole Debye circuit model is automatically Hilbert consistent.

We give an example for a multipole Debye model. The structure consists of a microstrip line over a dielectric layer with a ground plane underneath. The lossy dielectric layer is 1 mm thick and the zero thickness trace is 1 mm wide while the overall length of the structure is 10 cm.

We choose an FR-4 dielectric material in the example with a tan $\delta = 0.02$ and $\varepsilon_S = 4.7$, where $\tau_1 = 1.59$ ns in (10.27). Equation (10.28a) is used to determine the other time constants for the number of poles, which are $N_D = 4$. This results in the representation of the dielectric as

$$\varepsilon = \varepsilon_0 \left(\varepsilon_\infty + \sum_{k=1}^{N_D} \frac{\Delta\varepsilon_k}{1 + s\tau_k} \right), \tag{10.32}$$

where $\Delta\varepsilon_k = \varepsilon_{S,k} - \varepsilon_{\infty,k}$ and τ_k is again the time constant of the kth relaxation process, respectively. The parameters used are summarized in Table 10.1.

Figures 10.6 and 10.7 show the loss tangent and the magnitude of electrical permittivity where the number of poles is increased from 1 to 4. This illustrates how the number of poles to the model increases the frequency range for which the model is valid. The overall PEEC model mesh included 924 volume cells for the volume filament (VFI) skin-effect inductance model and 760 zero thickness surface cells, resulting in 924 currents and 168 nodes since some of the capacitive cells are joined at the nodes in this example.

TABLE 10.1 FR-4 Debye model parameters.

	$\varepsilon_{S,k}$	$\varepsilon_{\infty,k}$	τ_k (ns)
Pole 1	4.7	4.55	1.59
Pole 2	4.55	4.40	0.159
Pole 3	4.40	4.25	0.0159
Pole 4	4.25	4.10	0.00159

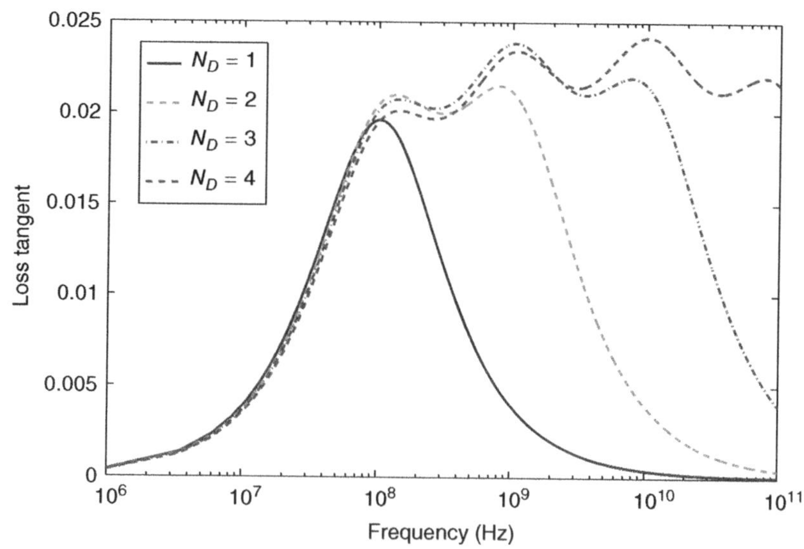

FIGURE 10.6 FR-4 loss tangent representation for an increasing number of poles (example in Section 10.5.2).

10.3.1 Combined Debye and Lorentz Dielectric Model

This model is added mostly for generality since its complexity contributes to very long compute times. Of course, the models are very general in representing the behavior of different materials by the sum of Debye and Lorentz terms. We assume that N_D Debye and N_L Lorentz terms are combined in the form of (10.15)

$$\varepsilon(s) = \varepsilon_0 \left[\varepsilon_\infty + \sum_{m=1}^{N_D} \frac{(\varepsilon_{DS,m} - \varepsilon_{D\infty,m})}{1 + s\tau_m} \right. \\ \left. + \sum_{n=1}^{N_L} \frac{(\varepsilon_{LS,n} - \varepsilon_{L\infty,n})\omega_{0,n}^2}{s^2 + 2s\delta_n + \omega_{0,n}^2} \right], \quad (10.33)$$

where ε_∞ is the permittivity at infinite frequency, $\varepsilon_{DS,m}$ and $\varepsilon_{LS,m}$ are the static dielectric constants for the mth Debye and Lorentz medium, $\omega_{0,m}$ is the resonance frequency of the mth Lorentz medium. The loss constant of the mth Debye term is τ_m. The admittance corresponding to the capacitance is

$$sC_e(s) = sC_\infty(s) + \sum_{m=1}^{N_D} Y_D(s) + \sum_{m=1}^{N_L} Y_L(s), \quad (10.34)$$

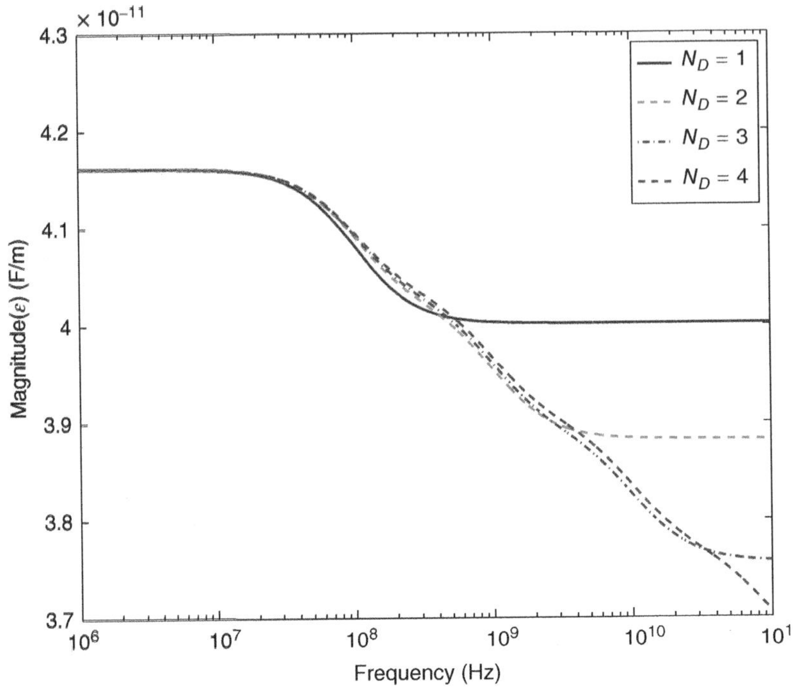

FIGURE 10.7 FR-4 permittivity representation for an increasing number of poles.

which is given by the synthesized circuit shown in Fig. 10.8. The techniques presented in these sections are applied in the rest of this chapter.

10.4 INCLUDING DIELECTRIC MODELS IN PEEC SOLUTIONS

Dielectric models have so far been presented independently of how we include them in a PEEC model. Different approaches will be presented below which show applications of the dielectric models presented so far in this chapter. The model we use may be problem specific and not general. Mainly, the compute time for general techniques could be excessive in some cases. For example, only a limited number of approaches apply to lossy full-wave (FW)PEEC models.

It is important that not all dielectric models result in an increase in the size of the PEEC circuit model. For example, the technique in Section 10.4.2 impacts the Green's function rather than the circuit topology of the PEEC model. This exemplifies the fact that a combination of circuits as well as EM techniques can be mixed for different solution methods.

10.4.1 Models for Uniform, Lossless Dielectrics

In the most simple situation, we can use a model where the entire problem is embedded in a uniform dielectric material, which, theoretically, extends to infinity in all directions. The relative dielectric constant $\varepsilon_r = 1$ is replaced by $\varepsilon_r > 1$ for all the partial potential coefficients. This uniformly results in a reduces velocity v everywhere $v = c_0/\sqrt{(\varepsilon_r)}$, where c_0 is the speed of light in air.

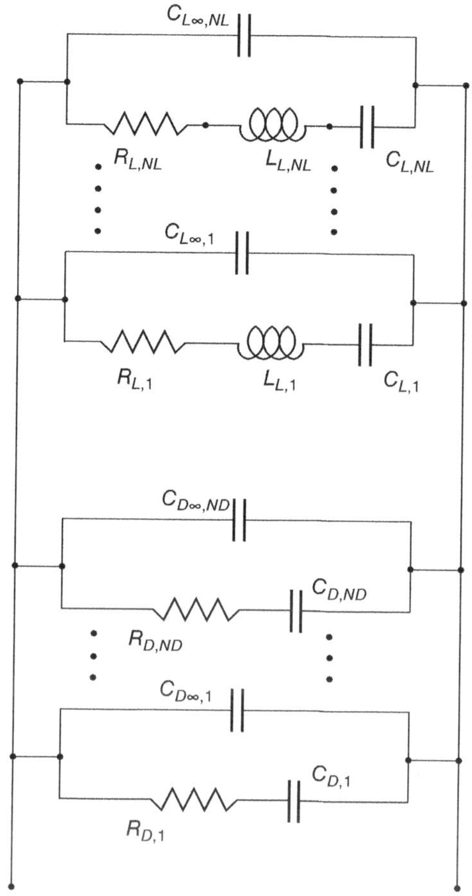

FIGURE 10.8 Equivalent circuit for a general dispersive dielectric.

This can result in very effective solutions for many problems. While the implementation in PEEC is simple, we notice that the meshing also needs to be made more dense to achieve the same accuracy. The mesh size has to be reduced by $\sqrt{\varepsilon_r}$. This can result in a significant increase in compute time for very high values of ε_r. Also, the inclusion of lossy dielectric regions with $\varepsilon_r^L = \varepsilon_r - j\sigma/(\omega\varepsilon_0)$ can be used to make the problem lossy.

10.4.2 Green's Functions for Dielectric Layers Based on the Image Theory

Dielectric layers can be taken into account in a different way. First, we show that they can be included directly in the Green's function by a combination of free-space type Green's functions. The capacitance and inductance models are de-coupled, and the Green's functions are not classified by the direction of the current, etc., as is the case for the ones considered in Chapter 3 and Section 3.4.

The details of the solution are determined by the location of the source and observation points in the layer structure. For this reason, we label the Green's functions according to the layer region where the observation point and the field points are located. For example,

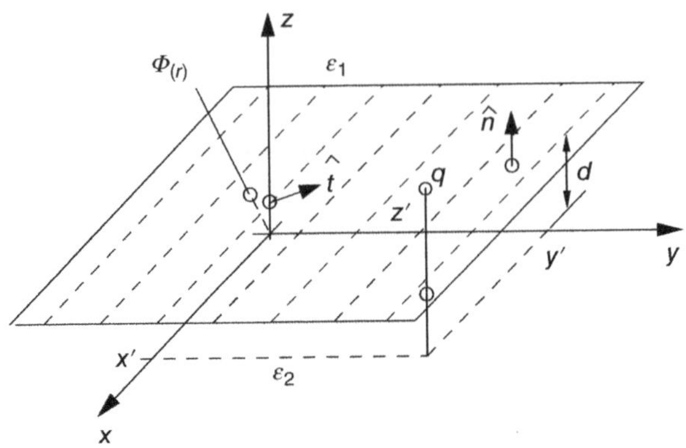

FIGURE 10.9 Two infinite dielectric layers geometry for images.

for the Green's function g_{12} the observation point is in region 1 while the source point is in region 2.

The method of images we use has a long history, e.g., [18, 19]. The image planes are theoretically infinite in size since the boundary conditions are matched between the entire planar layers.

We start with the two infinite dielectric half-spaces where the dielectric half-space above the boundary is filled with a dielectric constant of ε_1 while the dielectric constant below the interface is filled with ε_2 as shown in Fig. 10.9. We use the notation that $\varepsilon_k = \varepsilon_0\ \varepsilon_{r,k}$ for layer k. For clarity, we only show a finite section of the infinite dielectric layer in Fig. 10.9 and all the subsequent figures for layered Green's functions. They extend to infinity in the x and y direction.

The interface between two dielectrics is in the x, y plane and it is placed at a distance $z = d$ from the $z = 0$ plane. We are interested in computing the potential at $\boldsymbol{r} = x\ \hat{\boldsymbol{x}} + y\ \hat{\boldsymbol{y}} + z\ \hat{\boldsymbol{z}}$ which may be located in one of the two dielectric regions. For the Green's function, we assume that a unit charge $q(\boldsymbol{r}')$ is placed at $\boldsymbol{r}' = x'\hat{\boldsymbol{x}} + y'\hat{\boldsymbol{y}} + z'\hat{\boldsymbol{z}}$ which can be located anywhere.

For the case when both observation point $\boldsymbol{r}$ and the source point $\boldsymbol{r}'$ are located in the same dielectric region, the observation point sees two effects. One is the direct field contribution from the source direction through a distance vector $\boldsymbol{r} - \boldsymbol{r}'$ and another one is the reflected contribution by the source. The reflection happens at the interface and its equivalence consists of a weighted image source. The total contribution of these two is added up for the Green's function g_{11}.

If the observation point is in region 1 and the source point is in region 2, then the situation is different. A new weighted source has to be computed to find the Green's function g_{12}. Obviously, there are two other cases to be considered when the observation point is located in region 2 instead. Fortunately, the above results can be used where the corresponding Green's functions are g_{22} and g_{21}, respectively.

In all cases, the solutions must satisfy the boundary conditions at the interface(s). One of the required boundary conditions is that the tangential electric fields are matched (3.14a)

$$\hat{\boldsymbol{t}} \cdot \boldsymbol{E}_1 = \hat{\boldsymbol{t}} \cdot \boldsymbol{E}_2 \tag{10.35}$$

on the two sides of the interface where $\hat{\boldsymbol{t}}$ is the tangential unit vector shown in Fig. 10.9. Since this boundary condition has to be satisfied at any point on the interface, we can put a tangential vector of unit length $\hat{\boldsymbol{t}} = a\hat{\boldsymbol{x}} + b\hat{\boldsymbol{y}}$ in any direction on the plane. For the unit length to be satisfied, we require that $\sqrt{a^2 + b^2} = 1$. Without loss of generality, we place the tangential vector at the origin in Fig. 10.9.

The second boundary condition which has to be met is the continuity of the normal electric displacement flux density for $\boldsymbol{D}$ in (3.14c)

$$\hat{\boldsymbol{n}} \cdot \boldsymbol{D}_1 = \hat{\boldsymbol{n}} \cdot \boldsymbol{D}_2 \tag{10.36}$$

where $\hat{\boldsymbol{n}}$ is the unit normal vector on the dielectric interface and points into region 1 as shown in Fig. 10.9.

10.4.3 Green's Function for One Dielectric Interface

All the preliminaries for computing the Green's functions have been covered in the last section. Assume the observation point $\boldsymbol{r}$ as well as the source point $\boldsymbol{r}'$ are in region 1. The contribution from the source charge q located at $\boldsymbol{r}'$ is

$$\boldsymbol{E}(\boldsymbol{r}) = \frac{q}{4\pi\varepsilon_1} \frac{(x-x')\hat{\boldsymbol{x}} + (y-y')\hat{\boldsymbol{y}} + (z-z')\hat{\boldsymbol{z}}}{[(x-x')^2 + (y-y')^2 + (z-z')^2]^{3/2}}. \tag{10.37}$$

Based on the image theory we introduced at the beginning of this section, we can set up a proposed solution using one image source q_1 and then match the boundary conditions at the interface. Figure 10.10 shows the image solution valid for the region 1 case where both source and field positions are above the dielectric interface. If the image source is properly constructed, the electric field can be finally computed from the following equation

$$\begin{aligned} \boldsymbol{E}_1(\boldsymbol{r}) = & \frac{q}{4\pi\varepsilon_1} \frac{(x-x')\hat{\boldsymbol{x}} + (y-y')\hat{\boldsymbol{y}} + (z-z')\hat{\boldsymbol{z}}}{[(x-x')^2 + (y-y')^2 + (z-z')^2]^{3/2}} \\ & + \frac{q_1}{4\pi\varepsilon_1} \frac{(x-x')\hat{\boldsymbol{x}} + (y-y')\hat{\boldsymbol{y}} + (z+z'-2d)\hat{\boldsymbol{z}}}{[(x-x')^2 + (y-y')^2 + (z+z'-2d)^2]^{3/2}}. \end{aligned} \tag{10.38}$$

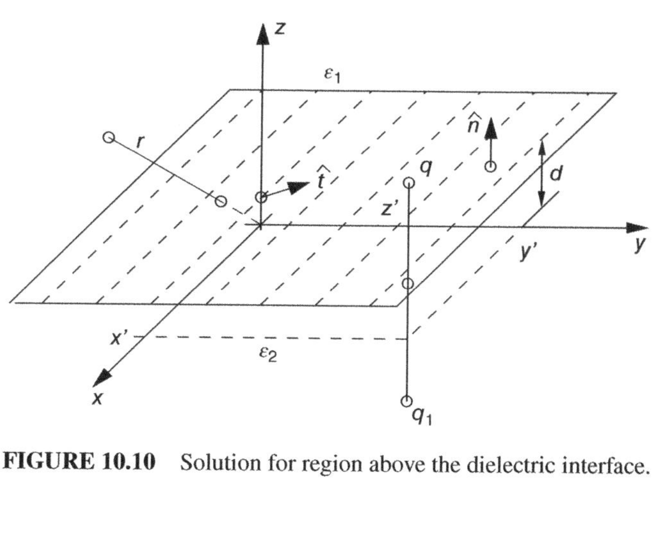

FIGURE 10.10 Solution for region above the dielectric interface.

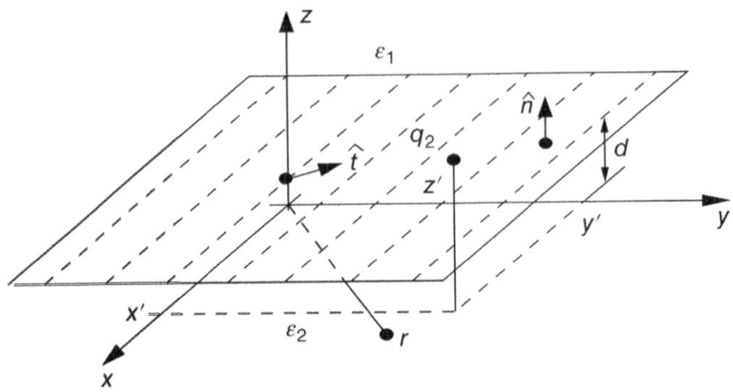

FIGURE 10.11 Solution for region below the dielectric interface.

It is noted that the source charge q and the image charge q_1 are the same distance to the other side of the common interface.

If the observation field position is in region 2 while the source q is still $\boldsymbol{r}'$ in region 1, we use an image source q_2 located at $\boldsymbol{r}'$ that is working in a homogeneous medium with the same permittivity $\varepsilon_2 = \varepsilon_0\varepsilon_{r2}$ as region 2. The situation is shown in Fig. 10.11, where the electric field is

$$\boldsymbol{E}_2(\boldsymbol{r}) = \frac{q_2}{4\pi\varepsilon_2}\frac{(x-x')\hat{\boldsymbol{x}} + (y-y')\hat{\boldsymbol{y}} + (z-z')\hat{\boldsymbol{z}}}{[(x-x')^2 + (y-y')^2 + (z-z')^2]^{3/2}} \tag{10.39}$$

where q_2 is also unknown and the entire space is filled with ε_2. The two unknown charges q_1 and q_2 have to be found by matching the boundary conditions (10.35) and (10.36).

To match the tangential components starting with (10.38), the dot product between with $\hat{\boldsymbol{t}}$ results in

$$\hat{\boldsymbol{t}}\cdot\boldsymbol{E}_1(\boldsymbol{r}) = \frac{q}{4\pi\varepsilon_1}\frac{a(x-x') + b(y-y')}{[(x-x')^2 + (y-y')^2 + (d-z')^2]^{3/2}} + \frac{q_1}{4\pi\varepsilon_1}\frac{a(x-x') + b(y-y')}{[(x-x')^2 + (y-y')^2 + (d+z'-2d)^2]^{3/2}}. \tag{10.40}$$

Also for (10.39), we obtain

$$\hat{\boldsymbol{t}}\cdot\boldsymbol{E}_2(\boldsymbol{r}) = \frac{q_2}{4\pi\varepsilon_2}\frac{a(x-x') + b(y-y')}{[(x-x')^2 + (y-y')^2 + (d-z')]^{3/2}}. \tag{10.41}$$

Equating (10.40) to (10.41), we obtain one of the equations for the determination of the charges

$$\frac{q+q_1}{\varepsilon_1} = \frac{q_2}{\varepsilon_2} \tag{10.42}$$

For the electric displacement $\boldsymbol{D} = \varepsilon\boldsymbol{E}$ normal boundary condition, we apply $\hat{\boldsymbol{n}} = \hat{\boldsymbol{z}}$ to (10.38) and (10.39), which results in the displacement flux density on the interface but in region 1 ($z = d$):

$$\hat{n} \cdot \boldsymbol{D}_1(\boldsymbol{r}) = \frac{q}{4\pi} \frac{d - z'}{[(x - x')^2 + (y - y')^2 + (d - z')^2]^{3/2}} + \frac{q_1}{4\pi} \frac{d + z' - 2d}{[(x - x')^2 + (y - y')^2 + (d + z' - 2d)^2]^{3/2}}. \tag{10.43}$$

Similarly, the normal dot product of the field on the interface and in region 2 is

$$\hat{n} \cdot \boldsymbol{D}_2(\boldsymbol{r}) = \frac{q_2}{4\pi} \frac{d - z'}{[(x - x')^2 + (y - y')^2 + (d - z')^2]^{3/2}}. \tag{10.44}$$

For the evaluation of (10.37), matching (10.44) and (10.45) on the boundary $z = d$, we get

$$-q + q_1 = -q_2 \tag{10.45}$$

Since q is the original known charge, from the (10.42) and (10.45), we can solve q_1 and q_2 to be

$$q_1 = \frac{\varepsilon_1 - \varepsilon_2}{\varepsilon_1 + \varepsilon_2} q := \alpha q \quad \text{and} \quad q_2 = \frac{2\ \varepsilon_2}{\varepsilon_1 + \varepsilon_2} q = (1 - \alpha) q \tag{10.46}$$

where

$$\alpha = \frac{\varepsilon_1 - \varepsilon_2}{\varepsilon_1 + \varepsilon_2}. \tag{10.47}$$

Using q_1 and q_2, we can revise the homogeneous Green's function to adapt the effects of the two dielectrics. As we considered before, there are four Green's functions that can be divided into two cases: two for the source charge point are in the upper layer Fig. 10.10 and two for the source point in the lower region.

The Green's function whose source and observation points are both in region 1 is found from (10.38) and the coefficient α as

$$g_{11}(\boldsymbol{r},\ \boldsymbol{r}') = \frac{1}{4\pi\varepsilon_1} \left[\frac{1}{[\rho^2 + (z - z')^2]^{1/2}} + \frac{\alpha}{[\rho^2 + (z + z' - 2d)^2]^{1/2}} \right] \tag{10.48}$$

where we define to simplify $\rho^2 = (x - x')^2 + (y - y')^2$.

Next, the Green's function whose observation point is in region 1 and the source charge is in region 2 is

$$g_{12}(\boldsymbol{r},\ \boldsymbol{r}') = \frac{1}{4\pi\varepsilon_1} \frac{(1 + \alpha)}{[\rho^2 + (z - z')^2]^{1/2}} \tag{10.49}$$

The third case is for the observation point in region 2 and the source point in region 1. The result is

$$g_{21}(\boldsymbol{r},\ \boldsymbol{r}') = \frac{1}{4\pi\varepsilon_2} \frac{(1 - \alpha)}{[\rho^2 + (z - z')^2]^{1/2}}. \tag{10.50}$$

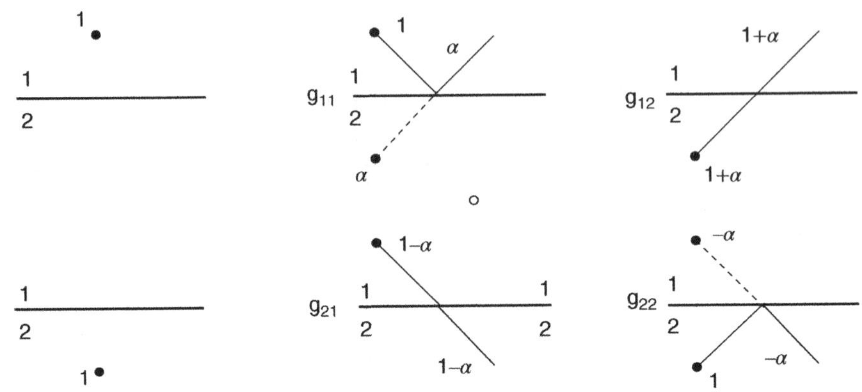

FIGURE 10.12 Reflection and image for the four Green's functions.

The last case is given when both the observation and the source point are region 2 with a dielectric constant ε_2

$$g_{22}(\boldsymbol{r}, \boldsymbol{r}') = \frac{1}{4\pi\varepsilon_2}\left[\frac{1}{[\rho^2 + (z - z')^2]^{1/2}} - \frac{\alpha}{[\rho^2 + (z - z' + 2d)^2]^{1/2}}\right]. \tag{10.51}$$

Working out the proper images for the three dielectric layers considered in the next section is quite challenging. However, we can use the two layer results to significantly simplify the process since meeting local boundary conditions is the same for all cases of interest. We summarize the four Green's function in Fig. 10.12 where we show the location of the image as well as the reflection needed for the next section.

10.4.4 Three Dielectric Layers Green's Functions

Another class of problems which can be solved with the image method consists of three dielectric layers as shown in Fig. 10.13. Even for the more complicated problems, we can take advantage of what we did learn about the two layer dielectric problems. For the three dielectric layer problems, the observation point $\boldsymbol{r}$ as well as the source point $\boldsymbol{r}'$ can be located in each of the three layer. This results in nine Green's functions for all the situations. However, fortunately only five Green's functions are needed to cover the fundamentally different functions. The rest of the equations can be obtained from the formulas in this section by a simple exchange of variables.

For the three layer case considered, we have two reflection coefficients, α_1 for the upper dielectric interfaces and α_2 for the lower case. By applying (10.47) to the three layer situation in Fig. 10.13 we obtain

$$\alpha_1 = \frac{\varepsilon_1 - \varepsilon_2}{\varepsilon_1 + \varepsilon_2} \quad \text{and} \quad \alpha_2 = \frac{\varepsilon_2 - \varepsilon_3}{\varepsilon_2 + \varepsilon_3} \tag{10.52}$$

It is noticed from the above example in Fig. 10.12, that in general the regions where the observation point is located does not contain any image charges.

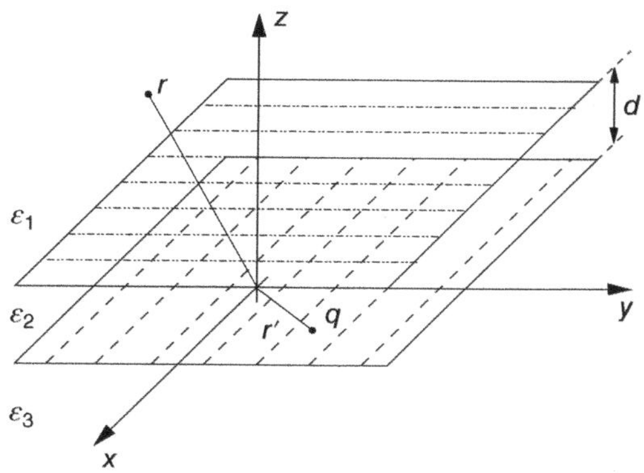

FIGURE 10.13 Example geometry for three dielectric layers.

The locations of images are always equidistant from the interface. The strength of the images requires some thought in that the reflections must be considered carefully. Because there are multi-reflections due to the existence of two parallel dielectric boundaries, the coefficients α_i in (10.52) have to be determined by multiple applications of the α's in (10.53). A summary of diagrams in Fig. 10.12 is very helpful for the determination of the strength as well as the location of the images.

We start with g_{11} where both the source and the observation points are located in region 1 shown in Fig. 10.14.

$$g_{11}(\boldsymbol{r},\ \boldsymbol{r}') = \frac{1}{4\pi\varepsilon_1}\left[\frac{1}{[\rho^2 + (z - z')^2]^{1/2}} + \frac{\alpha_1}{[\rho^2 + (z + z' - 2d)^2]^{1/2}} + (1 - \alpha_1^2)\sum_{k=0}^{\infty}\frac{(-\alpha_1)^k(\alpha_2)^{k+1}}{[\rho^2 + (z + z' + 2\ k\ d)^2]^{1/2}}\right] \tag{10.53}$$

Each of the situations lead to as new formulation. Also, g_{33} can be derived from g_{11} by a change of variables.

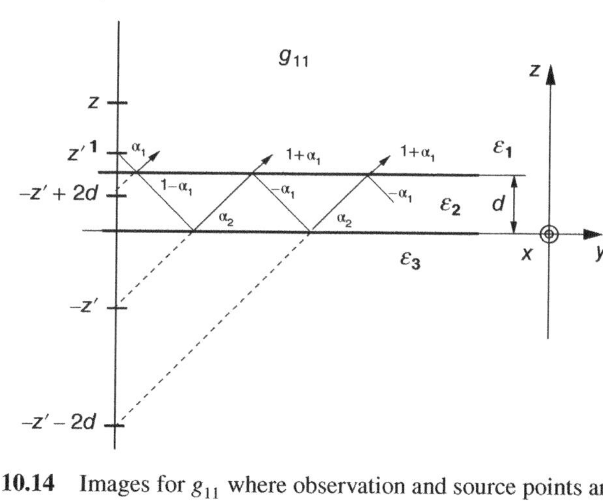

FIGURE 10.14 Images for g_{11} where observation and source points are in layer 1.

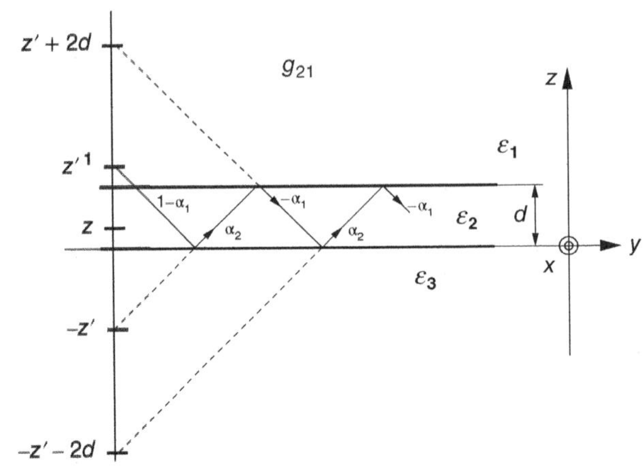

FIGURE 10.15 Images for g_{21} where the source point is in region 1 and the observation point is in region 2.

For the second Green's function, the source is still located in layer 1 while the observation point is in the middle layer 2, as shown in Fig. 10.15. By a change of variables we can construct G_{23} from this result since the observation point is staying in layer 2.

$$g_{21}(\boldsymbol{r},\ \boldsymbol{r}') = \frac{1-\alpha_1}{4\pi\varepsilon_1}\left[\sum_{k=0}^{\infty}\frac{(-\alpha_1)^k\alpha_2^{k+1}}{[\rho^2+(z+z'+2\ k\ d)^2]^{1/2}} + \sum_{k=0}^{\infty}\frac{(-\alpha_1\alpha_2)^k}{[\rho^2+(z-z'-2\ k\ d)^2]^{1/2}}\right] \tag{10.54}$$

The third Green's function for which the source is in layer 1 is g_{31}. The observation point is moved to layer or region 3, as shown in Fig. 10.16. We should note that by a change of

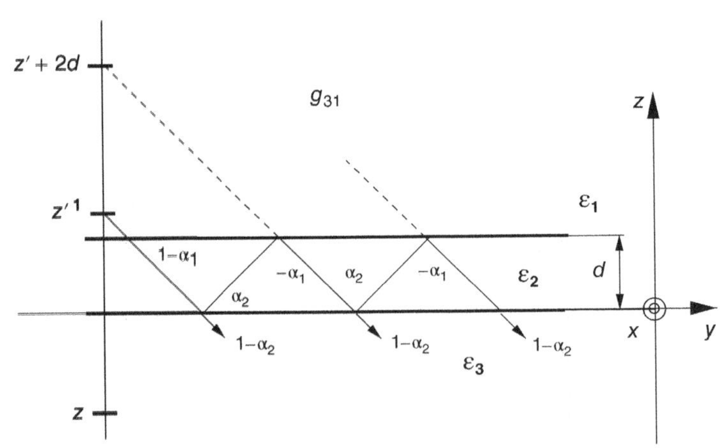

FIGURE 10.16 Images for g_{31} where the source point is in region 1 and the observation point is in region 3.

variables in the solution can also be used to find the result for g_{13}. Hence, both the source point and observation points are in the outside layers.

$$g_{31}(\boldsymbol{r},\ \boldsymbol{r}') = \frac{(1-\alpha_1)(1-\alpha_2)}{4\pi\varepsilon_3}\left[\sum_{k=0}^{\infty}\frac{(-\alpha_1\alpha_2)^k}{[\rho^2+(z-z'-2\ k\ d)^2]^{1/2}}\right] \tag{10.55}$$

For the case where the source point is in layer 2 and the observation point is in Layer 1 which is g_{12} is treated next. We should observe that this solution can also used to obtain g_{32} since the observation point is just in the other outer layer while the source point is in the same layer.

$$g_{12}(\boldsymbol{r},\ \boldsymbol{r}') = \frac{(1+\alpha_1)}{4\pi\varepsilon_1}\left[\sum_{k=0}^{\infty}\frac{(-\alpha_1\alpha_2)^k}{[\rho^2+(z-z'+2\ k\ d)^2]^{1/2}} + \sum_{k=0}^{\infty}\frac{(-\alpha_1)^k(\alpha_2)^{k+1}}{[\rho^2+(z+z'+2\ k\ d)^2]^{1/2}}\right] \tag{10.56}$$

We give the reflection diagrams for g_{12} in Fig. 10.17. The resultant Green's function is given in (10.57).

Finally, the Green's function for the case where both the source and observation point are in layer 2 is g_{22}. In this case, images occur on both sides. The refections are shown in Fig. 10.18.

$$g_{22}(\boldsymbol{r},\ \boldsymbol{r}') = \frac{1}{4\pi\varepsilon_2}\left[\frac{1}{[\rho^2+(z-z')^2]^{1/2}} + \sum_{k=1}^{\infty}\frac{(-\alpha_1)^k\alpha_2^{k-1}}{[\rho^2+(z+z'-2kd)^2]^{1/2}} + \sum_{k=1}^{\infty}\frac{(-\alpha_1\alpha_2)^k}{[\rho^2+(z-z'+2kd)^2]^{1/2}}\right.$$

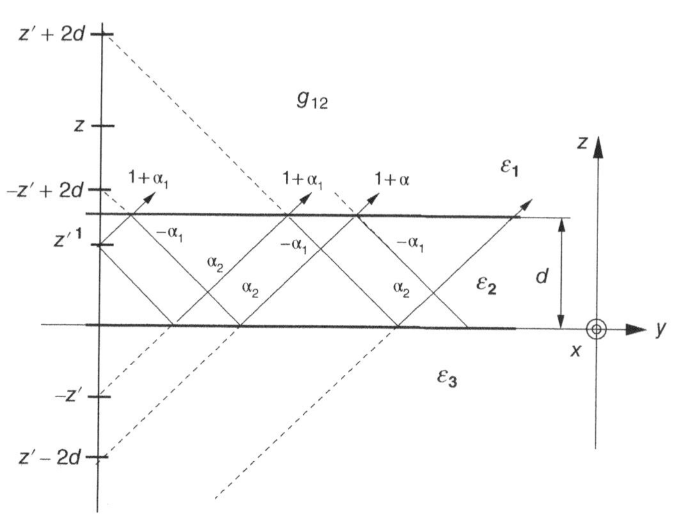

FIGURE 10.17 Images for g_{12} where the source point is in region 2 and the observation point is in region 1.

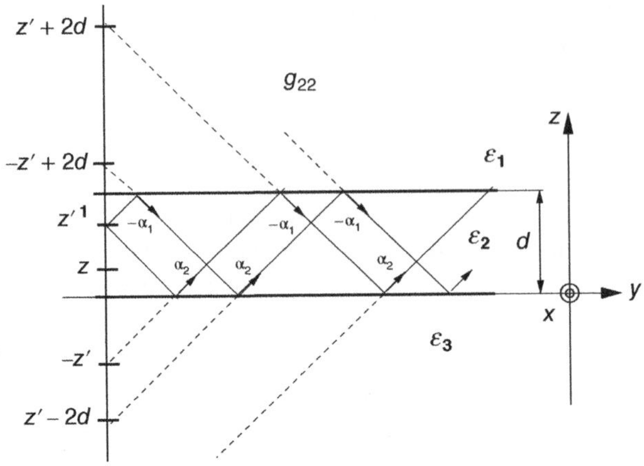

FIGURE 10.18 Images for g_{22} where the observation and source points are in region 2.

$$+\sum_{k=1}^{\infty}\frac{(-\alpha_1)^k\alpha_2^{k+1}}{[\rho^2+(z+z'+2kd)^2]^{1/2}}+\sum_{k=1}^{\infty}\frac{(-\alpha_1\alpha_2)^k}{[\rho^2+(z-z'-2kd)^2]^{1/2}}\Bigg] \quad (10.57)$$

This includes the images which are needed to construct all cases required. Note that we also included the permitivtiy needed for each Green's function for clarity.

10.4.5 Dielectric Model Based on the Volume Equivalence Theorem

The technique presented in this section is based on the *volume equivalence theorem* in Section 3.5.1. The theorem allows the replacement of a finite dielectric body by charges and currents. The volume PEEC finite dielectric bodies was originally added to the classical PEEC conductor model in Refs [25, 21]. The model was extended to nonorthogonal-shaped bodies [22]. Surface PEEC models have also been added to the classical PEEC volume approach [23, 24], which includes finite dielectric bodies. This different approach is given in Section 11.4.2.

The volume model considered in this section transforms the dielectric materials to bound charges located in a uniform free space with a dielectric constant ε_0. Hence, this approach results in a very convenient model. Importantly, the technique can be used for both quasistatic (QS)PEEC models and full-wave (FW)PEEC models. Also, lossy and dispersive dielectrics are included in the model in Section 10.4.7.

We start the derivation of the volume equivalence model with a simple problem consisting of a single dielectric block between two zero thickness conducting sheets, which is shown in Fig. 10.19.

As always, we consider a problem geometry, which may be part of a larger one, with other conductors or dielectrics. Here, we take a finite dielectric part only since the PEEC formulation for conductors is treated carefully in Chapter 6. This is permissible if we remember to take the couplings into account between all the different parts of the overall problem.

So far, we have dealt with the free charge density $q^F(\boldsymbol{r})$ on conductors. To represent the dielectrics using the volume equivalence theorem given in Section 3.5.1, we use bound charge density $q^B(\boldsymbol{r})$ that describes the dielectric charge where $\varepsilon_r > 1$ is handled separately

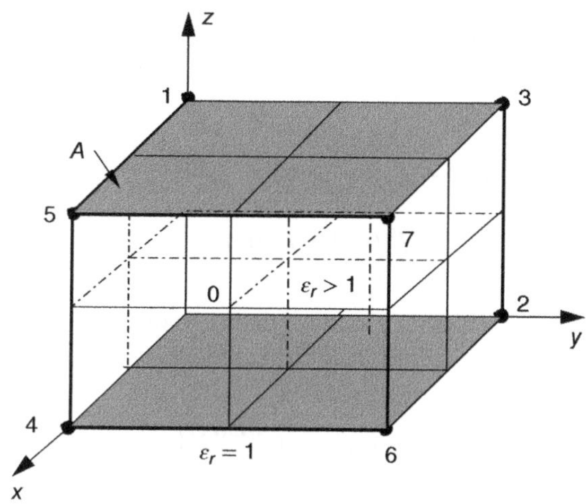

FIGURE 10.19 Dielectric block between two zero thickness metal plates.

from the conducting currents that are due to free charge q^F. Maxwell's equation (3.1d) defines the relationship between the displacement flux density and the charge densities as

$$\nabla \cdot \boldsymbol{E} = \frac{q^F + q^B}{\varepsilon_0}, \tag{10.58}$$

where q^F is the free charge and q^B is the bound charge.

We purposely chose an example in Fig. 10.19 where the side surfaces have a bound charge only, since the surfaces are dielectric–air interfaces. The top and bottom surfaces have dielectric bounded polarization charge densities q^B as well as free charge densities q^F due to the conductors.

The volume equivalence formulation is derived by adding and subtracting the displacement current $d\boldsymbol{E}(\boldsymbol{r},t)/dt$ in the free space through the Ampere's law for $\boldsymbol{H}$ in Section 3.1.1 such that the formulation results

$$\begin{aligned} \nabla \times \boldsymbol{H}(\boldsymbol{r},t) &= \boldsymbol{J}^C(\boldsymbol{r},t) + \varepsilon_0\varepsilon_r \frac{\partial \boldsymbol{E}(\boldsymbol{r},t)}{\partial t} \\ &= \boldsymbol{J}^C(\boldsymbol{r},t) + \varepsilon_0(\varepsilon_r - 1)\frac{\partial \boldsymbol{E}(\boldsymbol{r},t)}{\partial t} + \varepsilon_0 \frac{\partial \boldsymbol{E}(\boldsymbol{r},t)}{\partial t} \end{aligned} \tag{10.59}$$

in the time domain. Of course, the formulation is the same if we replace the time derivative d/dt with the Laplace variable s for a frequency domain solution. Thus, the total current in (10.59) takes into account both the conduction electric current related to the conductivity of the medium and the polarization current due to the dielectrics

$$\begin{aligned} \boldsymbol{J}^T(\boldsymbol{r},t) &= \boldsymbol{J}^C(\boldsymbol{r},t) + \varepsilon_0(\varepsilon_r - 1)\frac{\partial \boldsymbol{E}(\boldsymbol{r},t)}{\partial t} + \varepsilon_0 \frac{\partial \boldsymbol{E}(\boldsymbol{r},t)}{\partial t} \\ &= \boldsymbol{J}^C(\boldsymbol{r},t) + \boldsymbol{J}^{eq}(\boldsymbol{r},t) + \boldsymbol{J}^D(\boldsymbol{r},t). \end{aligned} \tag{10.60}$$

Hence, considering that the last term is related to displacement current density, which is modeled by time variation of charges residing only on the surface of the objects, the total current within the dielectric is

$$\begin{aligned} \boldsymbol{J}_d^T(\boldsymbol{r},t) &= \boldsymbol{J}^C(\boldsymbol{r},t) + \varepsilon_0(\varepsilon_r - 1)\frac{\partial \boldsymbol{E}(\boldsymbol{r},t)}{\partial t} \\ &= \boldsymbol{J}^C(\boldsymbol{r},t) + \boldsymbol{J}^{eq}(\boldsymbol{r},t), \end{aligned} \tag{10.61}$$

which suggests a parallel RC circuit modeling both conductive and polarization current densities. From (10.60), we can define an equivalent dielectric current density $\boldsymbol{J}^{eq}$ as

$$\boldsymbol{J}^{eq} = \varepsilon_0(\varepsilon_r - 1)\frac{\partial \boldsymbol{E}(\boldsymbol{r},t)}{\partial t}. \tag{10.62}$$

We can include this term in the electric field integral equation in the form

$$\boldsymbol{E}(\boldsymbol{r},t) = \frac{1}{\varepsilon_0(\varepsilon_r - 1)}\int_{-\infty}^{t} \boldsymbol{J}^{eq}(\boldsymbol{r},\tau)\, d\tau. \tag{10.63}$$

This leads to the electric field equation for points in the dielectric region

$$\boldsymbol{E}^{inc}(t) = \frac{1}{\varepsilon_0(\varepsilon_r - 1)}\int_{-\infty}^{t} \boldsymbol{J}^{eq}(\tau)\, d\tau + \frac{d\boldsymbol{A}(t)}{dt} + \nabla\Phi_e(t) \tag{10.64}$$

corresponding to the conventional EFIE for conductors (3.25).

10.4.6 Discretization of Dielectrics

The next step is again the discretization of the integral equation and to construct an equivalent circuit for the equation. Fortunately, we already have discretized two out of the three right-hand-side terms in (10.64). For this reason, we are not repeating the parts that are given in Sections 6.1, 6.2 and 6.3.3. We need to construct the PEEC model for the first term, which is (10.63).

The same averaging technique that is used for the resistance (6.7) is used for this term. As an example, we pick a z-directed cell in Fig. 10.19 for the derivation. Of course, all four vertical divisions are quarter cells. Then, the integration will be for the left-front cell

$$V_{5,4}(t) = \int_{z_4}^{z_5} \boldsymbol{E}(\boldsymbol{r},t)\cdot\hat{z}\, dz = \frac{1}{\varepsilon_0(\varepsilon_r - 1)}\int_{z_4}^{z_5}\int_{-\infty}^{t} J_z^{eq}(\boldsymbol{r},\tau)\, d\tau\, dz. \tag{10.65}$$

Using conventional current density averaging, we get

$$V_{5,4}(t) = \frac{1}{\mathcal{A}\,\varepsilon_0(\varepsilon_r - 1)}\int_{z_4}^{z_5}\int_{\mathcal{A}}\int_{-\infty}^{t} J_z^{eq}(\boldsymbol{r},\tau)\, d\tau\, d\mathcal{A}\, dz. \tag{10.66}$$

With the constant approximation in the z-direction, this simplifies to

$$V_{5,4}(t) = \frac{(z_5 - z_4)}{\mathcal{A}\,\varepsilon_0(\varepsilon_r - 1)} \int_{-\infty}^{t} I_z^{eq}(\tau)\,d\tau. \tag{10.67}$$

We note that the first part can be interpreted as the inverse of a parallel plate capacitance of the form $\varepsilon A/d$ or

$$C_e = \frac{\mathcal{A}\,\varepsilon_0(\varepsilon_r - 1)}{(z_5 - z_4)}. \tag{10.68}$$

However, we notice that the difference in the capacitance is proportional to $(\varepsilon_r - 1)$. We recognize that this capacitance is in excess of the free space values. For this reason, it is called excess capacitance [21, 25].

Definition 7 *The excess capacitance C_e of a dielectric cell is defined as*

$$C_e := \frac{\varepsilon_0(\varepsilon_r - 1)\mathcal{A}}{\ell}, \tag{10.69}$$

where ε_r is the relative dielectric constant of the cell. and where 1 is the plate-to-plate spacing and A is the area of the plates in Fig. 10.19

We further observe that for a capacitor, the current and the voltage can be written as

$$i(t) = C_e \frac{dv(t)}{dt} \qquad v(t) = \frac{1}{C} \int_{-\infty}^{t} i(t)\,dt. \tag{10.70}$$

It is clear from (10.70) that the model for the new term in (10.61) is interpreted as the equivalent circuit model in Fig. 10.20.

Some observations help to understand this important model.

- The excess capacitance C_e takes the dielectric constant above air $\varepsilon_r = 1$ into account.
- A parallel resistance could be used to include a model for losses due to finite conductivity of the dielectric material. This would include dielectrics with a *dc* resistance.

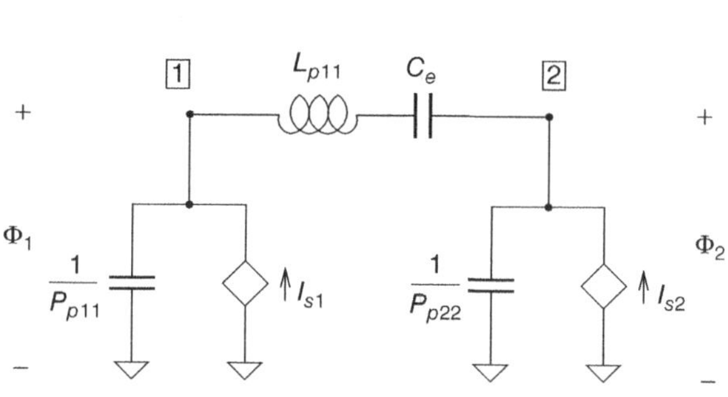

FIGURE 10.20 One section example of equivalent circuit for dielectric.

- Surprisingly, a partial inductance Lp_{11} has to be computed for the dielectric cell. Also, all the partial inductances in a PEEC model are coupled including the dielectric ones.
- Capacitive cells are only placed on surfaces with conductor–dielectric interfaces. Interfaces with dielectric cells only do not connect to the free-charge model.

We take a specific example to illustrate the last point. For example, node 5 in Fig. 10.19 includes several faces. The free side faces do not have free charge associated with them, whereas the top face does have a free charge associated with the conductor surface. However, bound charge or the excess capacitance C_e is associated with all faces. This happens automatically when we compute C_e for the model.

10.4.7 Dispersive Dielectrics Included in the Volume Equivalence Theorem Model

An important observation is that the models for dispersive dielectrics in Section 10.1 can directly be used in the volume equivalence theorem model. The implementation is simple since we can use any one of the dielectric models.

To give a specific example, we take the widely used multipole Debye model in (10.27) as part of the volume equivalence model. We consider the ε_r excess capacitance in Definition 7 and replace it with a model with the appropriate equivalent circuit.

The Debye model for the excess capacitance is given by

$$C_e(s) = \frac{\varepsilon_0 \mathcal{A}}{\ell} \left[\varepsilon_\infty - 1 + \sum_{i=1}^{N_D} \frac{(\varepsilon_{S,i} - \varepsilon_{\infty,i})}{1 + s\tau_i} \right], \tag{10.71}$$

where N_D, RC series branches are used in the circuit model. This model can be implemented in a solver with the equivalent circuit shown in Fig. 10.5.

10.4.8 Dispersive Dielectrics with Finite Electrical Conductivity

If a dispersive dielectric is characterized by a finite conductivity, then the corresponding equivalent circuit is obtained by placing a resistance representing the finite conductivity in parallel to the equivalent circuit for the dispersive behavior of the dielectric. An example of the equivalent circuit for a Debye medium with finite electrical conductivity is shown in Fig. 10.21. This model is based on the one-pole Debye model in Fig. 10.3 where as an example, a dc resistance R_{dc} is added to the equivalent circuit.

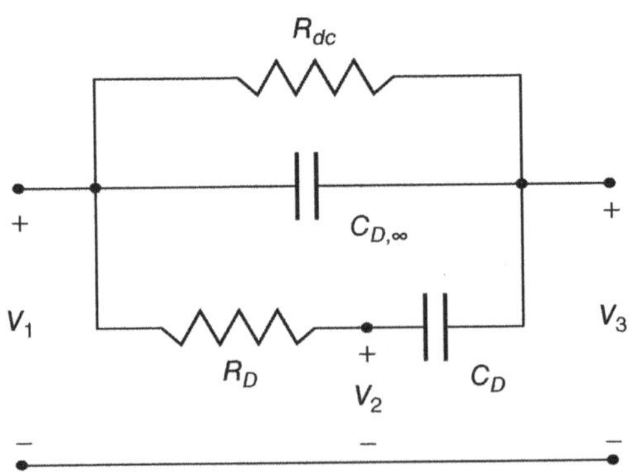

FIGURE 10.21 Equivalent circuit for a Debye medium with finite electrical conductivity.

10.4.9 Convolution Formulation for General Dispersive Media

The implementation of the frequency-dependent dielectrics is simple in the frequency domain since the dielectric model can directly be applied in (3.2b), $\boldsymbol{D}(\omega) = \varepsilon(\omega)\boldsymbol{E}(\omega)$. For the time domain, we use convolution as an alternative way to implement the dielectric model, which includes the susceptibility functions (10.11). The displacement vector (10.7) $\boldsymbol{D}$, for the linear dispersive medium is given by

$$\boldsymbol{D}(t) = \varepsilon_0\, \boldsymbol{E}(t) + \varepsilon_0 \int_0^t \boldsymbol{E}(t-\tau)\chi(\tau)\, d\tau, \tag{10.72}$$

in the time domain. Here, we simplified the notation by omitting the space variable dependence of the field quantities. Again, ε_0 is the permittivity of free space and the electric susceptibility is given by $\chi(t) = \varepsilon_r(t) - 1$.

A discrete time solution for the convolution at p time steps $t = p\Delta t$, is

$$\begin{aligned} \boldsymbol{D}(t) &= \boldsymbol{D}(p\Delta t) \\ &= \varepsilon_0 \boldsymbol{E}(p\Delta t) + \varepsilon_0 \int_0^{p\Delta t} \boldsymbol{E}(p\Delta t - \tau)\chi(\tau)\, d\tau. \end{aligned} \tag{10.73}$$

Fortunately, the recursive convolution algorithm from Section 2.10.2 can be used to speed up the computation. This has been done for a single-pole model [26] and also in PEEC framework [27].

As done in (2.80), the derivation can be based on a residue–pole model that directly fits with the Debye model for the volume dielectric model in Section 10.4.7, (10.71). Of course, it can also be used for a conventional Debye circuit model (10.27) if the number of elements is large.

$$\varepsilon(s) = \varepsilon_0 \left[\varepsilon_\infty + \sum_{i=1}^{N_D} \frac{\Delta\varepsilon_i}{1 + s\tau_i} \right], \tag{10.74}$$

which we can represent in the general form (2.83)

$$x(s) = \left[d + \sum_{m=1}^{M_r} \frac{r_m}{s - p_m} \right] y(s), \tag{10.75}$$

where r_m are the usual residues and p_m the poles.

The recursive convolution algorithm presented in Section 2.10.2 is

$$w(t_{p-1}) := \sum_{k=0}^{K_p} e^{p_m(t_{p-1}-t_k)} y(t_k). \tag{10.76}$$

We write (10.76) recursively as

$$w(t_{p-1}) = e^{p_m\Delta t} \sum_{k=0}^{K_p} e^{p_m(t_{p-2}-t_k)}\, y(t_k) = e^{p_m\Delta t}\, w(t_{p-2}) \tag{10.77}$$

using (2.85) for the time step t_{p-1} where $w(t_{k-2})$ is evident from (10.76) and by recursively computing $w(t_0), w(t_1), \ldots, w(t_{k-2})$. Using this approach, we can compute the solution as

$$x(t_p) = \Delta t\, y(t_p)\, d + \Delta t \sum_{m=1}^{M_r} r_m\, e^{p_m \Delta t} w(t_{p-1}) \tag{10.78}$$

by recursively updating. The last part in the convolution sum is evaluated by the recursive scheme [28]

$$v_c(n) = \frac{q_c(n)}{C_{e\infty}} + \Psi = \kappa q_c(n) + \Psi, \tag{10.79}$$

where $\Psi = q_c(n) f(0) + \sum_{m=1}^{n-1} q_c(n - m\Delta t) f(m)$, which takes into account the past history and can be efficiently computed by a recursive scheme [28].

Losses for the dielectric material are usually available in different forms. The use of table data as a function of frequency is a common approach. Alternatively, data is based on physics-based models such as the Lorentz or Debye models considered earlier. A fundamental requirement is that the real and imaginary parts of the data or model are *Hilbert consistent*. This implies that the requirements of the model for passivity are guaranteed, for example, Ref. [29].

10.5 EXAMPLE FOR IMPACT OF DIELECTRIC PROPERTIES IN THE TIME DOMAIN

In this section, we want to illustrate the impact of the dielectric loss on the solution results. Additional examples for the application of the dielectric models are given in Ref. [27]. It is clear that at high frequencies both the dielectric losses and the skin-effect losses modeled in Chapter 9, are of importance. For this reason, we want to give examples that include both types of loss problems. We should specify that we assume that the conductors in all examples are copper. The conductivity of copper is specified in Appendix A.

10.5.1 On-Chip Type Interconnect

We modeled the simple small interconnect connection over a ground plane shown in Fig. 10.22. The source resistance was $R_i = 10\ \Omega$ and the termination was $R_L = 100\ \Omega$. The highest frequencies in the spectrum of the current impulse waveform is in the 50–100 GHz range. We considered the lossless dielectric case as well as a Debye model with different loss tangents. The mesh discretization was chosen using the $\lambda_{min}/20$ rule

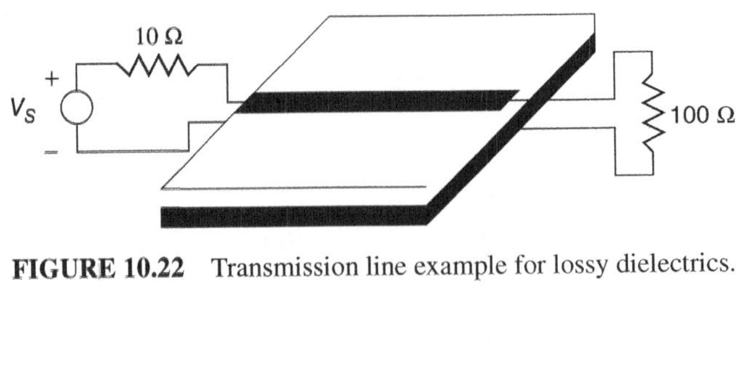

FIGURE 10.22 Transmission line example for lossy dielectrics.

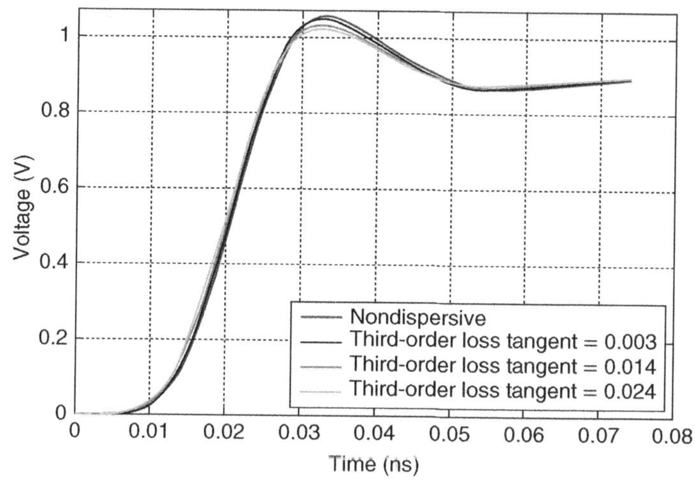

FIGURE 10.23 Time domain output port voltage with Debye model with $\varepsilon_S = 20, f_1 = 1$ GHz and different values of tan δ.

such that λ_{min} corresponds to 200 GHz. This resulted in the unknowns of 948 currents or partial inductances and 156 node potentials for the PEEC model.

We applied a 1 V voltage source with a 20 ps rise time in this example. The dielectric has been modeled by a third-order Debye model with increasing values of tan δ for fixed values of ε_S and f_1. The results of the analysis are shown in Fig. 10.23 for different dielectric loss tangent. We observe that the larger value of tan δ results in the reduction of the peak amplitude.

10.5.2 Microstrip Line with Dispersive, Lossy dielectric

In this section, we consider an example where the dielectric is represented by $N_D = 4$ poles. The dielectric material was assumed to be FR-4 with tan $\delta = 0.02$, $\varepsilon_S = 4.7$, $\tau_1 = 1.59$ ns. The dispersive, lossy behavior of the dielectric has been taken into account by means of a Debye model where

$$\varepsilon = \varepsilon_0 \left(\varepsilon_\infty + \sum_{k=1}^{N_D} \frac{\Delta\varepsilon_k}{1 + s\tau_k} \right) \tag{10.80}$$

and where $\Delta\varepsilon_k = \varepsilon_{S,k} - \varepsilon_{\infty,k}$ and τ_k are the value and the time constant for the kth relaxation process, respectively. The parameters used are given in Table 10.2.

Figures 10.24 and 10.25 give the loss tangent and also the magnitude of electrical permittivity as a function of an increasing number of poles. Clearly, adding poles to the model increases the accuracy for a wider range of frequencies. From this, it is evident that for accurate wide-band frequency responses, multipole models are required.

A microstrip transmission line on a dielectric was modeled using this dispersive model. The trace with a zero thickness was 1 mm wide and the dielectric thickness was also 1 mm. The ground plane underneath was 1 cm wide and 10 cm long. The modeling resulted in 924 inductance volume cells and 760 capacitance surface cells. This lead to a PEEC model with 924 current unknowns and 168 potential nodes.

TABLE 10.2 Parameters for an FR-4 Debye model.

Pole number	$\varepsilon_{S,k}$	$\varepsilon_{\infty,k}$	$\tau_k(ns)$
Pole 1	4.7	4.55	1.59
Pole 2	4.55	4.40	0.159
Pole 3	4.40	4.25	0.0159
Pole 4	4.25	4.10	0.00159

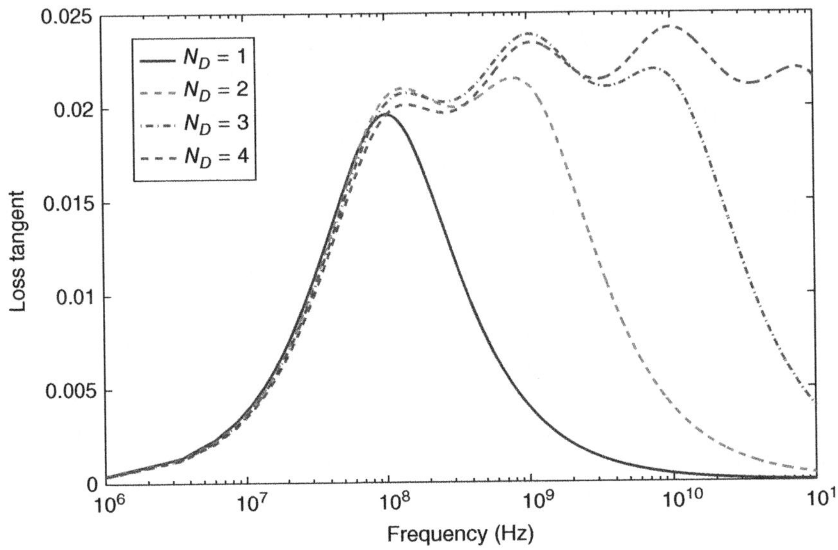

FIGURE 10.24 FR-4 loss tangent for increasing number of poles (example in Section 10.5.2).

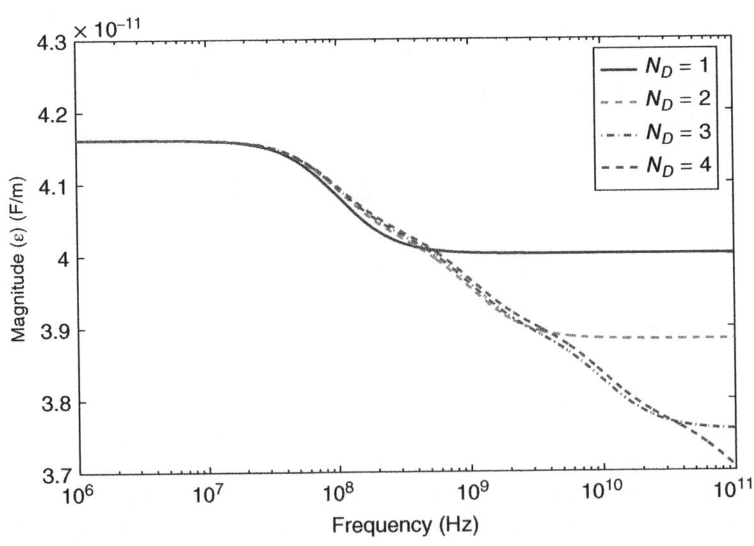

FIGURE 10.25 FR-4 permittivity for increasing number of poles (example in Section 10.5.2).

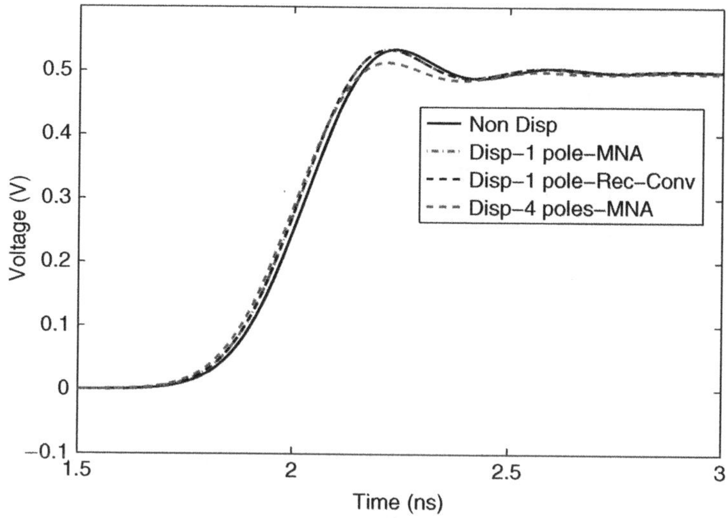

FIGURE 10.26 Output port voltage (example in Section 10.5.2).

Both the source and the load resistances for the microstrip are 50 Ω. The input voltage source is a 2 V volt step with a rise time of 35 ps. The results for this example are shown in Fig. 10.26. The voltage given is for a perfect dielectric (solid line, Non Disp) and for Debye models with a different number of poles. In this example, the dielectric is taken into account using recursive convolution for this time domain response. The damping due to dispersion is clearly visible in the broadband loss model.

Figure 10.27 shows the voltage magnitude of the spectrum at the output obtained by using the two solvers, implementing dispersive and lossy dielectrics by means of equivalent circuits. A satisfactory agreement is reached in the frequency range 0–1 GHz.

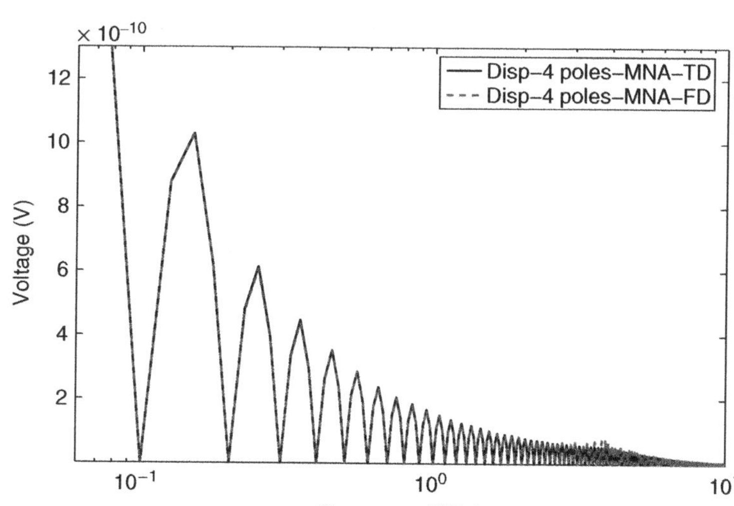

FIGURE 10.27 Magnitude spectrum of the output port voltage (example in Section 10.5.2).

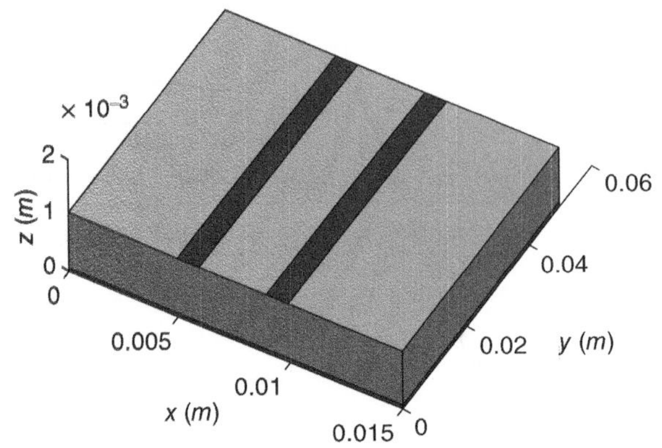

FIGURE 10.28 Coplanar microstrip line (example in Section 10.5.3).

10.5.3 Coplanar Microstrip Line Example

A last example with dielectrics is based on two coupled lines on a lossy dielectric block with a size of 50 × 15 mm with a thickness of 1 mm. The dielectric is again FR-4 and the loss is again represented with a fourth-order $N_D = 4$ Debye model in a (QS)PEEC solution. As shown in Fig. 10.28, two 1 mm wide zero thickness wires are spaced by 3 mm centered on the dielectric. Both conductors are terminated with 50 Ω resistors, while one of them is driven by a 1 V source with a 30 ps rise and fall time where the pulse is 5 ns wide. The PEEC model is discretized with 1008 volume and 796 surface cells, which leads to 1008 partial inductance with currents and 176 nodes.

The response for the model for the coplanar microstrip transmission line shown in Fig. 10.29 was validated using an finite integration technique (FIT)-based computer

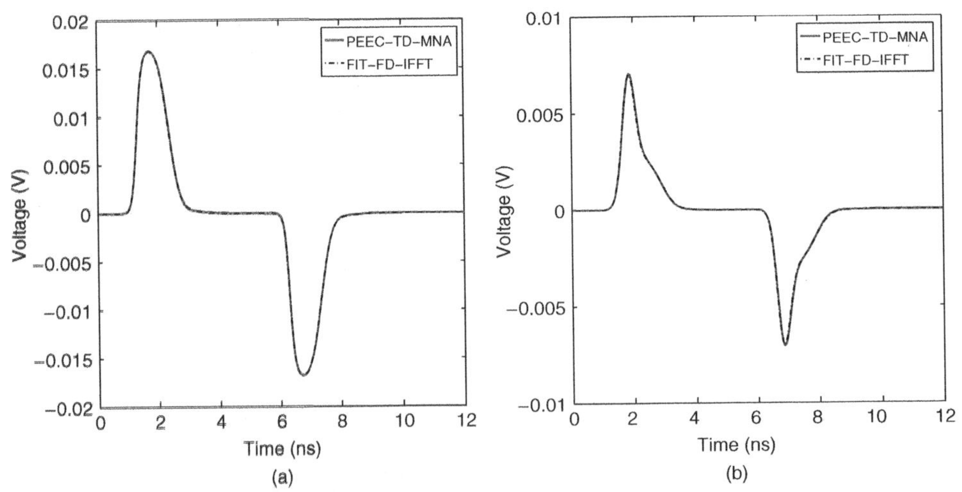

FIGURE 10.29 Near-end (a) and far-end (b) voltages for the coplanar lines. The solid line refers to the results obtained using the proposed methodology in the time domain (PEEC–TD–MNA). The dash–dot line refers to the results obtained using the finite integration technique (FIT) technique in the frequency domain via-inverse fast Fourier transform (IFFT) (FIT–FD–IFFT).

simulation technology (CST) solver [30] in the frequency domain, which has been transformed to the time domain using the inverse fast Fourier transform (IFFT). The agreement can be observed to be very good.

PROBLEMS

10.1 *Hilbert transform*
Verify that you understand the Hilbert transform given in (10.4a) and (10.4b). Note that they should be satisfied for the real and imaginary parts of any complex impedance function. An important issue is how we take care of the singular behavior when $x = \omega$. Describe how you numerically treat the singularity in the Hilbert integrals.

10.2 *Debye model evaluation*
Plot the magnitude of the permittivity and the loss tangent versus frequency for the fourth-order Debye model introduced in Section 10.1.5, (10.15). Do not use the Lorentzian terms in the model.

10.3 *Lorentz medium*
Derive an equivalent circuit for Lorentz medium using a complex pole model. (Hint: Complex pole models are discussed in Chapter 2.)

10.4 *Static Green's function*
Using Fig. 10.12 explain the contributions to the static Green's function in (10.53).

10.5 *Asymptotic form of the static Green's function*
Assume that a problem consists of air above a planar dielectric half space. Assume that the permittivity ε_r is close to infinite. Derive the static Green's function asymptotically based on the results from (10.48)–(10.51).

REFERENCES

1. P. G. Huray. *The Foundations of Signal Integrity*. John Wiley and Sons, Inc., New York, 2010.
2. C. A. Balanis. *Advanced Engineering Electromagnetics*. John Wiley and Sons, Inc., New York, 1989.
3. R. Achar and M. Nakhla. Simulation of high-speed interconnects. *Proceedings of the IEEE*, 89(5):693–728, May 2001.
4. E.-P. Li. *Electrical Modeling and Design for 3D System Integration*. John Wiley and Sons, Inc., New York, Hoboken, NJ, 2012.
5. K. M. C. Branch, J. Morsey, A. Cangellaris, and A. Ruehli. Physically consistent transmission line models for high-speed interconnects in lossy dielectrics. *IEEE Transactions on Advanced Packaging*, 25(2):129–135, May 2002.
6. M. J. Kobrinsky, S. Chakravarty, D. Jiao, M. Harmes, S. List, and M. Mazumder. Experimental validation of crosstalk simulations for on-chip interconnects at high frequencies using S-parameters. In *Digest of Electrical Performance of Electronic Packaging*, pp. 329–332, October 2003.
7. D. Jiao, C. Dai, S.-W. Lee, T. R. Arabi, and G. Taylor. Computational electromagnetics for high-frequency IC design. In *Proceedings of IEEE Antennas and Propagation Society International Symposium*, Volume 3, pp. 3317–3320, June 2004.

8. P. Triverio, S. Grivet-Talocia, M. S. Nakhla, F. G. Canavero, and R. Achar. Stability, causality, and passivity in electrical interconnect models. *IEEE Transactions on Advanced Packaging*, 30(4):795–808, November 2007.
9. E. C. Titchmarsh. *Introduction to the Theory of Fourier Integrals*. Clarendon Press, Oxford, 1967.
10. L. D. Landau and E. M. Lifshitz. *Electrodynamics of Continuous Media*. Addison Wesley Publishing Company, New York, London, Paris, 1960.
11. R. E. Diaz and N. G. Alexopoulos. An analytic continuation method for the analysis and design of dispersive materials. *IEEE Transactions on Antennas and Propagation*, 45(11):1602–1610, November 1997.
12. K. Inan and R. E. Diaz. Automatic analytic continuation method for the analysis and design of dispersive materials. *IEEE Transactions on Antennas and Propagation*, 59(4):1228–1236, April 2011.
13. K. S. Cole and R. H. Cole. Dispersion and absorption in dielectrics. *Journal of Chemical Physics*, 9(4):341–351, 1941.
14. D. W. Davidson and R. H. Cole. Dielectric relaxation in glycol, propylene glycol and n-propanol. *Journal of Chemical Physics*, 19(12):1484–1490, 1951.
15. S. Havriliak and S. Negami. A complex plane analysis of α dispersion in some polymer systems. *Journal of Polymer Science, Part C*, 14(1):99–117, 1966.
16. E. Tuncer and S. M. Gubanski. On dielectric data analysis. Using the Monte Carlo method to obtain relaxation time distribution and comparing non-linear spectral function fits. *IEEE Transactions on Dielectrics and Electrical Insulation*, 8(3):310–320, June 2001.
17. A. E. Engin, W. Mathis, W. John, G. Sommer, and H. Reichl. Time domain modeling of lossy substrates with constant loss tangent. In 6th IEEE Workshop on Signal Propagation on Interconnects, Heidelberg, Germany, pp. 151–154, May 2004.
18. E. Weber. *Electromagnetic Theory*. Dover Publications, New York, 1965.
19. S. Ramo, J. R. Whinnery, and T. Van Duzer. *Fields and Waves in Communication Electronics*. John Wiley and Sons, Inc., New York, 1994.
20. L. J. Jiang, W. C. Chew, and Y. C. Pan. Capacitance Extraction in the Multilayer Medium Using DCIM and SMFMA. *Journal of Electromagnetic Waves and Applications*, 19(14):1851–1864, 2005.
21. H. Heeb and A. Ruehli. Three-dimensional interconnect analysis using partial element equivalent circuits. *IEEE Transactions on Circuits and Systems*, 39(11):974–982, November 1992.
22. A. E. Ruehli, G. Antonini, J. Esch, J. Ekman, A. Mayo, and A. Orlandi. Non-orthogonal PEEC formulation for time and frequency domain EM and circuit modeling. In *Proceedings of the IEEE International Symposium on Electromagnetic Compatibility*, Volume 45, pp. 167–176, May 2003.
23. A. Rong and A. C. Cangellaris. Generalized PEEC models for three-dimensional interconnect structures and integrated passives of arbitrary shapes. In *Digest of Electrical Performance of Electronic Packaging*, Volume 10, Boston, MA, pp. 225–228, October 2001.
24. D. Gope, A. Ruehli, and V. Jandhyala. Surface-based PEEC formulation for modeling conductors and dielectrics in time and frequency domain combined circuit electromagnetic simulation. In *Digest of Electrical Performance of Electronic Packaging*, Volume 13, Portland, OR, pp. 329–332, October 2004.
25. A. E. Ruehli and H. Heeb. Circuit models for three-dimensional geometries including dielectrics. *IEEE Transactions on Microwave Theory and Techniques*, 40(7):1507–1516, July 1992.
26. G. Antonini. PEEC modelling of Debye dispersive dielectrics. In *Electrical Engineering and Electromagnetics* (Eds C. A. Brebbia and D. Polyak), WIT Press, Southampton, pp. 126–133, 2003.

27. G. Antonini, A. E. Ruehli, and A. Haridass. Including dispersive dielectrics in PEEC models. In *Digest of Electrical Performance of Electronic Packaging*, Princeton, NJ, pp. 349–352, October 2003.

28. K. S. Kunz and R. J. Luebbers. *The Finite Difference Time Domain Method for Electromagnetics*. CRC Press, Boca Raton, FL, 1993.

29. T. R. Arabi, A. T. Murphy, T. K. Sarkar, R. F. Harrington, and A. R. Djordjevic. On the modeling of conductor and substrate losses in multiconductor, multidielectric transmission line systems. *IEEE Transactions on Microwave Theory and Techniques*, 39(7):1090–1097, July 1991.

30. Computer Simulation Technology, *www.cst.com. CST Studio Suite, 2006B*, 2007.

11

PEEC MODELS FOR MAGNETIC MATERIAL

11.1 INCLUSION OF PROBLEMS WITH MAGNETIC MATERIALS

Models with magnetic materials can be a part of many different interesting PEEC solutions. This leads to a different class of issues. In some structures, magnetic bodies may be a small part of the overall problem. Examples are a small inductor on a chip, a filter on a printed circuit board (PCB), or a microwave filter. On the other hand, substantial parts of a problem can consist of magnetic materials. Other examples are parts of a motor or a very large transformer. Different solution approaches are necessary to cover such a wide range of problems. Hence, different models may be used to include magnetic materials in partial element equivalent circuit (PEEC) models.

The first approach considered in Section 11.1 deals with magnetically closed loops such that the magnetic part may be treated separately from the rest of the problem. For this class of problems, the solution may result in a simpler model. We consider how such a model is applied in a PEEC environment. In fact, integrating such a part may be very simple since it can be included as circuit models. The second class of problems consists predominantly of magnetic materials, where most of the solution is based on magnetic materials.

11.1.1 Magnetic Circuits for Closed Flux Type Class of Problems

We first consider perhaps one of the oldest solution methods for magnetic problems, which was and still is applied, including many situations such as power electronics problems and transformers or motors, for example, [1, 2]. The key assumption is that the majority of the magnetic flux is in the magnetic material rather than through the air. Today, this class of problems can also include chokes used with PCBs as well as other magnetic core structures such as magnetic memory cores. An interesting aspect is that we can solve this class of problems by using a circuit analog. To compare the corresponding statement for the

Circuit Oriented Electromagnetic Modeling Using the PEEC Techniques, First Edition.
Albert E. Ruehli, Giulio Antonini, and Lijun Jiang.

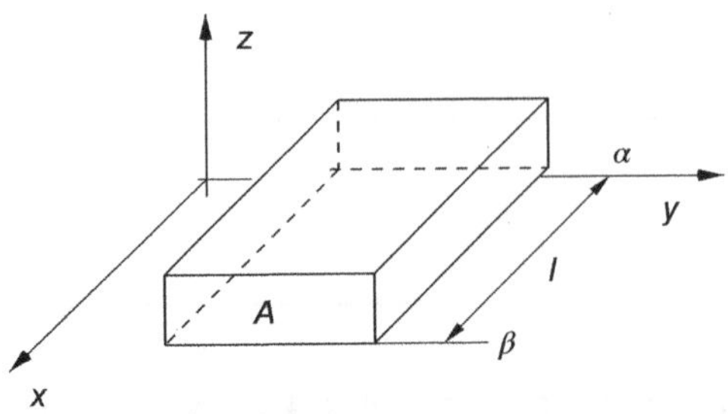

FIGURE 11.1 Simple example bar of magnetic material.

electrical and magnetic circuits, we display the electrical circuit equation on the left and the equivalent magnetic one on the right side in the derivation.

We first show in Fig. 11.1 part of a magnetic circuit to define the geometrical parameters such as the cross section $\mathcal{A}$ and the length ℓ. However, we should remark that the reluctance approach does not work well for open-loop problems such as the bar shown in Fig. 11.1. Hence, we assume that it is a portion of a loop without large gaps.

Corresponding to the electrical potential, we define an artificial magnetic potential Φ_m with the integral form

$$\boldsymbol{E} = -\nabla\Phi, \qquad \boldsymbol{H} = -\nabla\Phi_m \tag{11.1}$$

$$\Phi_{\alpha,\beta} = \int_{\alpha}^{\beta} \boldsymbol{E}\cdot d\ell, \qquad \Phi_{m,\alpha,\beta} = \int_{\alpha}^{\beta} \boldsymbol{H}\cdot d\ell. \tag{11.2}$$

α and β are the line integration starting position and the ending position respectively. The relationship between the current density and electrical fields is in terms of the conductivity σ for conductors that corresponds to the permeability μ in the magnetic material.

$$\boldsymbol{J} = \sigma\boldsymbol{E}, \qquad \boldsymbol{B} = \mu\boldsymbol{H}. \tag{11.3}$$

The relation between current and current density corresponds to the magnetic field and the magnetic flux

$$I = \int_{\mathcal{A}} \boldsymbol{J}\cdot\hat{\boldsymbol{n}}\, d\mathcal{A}, \qquad \Psi = \int_{\mathcal{A}} \boldsymbol{B}\cdot\hat{\boldsymbol{n}}\, d\mathcal{A}, \tag{11.4}$$

where I is the electrical current and Ψ is the magnetic flux in the magnetic equivalent circuit. It is equivalent to the electrical current for electrical circuits.

In this case, we use the *current flow* type notation $\mathcal{A}$ for the area on purpose for the magnetic flux since it corresponds to the current in the conventional circuit case. Again, Ψ corresponds to current and $\boldsymbol{B}$ corresponds to current density. With this, Ohm's law with the reluctance resistance corresponds to

$$\Phi = RI, \qquad \Phi_m = R_m\Psi \tag{11.5}$$

$$\mathrm{R} = \frac{\ell}{\sigma\mathcal{A}}, \qquad R_m = \frac{\ell}{\mu\mathcal{A}}. \tag{11.6}$$

Finally, we see a difference between the quasistatic equations for the electrical and magnetic fields where

$$\oint \boldsymbol{E} \cdot d\ell = 0, \qquad \oint \boldsymbol{H} \cdot d\ell = I_{tot}. \tag{11.7}$$

By comparing (11.2) and (11.7), we find for the closed loop that

$$\Phi_m = I_{tot}, \tag{11.8}$$

where I_{tot} is the total current penetrating the loop. We see that the magnetic equivalent circuit equation differs in an important way in that the closed loop in (11.7) leads to an equivalent voltage source corresponding to the driving voltage. This requires some adjustment in thinking. Note that the units of the magnetic potential is A or Amperes. We need to note that the current I_{tot} is the total *conventional* current. We see that for a multiturn loop with N-turns, the current may be entering the loop N times, which results in an N multiplication to get the total contributing current. The first equation of (11.7) results from Faraday's law and the second equation in (11.7) is the well-known Ampere's law.

11.1.2 Example for Inductance Computation

First, we apply the equivalent magnetic circuit Ohm's law equation (11.6) to an example of a magnetic core shown in Fig. 11.2 where the magnetic potential is

$$\Phi_m = R_m \Psi = \frac{\ell}{\mu A} \Psi. \tag{11.9}$$

The application of the currents as sources is not trivial. We start with (11.2), for a closed loop

$$\Phi_m = \int_\ell \boldsymbol{H} \cdot d\ell = \int_A \boldsymbol{J} \cdot \hat{\boldsymbol{n}}\, dA. \tag{11.10}$$

The important aspect is that the loop and surface for using Ampere's law must be chosen correctly. In the example in Fig. 11.2, the loop is given by the dashed line, which is a closed

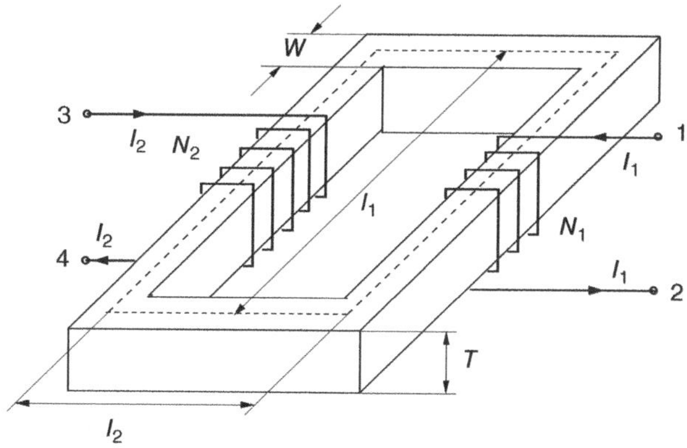

FIGURE 11.2 Magnetic transformer core example.

loop in this case. The currents that penetrate this loop are flowing in the winding wires, where each of them carries a current I. Therefore, the magnetic source is given by the total current entering the magnetic core

$$\Phi_m = N_{turns}\, I \tag{11.11}$$

FOR $N_{turn} = N$ turns. Note that the other parts of the windings are outside of the loop and they do not count.

Substituting (11.11) in (11.9), we get

$$N_{turns} I = \Phi_m = R_m \Psi \tag{11.12}$$

and from this, since the total flux is equal to the inductance L multiplied by the current I, or

$$N_{turns}\, \Psi = L\, I \tag{11.13}$$

we have with (11.12) results in the inductance

$$L = \frac{N^2}{R_m}. \tag{11.14}$$

Finally, from Fig. 11.2 we can see that the magnetic circuit length is $\ell = 2\,(\ell_1 + \ell_2)$ and the magnetic reluctance resistance is

$$R_m = \frac{2\,(\ell_1 + \ell_2)}{\mu\, T\, W}. \tag{11.15}$$

If we connect terminals 2 and 4 together, we have an inductor with the number of turns is $N = N_1 + N_2$. From (11.14) and (11.15), we can compute the inductance for this arrangement as

$$L_{tot} = \frac{\mu\, T\, W\, (N_1 + N_2)^2}{2\,(\ell_1 + \ell_2)}. \tag{11.16}$$

Of course, we could have left the second loop unconnected such that $I_2 = 0$ and then only the N_1 winding would have contributed to the inductance.

Note that we can expand (11.16) into

$$L_{tot} = k\, {N_1}^2 + 2\, k\, N_1\, N_2 + k\, {N_2}^2, \tag{11.17}$$

where $k = \mu\, T\, W/[2\,(\ell_1 + \ell_2)]$. We should note that this can be interpreted as

$$L_{tot} = L_{11} + 2\, L_{12} + L_{22}, \tag{11.18}$$

since the L matrix is symmetrical $L_{12} = L_{21}$. While it is clear for the self-inductances, we also see that the mutual coupling inductance is given by $L_{12} = kN_1N_2$. Also, the formula (11.17) is the conventional inductance for the series connected windings.

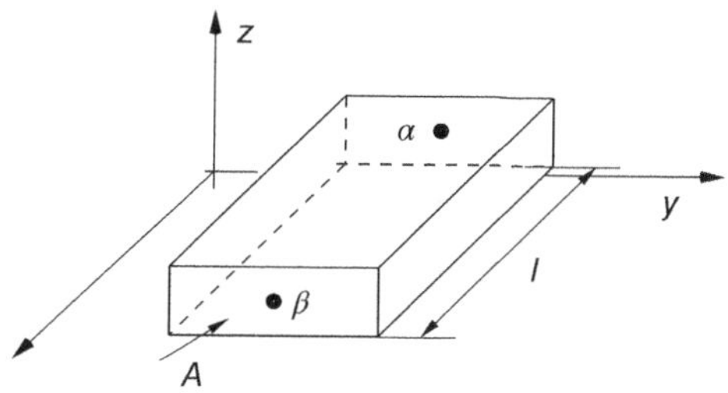

FIGURE 11.3 Example bar of magnetic material for reluctance computation.

11.1.3 Magnetic Reluctance Resistance Computation

Since the computation of the reluctance resistance is a key part of the magnetic circuit approach, we give a specific example for the computation of a piece of magnetic material. The flux direction is laminar from α to β as shown in Fig. 11.3. It is important to consider the details of computing the magnetic reluctance resistance (11.6), $R_m = \Phi_m / \Psi$, or

$$R_m = \frac{\int_\alpha^\beta \boldsymbol{H} \cdot d\ell}{\int_A \boldsymbol{B} \cdot d\mathcal{A}}. \tag{11.19}$$

Finally, the reluctance resistance for the simple geometry in Fig. 11.3 is obtained by approximating (11.19)

$$R_{m1} = \frac{H_1 \, \ell_1}{B_1 \, \mathcal{A}_1} = \frac{\ell_1}{\mu_1 \, \mathcal{A}_1}. \tag{11.20}$$

The last step is clearly dependent on a sufficiently high permeability μ_1 since the accuracy of the approximation depends on the condition that $\mu_1 \gg 1$.

11.1.4 Inductance Computation for Multiple Magnetic Paths

A key application of the magnetic circuit approach is the computation of the inductances for large permeability magnetic cores as we showed above. Further, more complex structures can be treated with a similar approach. The following example shows how the technique can be applied for a realistic case. We use the example in Fig. 11.4 to show how to formulate the solution of the problem. The solution of interest is an inductance model for the three transformer windings. We again assume that the permeability is $\mu \gg 1$.

The equivalent circuit for the example in Fig. 11.4 is given in Fig. 11.5. Some simple subdivisions for the resistances are given in Fig. 11.4 and the values are computed as discussed in the previous section. The easiest way to solve for the inductances is to write the loop equation for the two loops in Fig. 11.5. For the first loop, we get

$$-N_1 I_1 + R_1 \Psi_{L1} + R_4 \Psi_{L1} + R_3 \Psi_{L1} - N_2 I_2 + R_2(\Psi_{L1} - \Psi_{L2}) = 0 \tag{11.21}$$

and for the second loop

$$R_2(\Psi_{L2} - \Psi_{L1}) + N_2 I_2 + R_5 \Psi_{L2} - N_3 I_3 + R_7 \Psi_{L2} + R_6 \Psi_{L2} = 0. \tag{11.22}$$

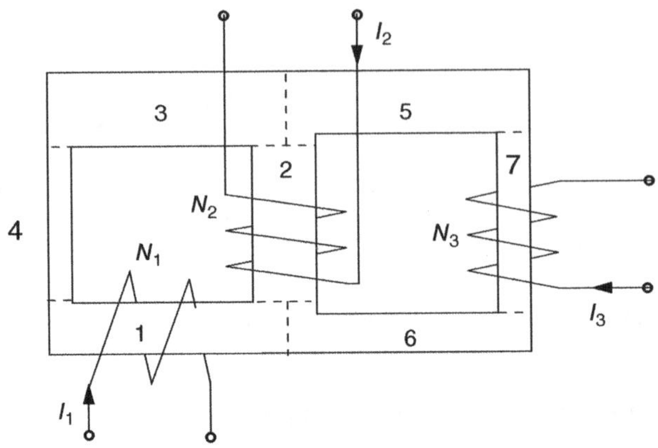

FIGURE 11.4 Example problem of a three-bar magnetic circuit.

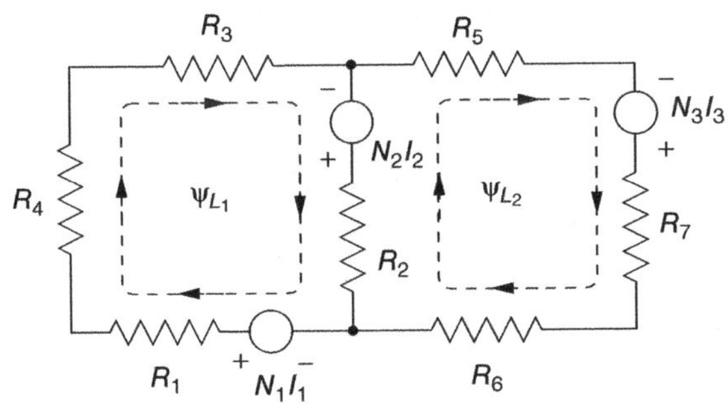

FIGURE 11.5 Magnetic equivalent circuit for three-bar problem.

This can be written in matrix form as follows:

$$\begin{bmatrix} \Psi_{L1} \\ \Psi_{L2} \end{bmatrix} = \frac{1}{\Delta} \begin{bmatrix} R_\beta & R_2 \\ R_2 & R_\alpha \end{bmatrix} \begin{bmatrix} N_1 I_1 + N_2 I_2 \\ -N_2 I_2 + N_3 I_3 \end{bmatrix}, \tag{11.23}$$

where we defined $R_\alpha = R_1 + R_2 + R_3 + R_4$ and $R_\beta = R_2 + R_7 + R_5 + R_6$ and where the determinant is $\Delta = R_\beta R_\alpha - R_2^2$.

As the last step, we want to use (11.23) to compute the inductances for the system. We start out by carefully expanding (11.13) into the matrix form since we need to compute the 3×3 inductance matrix for the coils.

$$\begin{aligned} N_1\, \Psi_{L1} &= L_{11}\, I_1 + L_{12}\, I_2 + L_{13}\, I_3 \\ N_2\, \Psi_{L2} &= L_{21}\, I_1 + L_{22}\, I_2 + L_{23}\, I_3 \\ N_3\, \Psi_{L3} &= L_{31}\, I_1 + L_{32}\, I_2 + L_{33}\, I_3. \end{aligned} \tag{11.24}$$

The interpretation of (11.24) for our problem is not straightforward since we need to include the appropriate couplings. It is best to consider Figs. 11.4 and 11.5. First, we want to compute L_{11}. An important observation is that the total flux picked up by this is $N_1\ \Psi_{L1}$. Hence, the interpretation of (11.24) for this case is

$$L_{11} = \frac{N_1\ \Psi_{L1}}{I_1}|_{(I_2=I_3=0)} = \frac{R_\beta\ N_1{}^2\ I_1}{\Delta\ I_1} = \frac{R_\beta\ N_1{}^2}{\Delta}. \tag{11.25}$$

We want to give another example of the computation of L_{13} since it involves both loops in the example.

$$L_{13} = \frac{N_1\ \Psi_{L1}}{I_3}|_{(I_1=I_2=0)} = \frac{R_\alpha\ N_1 N_3\ I_3}{\Delta\ I_3} = \frac{R_\alpha\ N_1 N_3}{\Delta}. \tag{11.26}$$

From this, it is evident how we compute all the elements of the inductance matrix. Also, the approach presented is general and it can be applied to other topologies.

11.1.5 Equivalent Circuit for Transformer-Type Element

The model presented in the previous section allows us to make an equivalent circuit for an inductor- or transformer-type structure [2]. If they are physically small compared to other parts of a problem, then the magnetic object can be modeled separately. Hence, we can embed the PEEC circuit model via external connections only and parasitic models can be included in the model. An obvious example is to add the *dc* resistance of the windings to the magnetic model.

As an example, we take the arrangements of the windings in Fig. 11.2. A transformer is formed between connections 1,2 and 3,4. The equations for this transformer are, if we also include the series resistance, as follows:

$$\begin{aligned} -V_{12} + R_1 I_1 + sL_{11} I_1 + sL_{12} I_2 &= 0 \\ -V_{34} + R_2 I_2 + sL_{22} I_1 + sL_{21} I_1 &= 0. \end{aligned} \tag{11.27}$$

We can easily verify that the equivalent circuit in Fig. 11.6 satisfies the usual equations (11.27).

Hence, it is evident that we could use the equivalent circuit in conjunction with an otherwise PEEC or conventional circuit model.

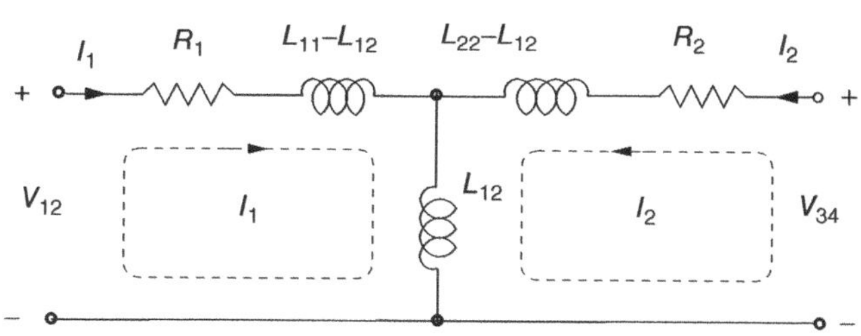

FIGURE 11.6 Equivalent circuit for transformer.

11.2 MODEL FOR MAGNETIC BODIES BY USING A MAGNETIC SCALAR POTENTIAL AND MAGNETIC CHARGE FORMULATION

11.2.1 Magnetic Scalar Potential

Assuming that if the current density is zero in a volume space and if the frequency is zero, then Ampere's law holds $\nabla \times \boldsymbol{H} = \boldsymbol{0}$. This allows us to introduce a scalar magnetic potential Φ_m in the same way we introduce the electrical potential as

$$\boldsymbol{H} = -\nabla\Phi_m. \tag{11.28}$$

Also in a linear medium $\nabla \cdot \boldsymbol{B} = \boldsymbol{0}$ or

$$\nabla \cdot (\mu\nabla\Phi_m) = 0. \tag{11.29}$$

If μ is piecewise constant within each region, then the scalar magnetic potential satisfies the Laplace equation

$$\nabla^2\Phi_m = 0. \tag{11.30}$$

The solution in the volume where μ is constant needs to be matched on its surfaces through appropriate boundary conditions for the magnetic field.

11.2.2 Artificial Magnetic Charge

Different models can be used to represent magnetic problems. In this section, we introduce an approach that is based on artificial magnetic volume bound charge density q_{vm} and magnetic surface density q_{sm}, with the artificial magnetic potential Φ_m.

Next, we introduce the magnetization $\boldsymbol{M}$ of the material [3]. For some magnetic materials, magnetization is independent of the field strength, at least for moderate field intensities. These materials are known as "*magnetically hard materials*". Where the magnetization $\boldsymbol{M}$ is known. Since we can assume that the electrical current density $\boldsymbol{J} = \boldsymbol{0}$ in these regions, we can use the scalar magnetic potential Φ_m.

The problems we consider here are based on so-called linear materials where $\boldsymbol{B} = \mu_0\mu_r\boldsymbol{H}$. Then, the conventional Maxwell's equation (3.1d), $\nabla \cdot \boldsymbol{B} = 0$ is used and from this

$$\nabla \cdot \boldsymbol{B} = \nabla \cdot (\mu_0\, \boldsymbol{H} + \mu_0\, \boldsymbol{M}) = 0 \tag{11.31}$$

and

$$\nabla \cdot \boldsymbol{H} = -\nabla \cdot \boldsymbol{M}, \tag{11.32}$$

which defines the magnetic field $\boldsymbol{B}$ in terms of the magnetic intensity $\boldsymbol{H}$ and the magnetization vector $\boldsymbol{M}$ to include the contribution of the magnetic material.

Using (11.28), (11.30) becomes the Poisson equation:

$$\nabla^2\Phi_m = -q_{vm}, \tag{11.33}$$

where the effective magnetic charge is defined as

$$q_{vm} = -\nabla \cdot \boldsymbol{M}. \tag{11.34}$$

This changes the magnetic charge definition to

$$\nabla \cdot \boldsymbol{H} = -\nabla \cdot \boldsymbol{M} = q_{vm}, \tag{11.35}$$

where the solution for Φ_m is given by

$$\Phi_m = -\frac{1}{4\pi}\int_{\mathcal{V}} \frac{\nabla \cdot \boldsymbol{M}(\boldsymbol{r}')}{|\boldsymbol{r}-\boldsymbol{r}'|} d\mathcal{V}. \tag{11.36}$$

The integration can be extended to ∞, since the magnetization $\boldsymbol{M}$ is typically continuous.

It can be convenient to represent the magnetization with a discontinuous distribution. For instance, if a *magnetically hard material* occupies a volume $\mathcal{V}$, closed by a surface $\mathcal{S}$, then magnetization can be defined only within $\mathcal{V}$ assuming it is zero outside $\mathcal{S}$. Applying the divergence theorem to (11.34), considering an elementary volume enclosed by a Gauss surface across the surface $\mathcal{S}$, it can be defined as a magnetic surface charge q_{sm} as

$$q_{sm} = \hat{\boldsymbol{n}} \cdot \boldsymbol{M}, \tag{11.37}$$

where $\hat{\boldsymbol{n}}$ is the outward normal to the surface $\mathcal{S}$. In this case, the solution for Φ_m becomes

$$\Phi_m(\boldsymbol{r}) = \frac{1}{4\pi}\int_{\mathcal{V}'} \frac{q_{vm}(\boldsymbol{r}')}{|\boldsymbol{r}-\boldsymbol{r}'|}\, d\mathcal{V}' + \frac{1}{4\pi}\int_{\mathcal{S}'} \frac{q_{sm}(\boldsymbol{r}')}{|\boldsymbol{r}-\boldsymbol{r}'|}\, d\mathcal{S}'. \tag{11.38}$$

For the case of interest where we assume that the magnetization is uniform inside $\mathcal{V}$, the first term in (11.38) is zero being $q_{vm} = 0$ and only the contribution of surface magnetic charge q_{sm} holds.

We get the magnetic field $\boldsymbol{H}(\boldsymbol{r}) = -\nabla\Phi_m(\boldsymbol{r})$, from (11.38)

$$\boldsymbol{H}(\boldsymbol{r}) = \frac{1}{4\pi}\int_{\mathcal{S}'} q_{sm}(\boldsymbol{r}')\frac{(\boldsymbol{r}-\boldsymbol{r}')}{|\boldsymbol{r}-\boldsymbol{r}'|^3}\, d\mathcal{S}' + \boldsymbol{H}^{inc}(\boldsymbol{r}), \tag{11.39}$$

where $\boldsymbol{H}^{inc}(\boldsymbol{r})$ is a potential externally applied magnetic field.

11.2.3 Magnetic Charge Integral Equation for Surface Pole Density

We show the interface between two materials with a different permeability in Fig. 11.7. The continuity of the magnetic field across the interface yields

$$\mu_2\, H_{n2} - \mu_1\, H_{n1} = 0. \tag{11.40}$$

The unit vectors $\hat{\boldsymbol{n}}_2$ and $\hat{\boldsymbol{n}}_1$ are normal to the interface between the two materials with the permeability μ_2 and μ_1. The surface S is assumed to be sufficiently smooth. We subdivide

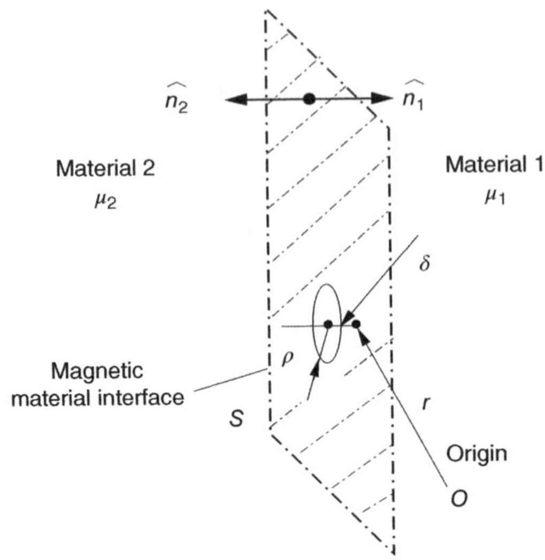

FIGURE 11.7 Interface between the two regions with different permeability.

the surface as shown in Fig. 11.7 into a small disk of radius ρ with the center at the projection of r onto the surface S. The radius of the disk $\rho \to 0$ such that $\delta = o(\rho)$. Here, δ is the normal distance between the point r and the center of the disk. In the limit as $\delta \to 0$, the total magnetic field on either side of the interface S is given by

$$H_n(\boldsymbol{r}) = \pm H_i(\boldsymbol{r}) + \frac{1}{4\pi}\int_{\mathrm{PV}} q_{sm}(\boldsymbol{r}')\frac{(\boldsymbol{r}-\boldsymbol{r}')}{|\boldsymbol{r}-\boldsymbol{r}'|^3}\,dS' + H^{inc}, \tag{11.41}$$

where H^{inc} is the usual incident field.

The contribution of the magnetic surface charge close to the surface is dominated by the *local normal field* in both directions given by $H_i(\boldsymbol{r}) = q_{sm}/2$. We observe that the sign depends on which side of the boundary we are located. This part is used in a numerical implementation to include the contribution of the local discretization cell. Next, we finally substitute (11.41) into the local boundary condition (11.40) to get the integral equation. With the definition for the reflection coefficient $R = (\mu_2 - \mu_1)/(\mu_2 + \mu_1)$, the following equation results

$$-\,q_{sm}(\boldsymbol{r}) = \frac{R}{2\pi}\int_{S'} q_{sm}(\boldsymbol{r}')\frac{(\boldsymbol{r}-\boldsymbol{r}')\cdot\hat{\boldsymbol{n}}_1}{|\boldsymbol{r}-\boldsymbol{r}'|^3}\,dS' + 2\,R\,H_{n2}. \tag{11.42}$$

This integral equation of second kind has been used to compute magnetic fields for thin magnetic shields [4] in the presence of an external field. Of course, for conductors outside of the magnetic material, inductance problems can be solved [5, 6]. This is an assumption stated at the beginning of Section 11.2.2.

11.2.4 Magnetic Vector Potential

Following Ref. [3], we want to consider the magnetic vector potential in terms of the magnetization vector $\boldsymbol{M}$. Since the magnetic flux density $\boldsymbol{B}$ is a solenoidal field ($\nabla \cdot \boldsymbol{B} = 0$), we use the definition in (3.18) where

$$B = \nabla \times A. \tag{11.43}$$

Thus, the second Maxwell curl equation can be written using (11.32) as follows:

$$\nabla \times H = \nabla \times (B/\mu_0 - M). \tag{11.44}$$

Combining (11.44) with (11.43) and assuming a low-frequency regime, we obtain the Poisson equation for the magnetic vector potential A using (3.37) to get

$$\nabla^2 A = -\mu_0 \nabla \times M. \tag{11.45}$$

For this, the solution for the magnetic vector potential $A(r)$ is

$$A(r) = \frac{\mu_0}{4\pi} \int_{\mathcal{V}'} \frac{M(r') \times (r - r')}{|r - r'|^3} \, d\mathcal{V}', \tag{11.46}$$

where the integral is to be computed over the entire free space since the magnetization is continuous in space. If we assume the magnetization confined into a volume $\mathcal{V}'$ and, thus, we assume it discontinuous across its surface $\mathcal{S}'$, it can be shown that the magnetic vector potential can be written as [3]

$$A(r) = \frac{\mu_0}{4\pi} \int_{\mathcal{V}'} \frac{\nabla' \times M(r')}{|r - r'|} \, d\mathcal{V}' + \frac{\mu_0}{4\pi} \int_{\mathcal{S}'} \frac{M(r') \times \hat{n}'}{|r - r'|} \, d\mathcal{S}', \tag{11.47}$$

where $\hat{n}'$ is the outward normal to the surface $\mathcal{S}'$. However, we use the volume magnetization formulation (11.46) in the derivation in the following section.

11.3 PEEC FORMULATION INCLUDING MAGNETIC BODIES

A realistic simple example of a problem with conducting bodies as well as a magnetic one is shown in Fig. 11.8. The treatment of the conductor 1 in PEEC with internal subdivisions leads to the conventional volume filament (VFI) skin-effect model in Chapter 9. Hence, the conductor part will lead to the conventional PEEC VFI skin-effect model with capacitive surface cells. Therefore, the key problem is the inclusion of magnetic material as shown by bar 2 in Fig. 11.8 to a conventional PEEC model.

Here, we assume that the magnetic part of the body can either be conductive or an insulator. The units of the quantities are evident from (11.32). The units for both the magnetic field intensity H and the magnetization M are A/m.

11.3.1 Model for Magnetic Body

We apply the derivation in Chapter 6 to form the PEEC equation for the magnetic body 2. Since we operate in the circuit domain, we can view the magnetic material as the addition

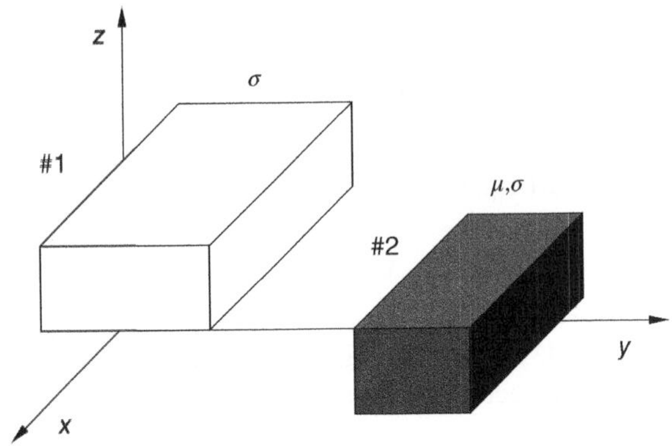

FIGURE 11.8 Example geometry for conductor and magnetic material problem.

of other circuit elements – or circuit element stamps – to the existing quasi-static PEEC model. We start with the usual integral equation (6.55), or

$$\boldsymbol{E}^{inc}(\boldsymbol{r}, s) = \frac{\boldsymbol{J}(\boldsymbol{r}, s)}{\sigma} + s\,\boldsymbol{A}(\boldsymbol{r}, s) + \nabla\phi(\boldsymbol{r}, s) \tag{11.48}$$

for the derivation where s is the Laplace variable. In the following, we do not rederive PEEC circuit elements that we considered in other chapters where the external field $\boldsymbol{E}^{inc}$ is given in Chapter 12 and the capacitance models in Chapter 6 as well as conventional partial inductance models.

We distinguish between the current in conductors 1 and in the magnetic material block in 2. The fundamental issue is rather straightforward. An additional vector potential needs to be included with the magnetic materials. This leads to

$$\boldsymbol{E}^{inc}(\boldsymbol{r}, s) = \frac{\boldsymbol{J}(\boldsymbol{r}, s)}{\sigma} + s\,\boldsymbol{A}_c(\boldsymbol{r}, s) + s\,\boldsymbol{A}_m(\boldsymbol{r}, s) + \nabla\phi(\boldsymbol{r}, s), \tag{11.49}$$

where $\boldsymbol{A}_c(\boldsymbol{r})$ results in the conventional partial inductances dependent on the electrical currents and where the magnetic vector potential $\boldsymbol{A}_m(\boldsymbol{r})$ is due to the magnetic material in Fig. 11.8 with its source, the magnetization $\boldsymbol{M}(\boldsymbol{r}, s)$.

To find the contribution of the magnetization to the vector potential, we start with (11.46)

$$\boldsymbol{A}_m(\boldsymbol{r}) = \frac{\mu_0}{4\pi}\int_{\mathcal{V}_{m'}} \frac{\boldsymbol{M}(\boldsymbol{r}') \times (\boldsymbol{r} - \boldsymbol{r}')}{|\,\boldsymbol{r} - \boldsymbol{r}'|^3} d\mathcal{V}', \tag{11.50}$$

which is used in the following section to evaluate the so-called magnetic inductance. It is clear that the second circuit equation in the modified nodal analysis (MNA) formulation (6.54) now has an additional term for the magnetic inductance besides the conventional partial inductances, which results in

$$\mathbf{A}_\ell^T\,\mathbf{\Phi}_n - (\mathbf{R} + s\,\mathbf{Lp})\mathbf{I} - s\,\mathbf{Lm}\,\mathbf{M} = -\mathbf{N}\,\mathbf{V}_i, \tag{11.51}$$

where the magnetic inductance term function $\mathbf{f}(\mathbf{M})$ is considered in more detail in the following section. We should note that $\mathbf{A}_\ell$ is the circuit incidence matrix or the matrix Kirchhoff's current law (KCL) (Section 2.7.1) in Chapter 2.

11.3.2 Computation of Inductive Magnetic Coupling

Figure 11.9 shows the conventional discretization we use for the magnetic material. Exactly like the conductors, we subdivide the volume into rectangular bars or cells for Manhattan coordinates. The magnetic inductances couple only to magnetization cells. However, they contribute to the vector potential for all cells in the system.

Expanding the cross-product in the numerator under the integral in (11.50), we get

$$\begin{aligned} \boldsymbol{M}(\boldsymbol{r}') \times (\boldsymbol{r} - \boldsymbol{r}') = \hat{\boldsymbol{x}}\, [M_y\,(z - z') - M_z\,(y - y')] \\ - \hat{\boldsymbol{y}}\, [M_x\,(z - z') - M_z\,(x - x')] \\ + \hat{\boldsymbol{z}}\, [M_x\,(y - y') - M_y\,(x - x')]. \end{aligned} \tag{11.52}$$

Using (11.52) and substituting it into (11.50) and by averaging as we did for the partial inductance in (5.18), a magnetic inductive coupling matrix **Lm** can be defined for the example given in Fig. 11.10.

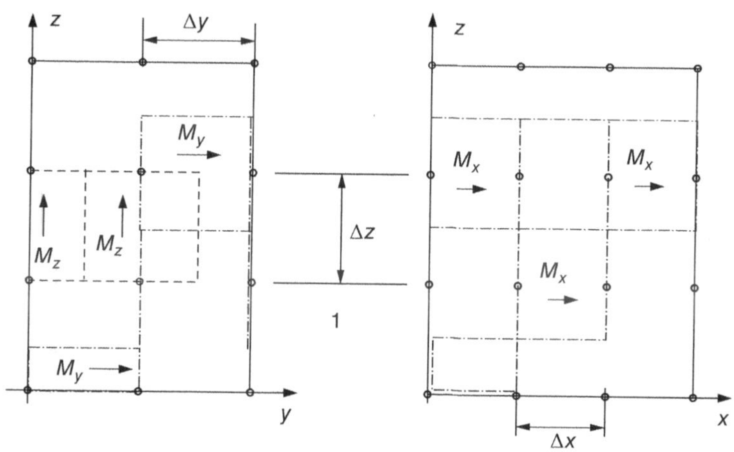

FIGURE 11.9 Meshing of magnetic body into conventional PEEC cells.

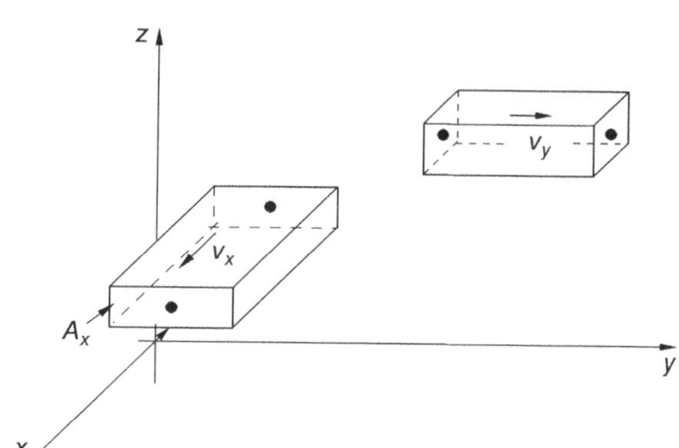

FIGURE 11.10 Example of two cell volumes for inductive coupling for Lm_{xy}.

The contribution of M_y to x is

$$Lm_{km}^{xy} = \frac{\mu_0}{4\pi}\frac{1}{\mathcal{A}_x}\int_{\mathcal{V}_{x,k}}\int_{\mathcal{V}'y,m}\frac{(z-z')}{|\boldsymbol{r}-\boldsymbol{r}'|^3}\,d\mathcal{V}'_y\,d\mathcal{V}_x, \tag{11.53}$$

where we need to find each of the magnetic inductive couplings for all three magnetization directions. The contribution of M_z to x is

$$Lm_{km}^{xz} = \frac{\mu_0}{4\pi}\frac{1}{\mathcal{A}_x}\int_{\mathcal{V}_{x,k}}\int_{\mathcal{V}'_{z,m}}\frac{(y-y')}{|\boldsymbol{r}-\boldsymbol{r}'|^3}\,d\mathcal{V}'_z\,d\mathcal{V}_x. \tag{11.54}$$

The contribution of M_x to y is

$$Lm_{km}^{yx} = \frac{\mu_0}{4\pi}\frac{1}{\mathcal{A}_y}\int_{\mathcal{V}_{y,k}}\int_{\mathcal{V}'_{x,m}}\frac{(z-z')}{|\boldsymbol{r}-\boldsymbol{r}'|^3}\,d\mathcal{V}'_x\,d\mathcal{V}_y. \tag{11.55}$$

The contribution of M_z to y is

$$Lm_{km}^{yz} = \frac{\mu_0}{4\pi}\frac{1}{\mathcal{A}_y}\int_{\mathcal{V}_{y,k}}\int_{\mathcal{V}'_{z,m}}\frac{(x-x')}{|\boldsymbol{r}-\boldsymbol{r}'|^3}\,d\mathcal{V}'_z\,d\mathcal{V}_y. \tag{11.56}$$

The contribution of M_x to z is

$$Lm_{km}^{zx} = \frac{\mu_0}{4\pi}\frac{1}{\mathcal{A}_z}\int_{\mathcal{V}_{z,k}}\int_{\mathcal{V}'_{x,m}}\frac{(y-y')}{|\boldsymbol{r}-\boldsymbol{r}'|^3}\,d\mathcal{V}'_x\,d\mathcal{V}_z. \tag{11.57}$$

Therefore, the contribution of M_y to z is

$$Lm_{km}^{zy} = \frac{\mu_0}{4\pi}\frac{1}{\mathcal{A}_z}\int_{\mathcal{V}_{z,k}}\int_{\mathcal{V}'_{y,m}}\frac{(x-x')}{|\boldsymbol{r}-\boldsymbol{r}'|^3}\,d\mathcal{V}'_y\,d\mathcal{V}_z. \tag{11.58}$$

Finally, according to (11.52) and the inductive couplings, the contribution to the coupling in (11.51) is

$$\mathbf{Lm\,M} = \begin{bmatrix} \mathbf{0} & \mathbf{Lm}^{xy} & -\mathbf{Lm}^{xz} \\ -\mathbf{Lm}^{yx} & \mathbf{0} & \mathbf{Lm}^{yz} \\ \mathbf{Lp}^{zx} & -\mathbf{Lm}^{zy} & \mathbf{0} \end{bmatrix}\begin{bmatrix} \mathbf{M}_x \\ \mathbf{M}_y \\ \mathbf{M}_z \end{bmatrix}. \tag{11.59}$$

The submatrices in (11.59) include all the coupling elements with the same orientation in space. At this point, we consider the contributions to (11.51). In the following section, we need to establish how the magnetization relates to the electrical currents.

11.3.3 Relation Between Magnetic Field, Current, and Magnetization

The discretization of the magnetization $\boldsymbol{M}$ in the magnetic material is shown in Fig. 11.9.

The following relation between the magnetization $\boldsymbol{M}$ and the magnetic field intensity $\boldsymbol{B}$ can be established as follows:

$$\boldsymbol{B} = \mu_0 \mu_r \boldsymbol{H} = \mu_0 \, (\boldsymbol{H} + \boldsymbol{M}), \tag{11.60}$$

or $\boldsymbol{M} = (\mu_r - 1)\boldsymbol{H}$. Solving for $\boldsymbol{H}$ and by multiplication by μ, we get the relation needed

$$\boldsymbol{B} = \mu_0 \mu_r \boldsymbol{H} = \frac{\mu_0 \mu_r}{\mu_r - 1} \boldsymbol{M} = \nabla \times \boldsymbol{A}. \tag{11.61}$$

Here, we assume that μ_r is a constant, which is not always the case. We use (11.61) as an additional equation relating the unknowns in the following form:

$$\boldsymbol{B}^{inc}(\boldsymbol{r}, t) + \nabla \times \boldsymbol{A}_c(\boldsymbol{r}, t) + \nabla \times \boldsymbol{A}_m(\boldsymbol{r}, t) = \frac{\mu_0 \mu_r}{\mu_r - 1} \boldsymbol{M}(\boldsymbol{r}, t), \tag{11.62}$$

where $\boldsymbol{B}^{inc}(\boldsymbol{r}, t)$ includes all types of magnetic field sources.

Three equations result for the Manhattan rectangular coordinates, $\hat{\boldsymbol{x}}, \hat{\boldsymbol{y}}$, and $\hat{\boldsymbol{z}}$. Hence, we only consider the x-component for the derivation. The vector potential for the electrical current is

$$\beta_{c,\ell} I_\ell \nabla \times \boldsymbol{A}_c(\boldsymbol{r}, t) = \frac{\mu_0}{4\pi} \frac{I_\ell}{A_\ell} \int_{\mathcal{V}c,\ell} \nabla \times \frac{\hat{\boldsymbol{x}}}{|\boldsymbol{r} - \boldsymbol{r}'|} \, d\mathcal{V}'. \tag{11.63}$$

The x-component of the vector potential for the magnetic material in (11.50) is

$$\beta_{m,\ell} M_\ell \nabla \times \boldsymbol{A}_m(\boldsymbol{r}, t) = \frac{\mu_0}{4\pi} M_\ell \int_{\mathcal{V}m,\ell} \nabla \times \frac{[\hat{\boldsymbol{x}} \times (\boldsymbol{r} - \boldsymbol{r}')]}{|\, \boldsymbol{r} - \boldsymbol{r}' \,|^3} \, d\mathcal{V}', \tag{11.64}$$

where I_ℓ and M_ℓ are the magnitude of the electrical current and the magnetization in the cell ℓ, respectively. The cell or bar in the magnetic material is x-directed in the example in (11.63) and (11.64), but it applies the same way for y and z.

With this, we can evaluate each term of (11.62). This is accomplished by integrating the equation over the volume of each cell m. However, if the aspect ratio of the dimensions of all cells is not large for all cells in the problem, we can evaluate (11.62) at the center of all cells. Hence, the observation point also in $\boldsymbol{r}$ in (11.63) and (11.64) is evaluated at the center of cell m. This leads to three equations for each point in cell m. This corresponds to a set coupled source of the form

$$-\boldsymbol{B}^{inc} = \mathcal{B}_c \mathbf{I} + \mathcal{B}_m \mathbf{M} \; - \alpha \mathbf{M}, \tag{11.65}$$

where

$$\alpha = \frac{\mu_0 \mu_r}{\mu_r - 1},$$

which has the same units as μ. Hence, we can view this equation as the coupling that exists between the conduction current, the magnetization, and a potential source of magnetic fields.

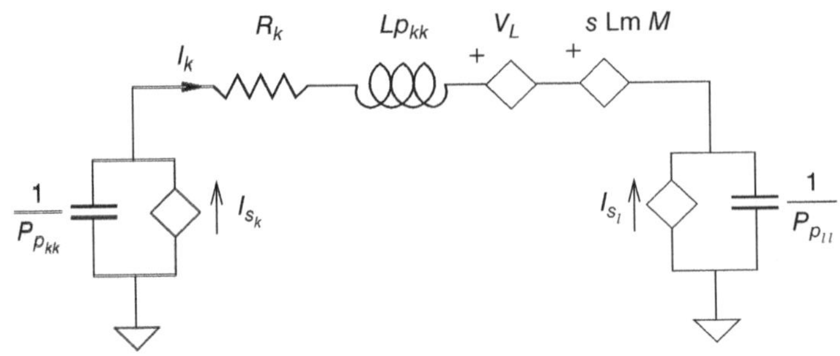

FIGURE 11.11 Basic PEEC loop that includes the coupling source for the magnetic coupling.

All the terms in (11.65) are based on the evaluation of the appropriate terms at the center points $\boldsymbol{r}$. We should point out that the field $\boldsymbol{B}^{inc}$ can consist of many different source types such as external magnetic fields or currents.

At this point, we can set up the augmented MNA system of equations by starting with the conventional form that also includes conductors as is shown in Fig. 11.8. Hence, we include the formulation in (6.55) with the additional magnetic inductance **Lm** and the source equation (11.65). The unknowns are $\boldsymbol{\Phi}$, **I**, and **M**

$$\begin{bmatrix} s\,\mathbf{Pp}^{-1} & -\mathbf{A}_\ell & \mathbf{0} \\ \mathbf{A}_\ell^T & -\mathbf{R} - s\,\mathbf{Lp} & -s\,\mathbf{Lm} \\ 0 & \mathcal{B}_c & \mathcal{B}_m - \alpha\hat{\mathbf{I}} \end{bmatrix} \begin{bmatrix} \boldsymbol{\Phi} \\ \mathbf{I} \\ \mathbf{M} \end{bmatrix} = \begin{bmatrix} \mathbf{I}_0 \\ -\mathbf{V}_0 \\ -\boldsymbol{B}^{inc} \end{bmatrix}. \tag{11.66}$$

where $\hat{\mathbf{I}}$ is again the identity matrix to distinguish it from a current.

We finally consider the equivalent circuit that corresponds to (11.66). The standard PEEC basic loop equivalent circuit in Fig. 11.11 includes an additional dependent voltage source. Further, the electrical and magnetic current densities equations are included, which control the dependent sources. Also, independent magnetic field sources are included.

It is to be pointed out that we present a quasistatic version. However, it takes into account the electrical field coupling. This makes it suitable for a large variety of applications ranging from power systems to radio frequency integrated circuits (RFIC). Further, the model is well suited to be extended to incorporate nonlinear and materials with hysteresis. Models with complex materials having special characteristics, for example, meta-materials with negative permeability can be included. A formulation for current free conductors with surface magnetic charges only is presented in [18].

11.4 SURFACE MODELS FOR MAGNETIC AND DIELECTRIC MATERIAL SOLUTIONS IN PEEC

So far, several approaches for solving models with magnetic parts have been presented. Another class of methods is considered, which is based on surface integral equation methods. These methods are considered in detail in several books [7–9]. It is evident that the solution of general problems which includes magnetic materials is not trivial. However, the class of methods that are based on a combination of electrical as well as magnetic field integral methods can include both magnetic and dielectric materials for full-wave solutions. In PEEC, as we see, we can represent the method in terms of circuit elements using some of the conventional models.

11.4.1 PEEC Version of Magnetic Field Integral Equation (MFIE)

Chapter 6 has introduced the development of the electrical field charge/potential version of an electrical field integral equation (EFIE) integral equation. In this section, we introduce a magnetic surface current and surface charge-based version of a magnetic field integral equation (MFIE). This formulation is required for the surface integral equation, for example, Ref. [10]. The basic equation is also derived in Section 3.2.2. Here, we use the dual magnetic quantities that correspond to the electrical ones from Maxwell's equations given in Table 11.1.

This leads to a simple way to define an equivalent MFIE. We start out with (3.30), the dual of (3.22) using Table 11.1 to get

$$\boldsymbol{H}^{scatt}(\boldsymbol{r}) = -j\omega \boldsymbol{F}(\boldsymbol{r}) - \nabla \phi_m(\boldsymbol{r}). \tag{11.67}$$

This result is also given in (3.30), where the electric vector potential $\boldsymbol{F}$ is given using the notation in the text [8]

$$\boldsymbol{F}(\boldsymbol{r}, t) = \epsilon_m \int_{S'} g(\boldsymbol{r}, \boldsymbol{r}')\, \boldsymbol{J}_{sm}(\boldsymbol{r}', t')\, dS', \tag{11.68}$$

where $\boldsymbol{J}_{sm}$ is the magnetic surface current density and where $t' = t - \tau$ with the delay or retardation τ is considered in Section 2.11.1. Also, retardation in the frequency domain is considered in Section 5.8.

The magnetic scalar potential Φ_m is

$$\phi_m(\boldsymbol{r}, t) = \frac{1}{\mu_m} \int_{S'} g(\boldsymbol{r}, \boldsymbol{r}')\, q_{sm}(\boldsymbol{r}', t')\, dS', \tag{11.69}$$

where q_{sm} is the magnetic surface charge density. One of the important problems with the MFIE is that the evaluation surface S' must be closed. The issue is that the magnetic field part of the formulation must include the discontinuity in the local field as is given in (11.41). As is explained in Ref. [11], the discontinuity is responsible for the requirement that the MFIE formulation must be applied to closed surface bodies. For the Poggio–Miller–Chang–Harrington–Wu–Tsai (PMCHWT) formulation considered below, the discontinuity terms cancel. For this reason, we do not include them here.

TABLE 11.1 Equivalent variables for the electric and magnetic integral equation.

EFIE	$\boldsymbol{E}$	ϕ_e	J_{se}	ϵ_e	$\frac{1}{\mu_e}$
MFIE	$\boldsymbol{H}$	ϕ_m	J_{sm}	$\frac{1}{\mu_m}$	ϵ_m

As a next step, we observe that the basic EFIE and MFIE are duals as is evident from Table 11.1. If we replace the quantities according to Table 11.1, we can use the same parts of a computer program for both cases that results in a considerable reduction in implementation work. Hence, it suffices to use the EFIE in Chapter 6 with surface cells only. Similarly, the nonorthogonal techniques in Chapter 7 can be used. The meaning of the variables is clear from Table 11.1. We should note that the discretization of the integrals is exactly the same as in Chapter 6. Here, we do not have to repeat the formulation of the integral discretization step.

11.4.2 Combined Integral Equation for Magnetic and Dielectric Bodies

The combined solution of both electrical and the MFIE leads to an approach for which both the relative permittivity ϵ_r and the permeability μ_r can be different from 1 for multiple regions. Different methods of combined approaches have been devised that can be found in Refs [9, 12]. One of the predominant implemented formulation is the so-called PMCHWT approach [10, 13]. Here, we consider the PEEC version of the PMCHWT technique, which was introduced in Ref. [14].

The boundary conditions are based on the equivalence principle presented in Chapter 3, Section 3.5.2. Since the approach requires that the objects are included by closed surfaces, this also means that the number of surface cells is large enough such that the cell size is sufficiently small compared to the size of the object.

Depending on the problem solved, the surface approach may cover all aspects of the geometry at hand. In some cases, the magnetic or dielectric regions also include smaller conducting objects, which we can embed inside the surface regions using a conventional volume PEEC formulation inside the regions [15]. However, this approach requires the computation of the additional appropriate volume-to-surface coupling integrals between the two formulations. For meshing, the techniques in Chapter 8 apply for the surfaces in the regions. Of course, we can 3D-MESH conductors which is necessary if they are VFI volume skin-effect PEEC models presented in Chapter 9.

The combined PMCHWT approach considered here uses the surface electrical vector potential $\boldsymbol{A}$, or

$$\boldsymbol{A}(\boldsymbol{r}, t) = \mu_{en} \int_{S'} g(\boldsymbol{r}, \boldsymbol{r}') \boldsymbol{J}_{se}(\boldsymbol{r}', t')\, dS', \tag{11.70}$$

where the surface electrical current is $\boldsymbol{J}_{se}$. $\boldsymbol{F}$, the magnetic vector potential given in Section 11.4.1, is repeated for convenience

$$\boldsymbol{F}(\boldsymbol{r}, t) = \epsilon_{mn} \int_{S'} g(\boldsymbol{r}, \boldsymbol{r}')\, \boldsymbol{J}_{sm}(\boldsymbol{r}', t')\, dS', \tag{11.71}$$

where $\boldsymbol{J}_{sm}$ is used for the surface magnetic current densities. It is clear that the permittivity ϵ_{en} and the permeability μ_{mn} are the appropriate values for the material region n in the problem illustrated in Fig. 11.12.

Further, ϕ_e and ϕ_m are the electrical and magnetic scalar potentials are in the time domain

$$\phi_e(\boldsymbol{r}, t) = \frac{1}{\epsilon_{en}} \int_{S'} g(\boldsymbol{r}, \boldsymbol{r}')\, q_{se}(\boldsymbol{r}', t')\, dS', \tag{11.72}$$

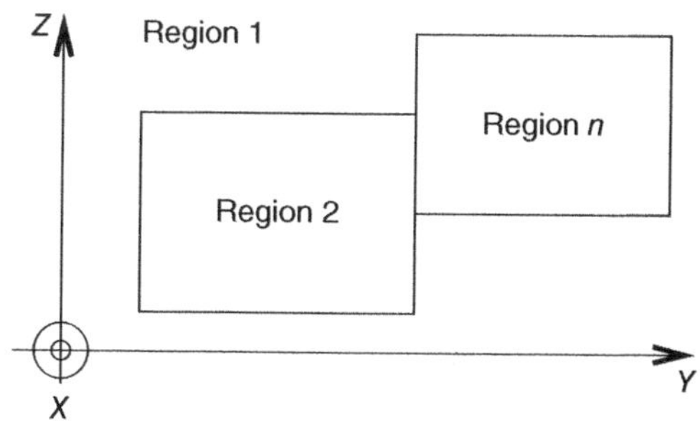

FIGURE 11.12 Side view of two regions.

and

$$\phi_m(\boldsymbol{r},t) = \frac{1}{\mu_{mn}} \int_{S'} g(\boldsymbol{r},\boldsymbol{r}')\, q_{sm}(\boldsymbol{r}',t')\, dS', \tag{11.73}$$

where again, q_{se} and q_{sm} are electrical and magnetic surface charges. The scalar Green's functions g are considered in Section 3.4 for simple cases. As always, the derivations apply to both the time and frequency domains. Only very few aspects are difficult to convert from one domain to the other one.

The central part of the surface formulation is the boundary conditions at the interfaces. Reliable electromagnetic solvers based on such surface formulations have been successfully used, for example, Ref. [16]. For our derivation, we only need to consider a two region problem where we call the outside region 1 while for the inside region we use 2 as shown in Fig. 11.12. Then, the boundary condition for the electric current densities is given by $\boldsymbol{J}_{e1}$ and the magnetic current density by $\boldsymbol{J}_{m1}$. The current densities at the inside surface of region 2 will then be $\boldsymbol{J}_{e2} = -\boldsymbol{J}_{e1}$ and $\boldsymbol{J}_{m2} = -\boldsymbol{J}_{m1}$. The other boundary conditions that are important are given by the continuity of the tangential electrical and magnetic fields for points on the interface, or

$$\hat{\boldsymbol{t}} \cdot [\boldsymbol{E}_1 - \boldsymbol{E}_2] = \hat{\boldsymbol{t}} \cdot \boldsymbol{E}^{inc}, \quad \hat{\boldsymbol{t}} \cdot [\boldsymbol{H}_1 - \boldsymbol{H}_2] = \hat{\boldsymbol{t}} \cdot \boldsymbol{H}^{inc}, \tag{11.74}$$

where $\hat{\boldsymbol{t}}$ is a unit vector that is tangential to the interface and $\boldsymbol{E}^{inc}$ and $\boldsymbol{H}^{inc}$ are possible incident fields.

A large part of the models used for the surface integral equation is similar to the volume formulation. The starting point is again the total electrical field at a point $\boldsymbol{r}$ in region $n = 1, 2$ given by

$$\boldsymbol{E}_n^{inc}(\boldsymbol{r},t) = \frac{\partial \boldsymbol{A}_n(\boldsymbol{r},t)}{\partial t} + \nabla\phi_{e,n}(\boldsymbol{r},t) + \frac{1}{\epsilon_n}\nabla \times \boldsymbol{F}_n, \tag{11.75}$$

where all the quantities are defined above. The MFIE is similarly in region n given by

$$\boldsymbol{H}_n^{inc}(\boldsymbol{r},t) = \frac{\partial \boldsymbol{F}_n(\boldsymbol{r},t)}{\partial t} + \nabla\phi_{m,n}(\boldsymbol{r},t) - \frac{1}{\mu_n}\nabla \times \boldsymbol{A}_n, \tag{11.76}$$

where the quantities are defined above. Two coupled circuits result once the boundary conditions for the currents and for fields (11.74) are applied. The equivalent circuit in Fig. 11.13

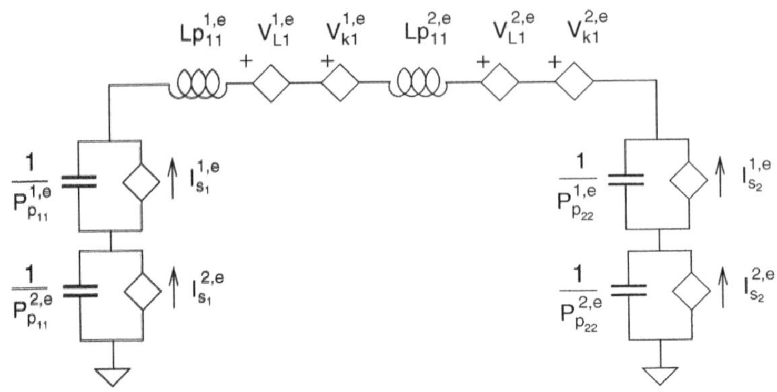

FIGURE 11.13 PEEC equivalent circuit for electrical surface equation.

is constructed by applying (11.75) to the interface between regions 1 and 2 with the boundary conditions as

$$\hat{t} \cdot [\boldsymbol{E}_1^{inc}(\boldsymbol{r},t)] = \hat{t} \cdot \left[\frac{\partial \boldsymbol{A}_1(\boldsymbol{r},t)}{\partial t} - \frac{\partial \boldsymbol{A}_2(\boldsymbol{r},t)}{\partial t}\right]$$
$$+\hat{t} \cdot [\nabla \phi_{e,1}(\boldsymbol{r},t) - \nabla \phi_{e,2}(\boldsymbol{r},t)]$$
$$+\hat{t} \cdot \left[\frac{1}{\epsilon 1} \nabla \times \boldsymbol{F}_1 - \frac{1}{\epsilon 2} \nabla \times \boldsymbol{F}_2\right]. \qquad (11.77)$$

We note that the coupling between the electrical and the magnetic circuit is given by the magnetic vector potential in (11.77), which results in voltage sources in the equivalent circuit in Fig. 11.13. It is evident that this results in a coupled model for the electrical type.

Similarly, we also have to take care of the MFIE based on the magnetic current and the field boundary condition in (11.74), or

$$\hat{t} \cdot [\boldsymbol{H}_1^{inc}(\boldsymbol{r},t)] = \hat{t} \cdot \left[\frac{\partial \boldsymbol{F}_1(\boldsymbol{r},t)}{\partial t} - \frac{\partial \boldsymbol{F}_2(\boldsymbol{r},t)}{\partial t}\right]$$
$$+\hat{t} \cdot [\nabla \phi_{m,1}(\boldsymbol{r},t) - \nabla \phi_{m,2}(\boldsymbol{r},t)]$$
$$-\hat{t} \cdot \left[\frac{1}{\mu_1} \nabla \times \boldsymbol{A}_1 - \frac{1}{\mu_2} \nabla \times \boldsymbol{A}_2\right], \qquad (11.78)$$

where $\hat{t}$ is again the tangential unit vector.

We show the PEEC circuit model corresponding to (11.78) for the magnetic circuit in Fig. 11.14. It is evident that we can choose the same ground node at infinity for both the electric and magnetic interface PEEC model. Further, we did show that the MFIE based in (11.77) is solved by multiplying (11.78) by μ/ϵ. This corresponds to the use of reciprocal medium in the formulation where μ is replaced by ϵ and vice versa.

The surface formulation requires a few observations. First, it is evident from (11.77) that region 1 and region 2 equivalent circuits are connected in series. The inductive surface

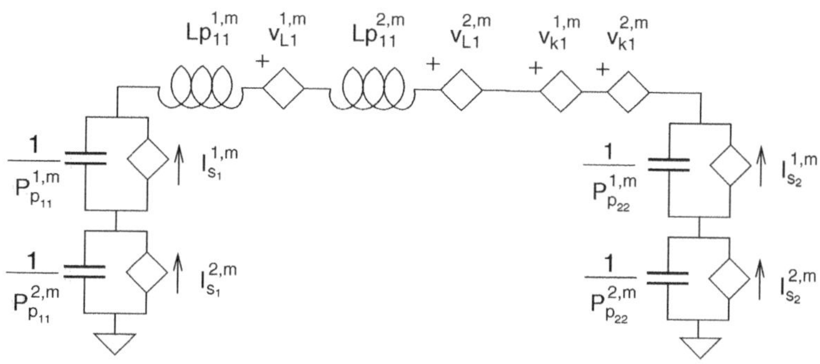

FIGURE 11.14 PEEC equivalent circuit for magnetic surface equation.

cells lead to the partial self-inductances for region 1 and the source v_{L1}^{1} is a reminder that the partial mutual inductances are coupled only to cells in region 1 and *not* region 2. Similarly, the circuit elements for region 2, which are indicated with a superscript 2, restrict the capacitive and inductive couplings to region 2. Finally, the coupled sources v_{k1}^{n} are restricted to region 1. This can be very helpful for large problems since the coupling is more contained in the regions.

The dependent voltage source represents the coupling between the electrical and the magnetic circuits, which is the $\nabla \times \boldsymbol{F}$ terms in (11.77). This leads to an interesting feedback-like coupling situation between the electrical and magnetic circuits.

We end the section with a small example that was computed with the PEEC-PMCHWT approach. The geometry is given in Fig. 11.15, which is part of the example given in Ref. [5, 17].

The width of the model is $W = 20$ μm, $T = 1$ μm, $D = 20$ μm, and $L = 300$ μm and $\epsilon_r = 10$. Here, we chose an example where the number of cells is rather small. This is a more

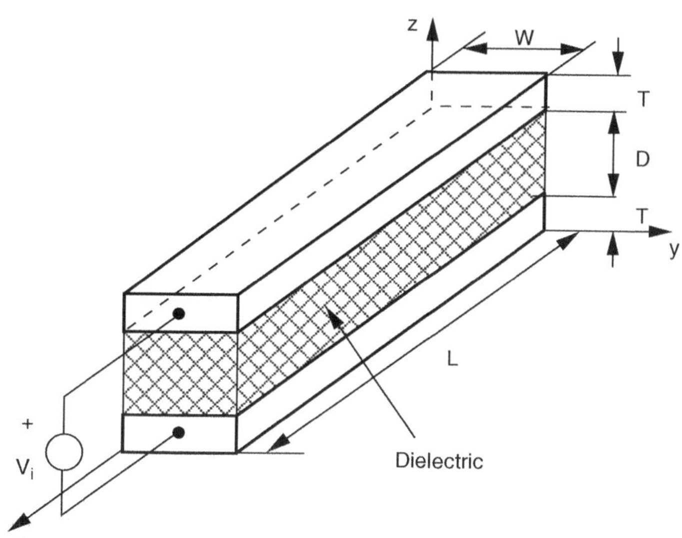

FIGURE 11.15 Short transmission line model with finite dielectric block.

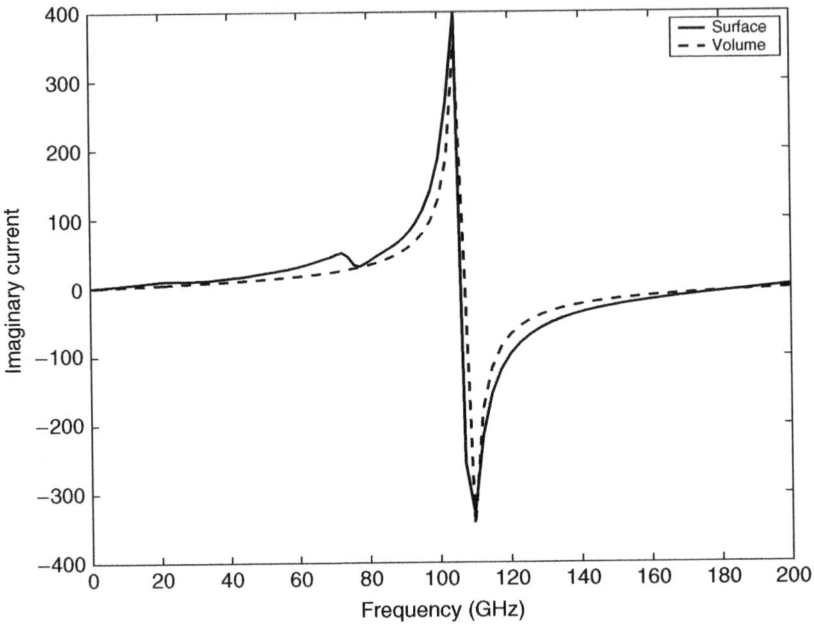

FIGURE 11.16 Imaginary part of input current.

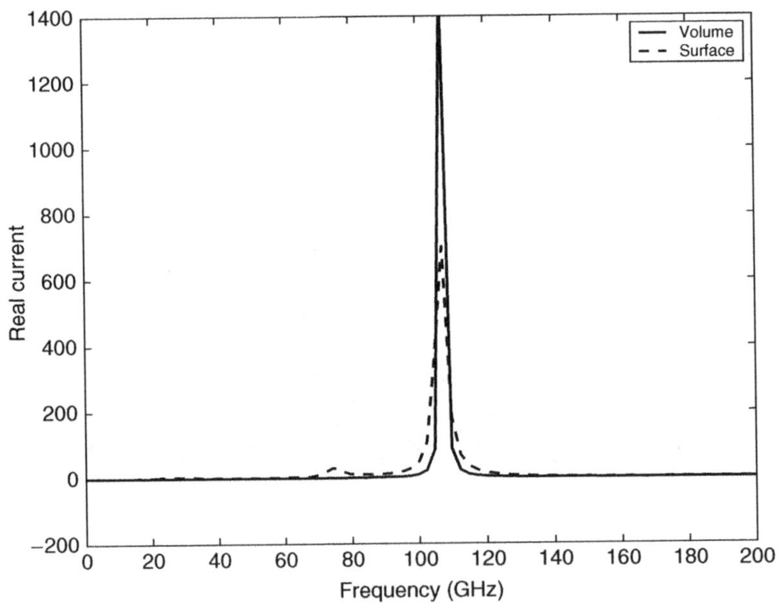

FIGURE 11.17 Real part of input current.

interesting case since important geometries may involve a multitude of such conductors and dielectrics. For both the volume and surface models, the conductors are subdivided using the usual uniform PEEC meshing. The width W, thickness T, and length L of conductors are subdivided into 5,2, and 10 divisions, respectively. The dielectric subdivisions are the same for the width and thickness (5 and 2), while the dielectric thickness divisions are chosen to

TABLE 11.2 Run time for volume compared to surface PEEC formulations.

Subdivision	Volume solution	Volume solution	Surface solution	Surface solution
Cell divisions	Number of unknowns	Time (min)	Number of unknowns	Time (min)
4–2–2	553	0.23	756	0.61
6–3–3	1348	3.7	1476	3.8
8–3–4	2109	14	2084	10
10–3–4	2997	46	2488	22
12–4–5	4448	141	3708	43

be 5. Figures 11.16 and 11.17 give the imaginary and real parts, respectively, for the current at the terminals of the *ac* 1 V voltage source. As is apparent, the results compare well.

For a larger number of cells, the surface formulation takes less computation time. The crossover point from a faster volume cell result to the surface model is shown in Table 11.2. The cell divisions for the dielectric part are given as *L–W–D* for the dielectric in Fig. 11.15. We show that the crossover occurs for 6–3–3 divisions. This is a good comparison since both formulations are using most of the same code and computer. Also, the same implementation of the integrals is used. The exception is the integral for the curl in (11.77) and in (11.77), which are only needed for the surface formulations.

PROBLEMS

11.1 *Resistance circuit*
Use the example of the transformer given in Fig. 11.4 to compute the voltage at each of the three coils in the figure, V_1, V_2, and V_3. Assume that the only current is $I_1 = 1$ A. Use the reluctance resistors in (11.20), where each branch has a length of 2 cm, a cross section of 5×5 mm, and a $\mu_r = 850$. The number of windings are $N_1 = 15$, $N_2 = 30$, $N_3 = 22$. Write a small Matlab program using equations (11.24).

11.2 *Inductance computation*
For the above problem, compute the 3×3 inductance matrix.

11.3 *PEEC formulation*
An auxiliary equation in (11.66) is used to include magnetic bodies in PEEC. Verify your understanding of the formula in the third row which represents the magnetic body.

11.4 *Single-loop inductance*
Write a Matlab program to compute the inductance of a square-shaped single loop borrowed from the problems in Chapter 5 of a size of 3 cm $\times$ 5 cm with a cross section 0.1 mm $\times$ 0.1 mm. A magnetic body with $\mu_r = 160$ is placed at a distance of 1 mm under the loop. The centered magnetic body also of size 3 cm $\times$ 5 cm and a thickness of 1 mm is placed under the loop. Compute the inductance of the loop with the magnetic sheet.

11.5 *PEEC model for MFIE*
Explain each electrical component given in the PEEC circuit model shown in Fig. 11.14 for the magnetic field integral equation (MFIE).

REFERENCES

1. L. V. Bewley. *Flux Linkages and Electromagnetic Induction*. Dover Publications, New York, 1964.
2. J. A. Brandao Faria. *Electromagnetic Foundations of Electrical Engineering*. John Wiley and Sons, Ltd, Hoboken, NJ, 2nd edition, 2008.
3. J. R. Reitz and F. J. Milford. *Foundation of Electromagnetic Theory*. Pergamon, New York, 1960.
4. A. E. Ruehli and D. Ellis. Numerical calculation of magnetic fields in the vicinity of a magnetic body. *IBM Journal of Research and Development*, 15(6):478–482, November 1971.
5. Y. Massoud, J. Wang, and J. White. Accurate inductance extraction with permeable materials using qualocation. *Proceedings of the 2nd International Conference of Modeling and Simulation of Micro Structure*, Volume 2, Puerto Rico, pp. 151–154, April 1999.
6. Y. Yi, V. Sarin, and W. Shi. An efficient inductance extraction algorithm for 3-D interconnects with frequency dependent nonlinear magnetic materials. In *Digest of Electrical Performance of Electronic Packaging*, San Jose, CA, October 2008.
7. W. C. Chew, J.-M. Jin, E. Michielssen, and J. Song. *Fast and Efficient Algorithms in Computational Electromagnetics*. Artech House, Boston, MA, 2001.
8. C. A. Balanis. *Advanced Engineering Electromagnetics*. John Wiley and Sons, Inc., New York, 1989.
9. B. H. Jung, T. K. Sarkar, S. W. Ting, Y. Zhang, Z. Mei, Z. Ji, M. Yuan, A. De, M. Salazar-Palma, and S. M. Rao. *Time and Frequency Domain Solutions of EM Problems Using Integral Equations and a Hybrid Methodology*. John Wiley and Sons, Inc., New York, 2010.
10. K. Umashankar, A. Taflove, and S. Rao. Electromagnetic scattering by arbitrary shaped three-dimensional homogeneous lossy dielectric objects. *IEEE Transactions on Antennas and Propagation*, 34(6):758–766, June 1996.
11. C. A. Balanis. *Advanced Engineering Electromagnetics*. John Wiley and Sons, Inc., New York, 2012.
12. B. M. Kolundzija and A. R. Djordjevic. *Electromagnetic Modeling of Composite Metallic and Dielectric Structures*. Artech House, Boston, MA, 2002.
13. S. M. Rao, C. C. Cha, R. L. Cravey, and D. R. Wilkes. Electromagnetic scattering from arbitrary shaped conducting bodies coated with lossy materials of arbitrary thickness. *IEEE Transactions on Antennas and Propagation*, 39(5):627–631, May 1991.
14. D. Gope, A. Ruehli, and V. Jandhyala. Surface-based PEEC formulation for modeling conductors and dielectrics in time and frequency domain combined circuit electromagnetic simulation. In *Digest of Electrical Performance of Electronic Packaging*, Volume 13, Portland, OR, pp. 329–332, October 2004.
15. A. E. Ruehli, D. Gope, and V. Jandhyala. Mixed volume and surface PEEC modeling. In *Proceedings of IEEE Antennas and Propagation Society International Symposium*, Volume 11, Monterey, CA, July 2004.
16. H. Singer, H.-D. Brüns, and G. Bürger. State of the art in the moment method. In *Proceedings of the IEEE International Symposium on Electromagnetic Compatibility*, Santa Clara, CA, pp. 122–227, August 1996.
17. A. E. Ruehli and G. Antonini. On modeling accuracy of EMI problems using PEEC. In *Proceedings of the IEEE International Symposium on Electromagnetic Compatibility*, Boston, MA, August 2003.
18. Y. Hackl, P. Scholz, W. Ackermann, and T. Weiland. Efficient simulation of magnetic components using the MagPEEC-Method. *IEEE Transactions on Magnetics*, 53(3), March 2017.

12

INCIDENT AND RADIATED FIELD MODELS

In general, electromagnetic problems are of two fundamental classes – *coupling* or *scattering*. In *coupling*, coherent equivalent sources are generated on the surfaces and in the element volumes if two objects are in close proximity. In the partial element equivalent circuit (PEEC) formulation, we can subdivide the coupling into inductive and capacitive coupling. Further, resistive coupling can exist between galvanically coupled bodies. However, this type of coupling is usually absent unless the source and observation parts are part of the same system. In most systems, coupling will become stronger for higher frequencies.

The second class is *scattering*. If electromagnetic waves originate from a field, an object or body, and propagate to another object or interface with certain material property discontinuity, the energy will be distributed in various directions, which could be to infinity. This phenomenon is called scattering. While coupling is the key factor in circuits and some electromagnetic compatibility (EMC) applications, scattering is dominant in antenna designs, radar cross sections (RCS), and some important electromagnetic interference (EMI) issues.

Integral equation-based techniques, for example, Ref. [1] are favored for scattering and radiation problems since they automatically take care of open spaces without requiring other techniques such as absorbing boundary conditions (ABCs). Finite element methods (FEMs) [2] are another popular choice if the ABC or perfectly matched layer (PML) [3] is used to completely wrap up the radiating region.

The calculation of external fields and radiation for the PEEC method is accomplished for external field calculations with only a moderate additional solution effort. In its beginning, the PEEC method was primarily used for solving electromagnetic coupling issues. However, scattering or radiation problems have also been solved by the PEEC method.

For quasistatic problems, a SPICE circuit solver can be used for nonradiation problems to find currents and surface charge distributions. However, a separate equivalent component should be added for incident field and radiation computations. The computation of the

Circuit Oriented Electromagnetic Modeling Using the PEEC Techniques, First Edition.
Albert E. Ruehli, Giulio Antonini, and Lijun Jiang.

external fields is essentially a postprocessing step once the current and charges or currents and potentials are computed.

Hence, the postprocessing of calculating the radiated field can be similar to that of conventional electric field integral equation (EFIE) formulations, especially if the low-frequency or *dc* solution is not important. We should note that, for example, the coupling to an antenna is a high-pass filter situation since low-frequency coupling between the source and the target object is absent.

In this chapter, we first consider the case of an external incident field impinging on a PEEC circuit. The first PEEC work in this area was done in Ref. [4]. Some early work on this subject was also done by M. Mandallena and by Pinello et al. [5].

We introduce two methods for the evaluation of the far-field radiation based on a solution obtained with a PEEC solver. The first one is based on placing artificial electric or magnetic sensors that extend the PEEC coupling concept to the far field. This was originated with the work in Ref. [6]. The approach includes the far-field sensors as part of the PEEC solving process. The PEEC model can include multiple field sensors – an approach that is presented in this chapter.

The second method is similar to the postprocessing used for conventional EFIE (3.67) solvers. Starting with the PEEC model solution vector, the radiated field is directly computed based on the auxiliary potentials. Far field approximation can be applied and the Sommerfeld radiation boundary condition can be clearly visualized from the results [7]. The radiation pattern can be computed from this process. This second method is more conventional than the first one.

Again, a key requirement for the modeling in this section is that we use full-wave (FW)PEEC models in Chapter 6, which include retardations or time delays. Neglecting retardation would not represent the far-field radiation, especially for higher frequencies above the quasistatic range such that the delays or retardation times are significant. Only the real part of the far field contributes to the far field radiation [14, 15].

We consider the derivation in the time domain. However, the solution method is fundamentally the same in the frequency domain. We start out with the electric field formulation (3.25) from Chapter 3,

$$\boldsymbol{E}^{inc}(\boldsymbol{r},t) = \frac{\boldsymbol{J}_{ve}(\boldsymbol{r},t)}{\sigma} + \frac{\partial}{\partial t}\frac{\mu}{4\pi}\int_{\mathcal{V}'}\frac{\boldsymbol{J}_{ve}(\boldsymbol{r}',t')}{|\boldsymbol{r}-\boldsymbol{r}'|}d\,\mathcal{V}' + \nabla\Phi_e(\boldsymbol{r},t), \tag{12.1}$$

where we simply can add a term for the case where finite dielectric bodies are involved in the model according to Section 10.4.5. Hence, the current contribution due to dielectric bodies can be added to the result.

In this chapter, we simply assume that the PEEC model solution has been set up according to Chapter 7 for nonorthogonal structures and in Chapter 6 for orthogonal structures. We simply add the parts of the models necessary for the incident and radiated field computations.

We should mention that radiation calculations have many applications in electronic systems such as, for example, to find the radiation from connectors [8]. Recently, more progress has been made in the computation of the radiated power [9, 10] and [13–15] using PEEC.

12.1 EXTERNAL INCIDENT FIELD APPLIED TO PEEC MODEL

We start this section with an independent external field applied in the presence of a PEEC model structure.

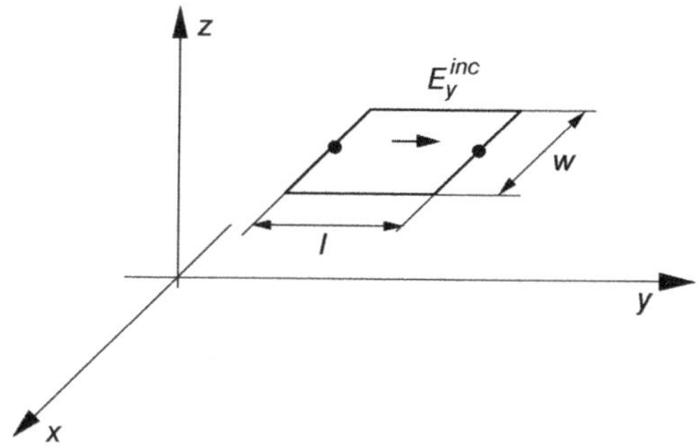

FIGURE 12.1 Example cell with component of incident field.

The new part of the incident field formulation is the collection of the electric field by the left-hand side of (12.1). Essentially, we need to add the source voltage along the inductive branch of the cell due to the incident electric field $\boldsymbol{E}^{inc}(\boldsymbol{r}, t)$. We integrate the external fields over the cell shown in Fig. 12.1, or

$$V_e(t) = \frac{1}{\mathcal{W}} \int_{\mathcal{W}} \int_{\ell} \boldsymbol{E}^{inc}(\boldsymbol{r}, t) \cdot d\ell \; d\mathcal{W}, \tag{12.2}$$

where

$$\boldsymbol{E}^{inc}(\boldsymbol{r}, t) = E_x^{inc}(\boldsymbol{r}, t)\ \hat{x} + E_y^{inc}(\boldsymbol{r}, t)\ \hat{y} + E_z^{inc}(\boldsymbol{r}, t)\ \hat{z}. \tag{12.3}$$

These sources are simply added to the basic PEEC loop, which is an addition to the circuit based on part of the Kirchhoff's voltage law (KVL). Hence, it is obvious that the extension of a PEEC model with the electric field is straightforward (Fig. 12.2). The right hand side basic loop in Fig. 12.2 represents the PEEC model for a dielectric with the incident field applied.

In the modified nodal analysis (MNA) equations, the independent field sources will appear in the right-hand-side source vector of the MNA equations as shown in Fig. 6.10 and (6.55).

As an example, we experiment a simple problem that was solved using both PEEC [4] and an approximate transmission line solution [11]. The example problem consists of a two-wire transmission line, which is 1 m long and the conductors are spaced by 5 cm. The wires that are round with a diameter of 0.2 mm are in the PEEC model approximated with

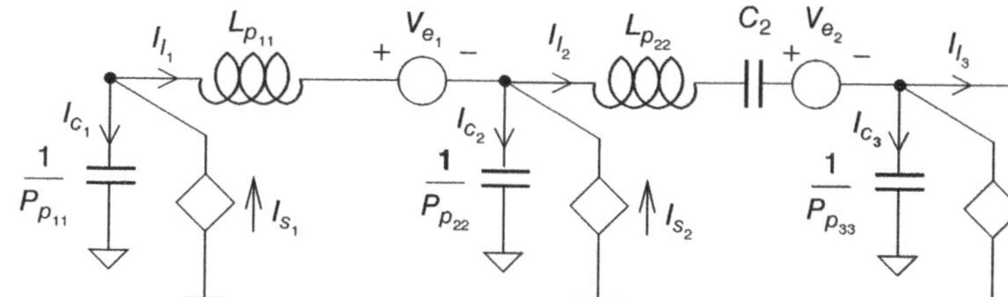

FIGURE 12.2 PEEC model with voltage sources for an incident electric field.

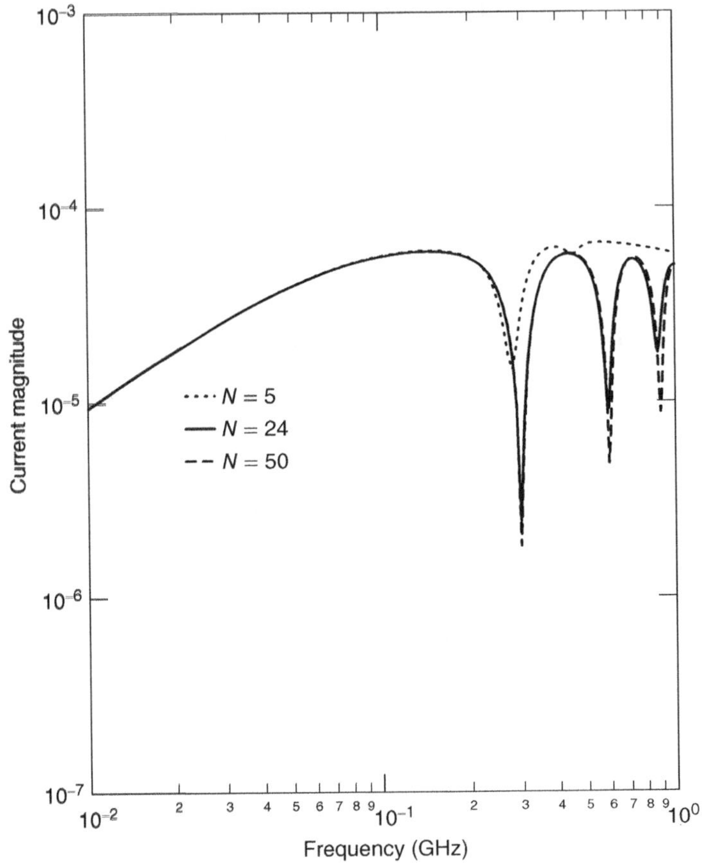

FIGURE 12.3 PEEC model with current induced by an incident electric field.

a zero thickness sheet cells with a width of 0.2 mm. Both ends of the transmission line are simply terminated with 0.5522 kΩ resistors.

For this problem, the result for the PEEC model is shown in Fig. 12.3 for three different numbers of subdivision along the length. As is evident, $N = 50$ cells leads to a good solution up to 1 GHz. The incident field is in the z-direction where $E_z^{inc} = 1$ mV/m.

12.2 FAR-FIELD RADIATION MODELS BY USING SENSORS

The computation of radiated fields emanating from a PEEC structure is important. Applications are the computation of radiated fields for EMI and antenna. As we mentioned in the introduction, the sensor technique first presented in Ref. [6], which is an interesting addition to the PEEC methods.

We evaluate the electric field with an E-field sensor based on the PEEC EFIE (12.1) solution. We assume that no direct galvanic coupling exists to the location of the observation points. Then, we only have to evaluate the contributions of the inductive and capacitive terms, as is done in the following two sections.

We first want to remind the reader that we do need to include the delays, or the retardation times in the field calculations to the sensors. The delay time $\tau_{r,r'}$ is the delay between the locations of the observation point $\boldsymbol{r}$ and the source point $\boldsymbol{r}'$, respectively. As considered in (2.11) for (FW)PEEC models

$$\tau_{r,r'} = \frac{|\boldsymbol{r} - \boldsymbol{r}'|}{c}, \tag{12.4}$$

where c is the speed of light.

12.2.1 Radiated Electric Field Calculations Using Sensors

The *eSensor* approach is based on work presented in Ref. [6]. An electric field sensor used for this approach is shown in Fig. 12.4. Essentially, the potential difference between two end points in the far field region is computed to find the field. The evaluation of the field is based on the conventional orthogonal PEEC models in Chapter 6 and/or the nonorthogonal ones in Chapter 7. The sensor is located at a point x, y, z, where the field needs to be computed. We only need to have three orthogonal sensors aligned in $\hat{\boldsymbol{x}}$, $\hat{\boldsymbol{y}}$, and the $\hat{\boldsymbol{z}}$ directions for the general case.

Further, a useful observation is that the sensor solution is independent of the rest of the PEEC model. Only a one-way coupling exists from the model to the sensors. This means that we can execute a PEEC solution independent of the sensor placement. Also, we can change the sensor locations to compute the field without the need to solve the problem again. Another positive issue for the solution is that multiple sensors, for different locations, do not interact with each other.

Another useful observation is that the structure of the coupling mechanisms from the PEEC circuit to the eSensor is similar to the coupling inside a PEEC model. Specifically, the partial element calculation facilities, which are part of the solver, can be used for the PEEC to sensors computations while the back coupling is set to zero.

The electric field sensors shown in Fig. 12.4 use three bars, one for the field in each direction. Since the method is the same for all three bars, it is sufficient to describe the sensor solution for 1 bar only. We chose the derivation for the z-directed sensor for E_z.

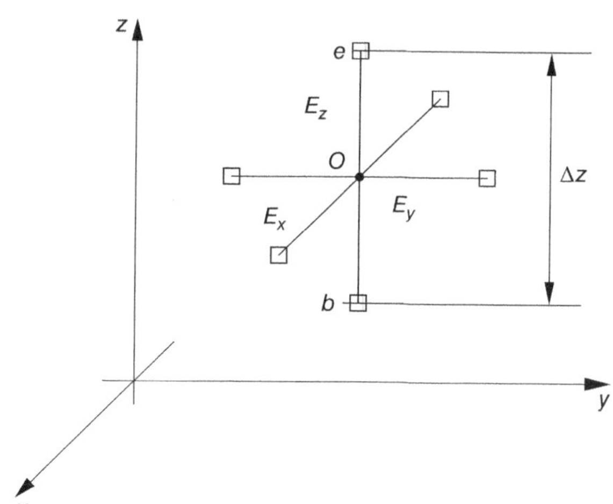

FIGURE 12.4 Electric field sensors for all three directions.

We should remember that the integral equation for PEEC is based on (3.25), which computes the $\boldsymbol{E}$-field. We use this to specifically compute electric fields, for the z-directed sensor bar as

$$(\boldsymbol{E} \cdot \hat{z})\Delta z = (\boldsymbol{E}^{inc} \cdot \hat{z})\Delta z - \sum_k Lp_{oz,k} \frac{di_k(t - \tau_{oz,k})}{dt} - \sum_m \int_z (\nabla \Phi_e \cdot \hat{z})\, dz, \tag{12.5}$$

where the k partial inductive coupling $Lp_{oz,k}$ terms consider the inductive couplings to the z-directed sensor element. The first right-hand side term in the field represents (12.1) is not coupled since we do not have a galvanic connection. Therefore, we did not include it in the sensor equations (12.5). If necessary, this term can be included directly in the field calculation.

12.2.2 Evaluation of z- Direction Inductive Coupling Term for the E-Field Sensor

The second right hand side term represents (12.1) the coupled voltage by the partial inductances. However, the very interesting situation is that we do not need a partial self-inductance coupling at the sensor to compute the inductively induced voltages. We must also compute the coupling to the z-directed part of the sensor in Fig. 12.4 for the inductive coupled voltage to a zero thickness sensor filament of a length Δz centered at the observation point O in Fig. 12.4. This represents one branch for the coupling of a partial mutual inductance.

The contribution of the inductively coupled voltage from the cell k to the sensor is similar to (6.9) as

$$Ve_z = E_z \Delta z = \frac{dI'_z(t - \tau_{oz,k})}{dt} \frac{\mu}{4\,\pi\, A'_k} \int_b^e \int_{A'_k} \int_{z_{k'}} \frac{1}{|\boldsymbol{r} - \boldsymbol{r}'|} dz'_k\, d\, A'_k\, dz, \tag{12.6}$$

where I'_z is the current in the z-directed partial element of the PEEC model. We recognize that the integrals on the right-hand side represent the coupling using the contribution by a partial inductance coupled to a z-directed filament representing the sensor.

The remaining issue in (12.6) is the computation of the time derivative of the transient currents $I'_z(t)$. Details for an accurate time domain solution are considered in Section 2.4.

For the time domain E-field computation, we have a postprocessing step since all the currents are already precomputed in the solution. Following the notation in Section 2.4, we call the present time step t_p as usual for which the solution has been obtained. The solution at the present time step $t = t_p$ which is given by $t_p = p\,\Delta t$, where Δt is a uniform time step.

To create a simple example, we assume that the delay from the inductive PEEC cell to the eSensor origin O is $\tau_{oz,k} = n\,\Delta t$. Then we can compute the current derivative as

$$\frac{di(t - \tau)}{dt} = \frac{i_{(p-n)} - i_{p-(n+1)}}{\Delta t}. \tag{12.7}$$

If the delay is not a multiple of the step delay Δt, the linear interpolations can be used. Note that these issues are the same as the updating of the right-hand-side elements for the time domain MNA equation solution (2.28) for systems with delay.

The computation of the delayed currents in the frequency domain is rather straightforward. The equivalent of (12.6) in the frequency domain can be approximated as

$$Ve_z = E_z \Delta z = sI_{z'}(s)e^{-s\ \tau(oz,k)} \frac{\mu}{4\ \pi\ A'_k} \int_b^e \int_{A'_k} \int_{z_{k'}} \frac{1}{|\boldsymbol{r} - \boldsymbol{r}'|} dz'_k\ dA'_k\ dz, \tag{12.8}$$

where $\tau_{oz,k}$ is the same delay as in the time domain. Recognizing that we can use a partial inductance that couples to the filament in Fig. 12.4, (12.8) can be simply evaluated as

$$Ve_z = E_z \Delta z = sI_{z'}(s)e^{-s\ \tau(oz,k)} Lp_{oz,k}. \tag{12.9}$$

It is clear that, in general, the length of the sensor Δz is chosen to be small compared to the shortest wavelength λ_{min}, like $\lambda_{min}/40$. But on the other hand it shall be large enough not to lead to numerical errors.

If finite dielectric bodies are present as considered in (10.62), the currents and the inductive couplings also need to be taken into account in exactly the same way. The equations for the contributions Ve_z for the dielectrics are the same as (12.6) for the time domain or (12.9) in the frequency domain. This is due to the fact that finite dielectric models also include partial inductances.

12.2.3 Potential Coefficient Coupling Contribution

We evaluate the last part in (12.5), $\nabla \Phi_e(\boldsymbol{r}, t)$. The electric field part of the sensed field is given by

$$E_z\ \Delta z = -\int_b^e \nabla \Phi_e \cdot \hat{z}\ dz, \tag{12.10}$$

where the potential evaluation points are shown as squares at the ends of the sensor in Fig. 12.4. The gradient is found by evaluating the potential Φ at the two end point squares of the sensor. The gradient is numerically approximated as

$$E_z(x, y, z, t) = \frac{\Phi(+) - \Phi(-)}{\Delta z}. \tag{12.11}$$

We note again that the sensors only involve mutual couplings. We have to establish a coupling mechanism from all the capacitive cells in the PEEC model to evaluate the potentials $\Phi(+)$ and $\Phi(-)$ at the two end points. To accomplish this, we have to compute the coupled potentials based on Section 6.3.4 for the m capacitive cells in the problem at hand:

$$\Phi(+) = \sum_m \frac{Pp_{ks+}}{Pp_{kk}} \Phi(t - \tau_{k,s+}) \tag{12.12}$$

and

$$\Phi(-) = \sum_m \frac{Pp_{ks-}}{Pp_{kk}} \Phi(t - \tau_{k,s-}), \tag{12.13}$$

where the delays $\tau_{k,s+}$ and $\tau_{k,s-}$ both are the delays from the capacitive PEEC cells to the eSensor squares. Here, the sums include all the capacitive cells in the problem. This includes all the contributions necessary for the E-field computation in (12.14) or (12.15).

It is evident that we need to add all the contributions from each of the terms of the equation to get the total E-field. This has to be done at each time point and the contributions at field sensor observation point $E_z(\boldsymbol{r})$ are found. Again, summing up all potentials will lead to the total radiated E-field.

12.2.4 Summary of E-Field Calculation with eSensor

Finally, all the components have been prepared for the E-field sensor. This leads to the following equation for the field evaluation:

$$(\boldsymbol{E}\cdot\hat{z})\Delta z = (\boldsymbol{E}^{inc}\cdot\hat{z})\Delta z - \sum_k Lp_{oz,k}\frac{di_k(t-\tau_{oz,k})}{dt} - \sum_m \frac{Pp_{ks+}}{Pp_{kk}}\Phi^+(t-\tau_{k,s+}) + \sum_m \frac{Pp_{ks-}}{Pp_{kk}}\Phi^-(t-\tau_{k,s-}), \quad (12.14)$$

where the potentials are evaluated at each end of the sensor filament. The E-field in the frequency domain is very similar to the time domain. Essentially, we need to transform (12.14) to the frequency domain for the direct evaluation of the field to get

$$(\boldsymbol{E}\cdot\hat{z})\Delta z = (\boldsymbol{E}^{inc}\cdot\hat{z})\Delta z - \sum_k Lp_{oz,k}\, e^{-s\tau_{oz,k}}\, I_k - \sum_m \frac{Pp_{ks+}}{Pp_{kk}} e^{-s\tau_{k,s+}}\,\Phi^+(s) + \sum_m \frac{Pp_{ks-}}{Pp_{kk}} e^{-s\tau_{k,s-}}\,\Phi^-(s), \quad (12.15)$$

where the delays are given by the physical location of the sensors. It is clear that the evaluation of the E-field in the frequency domain is similar to the time domain since the same small dipole arrangement is used at the observation point O.

12.2.5 Magnetic Field Calculation Using Sensors

The magnetic field (H-field) is calculated differently from the E-field in the previous sections. There are several ways to compute the H-fields where some of them do not lead to reliable numerical results, or where they do not work well for a wide frequency range. The hSensor approach we present is also from Ref. [6]. It is based on utilizing partial inductances from the PEEC model.

In Fig. 12.5, we show the hSensor, which consists of three small field test loops, where the total field is

$$\boldsymbol{H}(\boldsymbol{r},t) = H_x(\boldsymbol{r},t)\,\hat{x} + H_y(\boldsymbol{r},t)\,\hat{y} + H_z(\boldsymbol{r},t)\,\hat{z}. \quad (12.16)$$

Each of the loops senses the field for one of the orthogonal directions. They measure the field at the observation point O. We again assume that the sensors are small compared to the smallest wavelength in the spectrum.

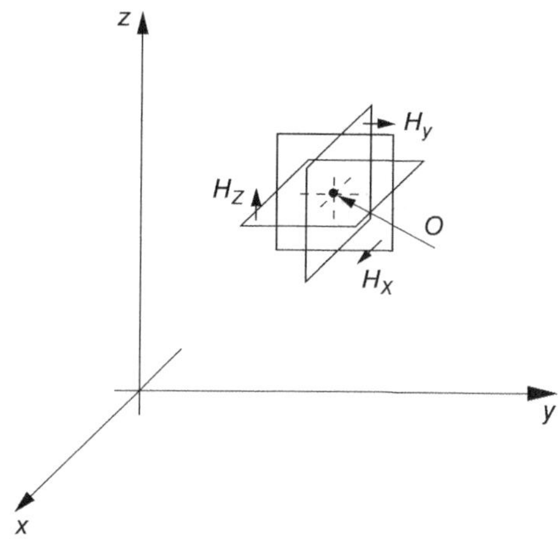

FIGURE 12.5 Magnetic field sensors loops for all three directions.

12.2.6 Time Domain Solution for *H*-Field Sensor

We assume that all sides of the sensor arrangements in Fig. 12.5 are the same, $\Delta x = \Delta y = \Delta z = w$. Since all three loops are fundamentally equivalent, it is sufficient to consider a single loop to illustrate the approach. Our example is a loop in the x–y plane to compute H_z shown in Fig. 12.6.

The voltage induced in the loop from the PEEC circuit is

$$v_z = \frac{d}{dt}\int_S \boldsymbol{B} \cdot \hat{z}\, dS \approx w^2 \frac{d}{dt} B_z(t), \tag{12.17}$$

where the normal component of $\boldsymbol{B}$ is integrated over the area w^2 of the sensor loop. With $\boldsymbol{B} = \mu \boldsymbol{H}$ given by (3.1d), we can solve (12.17) for H_z

$$H_z(t) = \frac{1}{\mu w^2} \int_0^t v_z(t)\, dt, \tag{12.18}$$

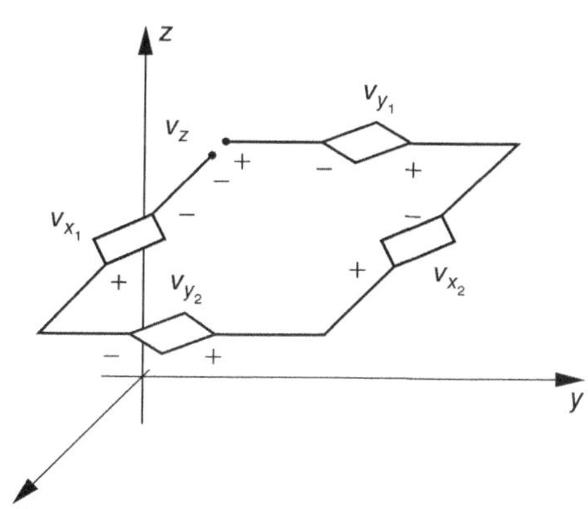

FIGURE 12.6 Example z-loop for inductive coupled voltages.

where the induced unknown voltage time integral of v_z in Fig. 12.5 needs to be computed to find $H_z(t)$.

Thin filament conductors are used to form the loop in Fig. 12.6 for the computation of the coupled induced voltage. To compute the total induced voltage for the z-loop in the example Fig. 12.6, we need to compute all partial mutual inductances that couple from the inductive PEEC cells in the problem to the loop filament conductors. Of course, the couplings are direction dependent and can be sparse, especially for rectangular Manhattan problems. Then, the voltage induced in test loop in Fig. 12.6 is

$$V_z = \sum_{m=1}^{4} \sum_{k=1}^{N_k} Lp_{mk} \frac{di_k(t - \tau_{mk})}{dt}, \tag{12.19}$$

where the sum is over all conductors N_k that are not orthogonal to the sensor filament considered. All three components of the field in (12.16) can be computed.

We observe that all sensors do not involve partial self-inductances. Hence, sensors do not have an impact on the solution and the above techniques can be applied as a postprocessing step once the solution vector $\mathbf{x}(t)$ is known. Rectangular loops can always be used even if the rest of the problem consists of nonorthogonal cells.

12.2.7 Frequency Domain Solution for *H*-Field Sensor

We consider the time domain computation $\boldsymbol{H}(t)$ and the frequency domain computation $\boldsymbol{H}(s)$. We note that the key equation (12.18) can easily be transformed into the frequency domain to

$$H_n(t) = \frac{1}{\mu w^2} \frac{V(s)}{s}, \tag{12.20}$$

since integration in the time domain is equivalent to the division by s in the Laplace domain. Due to the $1/s$ behavior, it is clear that the application for low frequencies loses accuracy.

All the other solution steps are the same as that of the time domain solution in the previous section. Of course, the coupling (12.19) with retardation transforms to

$$V_z = \sum_{m=1}^{4} \sum_{k=1}^{N_k} Lp_{mk} e^{-s\tau_{mk}} s I_k(s). \tag{12.21}$$

Finally, the computing of the contributions by the sides of the loop in Fig. 12.5 are very similar in both domains. In the frequency domain, we utilize the calculated currents at each frequency in a straightforward way. This is in contrast to the time domain, where we utilize all the current waveforms from the transient analysis over the time interval.

12.3 DIRECT FAR-FIELD RADIATION COMPUTATION

In this section, we present the second approach for the computation of far fields at points outside of the structures. This approach is easier to apply in the frequency domain rather than in the time domain. We start the derivation assuming that a single frequency time harmonic field in air at an angular frequency ω or a wave number β is applied as considered in Section 3.4.1.

12.3.1 General Radiated Field

Following the formulations in Chapter 3, the magnetic field can be represented by (3.18) based on the auxiliary vector potential

$$\boldsymbol{B}_A = \nabla \times \boldsymbol{A} \tag{12.22}$$

and the electric field (3.22) using also the scalar potential

$$\boldsymbol{E} = -j\omega \boldsymbol{A} - \nabla \Phi, \tag{12.23}$$

where the vector potential can be generally computed from the currents from the PEEC solution vector as (3.50)

$$\boldsymbol{A} = \mu \int_{\mathcal{V}} g(\boldsymbol{r}, \boldsymbol{r}') \boldsymbol{J}(\boldsymbol{r}')\, d\mathcal{V}'. \tag{12.24}$$

The scalar potential (3.52) is calculated from the solved charge distribution (3.52)

$$\Phi(\boldsymbol{r}) = \frac{1}{\epsilon} \int_{S'} g(\boldsymbol{r}, \boldsymbol{r}') q(\boldsymbol{r}')\, dS', \tag{12.25}$$

where $g(\boldsymbol{r}, \boldsymbol{r}')$ is the usual free space Green's function (3.63)

$$g(\boldsymbol{r}, \boldsymbol{r}') = \frac{e^{-j\beta|\boldsymbol{r}-\boldsymbol{r}'|}}{4\pi|\boldsymbol{r}-\boldsymbol{r}'|} = \frac{e^{-j\beta R}}{4\pi R} \tag{12.26}$$

where $R = |\boldsymbol{r} - \boldsymbol{r}'|$ and $\beta = \omega / c$.

Based on the gradient of Green's function

$$\nabla g(\boldsymbol{r}, \boldsymbol{r}') = -\hat{\boldsymbol{R}}\, \frac{1 + j\beta R}{R}\, g(\boldsymbol{r}, \boldsymbol{r}'), \tag{12.27}$$

by using (12.27) and (12.24) in (12.22), we find

$$\boldsymbol{B}(\boldsymbol{r}) = -\int_{\mathcal{V}} \frac{1 + j\beta R}{R}\, g(\boldsymbol{r}, \boldsymbol{r}')\, \hat{\boldsymbol{R}} \times \boldsymbol{J}(\boldsymbol{r}')\, d\mathcal{V}', \tag{12.28}$$

or

$$\boldsymbol{B}(\boldsymbol{r}) = -\int_{\mathcal{V}} \frac{1}{R}\, g(\boldsymbol{r}, \boldsymbol{r}')\, \hat{\boldsymbol{R}} \times \boldsymbol{J}(\boldsymbol{r}')\, d\mathcal{V}' - \int_{\mathcal{V}} j\beta\, g(\boldsymbol{r}, \boldsymbol{r}')\, \hat{\boldsymbol{R}} \times \boldsymbol{J}(\boldsymbol{r}')\, d\mathcal{V}', \tag{12.29}$$

which is based on [7]. From this equation, it is clear that the far field $\boldsymbol{B}$ can only have the tangential components relative to the longitudinal direction $\hat{\boldsymbol{R}}$. At infinite, according to the

Sommerfeld radiation boundary condition [7], the radiated waves are plain waves. Hence, the electric field will also have tangential components only.

Even though the electric field intensity can be computed directly based on (12.23), it is more consistent by using Ampere's circuital law at the observation position to avoid any numerical inconsistency between the computed current and charge distributions. In the frequency domain, at the field observation point where $\boldsymbol{J} = 0$, we have

$$\boldsymbol{E}(\boldsymbol{r}) = \frac{1}{j\omega\epsilon}\nabla \times \boldsymbol{H}. \tag{12.30}$$

We can compute the electric field $\boldsymbol{E}$ by substituting (12.28) into (12.30).

$$\begin{aligned}\boldsymbol{E}(\boldsymbol{r}) = &-\frac{1}{j\omega\epsilon\mu}\nabla \times \int_{\mathcal{V}} \frac{1}{R}\, g(\boldsymbol{r},\boldsymbol{r}')\,\hat{\boldsymbol{R}} \times \boldsymbol{J}(\boldsymbol{r}')\, d\mathcal{V}' \\ &- \frac{1}{j\omega\epsilon\mu}\nabla \times \int_{\mathcal{V}} j\beta\, g(\boldsymbol{r},\boldsymbol{r}')\,\hat{\boldsymbol{R}} \times \boldsymbol{J}(\boldsymbol{r}')\, d\mathcal{V}',\end{aligned} \tag{12.31}$$

12.3.2 Radiated Field Computation Based on the PEEC Computation Results

For the solution we use for weighted residuals method (WRM) approach, where the capacitive cells have the conventional constant charge density. The currents in rectangular cells are in one direction. Therefore, the current in each inductive cell is constant in both value and direction. Therefore, if the cross section of the filament is $\mathcal{A}$ and its length is ℓ, the resultant current and the current density in this rectangular filament is $\boldsymbol{I} = \boldsymbol{J}_s\mathcal{A}$. The current $\boldsymbol{I}$ is a vector with a given direction. Assuming that the problem has a total of N inductive current cells and we apply the discretization to (12.28), we have

$$\boldsymbol{B}(\boldsymbol{r}) = -\sum_{i=1}^{N} \int_{\ell_i} \frac{1 + j\beta R}{R}\, g(\boldsymbol{r},\boldsymbol{r}')\,\hat{\boldsymbol{R}} \times \boldsymbol{I}_i(\boldsymbol{r}')\, d\,\ell', \tag{12.32}$$

where ℓ' is along the longitudinal direction of the cell filament. Based on this assumption, the current density is replaced by the current for the constant current density in each cell. Accordingly, the electric field can be computed similar to (12.30) and (12.28).

12.3.3 Approximate Computation of Far Fields

As shown in Fig. 12.7, $\boldsymbol{r}$ is a vector to the distant observation point location. A condition for a far-field approximation is that the observation point is far in comparison to the relative size of the object $r = |\boldsymbol{r}| \gg D$. Also, the wavelength λ is assumed to be smaller than $|\boldsymbol{r}|$. On the other hand, the point vector $|\boldsymbol{r}'|$ for a point on the object close to the origin is much smaller than r. We call the angle at the origin α. Under these assumptions, it is possible to use the far-field approximation to simplify the field calculation [12]. It is well known from the geometry that we can approximate the distance R as

$$R = \sqrt{r^2 + r'^2 - 2rr'\cos\alpha} \approx r - r'\cos\alpha = r - (\hat{\boldsymbol{r}} \cdot \boldsymbol{r}'). \tag{12.33}$$

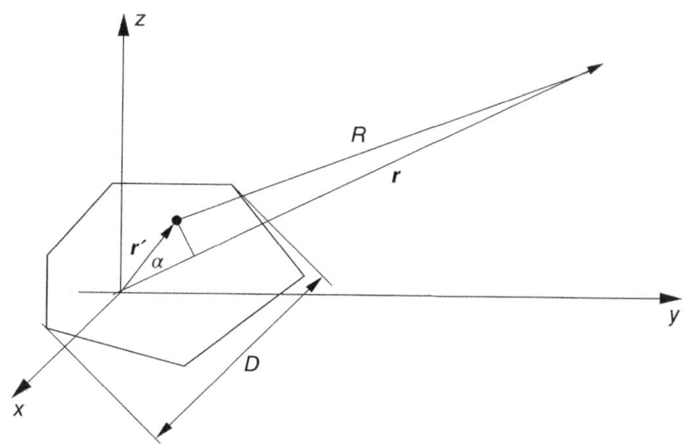

FIGURE 12.7 Example of approximation for far-field computation.

The condition for the approximation to be applied can be expressed as

$$r \geq \frac{2D^2}{\lambda}, \tag{12.34}$$

where again r is the length of the vector $\boldsymbol{r}$ and D is the maximum dimension of the radiator model. We can use it in (12.28), considering the fact that at infinity $r \gg r'$, $\hat{\boldsymbol{R}} = \hat{\boldsymbol{r}}$, $\boldsymbol{\beta} = \beta\hat{\boldsymbol{R}}$. We can assume that the components of the far field such as the $1/r^2$ or higher order term decay fast and can be ignored. This results in

$$\boldsymbol{B}(\boldsymbol{r}) = -j\beta\, g(\boldsymbol{r}, \boldsymbol{0})\, \hat{\boldsymbol{r}} \times \int_{\mathcal{V}} e^{-j\boldsymbol{\beta}\cdot\boldsymbol{r}'} \boldsymbol{J}(\boldsymbol{r}')\, d\mathcal{V}', \tag{12.35}$$

where starting from (12.26) Green's function used is

$$g(\boldsymbol{r}, \boldsymbol{0}) = \frac{1}{4\pi|\boldsymbol{r}|} e^{-j\beta|\boldsymbol{r}|}. \tag{12.36}$$

An approximation can be obtained for the far field starting with (12.23). The scalar potential primarily contributes to the longitudinal field components, while it cancels the contribution to the longitudinal components of the vector potential. Hence, we only need to compute the tangential components of $\boldsymbol{A}$ to approximate the far field of $\boldsymbol{E}$. As a result, the far field of $\boldsymbol{E}$ is derived to be

$$\boldsymbol{E}(\boldsymbol{r}) = j\omega\mu\, g(\boldsymbol{r}, \boldsymbol{0})\, \hat{\boldsymbol{r}} \times \hat{\boldsymbol{r}} \times \int_{\mathcal{V}} e^{-j\boldsymbol{\beta}\cdot\boldsymbol{r}'} \boldsymbol{J}(\boldsymbol{r}')\, d\mathcal{V}'. \tag{12.37}$$

By taking the ratio of $|\boldsymbol{E}|/|\boldsymbol{H}|$, we can find the intrinsic impedance $\eta = \sqrt{\mu/\epsilon}$. It is understood that the solution is a plain wave satisfying the Sommerfeld radiation condition at ∞ [7].

PROBLEMS

12.1 *External incident electrical field*
We assume that an electric field is incident to a PEEC cell in the x–y plane. Explain and formulate the contribution of the external incident electric field and derive the equivalent circuit source. Consider both the polarization and the orientation of the wave compared to the orientation of the mesh cell.

12.2 *Circular polarization incidence*
Derive the PEEC model when the external incident wave is a circularly polarized plane wave.

12.3 *Dipole radiation field*
Assume that a half-wavelength dipole is made of two strips that are perfect electrical conductors (PECs) of zero thickness. The length of the dipole is 1.5 m. Compute the radiated magnetic field at a distance of 50 m at a frequency of 300 MHz using the electric field sensors introduced in Section 12.2.

12.4 *Loop radiation field*
The problem consists of a square loop with an edge length of 1 m. The cross section of the loop is 5 mm × 5 mm being a PEC. Compute the radiated magnetic field at a distance of 50 m at a frequency of 300 MHz using the magnetic field sensors introduced in Section 12.2.

12.5 *Far-field zone*
The approximate computation of the far field is given in Section 12.3.3. In the derivation, the formula (12.34) is given. Give an explanation of the physical meaning of the far-field zone limit given in this equation.

REFERENCES

1. R. F. Harrington. *Field Computation by Moment Methods*. Macmillan, New York, 1968.
2. J. M. Jin. *The Finite Element Method in Electromagnetics*. John Wiley and Sons, Inc., New York, 2nd edition, 2002.
3. A. Taflove Ed. *Computational Electrodynamics*. Artech House, Inc., Norwood, MA, 1998.
4. A. E. Ruehli, J. Garrett, and C. R. Paul. Circuit models for 3D structures with incident fields. In *Proceedings of the IEEE International Symposium on Electromagnetic Compatibility*, Dallas, TX, pp. 28–31, August 1993.
5. W. Pinello, A. C. Cangellaris, and A. Ruehli. Hybrid electromagnetic modeling of noise interactions in packaged electronics based on the partial-element equivalent circuit formulation. *IEEE Trans MTT*, Volume 45(10):1889–1896, October 1997.
6. G. Wollenberg and A. Görisch. Analysis of 3-D interconnect structures with PEEC using SPICE. *IEEE Transactions on Electromagnetic Compatibility*, 41(2):412–417, November 1999.
7. W. C. Chew. *Waves and Fields in Inhomogeneous Media*. IEEE Press, New York, 1995.
8. B. Archambeault, S. Connor, M. S. Halligan, J. L. Drewniak, and A. E. Ruehli. Electromagnetic radiation resulting from PCB/high density connector interfaces. *IEEE Transactions on Electromagnetic Compatibility*, 55(4):614–623, August 2013.
9. J. Nitsch, F. Gronnwald, and G. Wollenberg. *Radiating Nonuniform Transmission-Line Systems and the Partial Element Equivalent Circuit Method*. John Wiley and Sons, Inc., New York, 2009.

10. Y. S. Cao, L. J. Jiang, and A. E. Ruehli. Distributive radiation and transfer characterization based on the PEEC method. *IEEE Transactions on Electromagnetic Compatibility*, 57(4):734–742, August 2013.

11. R. T. Abraham and C. R. Paul. Coupling of electromagnetic fields to transmission lines. In *Proceedings of the IEEE International Symposium on Electromagnetic Compatibility*, pp. 164–170, August 1981.

12. C. A. Balanis. *Advanced Engineering Electromagnetics*. John Wiley and Sons, Inc., New York, 1989.

13. L. K. Yeung and Ke-Li Wu. PEEC modeling of radiation problems for microstrip structure. *IEEE Transactions on Antennas and Propagation*, 61(7):3648–3655, July 2013.

14. C. C. Chou, W. C. Lee, and T. L. Wu. A rigorous proof on the radiation resistance in generalized PEEC models. *IEEE Transactions Microwave Theory and Techniques*, 64(12):4091–4097, December 2016.

15. Y. S. Cao, L. J. Jiang, and A. E. Ruehli. Distributive radiation and transfer characterization based on the PEEC method. *IEEE Transactions Electromagnetic Compatibility*, 57(4):134–142, August 2015.

13

STABILITY AND PASSIVITY OF PEEC MODELS

This chapter is dedicated to important issues related to the quality of the solutions obtained with a partial element equivalent circuit (PEEC) solver. Issues are considered such as the passivity in both the time and frequency domain responses and their stability in the time domain.

If an exact solution of Maxwell's equations could be obtained with infinitesimal size cell divisions, it would be passive. The necessary discretization of Maxwell's equations for a practical numerical solution introduces errors by the WRM solution approach. It is clear that the properties of the solution could be improved if we could apply an extremely small mesh size.

Historically, electromagnetic (EM) solver stability problems became apparent with the first-time domain computer solutions for the class of integral equation solvers [1–4] and references therein. Examples of fundamental investigations that lead to more insights for PEEC solvers are presented in several publications, for example, Refs [5–8]. We should note that time domain stability issues for integral equation solutions have been considered by many researchers, and problems are widely investigated, for example, Ref. [9]. Some challenging problems for time domain EM solvers lead to stability and passivity problems. As is apparent from this chapter, passivity inaccuracies are harder to observe in both domains.

In general, we prefer solver approaches that work in both the time and the frequency domains. For this reason, we formulated all chapters as much as possible for both domains. Of interest are circuit solver approaches [7] that utilize SPICE concepts [10, 11]. They incorporate some of the advantages of the approach with solutions in both domains. However, this implies that passivity should be obtained in both domains.

We would like to consider techniques for the improvement of stability as well as passivity for the solution obtained with (EM)PEEC methods. Unfortunately, a careful investigation is needed to show why the lack of time domain stability results in the false exponential

Circuit Oriented Electromagnetic Modeling Using the PEEC Techniques, First Edition.
Albert E. Ruehli, Giulio Antonini, and Lijun Jiang.

increase in the solution variables such that the real solution is totally masked. Also, nonpassivity can at least result in the reduction of solution accuracy in addition to potential instability.

It is very important to understand that the cascading of nonpassive systems may also lead to unstable solutions even if the individual parts are stable, for example, Refs [12, 13]. The nonpassivity may result in stability issues since cascading of EM solutions and circuit waveforms is a common approach used in the time domain. An example of the stability issue is given in Section 13.1.5.

Stability and passivity are easier to prove for quasistatic (QS)PEEC models since they do not include delays. This case can be treated as a common circuit passivity problem. Maintaining passivity is a well-researched issue in the circuit domain, and mathematical concepts are available today. A difficult issue is the evaluation of the passivity of distributed full-wave (FW)PEEC models. It is obvious that the magnitude of the partial element couplings reduces with the distance between the elements. This leads to a decreasing influence as the elements are farther apart. This decrease in magnitude has been shown to help stability [6].

In contrast, the phase change in the partial element couplings does not decrease with the distance between the coupled elements. The phase $\exp(-j\,\omega d/c_o)$ changes with the physical distance d as considered in Section 2.11.1. Hence, the phase change is similar for a larger distance change $d + \Delta d$, which leads to $e^{-j\,\omega \Delta d/c_0}$ independent of the relative locations of the two objects. We show that this is a key issue.

A fundamental difference exists between techniques that require the knowledge of all circuit elements of the PEEC models or whether we simply know the behavior at designated ports. We need to assume that only port data is available due to the extremely large matrices that result from the large electromagnetic and circuit (EM/Ckt) problems. Hence, we have to depend on the information that can be obtained from terminal or port information only.

We have a fundamental understanding of the external operations of the solver as can be observed at the ports. The characteristic behavior of a modified nodal analysis (MNA)PEEC system [14] can be utilized to consider stability and passivity enhancements. Known problems with unstable and nonpassive circuits have yielded different tests for passivity called *passivity assessment*, for example Refs [12, 15].

Here, we consider a different step we call *solver passivity enhancement*, which was introduced in Refs [16, 17]. This approach attempts to improve the stability/passivity during the EM/Ckt solution if required. However, access to the inner workings of the solver is required such that the solution itself is improved.

We summarize the different relevant assessment, enhancement, and enforcement techniques, which we consider in this chapter:

- *Solver stability enhancement:* This approach enhances the stability in the time domain at the EM solver level.
- *Solver passivity enhancement:* The passivity is enhanced in one or both of the time and/or the frequency domain at the EM solver level.
- *Solution passivity assessment:* The solution is verified to be passive after a portion or the total EM solution is obtained from the solver.
- *Solution passivity enforcement:* The result is enforced after the EM solver result is obtained independent of the solver.

The last item in this list is based on the assumption that we do not have access to the formulation of the EM/Ckt solver. For this reason, a postprocessing step technique was developed called *solution passivity enforcement* [18–22], which is used to change the solver output response such that the final new result is passive. Here, we assume that we do have access to the internals of the PEEC solver and we will not consider the enforcement approach. Other works in the postprocessing *passivity enforcement* area are, for example, Refs [12, 23–27, 57].

13.1 FUNDAMENTAL STABILITY AND PASSIVITY CONCEPTS

A fundamental understanding is required to deal with the stability and passivity issues considered in this chapter. From a circuit's point of view, research in this area dates back to early work such as [28] and later [29] and other key works. Other papers have contributed to the understanding of the issues, for example, Ref. [12].

From the previous references, it is evident that EM time domain integral equation solutions have a long history. Time domain stability has also impacted differential equation-based EM techniques such as the finite difference time domain (FDTD) [30] or the finite integration technique (FIT) models [31], as well as FEM methods [32, 33]. However, time domain stability is a more challenging issue for FW integral equation formulations such as the (FW)PEEC methods. We start with an example of a stable as well as an unstable solution in Fig. 13.1. It is clear that the response of the unstable solution for a sufficiently large time will totally mask the actual response for the problem. A key observation is the fact that for the same example, a simple inspection of the frequency domain response would not reveal the problem that caused the instability. Importantly, stability is necessary for passivity. Hence, the observation of an unstable time solution can be viewed as a way to observe issues in the solution for both domains.

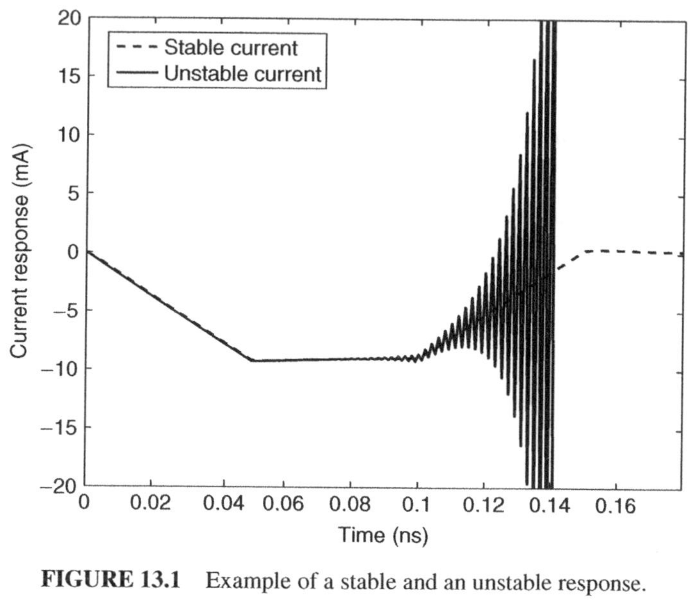

FIGURE 13.1 Example of a stable and an unstable response.

In the frequency domain, the source of error may be harder to detect since it does not completely distort the solution in an obvious way as can be the case in the time domain. To observe an internal passivity problem in the frequency domain is a more difficult problem, and it requires a more mathematical approach.

13.1.1 Time Domain Stability

Some definitions are required to clearly specify the concepts needed. In simple terms, stability is defined as a finite response of a circuit to a finite input signal. This specifies the bounded-input/bounded-output (BIBO) model. A more detailed definition is as follows:

Definition 1 ***(Stability)*** *A circuit is* stable *in* $t \in [0, \infty)$ *under the condition that all the inputs are reaching steady states for* $t < \infty$, *if the circuit solution vector* $x(t)$ *stays finite, or* $\lim_{t\to\infty} \|x(t) - x(t_0)\| = x_0$ *where* x_0 *is the appropriate stable final state, which may in general be different from zero by a* dc *offset.*

We have to be somewhat careful about the definitions of stability for circuits. We observe that the solution will go to ∞ if a constant current source is applied to the circuit such as the one in Fig. 6.9 – which we repeat here for the convenience of the reader in Fig. 13.2 – where we assume that a constant current source I_i is attached. Eventually, all the node potentials or nodal voltages go to ∞.

This issue is not related to the stability problem. For this reason, the definitions of stability have been extended to include such issues [34]. For example, for this specific case the definition is as follows:

Definition 2 ***(Open circuit stability)*** *A port of an n-port circuit is said to be* open circuit stable *if the circuit is stable for current source excitation.*

We observe that this issue is added for completeness but is not important for the stability/passivity issues we consider in this chapter.

13.1.2 Time Domain Passivity

When performing a transient solution, stability and passivity should be guaranteed. It is known that, while a passive system is also stable, the reverse may not be true [35].

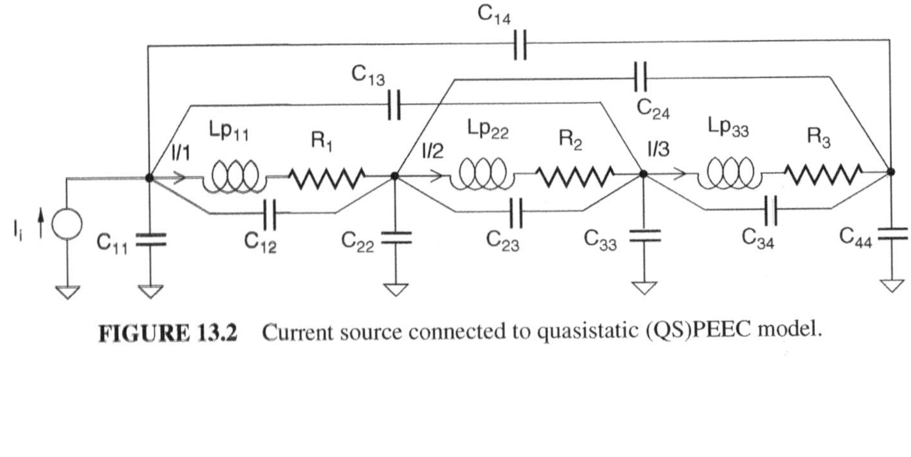

FIGURE 13.2 Current source connected to quasistatic (QS)PEEC model.

This is important when the model is utilized in a general-purpose analysis-oriented circuit solver. Passivity refers to the property of a system that cannot generate more energy than it absorbs through its electrical ports. Passivity (or positive realness) for a linear circuit can be described by an immittance matrix $\mathbf{Y}(s)$ or $\mathbf{Z}(s)$, which is defined as follows [36, 37]:

Definition 3 ***(Passivity)*** *A circuit is* passive *in* $t \in [t_0, t_{max}]$ *if the energy absorbed by the terminals obeys the relation* $\int_{t_0}^{t} \mathbf{v}^T(\tau)\mathbf{i}(\tau)\, d\tau \geq \epsilon(t) - \epsilon(t_0)$ *where* $\mathbf{v}, \mathbf{i}$ *are either the voltages and currents in an n-terminal circuit or the port voltages and currents in an n-port circuit and where* $\epsilon(t) - \epsilon(t_0)$ *is the net change in energy stored in the time interval.*

This definition implies that a circuit cannot be a source of energy in any interval of time $t \in [0, t_{max}]$ if it is passive. Of course, we have to subtract the potential initial energy that is already stored as system energy $\epsilon(t_0)$ at the initial time t_0.

In a practical situation, it is difficult to guarantee global passivity for each subcircuit or configuration in a system. For this reason, an effort is being made to prove passivity for all types of modeling procedures, for example, Refs [38, 39]. Unfortunately, passivity can be very restrictive and many useful algorithms do not lead to a passive model for all cases.

Definition 4 ***(Limited Passivity)*** *A circuit or system is* limited passive *if it is passive for a specific, relevant setup. It may be specified to be passive for a specified range of conditions. It may also be limited passive for a sufficiently large, yet finite, interval in time* $t \in [t_1, t_2]$.

This fundamental concept of *limited passivity* is quite general and it leads to a much larger set of acceptable circuits or systems. However, the conditions or the range of conditions under which it is passive need to be clearly specified. An essential idea behind limited passivity is that it is practically almost impossible to get all the data on the system in support of strict passivity. Hence, we would like to use more criteria to identify systems that are functional under more limited operating conditions.

The same concepts can be applied to stability. We again describe the load conditions, etc., for a system or circuit for which stability is achieved and call it *limited stable*. To give a specific example, we may need the solution only up to t_{max}. Hence, the solution for times $t > t_{max}$ may not be of importance for the particular application. The key idea behind these definitions is that a much larger class of circuits or systems can become useful. Or looking at it in reverse, a system may be stable and useful up to a time t_{max}. So, we do not need to test the system for conditions that are not of interest.

In general, passivity is an even more demanding property and more restrictive concept than stability as is evident from the above and other definitions, for example, Ref. [34]. However, the implications are that if we can prove passivity or limited passivity for the specific conditions or applications, then the solution is automatically stable for the specific conditions and we do not have to worry about stability.

13.1.3 Causality

Causality is another important concept that helps to confirm that a solution is valid. Essentially, the fundamental issue is very straightforward. It states that the output of a circuit cannot react in time before the input signal is applied. A more formal definition is as follows:

Definition 5 ***(Causality)*** *A circuit is causal if and only if for all input pairs* $\mathbf{x}_1(t)$ *and* $\mathbf{x}_2(t)$ *are such that for* t_0

$$\mathbf{x}_1(t_0) = \mathbf{x}_2(t_0), \tag{13.1}$$

the corresponding outputs satisfy for $t \in [t_0, t_1]$

$$\mathbf{x}_1(t) = \mathbf{x}_2(t). \tag{13.2}$$

We also find it is necessary to define strict or ideal causality. We usually associate this concept with an ideal transmission line (TL) model in a precise way.

Definition 6 ***(Strict or TL causality)*** *Causality defined for a system with an ideal delay such as an ideal transmission line, TL is called strict or ideal. This implies that in such a system, if a signal is applied at* t_0 *and it has a delay* τ*, then the output signal is zero up to* $t_0 + \tau$.

The motivations for this special causality definition is considered next. There are several practical issues related to causality that are important for PEEC and other equivalent circuit models as well as numerical solutions. Assume that we construct a QS lumped PEEC TL circuit that can be represented by several sections of the model shown in Fig. 13.2. We apply an input step current $I_1(t)$ at time t_0. Finite models such as a PEEC behave similar to a lumped circuit conventional R, L, C TL model with additional delays. It is clear that the more circuit sections we use, the more accurate the solution will be. However, the delay of the model is another issue. As with most numerical discrete models, as soon as the input is applied at $t = t_0$, a signal will appear that may be minute at the output node of the approximate nonideal, delay-less model. Practically, it is hard to observe the minute signals, and the system delay is approximated more or less well, representing the actual rise in the output signal that appears at $t_0 + Delay$.

From this it is clear that many practical systems do not fulfill the *strict TL causal* conditions such as the ideal transmission line delay. However, they are causal in the conventional circuit or system sense according to the above definition 6. For this reason, we do distinguish between the strict causality definition and the conventional definition of causality. Of course, for discrete PEEC and other numerical models, we are less interested in strict circuit causality.

We just learned an important lesson on causality. The conventional definition of causality may lead to a solution that starts before what we perceive as an ideal transmission line starting time $t = T_d$, where T_d is the delayed time of the arrival of the signal. However, as pointed out in Ref. [12], an approximate realistic casual signal may arrive as a very small signal before the appropriate delay.

13.1.4 Positive Real Function and Passivity

To consider circuit passivity, we need a definition for a *positive real* (PR) function for this theorem:

Definition 7 ***(Positive real PR)*** *A function $F(s)$ of the complex variable s is said to be positive real (PR) if $F(s)$ is real when s is real,* $\mathrm{Re}[F(s)] \geq 0$ *when* $\mathrm{Re}[s] = 0$, *and* $\mathrm{Re}[F(s)] > 0$ *when* $\mathrm{Re}[s] > 0$.

Next, we consider passivity for a linear model as is given in Ref. [40]. Passivity in the frequency domain for a transfer function (6.64) is based on several conditions.

A key condition for a transfer function in (6.64) or (6.67) is if the function is PR. A positive realness test is given in Ref. [40] as follows:

1. $\mathbf{H}(s^*) = \mathbf{H}^*(s)$
2. $\mathbf{H}(s)$ is analytic in $\mathrm{Re}[s] > 0$
3. $\mathbf{H}(s) + \mathbf{H}^*(s) \geq 0$ in $\mathrm{Re}[s] > 0$,

where "*" denotes the complex conjugate operator. In the case of lumped circuits, positive realness guarantees passivity, according to the following theorem [36]:

Theorem 1 ***(Passivity)*** *A system with rational transfer function* $\mathbf{H}(s)$ *is passive and stable if* $\mathbf{H}(s)$ *is PR.*

The PR condition 3. Can be restricted to the imaginary axis as a consequence of the following theorem [36].

Theorem 2 ***(Positive real on imaginary axis)*** *A rational transfer function* $\mathbf{H}(s)$ *is PR if and only if conditions (1) and (2) hold and if* $\mathbf{H}(j\omega) + \mathbf{H}(j\omega)^* \geq 0$ *for all* $\omega \in \Re$, *except for simple poles* $j\omega_0$ *of* $\mathbf{H}$, *where residue matrix must be nonnegative definite.*

If $\mathbf{H}$ has no poles on the closed right-hand half-plane, it is *PR* if and only if Theorem 3 holds.

Considering that $\mathbf{H}$ is reciprocal ($\mathbf{H} = \mathbf{H}^T$) checking condition (3) is equivalent to ensuring that

$$\lambda_i(\mathrm{Re}(\mathbf{H})) \geq 0, \quad \forall\, \omega \in \Re, \quad i = 1, \ldots, N_e, \tag{13.3}$$

where $\lambda_i(\cdot)$ denotes the eigenvalue of matrix in argument and N_e is the number of eigenvalues of the same matrix. Note that if $\mathbf{H}$ is a symmetric matrix, the eigenvalues are all real.

13.1.5 Example Circuit for Non- or Limited Passivity

In this section, we consider a conventional circuit model as a small example to illustrate the passivity issue. We want to show that the lack of passivity of a subcircuit can be impacted

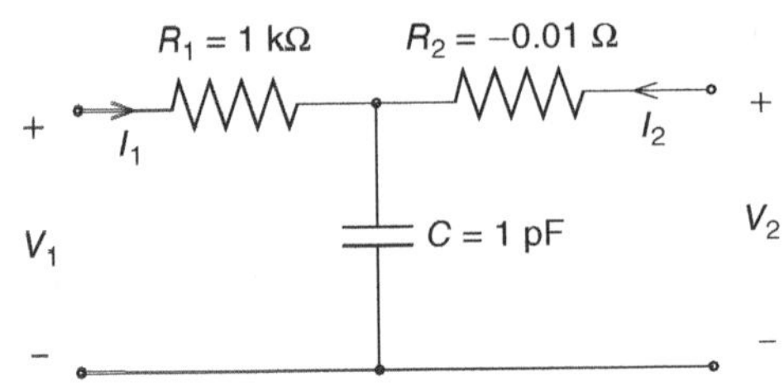

FIGURE 13.3 Two port non-passive circuit example.

by the series connection of two subsystems. In fact, today it is common to connect several subsystems in series for a SPICE type time domain solution. In Fig. 13.3, an example of a two-port circuit is shown which includes a negative resistor which leads to the input admittance

$$y_{11} = \frac{1}{R_1} \frac{s + \frac{1}{R_2 C}}{\left[s + \frac{1}{R_1 C}(1 + R_1/R_2)\right]}. \tag{13.4}$$

It is clear by considering y_{11} for the circuit that (13.4) is not passive. The demominator pole changes its sign if $R_1 < R_2$. However the numerator pole is still positive.

It is evident that the circuit can be part of a *limited passive* series connected circuit if the added circuit has a large enough positive input resistance. With this load, a stable response results. However, this is not the case for a capacitor termination of the example circuit. We observe, that the passivity problem is eliminated for reasonable resistive loads. So, this new combined circuit with the resistive load can be considered as *limited passive*.

13.2 ANALYSIS OF PROPERTIES OF PEEC CIRCUITS

In the previous section, we introduced the required definitions and fundamental concepts and theorems for stability and passivity. Next, we consider passivity of quasistatic (QS)PEEC models and we consider general properties of all PEEC models.

13.2.1 Ports and Nodal Potentials (Voltages)

An important issue for the following sections is the external connections to the systems under analysis. Fundamentally, the MNA matrix formulation for PEEC leads in general to a system of differential-algebraic equations [41]. It is evident from Chapter 2 that a pure PEEC-only model can be formulated without additional algebraic equations. However, circuits added to the system most likely add algebraic equations.

The MNA unknown vector of nodal voltages and inductive currents (2.11) gives access to all states in the system. This is an important issue for the observability considered below. In the same way, the PEEC circuit is controlled by the sources that are added to the right-hand side of the system as is exemplified in (13.6). Independent current sources can be attached to each node of the basic circuit loops. Independent voltage sources can be placed in series to the current branches as shown in (2.11). Hence, we can assume that the system is controllable and accessible if all these variables are available.

However, in realistic EM systems, access to the variables is mostly performed through ports. Port voltages are defined as differences of node potentials, or

$$v_{a,b}(t) = \Phi_a(t) - \Phi_b(t). \tag{13.5}$$

Of course, the same is true for the frequency domain. Also, the port currents are defined as $i_b(t) = -i_a(t)$. We observe that we assume that the port quantities are only dependent on the other port quantities through possible internal connections inside the PEEC model. The information obtained from ports is usually much less than the nodal voltage vector. It is clear that the connection to ports is the only choice for modeling some high-frequency microwave systems.

We also note that a port does not specify the relation of the port to ground. This assumes a ground independence of the ports. For this reason, ports can be used to connect, for example, a 2D transmission line model to a 3D PEEC model. High-frequency systems are usually connected through only a small number of ports. However, the number of nodal voltages in a system may be much larger. Hence, we can assume that very likely the system cannot be observed and controlled from the available ports. This case is considered in Section 13.3.

13.2.2 Passivity for Quasistatic PEEC Port Impedance

In this section, we show that the port impedance for a quasistatic (QS)PEEC model is passive. This represents a much simpler model that is given by (2.11) and (6.55) than the distributed full-wave (FW)PEEC model. Since we do not use a model with delays, the MNA equation simplifies to

$$\begin{bmatrix} \mathbf{G} + s\,\mathbf{Pp}^{-1} & \mathbf{A}_\ell \\ \mathbf{A}_\ell^T & -(\mathbf{R} + s\mathbf{Lp}) \end{bmatrix} \begin{bmatrix} \boldsymbol{\Phi} \\ \mathbf{I}_\ell \end{bmatrix} = \begin{bmatrix} -\mathbf{A}_i\mathbf{I}_i \\ 0 \end{bmatrix}, \tag{13.6}$$

where $\mathbf{G}$ is the simplest form to include a potential capacitive loss factor that is represented by a symmetric conductance matrix.

We consider the case where we have a current source excitation by the currents $\mathbf{I}_i$. We can subdivide the circuit matrix into the time- or frequency-dependent and independent parts as

$$\begin{bmatrix} \mathbf{Pp}^{-1} & 0 \\ 0 & -\mathbf{Lp} \end{bmatrix} \begin{bmatrix} s\,\boldsymbol{\Phi} \\ s\,\mathbf{I}_\ell \end{bmatrix} + \begin{bmatrix} \mathbf{G} & \mathbf{A}_\ell \\ \mathbf{A}_\ell^T & -\mathbf{R} \end{bmatrix} \begin{bmatrix} \boldsymbol{\Phi} \\ \mathbf{I}_\ell \end{bmatrix} = \begin{bmatrix} -\mathbf{A}_i\mathbf{I}_i \\ 0 \end{bmatrix}. \tag{13.7}$$

This can be written in the general form (13.8)

$$s\,\mathcal{C}\,\mathbf{X}(s) + \mathcal{G}\,\mathbf{X}(s) = \mathcal{B}\,\mathbf{U}(s) \tag{13.8}$$

with the exception of the different symbols used here from the standard symbols.

We follow the proof from Ref. [42] to show that the formulation (13.7) is passive based on the following:

Theorem 3 ***(Passivity for (QS)PEEC-MNA models)*** *A sufficient condition for a (QS)PEEC model to be passive is as follows:*

1. **G**, $\mathbf{Pp}^{-1}$, *are symmetric and nonnegative definite or all the eigenvalues are zero or larger.*
2. −**R**, −**Lp** *are symmetric and nonpositive definite or all eigenvalues are zero or less.*
3. *The off-diagonal coupling matrices are zero in the susceptance part* $\mathcal{C}$.
4. *The excitation is limited to currents as shown in (13.7).*

This test clearly is based on Section 13.1.4 for positive realness and also on **Theorem 2**. To show passivity of the system, the tests for positive realness theorem and the above passivity conditions must be fulfilled. It is clear from (5.16) for partial inductances and (4.31) for the partial coefficients of potential that they both are symmetric in that $Lp_{km} = Lp_{mk}$ and $Pp_{km} = Pp_{mk}$. Hence, both the **Lp** and **Pp** matrices are diagonally dominant with positive diagonal elements and they are positive definite.

We observe that we should always multiply the second row in the MNA matrix by −1. The result is that the MNA connection matrices will be **A** and $\mathbf{A}^T$. Also, the inductance and the capacitance matrices are independent from each other, without any couplings. Hence, the conditions for **Theorem 4** are met to prove the passivity.

13.3 OBSERVABILITY AND CONTROLLABILITY OF PEEC CIRCUITS

The fundamental properties of the full-wave (FW)PEEC circuit equations are more challenging to characterize. Observability and controllability are two of these interesting issues. Fortunately, we can learn from the information presented in the previous chapters. For convenience, we repeat the fundamental (MNA)PEEC circuit equation (6.54), (6.62), (13.8) in the time domain.

$$\mathcal{C}\,\dot{\boldsymbol{x}}(t) + \mathcal{G}\,\boldsymbol{x}(t) = \mathcal{B}\,\boldsymbol{x}_{in}(t) \tag{13.9a}$$

$$\boldsymbol{X}_{out}(t) = \mathcal{L}^T\boldsymbol{x}(t), \tag{13.9b}$$

where the ground node or reference potential is at infinity. As is considered in Section 13.2.1, we do not have access to all node potentials. The access we have consists of the voltage at a few ports. If we consider port k, we define the port voltage the quantity $Vp(k) = \Phi(m) - \Phi(n)$, where $\Phi(m)$ and $\Phi(n)$ are the nodal potentials at nodes m and n. Current and/or voltage sources are applied at ports at the node pairs. Of course, other input sources could be incident fields and output quantities could also be fields that we do not consider here.

We cannot assume that we know the detailed values of all partial elements of the large descriptor systems, which is in the general form of (13.9). All these issues must be taken into account for a realistic investigation of the passivity properties of a specific PEEC model. For our models, we need to distinguish between the frequency and time domain since the behavior of the circuit is substantially different in the two domains.

13.3.1 General Properties

Some fundamental properties of the resultant MNA descriptor matrices can be utilized in determining the behavior of the system even if we do not know the matrix elements in detail.

First, we consider the definition of observability, which is a fundamental property. The concept of observability is given in Ref. [43, p. 372] as follows:

Definition 8 ***(Observability)*** *A system (circuit) is said to be completely observable for the time interval $t_0 \le t \le t_1$ if, for every t_0 and some t_1, every state $X(t_0)$ can be determined from the knowledge if x_{out} on $t_0 \le t \le t_1$. Essentially, a system is completely observable if every transition of the system's state eventually impacts the system's output.*

Unfortunately, it is clear that for most realistic PEEC circuits, only a few states are directly observable due to the small number of ports compared to the number of unknowns. Hence, in general, the observability is clearly weak.

In PEEC circuits, some coupling exist among *all* elements since the majority of the partial inductances are coupled and so are the majority of the capacitive contributions. Therefore, multiple coupling paths exist to each node and all ports.

In Ref. [43], the definition of controllability also includes complete output controllability for the case when in (13.9) it is possible to construct an unconstrained input control vector $\mathbf{x}_{in}(t)$, which will transfer any given output $\mathbf{x}_{out}(t_o)$ to any final output $\mathbf{x}_{out}(t_1)$ in a finite time interval.

In a general PEEC circuit, a sparse input selector matrix $\mathcal{B}$ also leads to a very sparse access to ports. Hence, it is evident that this limits the input and outputs we can control. We next consider the concept of controllability.

Definition 9 ***(Controllability)*** *A system (circuit) is said to be completely output controllable if it is possible to construct an unconstrained input vector $\mathbf{x}_{in}(t)$ that will transfer any given initial output $\mathbf{x}_{out}(t_0)$ to any final output $\mathbf{x}_{out}(t_1)$ in a finite time interval $[t_0 - t_1]$.*

We conclude that for the few ports, limited observability exists due to the sparse output selector matrix $\mathcal{L}$. Also, controllability due to the sparse inputs or input selector matrix $\mathcal{B}$ is limited for practical PEEC circuits.

13.3.2 Passivity at Ports for PEEC Circuit in the Frequency Domain

We want to use the knowledge we gained in the previous section to consider the frequency domain behavior of typical (FW)PEEC circuits. We assume that we have typical large models with few input or output port connections. According to the previous section, we know that at least weak couplings exist throughout the system. Further, we do not have access to most nodal voltages and states. We consider two different situations based on this assumption.

In the first example, we assume that a small portion of the model is internally not passive. This may be caused by poor meshing or a locally poor or conventional delay model, which makes it a distributed system. However, the nonpassivity or problem inside of the model may not be observed due to the weak coupling path that exists to the small number of ports. Hence, we can conclude that in the frequency domain a large portion of the model is not observable and not controllable. It is unlikely that minor internal passivity issues can be observed at the ports.

A second frequency domain observation is based on the work in Ref. [44] where a very-high-frequency response is computed for example (Fig. 13.6). As a practical approach, the mesh density is chosen such that the accuracy is sufficient up to the maximum frequency of interest. This is especially true for the ever-increasing large problems to be solved. Hence, the challenge is to obtain a stable and passive solution without an expensive, excessively small mesh size.

Conventionally, the smallest mesh size is determined by the highest relevant frequency of interest f_{max}. A key observation gives insight into a fundamental problem. Assume that we chose the smallest mesh size for the PEEC model to be accurate up to f_{max}. Further, for a dielectric material, the frequency is even lower:

$$f_{max}[\text{GHz}] = \frac{1.5}{W[\text{cm}]\ \sqrt{\epsilon_r}}, \tag{13.10}$$

where W is the dimension of the largest cell in the geometry and where ϵ_r is the relative dielectric constant. As an example, we assume that the maximum cell size is $W = 0.1$ cm and $\epsilon_r = 4$. Then, by (13.10) the frequency $f_{max} = 7.5$ GHz. At this frequency, each cell is a small fraction of a wavelength. A key observation is that at a frequency that is 10 times higher, or 75 GHz, the cells will be subdivided only into the equivalent 2 cells/wavelength.

Hence, mostly the phase and also the amplitude response for the entire PEEC model are very poor at such a frequency. It is not surprising that at $f = 10 f_{max} = 75$ GHz, new nonphysical right-hand-side poles or resonances result – outside of the important frequency range –above f_{max}. Hence, at these frequencies, the entire system model will not be passive due the poor high-frequency response.

By measuring the impedance at a PEEC model node or a port in the circuit, the real part of the input impedance Re(Z_{in}) can be computed according to (6.68). It was shown in Ref. [44] that the solution becomes very poor for frequencies a decade above f_{max} where the meshing is only $\lambda/2$ and even less for two decades above f_{max}. Due to its importance, we summarize this issue as follows:

Observation 1 ***(High-frequency resonance problem)*** *Discretized models exhibit seemingly random resonances in the frequency range sufficiently above f_{max} where the cell subdivisions are much less than 20 cells/λ. These resonances or poles are a strong function of the cell subdivision size. We observed that most port impedance exhibit variable negative real parts at these high frequencies.*

To summarize, two problems may exist in the frequency domain. First, in the active frequency range below f_{max} some of the model may contain parts that are nonpassive, which may not be observable at the ports. Second, for frequencies higher than f_{max} – especially for full-wave (FW)PEEC models – we observe negative real parts in the impedances that lead to nonpassivity.

13.3.3 Time Domain Stability and Passivity Issues

The fundamental causes of time domain instabilities are related to the frequency domain passivity issues treated in the previous section. However, the problem manifests itself in a different way.

Generally, time domain excitation signals include a wide spectrum of frequencies. The very-high-frequency issue has the same source as the one described in the frequency domain given in the previous section. In the same way, the high-frequency content in the input signals corresponds to a mesh size where the cell size is very large, again like 2 cells/λ or worse. It is clear that this problem results in an unstable solution at the high-frequency content that corresponds to less than 10 cells/λ.

The PEEC circuit will have right-hand-side poles at these high frequencies that are excited by the input signal. An excited arbitrary pole/residue pair generates a transient response of the form

$$v(t) = r\, e^{\alpha\, t}, \tag{13.11}$$

where the residue r may be extremely small and α is positive for right-hand plane poles. Hence, even if the residue r is very small, eventually the exponential increase will dominate.

Importantly, the exponential increase will propagate throughout the system due to the fact that the entire circuit is coupled, even if the coupling is weak. This will cause an exponential increase of all nodal voltages and currents. Hence, the behavior at the global circuit level is much different in the time domain than the frequency domain.

Unfortunately, time domain stability can be compromised even if the high-frequency content of the input signal(s) is (are) weak. The exponential increase in (13.11) clearly spreads to all parts of the system. Hence, the instability in the time domain serves in some cases as an indicator of hidden passivity problems, which may be an issue in both domains.

13.4 PASSIVITY ASSESSMENT OF SOLUTION

In this section, we consider the area of *solution passivity assessment* mostly from a practical point of view. This step is clearly necessary for all approaches during or after the solver solution to confirm passivity. It is also a necessary step for the *solution passivity enforcement* approaches.

For the large systems, we need to consider passivity assessment approaches that do not require detailed information on circuit elements in the MNA equations. The Vector Fitting algorithm [18] has been used from the beginning as a way to model the behavior of the overall model with a limited number of ports [15, 19]. This approach insures that the fitted model is not very large due to the relatively small number of ports for realistic systems. In Ref. [45], the Vector Fitting algorithm is used with a Hamiltonian matrix check after the system is modeled with the vector fitting.

A large number of passivity assessment techniques are based on the assumption that the details of the solution system are known, including all the partial elements for a PEEC system [12, 18–27]. We again remind the reader that passivity assessment or limited passivity assessment given in Definition 4 is necessary step for passivity enforcement. Although we need this step for our work, we do not get into details for passivity enforcement.

13.4.1 Port-Based Passivity Assessment in Frequency Domain

In this section, we pursue a practical approach for the determination of limited passivity for full-wave (FW)PEEC circuits. We consider a practical approach since a strict passivity test is not possible to obtain with measurements.

Scattering parameters are very useful for the determination of passivity in both the frequency and the time domains. A theorem for the frequency domain determination of the passivity of a multiport system is given by Triverio et al. [12] as follows:

Theorem 4 ***(Passivity using scattering parameters)*** *A scattering matrix* $\mathbf{S}(j\,\omega)$ *represents a passive linear system if and only if*

1. *It is Hilbert consistent if the real and imaginary parts obey the equations (10.4);*
2. $\hat{\mathbf{I}} - \mathbf{S}^{\mathbf{T}}(-j\omega)\,\mathbf{S}(j\omega)$ *is nonnegative-definite matrix for all* ω*;*
3. $\mathbf{S}(-j\omega) = \mathbf{S}^{*}(j\omega)$*;*

where we again use the $\hat{\mathbf{I}}$ to represent the identity matrix to distinguish it from a current symbol. Step 1 in Theorem 5 can be checked by numerically evaluating the Hilbert consistency between the real and the imaginary parts. We note that this already leads to practical accuracy problems unless data from $\omega = 0$ to $\omega \rightarrow \infty$ is available. This is strictly impossible.

We consider an example for Step 2 of Theorem 5 for a port model that reduces to $S_{11}(s) = \rho(s)$, where $\rho(s) \leq 1$ for a single port. In Fig. 13.4, we show an equivalent circuit for the scattering port.

The reflection coefficient is computed according to (13.12) in the frequency domain as

$$\rho(s) = \frac{Z_{11}(s) - R_0}{Z_{11}(s) + Z_0} \leq 1, \tag{13.12}$$

where R_o is the reference resistance and $Z_{11}(s)$ is the impedance of the port under test. For limited passivity, we require that the PEEC model is passive up to a frequency such that the magnitude and phase are represented correctly at the highest frequency of interest, which should be somewhat higher than f_{max} in (13.10). We first want to show by example how harmful the very-high-frequency response for a FW model can be. The patch antenna in Fig. 13.5, which consists of two thin square patches with the dimension with $w = 4.5$ mm where $t = 50$ μm and a patch-to-patch spacing of $s = 0.5$ mm. For this geometry, f_{max} is designed to be 20 GHz, which is the highest frequency of operation. The result is shown in Fig. 13.6, where both sides of the antenna patches divided with six and seven subdivision cells. The large impact of the cell subdivisions for frequencies above f_{max} is apparent.

Next, we consider the frequency domain response for a loop using a scattering response again with the same port connection. The frequency domain reflection coefficient $\rho(s)$ or scattering response S_{11} is computed by using (13.12). We show results for a loop problem

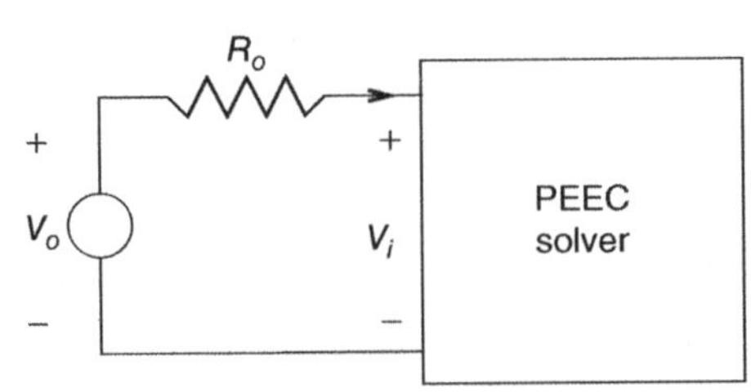

FIGURE 13.4 Port measurement for passivity for EM solver.

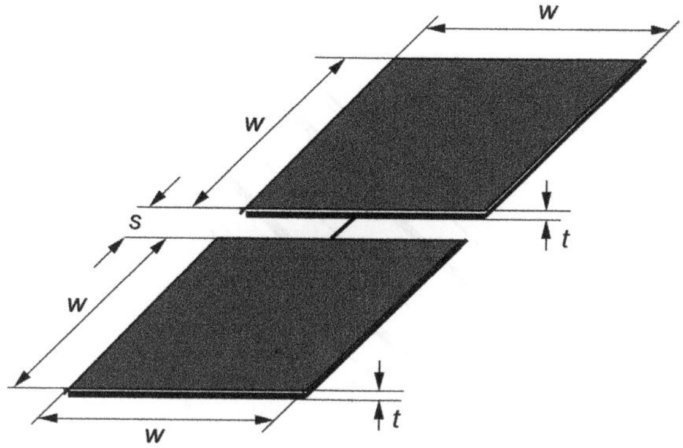

FIGURE 13.5 Small two-patch antenna where source is connected between the two patches.

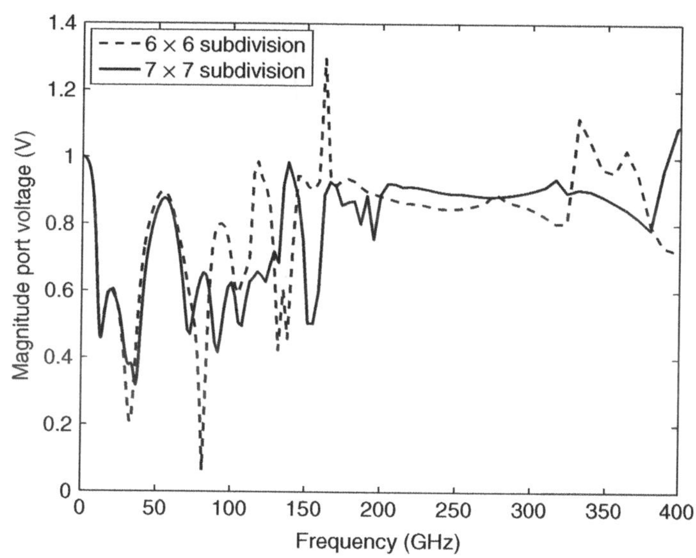

FIGURE 13.6 Very-high-frequency resonance using the PEEC model for patch antenna.

shown in Fig. 13.7, which is a longer version ($x_l = 0.53$ mm, $y_l = 50$ μm, $w = 10$ μm) of an example used in Ref. [16] and in Chapter 9. Here, we first present the solution without passivity enhancement.

First, we consider the real part of the input impedance of the loop as a function of frequency. The real part of the loop input impedance is given in Fig. 13.8. It is important that the response must be computed far above the maximum frequency of interest f_{max}. It is clear that the real part of the high-frequency response prevents strict passivity due to this response.

This result is confirmed by the scattering response in Fig. 13.9. Both responses indicate that this loop model is not passive. As can be observed, the real part turns negative in Fig. 13.8 while $|S_{11}| > 1$ for frequencies much higher than the 20 GHz operating frequency. This is a further indication that the resonances above f_{max} are leading to passivity problems.

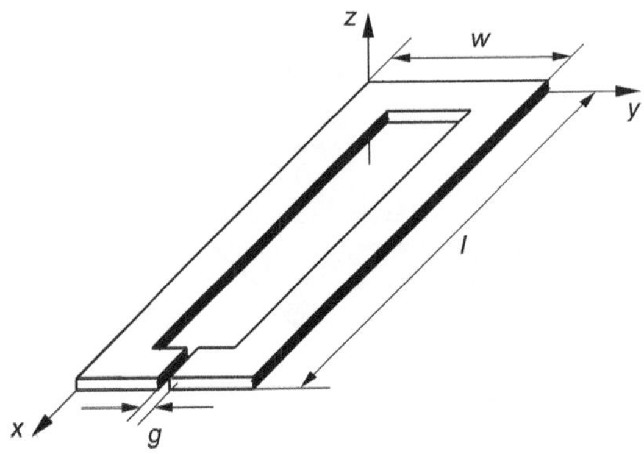

FIGURE 13.7 Example loop problem geometry for passivity study.

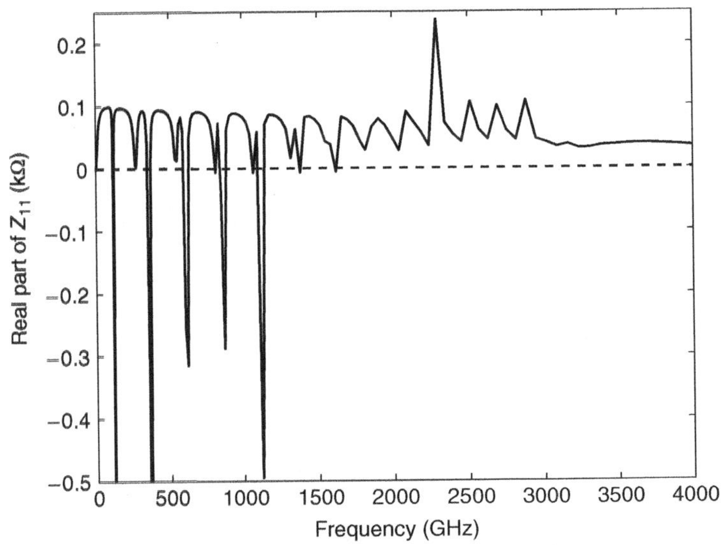

FIGURE 13.8 Real part of input impedance Z_{11} for loop problem.

13.4.2 Port-Based Passivity Assessment in Time Domain

In this section, we consider testing for passivity in the time domain. We can use the energy integral in Definition 10 in Section 13.1.2, which for the case that the initial energy is zero is

$$\int_{-\infty}^{t} \mathbf{v}^T(\tau)\mathbf{i}(\tau)d\tau \geq 0. \tag{13.13}$$

Similar definition exists for the time domain in terms of scattering parameters **a**, **b** [12] as

$$\int_{-\infty}^{t} [\mathbf{a}^T(\tau)\mathbf{a}(\tau) - \mathbf{b}^T(\tau)\mathbf{b}(\tau)]d\tau \geq 0. \tag{13.14}$$

Of course, if the initial energy is 0 at $t = 0$, then the integral can also start at 0.

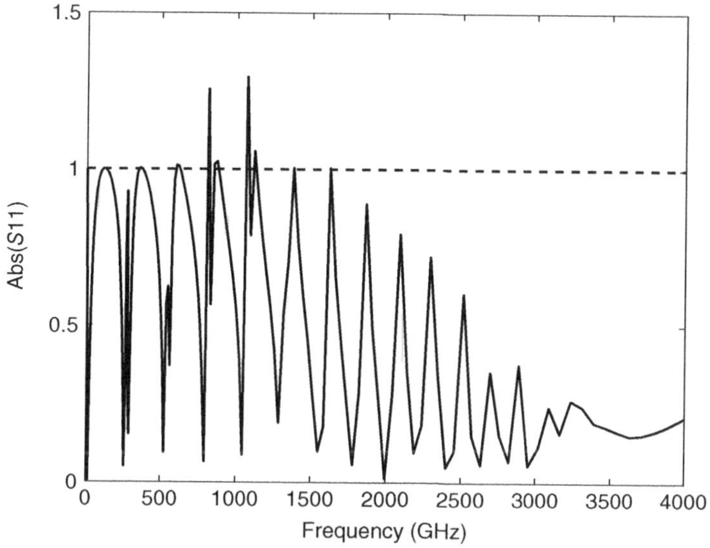

FIGURE 13.9 Scattering response $S11(\omega)$ for loop problem.

We consider an example for a single port. The scattering parameters for the single-port model are

$$a(t) = \frac{1}{2}\left[\frac{v(t)}{\sqrt{R_0}} + \sqrt{R_0}\, i(t)\right] \tag{13.15a}$$

$$b(t) = \frac{1}{2}\left[\frac{v(t)}{\sqrt{R_0}} - \sqrt{R_0}\, i(t)\right]. \tag{13.15b}$$

The energy integral (13.14) simplifies for the single port case to

$$\int_{-\infty}^{t} [a^2(\tau) - b^2(\tau)]d\tau \geq 0\,. \tag{13.16}$$

It is interesting to note that after substituting (13.15) into (13.16), the results will be the same as (13.13). Hence, we show an example for a single port for the energy integral (13.13), which is evaluated numerically. The result of this is shown in Fig. 13.10 for the very small time step of 0.05 ps. The energy integral turns negative, which shows that the model for the two-patch antenna is not passive. We also observe that the energy response does not change much with the change in the small time step.

For the case where the spectral content of the input signal is reduced, we increased both the time step and the input signal rise time. The time step Δt is increased from 0.5 to 10 ps and the input step rise time is increased from 0.05 to 1 ps. The result is shown in Fig. 13.11. The result for the very fast input signal in Fig. 13.10 is clearly not passive, whereas the slower signal response in Fig. 13.11 is limited passive while not being strictly passive. This is evident from the result.

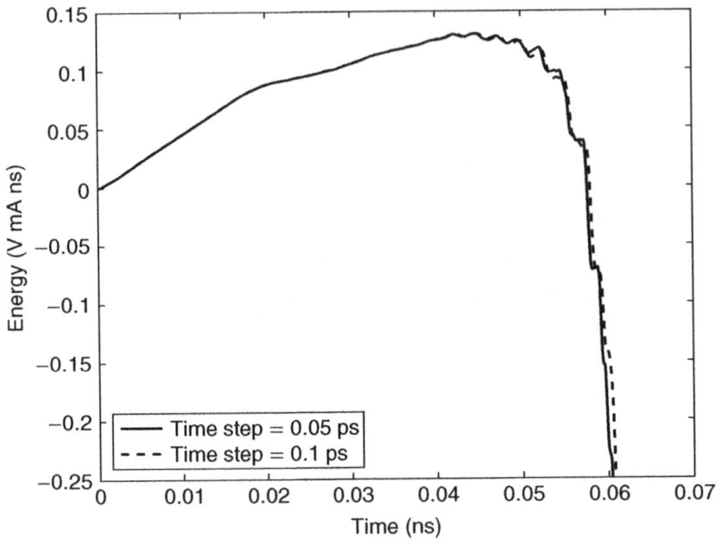

FIGURE 13.10 Result from energy integral for the two-patch antenna for fast input signal.

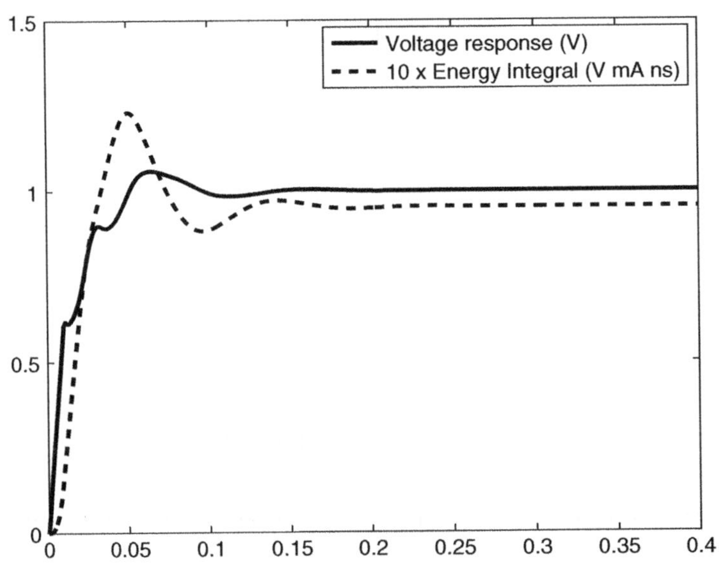

FIGURE 13.11 Response and energy integral results for the two-patch antenna for slower input signal.

13.5 SOLVER BASED STABILITY AND PASSIVITY ENHANCEMENT TECHNIQUES

It is clear that the stability/passivity assessment of the solver output should be performed for all results. In this section, we assume that we have access to the internal PEEC solver

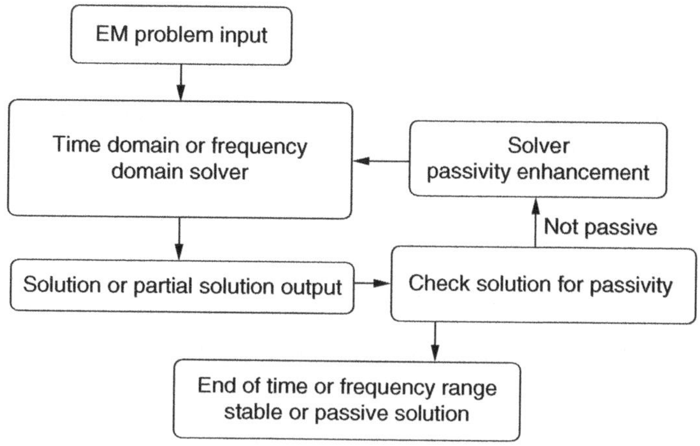

FIGURE 13.12 Example of EM solver with stability/passivity enhancement.

implementation. This access to the PEEC internals allows us to apply *solver passivity enhancement* techniques at the solver level. Stability and passivity enhancements are applied such that the EM solution assumes the required properties. An example of a control loop during the solution is shown in Fig. 13.12.

The monitoring of passivity during or at the end of the solution can be applied in some form depending on problem details. It does not make sense to continue the solution once we observe passivity or stability problems.

An advantage of the *solver passivity enhancement* in Section 13.5.1 is that we can improve the solution accuracy rather than perhaps a less accurate answer by postprocessing with *solution passivity enforcement*. The function of the control loop added to Fig. 13.12 is to make sure that the solution is stable or passive up to the point in the time or frequency of the solution. Different monitoring procedures in the PEEC solver can be used. It is clear that *solution passivity assessment* is a necessary first step for both the *solver passivity enhancement* and solution passivity enforcement.

13.5.1 Solver Enhancement Techniques for Time and Frequency Domains

It is shown in Section 13.2.2 that (QS)PEEC models are passive and also stable. Of course, this depends, as in all PEEC circuits, on the meshing being projection based, etc., according to Sections 4.4 and 8.4.2.

Techniques have been proposed for more than three decades to improve the time domain stability of FW integral equation EM and other solutions. Some stability enhancement techniques proposed for PEEC circuits are summarized in Ref. [7, p. 231]. Formally, the inclusion of delays in PEEC changes the type to a distributed model that may impact stability and passivity. Other issues are that often we do not include all the losses in the actual problem, which introduces damping.

Practical circuit-oriented steps are considered next, which help the passivity enhancement for full-wave (FW)PEEC solvers. One enhancement is based on the inclusion of the conductor and dielectric loss, which is also very important for the model accuracy. We show how we can combine loss models to improve the stability and accuracy of the overall solution. Fortunately, we have the (FW)PEEC circuits skin-effect conductor loss models from Chapter 9 and dielectric loss models from Chapter 10 for these more realistic models.

Approaches used to include delays in circuit models were introduced in Section 2.11.1. For many lower frequency applications, it is sufficient to take the delay or phase factor outside of the partial elements into account since the cell sizes are chosen to be sufficiently small and if the delays are moderate. The partial element size should be smaller than a fraction of the smallest wavelength like $\lambda_{min}/20$. As an example, for partial inductances, we use

$$Lp_{km}^{(FW)} = Lp_{km}^{(QS)} \; e^{-s\tau_{km}}, \tag{13.17}$$

where the delay τ_{km} is a center-to-center delay and where (5.18) is the usual (QS)PEEC partial inductance Lp_{km}.

We should note that this is an approximation to the delayed system in the frequency domain. However, it is equivalent to the solution used for the time domain, which is based on discrete time steps. These approximations do have an impact on the solution behaviors, which we consider next.

13.5.2 Passivity Enhancement by Subdivision of Partial Elements

The *passivity enhancement* approach considered in this section can be applied to both time and the frequency domain solutions. The approach has its origin in an observation presented in Ref. [44]. The key problem found was that the input impedance at a node becomes negative at some very high frequencies. The (FW)PEEC solution showed that arbitrary non-physical very-high- frequency poles or resonances were observed in Fig. 13.6. Related work [46] showed that the careful choice of mesh subdivisions had a positive impact on the solution stability. It is clear that the phase representation is very critical at high frequencies and it plays a very important role for the problem at hand.

Two cells of different sizes, **A** and **B**, are shown in an example in Fig. 13.13. We include a single inductive cell example for the small one while the larger cell is shown to be subdivided into 16 capacitive delay-only subcells. We should note that the subcells are used only for the evaluation of the phase or delay at the center of the partial inductance areas, while the current and charge for the "large" cell is computed based on the full cell such that no new unknowns are introduced.

We notice that for the partial inductance, the delay is evaluated at the **x** in the small inductive subcell in Fig. 13.13. The capacitive delay is evaluated differently from node to node on the capacitive subcells, which is at the center of the subdivisions only for full cells.

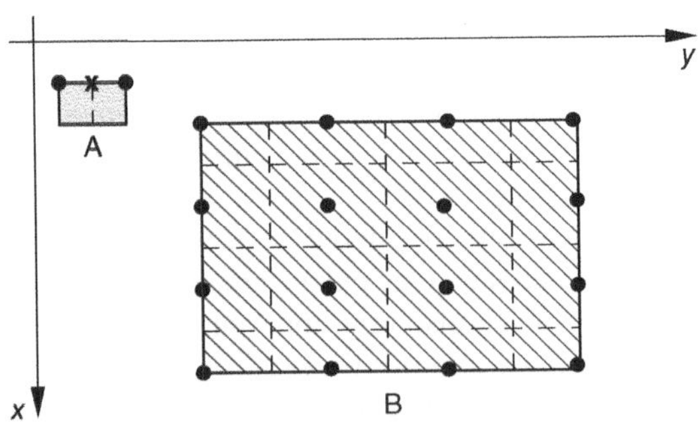

FIGURE 13.13 Example of surface cells with phase subdivisions.

In this example, the small **A** cell is not laid out with additional inductive delay subdivisions due to its smaller size, which is smaller than the subdivision on the large **B** cells.

We repeat that in this approach we do *not* add additional unknowns to the system of equations. The only difference is that delays are added to the subcells. Of course, these subdivisions with delays have a profound impact on the phase response at higher frequencies.

In our example, we present the coupling term between two cells **A** and **B** with a refined phase mesh for cell B. The refined potential coefficient Pp_{km} for an example where both cells are in an x–y plane is given by

$$Pp_{km} = \frac{1}{4\pi\epsilon_0} \frac{1}{S_k S_m} \int_{S_k} \int_{S_m} \frac{1}{R_{km}} \, dS_k \, dS_m \\ \sum_{n1=1}^{N_x} \sum_{n2=1}^{N_y} e^{j\beta \, d(n1,n2)}, \tag{13.18}$$

where the distance between two subcells 1 and 2 is

$$d(n1, n2) = \sqrt{[x(n1) - x(n2)]^2 + [y(n1) - y(n2)]^2}. \tag{13.19}$$

This represents the distance between the local points chosen on the subcells such as the ones shown in Fig. 13.13. In this example for cell B we have, $N_{x1} = 1$ and $N_{y1} = 1$ for cell A and for cell B we have $N_{x2} = 4$ and $N_{y2} = 4$. The example also illustrates a general case where the overall cell size is taken into consideration.

Next, we give the results for the patch antenna problem in Fig. 13.5 to illustrate the impact of the phase submeshing.

The frequency domain results for phase submeshing is given in Fig. 13.14. In this example, the equal size antenna patches are subdivided with $n = 1$, 2, and 3 subdivisions

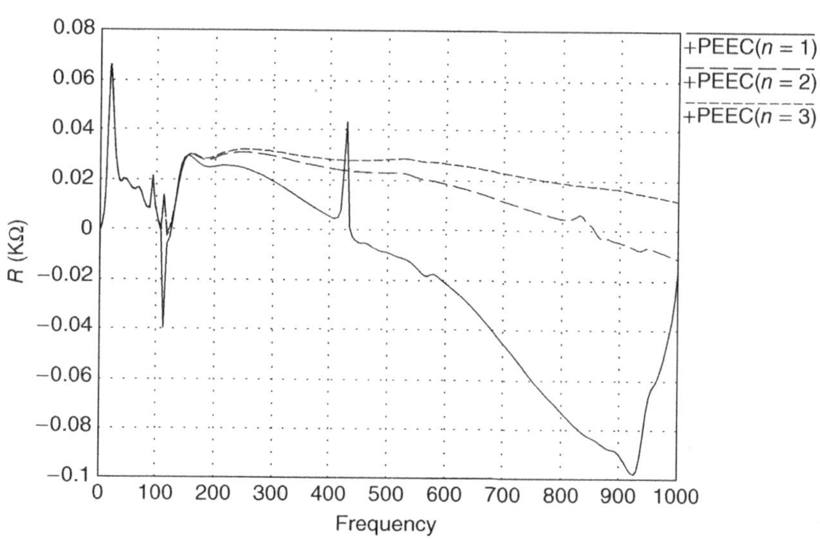

FIGURE 13.14 Real part of subsurface partitioning of path antenna for three subdivision cases.

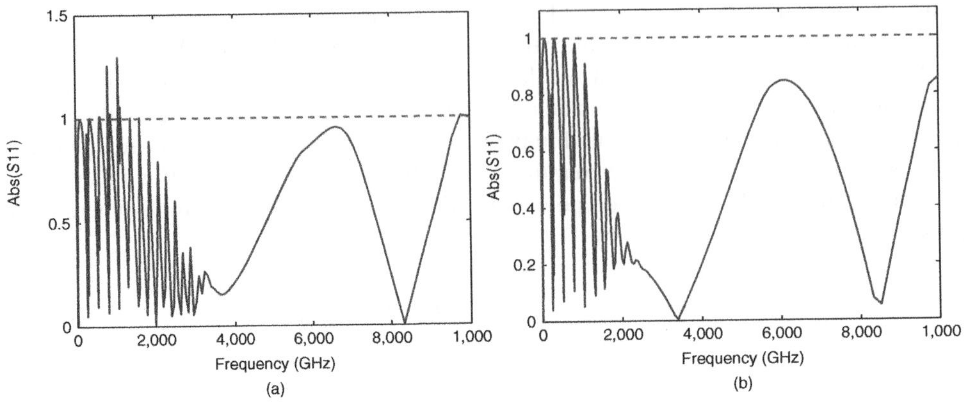

FIGURE 13.15 Results for reflection coefficient test for patch antenna. (a) Results without subdivisions. (b) Results with subdivisions.

both in the x and y directions. The result in Fig. 13.14 shows that the real part of the antenna impedance can be improved by using the cell subdivision approach.

We observe that only very high frequencies are impacted by the subdivision. Importantly, the negative real part of the input impedance leads to an unstable time domain solution since a fast input signal will contain the high frequencies. This explains why the response can be stable for a slower input signal but will be unstable for fast rise time input pulses with a higher frequency spectrum.

We also give results for the reflection coefficient test according to Section 13.4.1, where the results are shown in Fig. 13.15. It is evident that the passivity requirements for $\rho(s) \leq 1$ condition are satisfied for the lower frequencies. However, the problem is not passive for higher frequencies. After applying the cell subdivision technique given in this section, the passivity criterion is met for the total frequency region.

13.5.3 Passivity Enhancement Using Resistive Damping

The introduction of losses represents a general approach for the improvement of the solution stability and passivity. The results given in Section 13.5.2 show that nonphysical very-high-frequency poles exist in the right-hand half-plane. This clearly violates the positive realness requirement. The false response that results at sufficiently high frequencies above f_{max} needs to be eliminated. As a solution, resistive damping was introduced in Ref. [44] by parallel resistances to the partial inductances in the PEEC model.

An example for resistive damping is shown in Fig. 13.16. Damping resistors R_d are placed across the partial inductances such that the circuit turns resistive at very high frequencies since $2\pi f_{max} Lp_{kk} < R_{dk}$.

It is clear that the computational cost of such a change is minimal since it does not introduce additional unknowns in the MNA equations. Of course, each damping resistor has to be adjusted according to the partial inductance value and the resistances are required to be large enough such that the response below f_{max} is not altered by the resistance. Hence, we choose the damping resistors such that

$$R_{dk} = \alpha \, \omega_{max} \, Lp_{kk}, \tag{13.20}$$

where $\omega_{max} = 2\pi f_{max}$ and $\alpha = 10, \ldots, 100$.

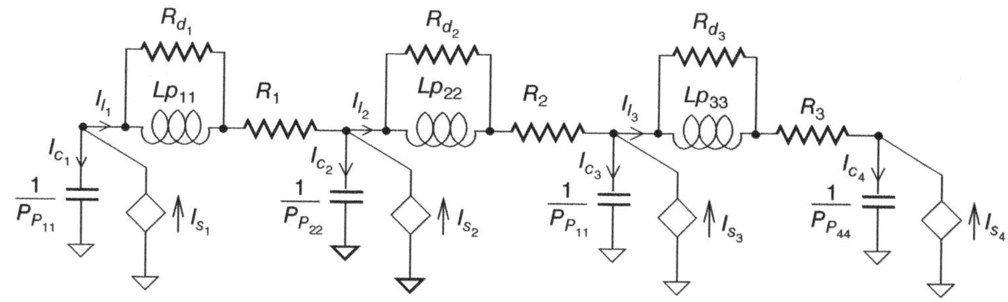

FIGURE 13.16 PEEC model example with damping resistors.

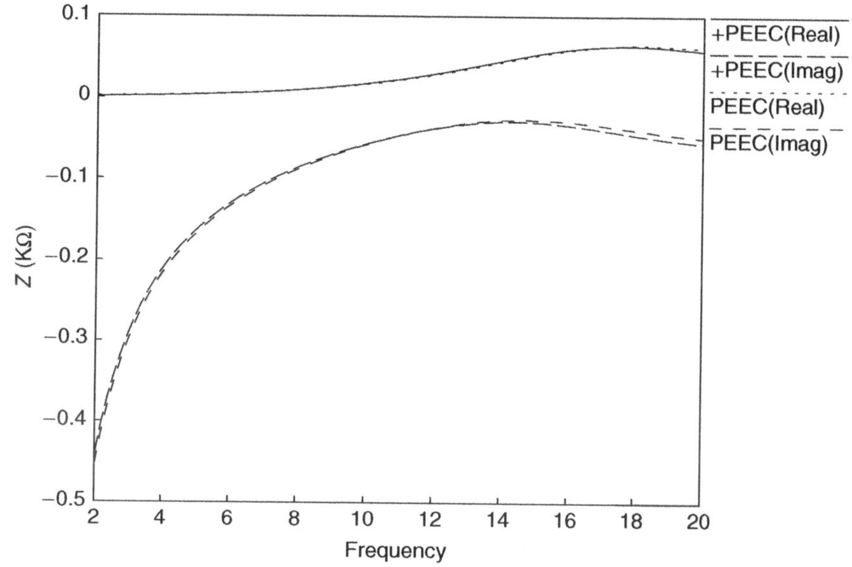

FIGURE 13.17 Real and imaginary parts of response below f_{max} for patch antenna with and without damping resistors.

To consider the impact of the damping resistors, we first want to show that the resistors do not impact the range below f_{max} as is shown in Fig. 13.17. The damping resistor R_d in Fig. 13.16 clearly needs to be chosen to be large enough such that the actual response is not distorted. We see a slight impact on imaginary part of the solution if R_d is too small.

Figure 13.18 shows the real part of the response of the small patch antenna above f_{max} up to 1000 GHz with several values of the damping resistors. It is clear that the high-frequency response with damping eliminates poles in the right-hand plane. This is especially evident from the imaginary part of the input impedance, which does not show artificial resonance behavior above f_{max}. This illustrates how very-high-frequency response can be controlled (Fig. 13.19).

A question is whether it is possible to add damping to the capacitance part of the PEEC model as well. We end this section with the observation that adding damping resistors parallel to capacitances [7, p. 213] may alter the *dc* solution. However, the *dc* solution should not be impacted by adding resistors in series to the capacitances. We experimentally added

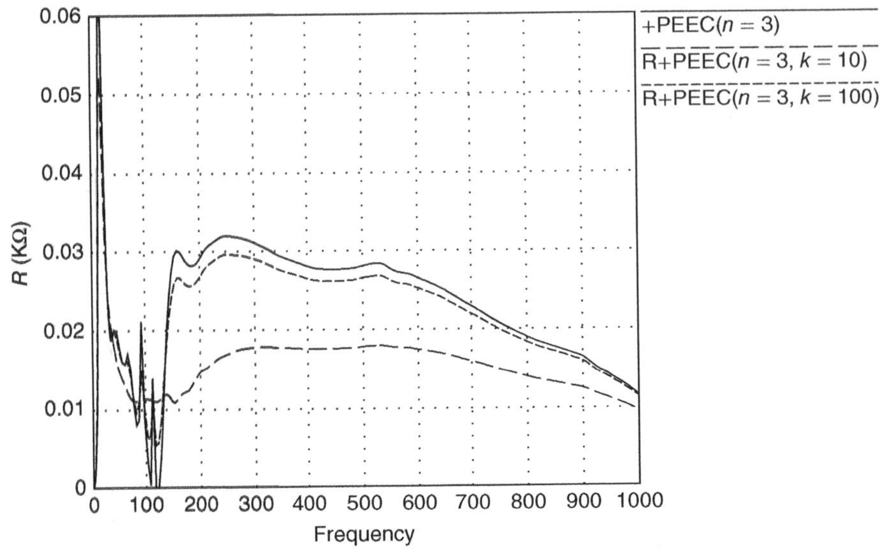

FIGURE 13.18 Real part Re[$Z(f)$] response above f_{max} with resistive damping.

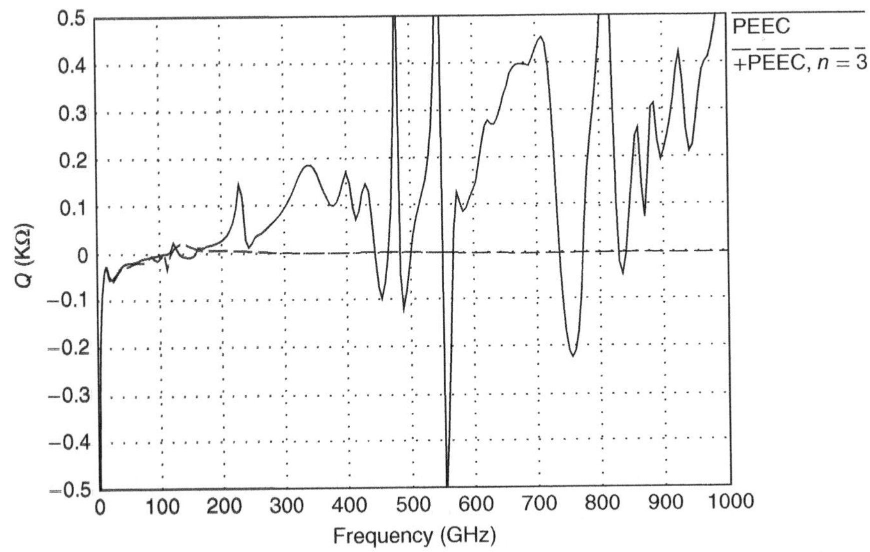

FIGURE 13.19 Imaginary part Im[$Z(f)$] above f_{max} with and without resistive damping.

circuit elements to the coefficient of potential models with series resistances. Some of these filters showed some promise.

13.5.4 Partial Elements Delay Macromodels for Passivity Enhancement

The approach in this section aims to modify partial elements of the (FW)PEEC model since taking the delay outside the integral somewhat reduces damping. The technique is based on synthesized macromodels used for the partial FW coefficients based on work in Refs [47, 48].

The approach pursued in Ref. [47] is illustrated for a partial inductance with a scalar Green's function, Section 3.4.2 function or

$$Lp_{km}^{(FW)}(s) = \frac{\mu}{4\pi \mathcal{A}_k \, \mathcal{A}_m} \int_{\mathcal{V}_k} \int_{\mathcal{V}_m} \frac{e^{-s|\boldsymbol{r}_k - \boldsymbol{r}_m|/v}}{|\boldsymbol{r}_k - \boldsymbol{r}_m|} \, d\mathcal{V}_k \, d\mathcal{V}_m, \tag{13.21}$$

where v is the speed of light in the material at hand. The model is also based on the center delay that is conventionally used. The delay τ_{km} is evaluated based on the center-to-center distance

$$Lp_{km}^{(FW)} = Lp_{km}^{(QS)} \, e^{-s\tau_{km}}, \tag{13.22}$$

where $Lp_{km}^{(QS)}$ is the conventional partial inductance. In Ref. [47], the term $Lp_{km}^{(QS)}$ is replaced with an impedance Zc_{km} to better approximate the result of (13.21). The equivalent circuit is based on a pole–residue representation from Section 2.9 as

$$Zc_{km}(s) = d_{km}^C + s \; e_{km}^C + \sum_{n=1}^{N_p^r} \frac{r_n^r}{s - p_n^r} + \sum_{n=1}^{N_p^c} \left(\frac{r_n^c}{s - p_n^c} + \frac{r_n^{c*}}{s - p_n^{c*}} \right), \tag{13.23}$$

such that

$$Vi_k(s) = sLp_{km}^{(FW)} I_m(s) = Zc_{km} \; e^{-s\tau_{km}} I_m(s), \tag{13.24}$$

where $Vi_k(s)$ is the voltage induced for k by the partial inductance and $I_m(s)$ is the current through branch m. As is presented in Chapter 2, an equivalent circuit can be found for both the partial self- and mutual inductances corresponding to the impedance Zc_{km}.

Vector fitting techniques [18] are used to determine the coefficients d_{km}^C, e_{km}^C and the values of the residues r_n^r and r_n^{c*}. Hence, the response of the original retarded partial inductance (13.21) is modeled with the macromodel. Further, the FW partial potential coefficients are modeled in the same way with the equivalent circuit and the Vector Fitting approach.

As an example for the application of the approach, we consider two identical elementary PEEC cells. On purpose, we chose the dimensions larger than usual, or $l = \lambda_{min}/10$, $w = \lambda_{min}/10$, and $t = \lambda_{min}/1000$, where λ_{min} is the minimum wavelength corresponding to the highest frequency, which is 12.5 GHz (Fig. 13.20).

In this example, the cells are coplanar and are spaced $0.2\lambda_{min}$ in the x- and y-directions. The result is computed with the numerical adaptive quadrature scheme (Adaptsim) in Ref. [49] as well as other numerical integration schemes. This numerically accurate result for (13.21) is compared to (13.22), the QS partial inductance. It is apparent that the magnitude decreases with frequency for the exact evaluation of the coefficient for a $\lambda_{min}/10$ cell (Fig. 13.21).

It is evident from this result that the magnitude or damping differs significantly, while the phase agrees well between the two models.

The Vector Fitting algorithm is used to extract the pole–residue representation of $Z_{L,mn}^{dl}$ in (13.23). Figure 13.22 shows the magnitude and phase spectra of the numerically computed $Z_{L,mn}^{dl}$ (original), the one obtained by the fitting procedure (fitted) based on two poles and the synthesized equivalent circuit.

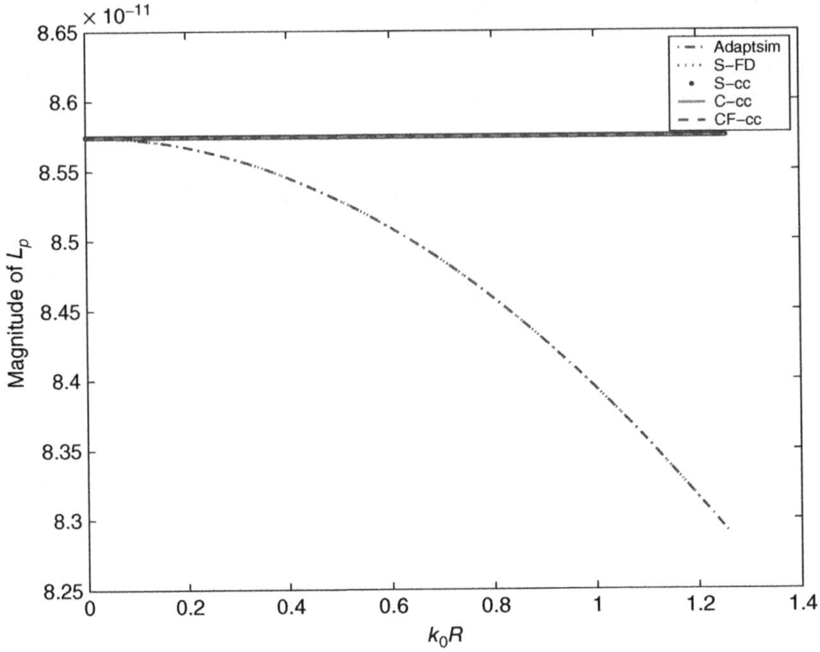

FIGURE 13.20 Magnitude example for retarded partial inductance Lp_{mn}.

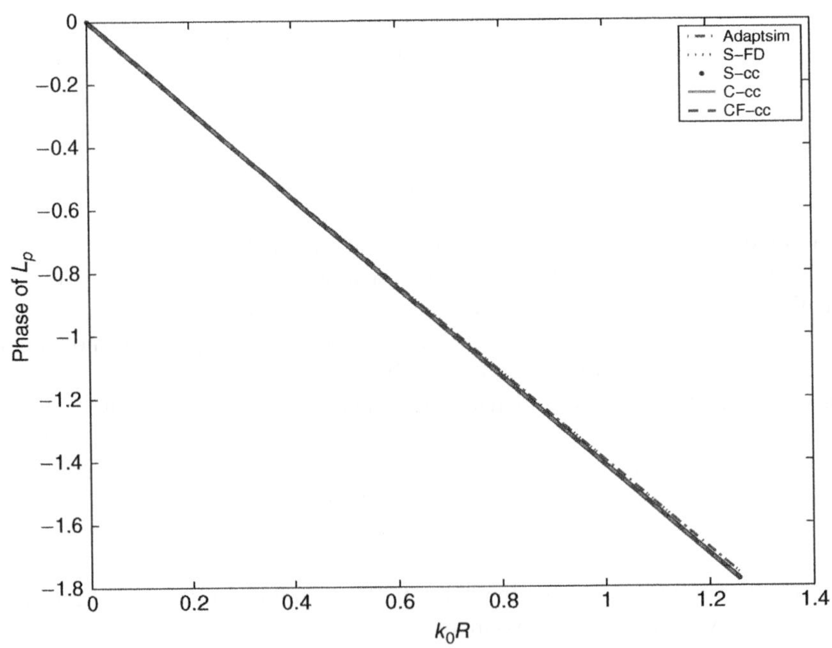

FIGURE 13.21 Phase example for retarded partial inductance Lp_{mn}.

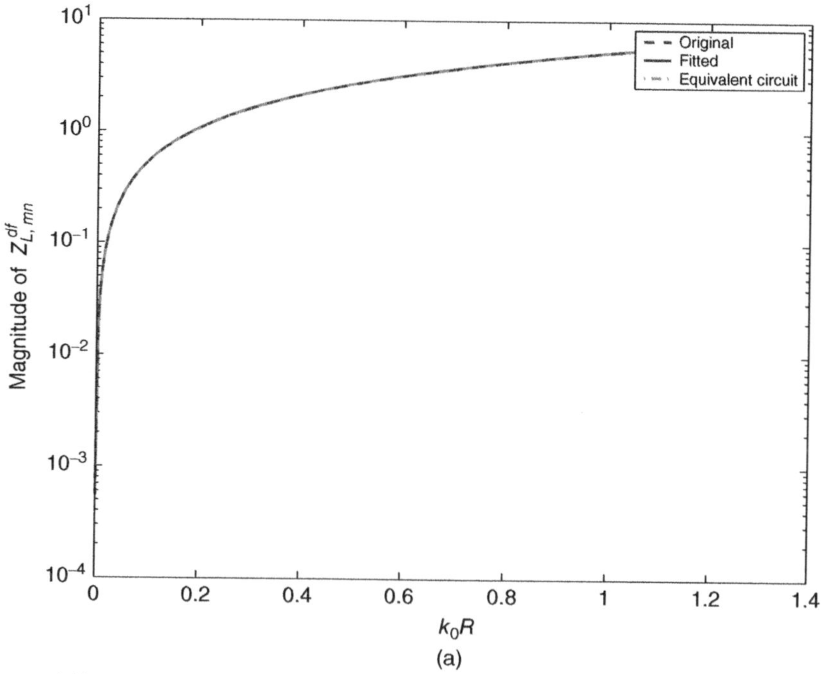

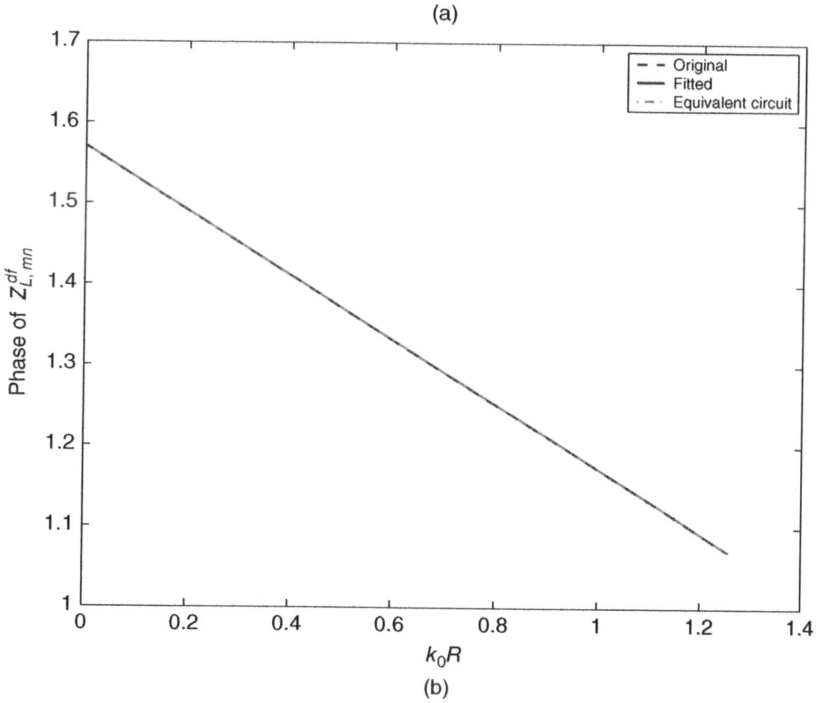

FIGURE 13.22 Pole–residue approximation of $Z^{dl}_{L,mn}$: near-field case. (a) Magnitude. (b) Phase.

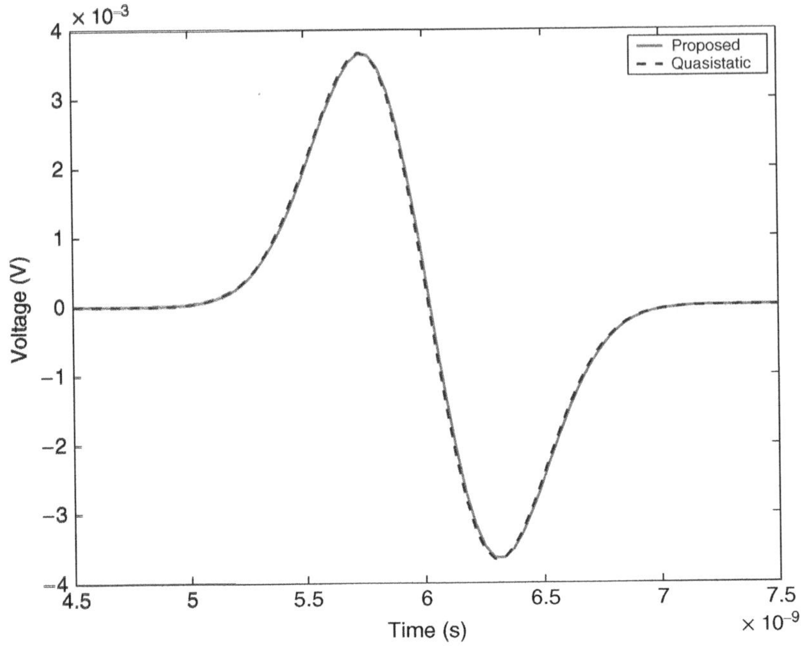

FIGURE 13.23 Induced voltage in the near-field case.

In a second test, an impulse current is injected into $Z^{dl}_{L,mn}$ and the induced voltage has been evaluated, accounting for the frequency dependence of mutual partial inductance Lp_{mn}.

The frequency content is significant up to 5 GHz, corresponding to an electrical distance $\beta_0 R = 0.502$, which is high enough to require a more accurate description of the frequency dependence of partial inductance. In Fig. 13.23, the induced voltage across the partial mutual inductance Lp_{mn} is plotted.

Due to the limited electrical distance at the maximum frequency of interest (~5 GHz), the result obtained by using the proposed enhanced (FW)PEEC model and the (QS)PEEC only differs by a small delay, as shown in Fig. 13.23.

The approach in Ref. [48] is also based on replacing the conventional partial elements with a filtered model. The purpose of the new model is not only a change in the phase representation but also to include a high-frequency damping filter. As an example, the model used for the partial inductance is given by

$$Lp^{(FW)}_{km} \approx Lp^{(QS)}_{km} D(j\omega) e^{-j\omega\tau_r}, \tag{13.25}$$

where $D(j\omega)$ is a damping factor and τ_r is a modified delay factor and $Lp^{(QS)}_{km}$ is the conventional QS partial inductance.

The damping factor is given by a low-pass filter

$$D(j\omega) = \left[\frac{n\omega_c}{j\omega + n\omega_c}\right]^n, \tag{13.26}$$

where n is 1 or 2, but $n \leq 4$ and ω_c is the filter cutoff frequency.

Next, we need to consider how the macromodel is tuned to represent a particular partial element. A single frequency evaluation of the exact FW partial inductance $Lp_{km}^{(FW)}$ is used to calibrate the macromodel at the angular frequency $\omega_g = min[\omega_{max}, \pi/(2\tau_{km})]$ where τ_{km} is the conventional center-to-center delay. Further details are given in Ref. [7].

13.5.5 Passivity Enhancement for Model with VFI Skin-Effect Models

Usually, a small cell size may be required to represent the volume filament (VFI) skin-effect models introduced in Chapter 9. Often, the cell size is dictated by the current and its redistribution with time or frequency. These subdivisions are usually much smaller than the conventional $\lambda_{min}/20$ cell size. Hence, the necessary finer subdivisions for the conductors result in better delay or retardation models. This leads to a natural refinement of the delays. It is clear that these smaller VFI cells increase the compute time. However, adding a separate center-to-center delay to each small cell is not adding much to the compute time.

The result of utilizing a subdivided VFI model with delays is shown in Fig. 13.24. It is clear from this study that the stability or the real part of the input impedance is considerably improved by including the delay for each of the subdivisions in a VFI model. It is important to remember that the active region, for which the model is designed is 20 GHz, is well below the frequency of 350 GHz where the impedance exhibits a negative real part.

13.5.6 Physics-Based Skin-Effect Macromodel for Partial Elements

An important observation is that natural skin-effect losses in the PEEC model introduce high-frequency damping, which increases strongly with frequency. From our studies in Section 13.5.1, we recognize that losses at high frequencies are important for the passivity of distributed PEEC circuits.

In this section, we consider passivity enhancement using physics-based equivalent circuit skin-effect models. We show how to model for the self-inductance or impedance term

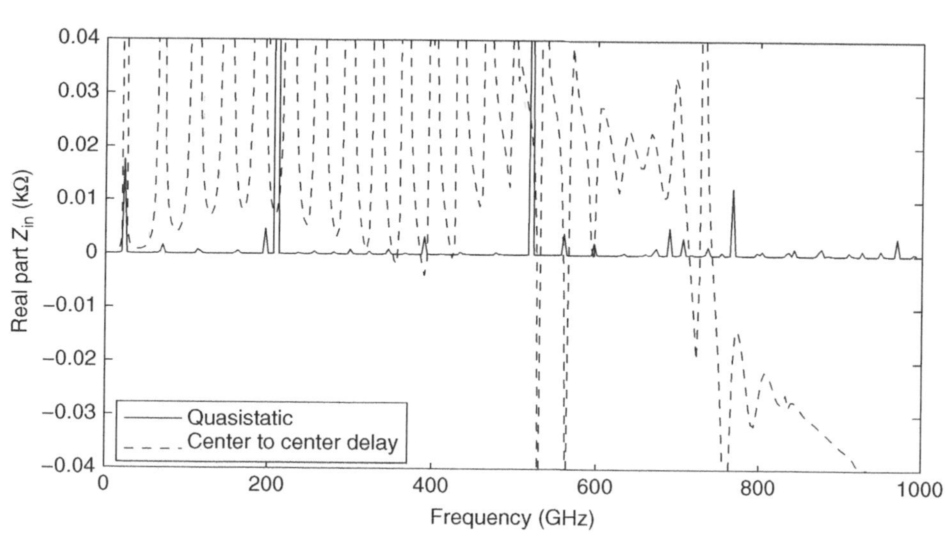

FIGURE 13.24 Response for VFI model with *Lp* center-to-center delays compared to quasistatic model.

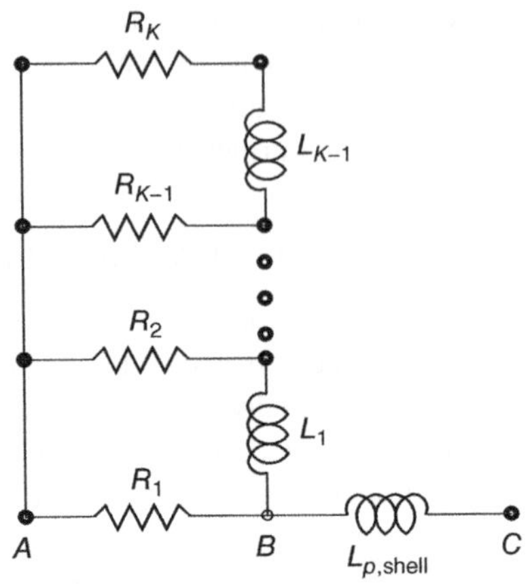

FIGURE 13.25 Skin-effect diffusion equivalent circuit for cylinder.

for skin-effect equivalent circuits. These losses also can enhance passivity. We use the round- wire geometry from Section 9.2.6 as an example where the diameter of the cylinder is much smaller than λ_{min}.

An analytical and a numerical solution for the skin effect can be used for the physics-based macromodel [50]. This skin-effect model is presented in Section 9.2.4. The equivalent circuit in Fig. 13.25 is a copy from Section 9.2.6.

So far, this model does not include the real part for the retardation term, which further increases losses especially for the self- inductance term. Cell sizes in the VFI skin-effect model are dictated by either the subdivisions for the current flow or frequency-dependent requirements. As considered in the previous section, smaller cells can lead to larger compute time requirements.

The inclusion of a good loss model does help to solve the high-frequency stability problem if we also take the retardation inside a conventional self-cell into account. As usual, the current flow in the partial self-inductance is predetermined. The partial inductance (5.64) computed with the retardation is again given by

$$Lp_{12}^R = \frac{1}{\mathcal{A}_1\,\mathcal{A}_2}\frac{\mu}{4\pi}\int_{\mathcal{V}_1}\int_{\mathcal{V}_2}\frac{e^{-j\beta R}}{R}\,d\mathcal{V}_1\,d\mathcal{V}_2. \tag{13.27}$$

Since the length of the wire segment is $\lambda/20$ or shorter at the largest frequency, we can expand the exponential in a series, or

$$e^{-j\beta R} = 1 - j\beta R + \frac{\beta^2 R^2}{2} - \cdots, \tag{13.28}$$

which is a well-known approach. We can view the third terms in (13.28) as an error term by using the first two terms in (13.27) to get

$$Lp_{11}^R = Lp_{11} - \frac{j\beta\mu}{4\pi}|z_e - z_s|^2. \tag{13.29}$$

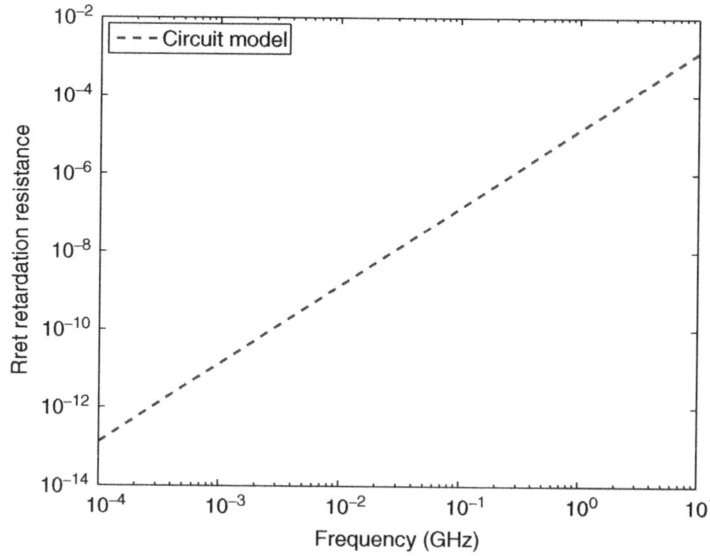

FIGURE 13.26 Loss resistance R_s for small cylinder conductor example in kiloohms.

Since the inductive impedance is given by $j\omega Lp^R_{11}$ and $\beta = \omega/v$, where v is the velocity in the material, we have

$$j\omega Lp^R_{11} = j\omega Lp_{11} + \frac{\omega^2\mu}{4\pi v}|z_e - z_s|^2, \tag{13.30}$$

where Lp_{11} is the QS partial inductance. Clearly, the second part is a frequency-dependent local resistance R_s due to the retardation is given by

$$R_s(\omega) = \frac{\omega^2\mu}{4\pi v}|z_e - z_s|^2. \tag{13.31}$$

The quadratic increase in the loss resistance is shown in Fig. 13.26 and the equivalent circuit, which includes the resistance, is given in Fig. 13.27. It is clear that R_s dominates at high frequencies. In fact, since the inductances will be open circuits at high frequencies, the very high frequency impedance will be $Z = R_1 + R_s$, which is resistive for this model. The real part for the combined equivalent circuit is shown in Fig. 13.28. The result compares well with the conventional FW partial inductance.

Finally, it is important that the inductance of the equivalent circuit shown in Fig. 13.29 remains small. This further shows that the impedance of the inductor model will be resistive, which is a more realistic high-frequency behavior. This supports the idea that the inclusion of skin-effect losses is important for stable and passive models. However, these equivalent circuit models are more problem-specific than the general skin-effect models such as the VFI model.

13.5.7 Mutual Coupling Inductance Terms with Retardation

An interesting question is how we can include the behavior of the circuit shown in the previous section into the partial mutual inductive couplings. It is clear that the large increase

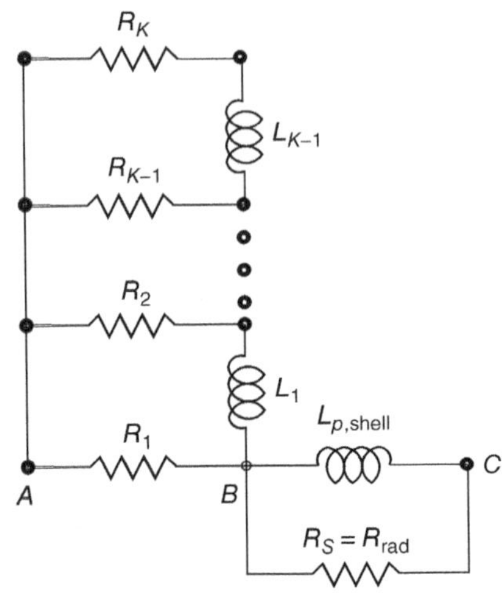

FIGURE 13.27 Modification of equivalent circuit to include the loss resistance R_s for self-term.

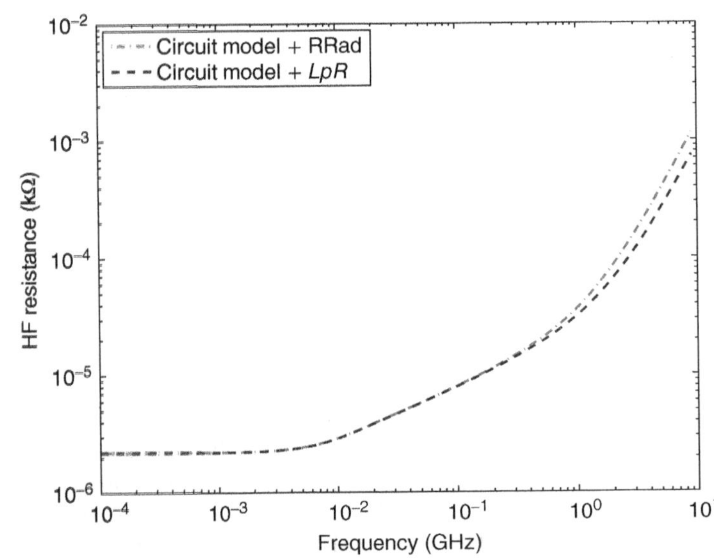

FIGURE 13.28 Total real part in kiloohms for high frequencies.

in the resistive part shown in Fig. (13.28) is very desirable from a stability and passivity point of view. This issue is particularly important for closely located coupling elements due to their larger impact.

In this section, we consider a specifice example based on the derivation in Section 5.8.1 for the partial inductances of coupling wire which are in line with retardation. The geometry for this case is again shown in Fig. 13.30. We are interested in an analytic solution for the

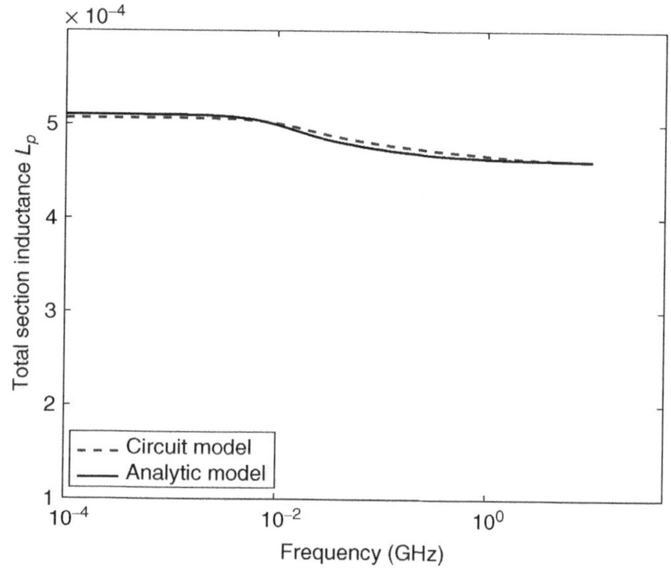

FIGURE 13.29 Inductance for equivalent circuit which includes R_c.

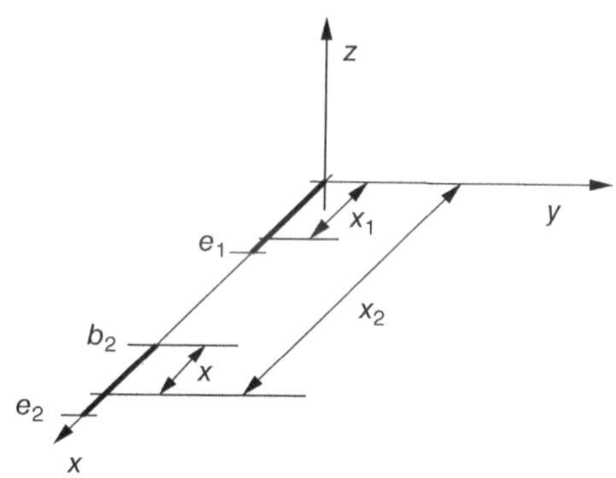

FIGURE 13.30 Two tube or wire segments in series for coupling calculations.

long objects to save evaluation time. Both wires are less than $\lambda/20$ in length for the highest frequency considered.

Starting from Section 5.8.1, we start by adding and subtracting $e^{-j\beta b_2}Lp_{12}$ to get

$$Lp_{12}^{R} = e^{-j\beta * b_2}Lp_{12} - e^{-j+\beta * b_2}Lp_{12} + Lp_{12}^{RM}. \tag{13.32}$$

where Lp_{12}^{RM} is given in (5.67). It is based on inductance for the case – where the delay is taken into account only external to the integral – as opposed to the case where the retardation is included in the partial element. This could be viewed as a *correction factor* for the external delayed partial inductance which is $e^{-j\beta b_2}Lp_{12}$.

Computing based on (5.67) we can evaluate the difference in (13.32) as

$$e^{-j\beta * b_2} Lp_{12}^{M} = e^{-j\beta * b_2} Lp_{12} - Lp_{12}^{RM} = \frac{\mu e^{-j\beta * b2}}{4\pi} \int_{x_1=0}^{e_1} \int_{x=b_2}^{e_2} \frac{1 - e^{-j\beta(x-x_1)}}{|b_2 + x - x_1|} dx dx_1 \quad (13.33)$$

which corresponds to the part within the brackets in (13.32). This process removes the common retardation factor $e^{-j\beta b_2}$, which can be very large, depending on the conductor distance. We re-iterate that the the segment size is such that each cell length, $e_2 - s_2$ and e_1 is less than $\lambda/20$. This allows us to evaluate the integrals analytically with the usual approximation of the exponent (13.28) as

$$\begin{aligned} Re(Lp_{12}^{M}) = & \frac{\mu e_1 (e_2 - b_2)\beta^2}{96\pi b_2^3} [7b_2^4 + 10 e_1 b_2^2 - 17 b_2^3 e_2 + \\ & +13 b_2^2 e_2^2 + 6 e_1 b_2 e_2^2 + 12 e_1 b_2^3 + 3 e_1^3 b_2 + \\ & -6 e_1^2 b_2 e_2 - 18 e_1 b_2^2 e_2 - 3 b_2 e_2^3] \end{aligned} \quad (13.34)$$

and the imaginary part is

$$\begin{aligned} Im(Lp_{12}^{M}) = & \frac{\mu e_1 (e_2 - b_2)}{96\pi b_2^3} [\beta(-12 e_1 e_2^2 + 6 e_2^3 - 26 b_2^3 + \\ & +36 e1 e_2 b_2 - 6 e_1^3 - 26 e_2^2 b_2 + 46 e_2 b_2^2 + \\ & -20 e_1^2 b_2 - 36 e_1 b_2^2 + 12 e_1^2 e_2) + \\ & +\beta^3 (2 e_1 b_2^4 + 2 e_1^2 b_2^3 - 3 b_2^4 e_2 + e1^3 b_2^2 + \\ & -b_2^2 e_2^3 + 3 b_2^3 e_2^2 - 2 e_1^2 b_2^2 e_2 + b_2^5 + \\ & +2 e_1 b_2^2 e_2^2 - 4 e_1 b_2^3 e_2)]. \end{aligned} \quad (13.35)$$

where the physical parameters are specified in Fig. 13.30. Finally, the result for the entire coupling factor will be

$$Lp_{12}^{R} = e^{-j\beta b_2} [Lp_{12} - Re(Lp_{12}^{M}) - j Im(Lp_{12}^{M})] \quad (13.36)$$

which results in an additional contribution to the quasi-static partial inductance computation.

Next, we give an example for the additional contribution of the magnitude of Lp_{12}^{M} in comparison to the quasi-static impedance ωLp_{12}. In this comparison, we ignore the impact of the external retardation since it represents the same multiplier in both cases. In the example, we chose $f = 10GHz$, where each wire is $\lambda/20 = 1.499mm$ and a distance of $(s_2 - e_1) = 29.979mm$. For this example, the partial quasi-static impedance is $\omega Lp_{12} = 4.48710^{-4}k\Omega$, and $Im(Lp_{12}^{M}) = -5.43810^{-4}k\Omega$ and $Re(Lp_{12}^{M}) = 5.45610^{-7} \mu H$.

We can conclude that the contribution of the retardation resistance dominates again in comparison to the contribution to the change in inductance with frequency. It is clear that these physics-based models for the inclusion of losses represent a very desirable approach where the losses can be used to dampen the high frequency response as well as representing the skin-effect loss more accurately.

13.6 TIME DOMAIN SOLVER ISSUES FOR STABILITY AND PASSIVITY

The numerical time domain integration method can contribute to the time domain stability and passivity of the solution. It can lead to instabilities even if the PEEC circuit model part is passive. On the other hand, the integration method can add some damping to the time solution. Numerical integration methods used for the time domain solution of the problems at hand are introduced in Section 2.6.

It is evident from the analysis as discussed in Section 13.3.3 that the time domain solution can immediately reveal the instability and, of course, nonpassivity of the solution. It is found in this chapter that the very high frequency response above the relevant frequencies can be responsible for instabilities. A fast time domain input pulse may contain very high frequencies well above f_{max}, the highest input frequency content of interest. We should note that even a minute high-frequency input signal may suffice to excite the resonances.

In this section, we want to identify how the numerical integration methods can contribute to the instability or stability of the time solution. Early on, the time domain solution was done with explicit numerical time integration methods [1–3] as is the case for most FDTD methods, for example, Ref. [51]. Explicit methods have the advantage that the solution of matrix systems is avoided. However, it is well known that the time step size for the region of stability needs to be very small. This step size limit for the explicit FDTD method is called the *Courant condition*. These issues are considered in the following section.

13.6.1 Impact of Time Integration on Stability

The numerical solution for the resultant differential equations for PEEC models is given in Section 2.6.1. It is essential to understand the impact of integration methods on the behavior of the solution. Fortunately, numerous works have been conducted on numerical integration methods, for example, Refs [52–55].

Originally, explicit integration methods such as the explicit forward Euler (FE) and higher order methods were used for IE-EM (TDIE) solvers. Interestingly, SPICE-type circuit solvers evolved in a similar way. Early SPICE-type circuit solvers before about 1970 also started with explicit integration methods. Today, implicit methods such as back Euler (BE) and higher order methods are used in SPICE and most time domain PEEC solvers.

To understand the issue, we need to consider the most simple differential equation with a single pole p, such as (2.7.1), as

$$\dot{x} = f(x) = p\,x \tag{13.37}$$

with the stable analytic solution $x = \exp(p\,t)$ where the pole p has a negative real part in the left-hand half s-plane.

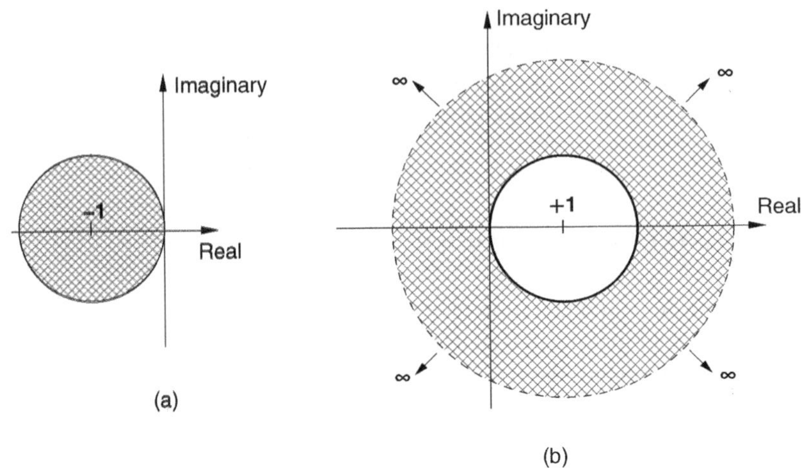

FIGURE 13.31 Absolute stability regions for (a) FE and (b) BE methods.

The FE numerical integration method given in Chapter 2 is

$$\frac{x_p - x_{p-1}}{\Delta t} = f(x_{p-1}) = p\, x_{p-1}, \tag{13.38}$$

or

$$x_p = (1 + \Delta t\, p)\, x_{p-1}, \tag{13.39}$$

where the contraction or attenuation factor needs to be $|1 + \Delta t\, p| \leq 1$. For the general case, we assume that the pole is located anywhere in the complex s-plane where $p = p_r + j\, p_i$. Hence, we can rewrite the attenuation factor as

$$(1 + \Delta t\, p_r)^2 + (\Delta t\, p_i)^2 \leq 1. \tag{13.40}$$

This equation represents a stability region that is a circle of radius 1 centered on the real axis at -1. This region of stability is shown in Fig. 13.31a. We clearly can see that the time step size Δt has to be limited especially for the high frequency poles such that $\Delta t\, p$ is small enough.

This small time step size Δt leads to other issues since the highest frequency f transmitted by a discrete time domain signal is given by $f = 1/(2\, \Delta t)$. Hence, the small time step also leads to more potential frequencies in the input signal and also higher frequency noise. It is clear from this study that explicit numerical integration methods are not well suited for an (FW)PEEC solution.

In PEEC, we used implicit integration methods from the start since the experience with circuit solvers was similar where a transition from explicit to implicit integration methods took place. To substantiate this issue, we apply the most simple (implicit) BE numerical integration method to the above problem. Applying the BE method to (13.37) leads to

$$\frac{x_p - x_{p-1}}{\Delta t} = f(x_p) = p\, x_p, \tag{13.41}$$

or

$$x_p = (1 - \Delta t\, p)^{-1}\, x_{p-1}, \tag{13.42}$$

where it is clear that the term $(1 - \Delta t\, p)^{-1}$ would require a matrix solution for a system of equations. Hence, the implicit methods require the solution of a system of equations. If we again consider the attenuation factor from x_{p-1} to x_p which is in this case given by

$$|(1 - \Delta t\, p_r)^2 + \Delta t\, p_i{}^2| \leq 1, \tag{13.43}$$

where we considered a complex pole p similar to (13.40). It is customary to define the function in (13.42)

$$R(\Delta t\, p) = \frac{1}{|1 - \Delta t\, p|} \leq 1, \tag{13.44}$$

the *stability function* such that $x_p = R(\Delta t\, p)\, x_{p-1}$. From (13.44), it is evident that the region of stability of BE is as shown in Fig. 13.31b.

We should note that a stable solution does not imply that the solution would not become unstable even for some poles in the right-hand half-plane. Further, we observe that Δt can assume any positive value and for stable left-hand-side poles. Hence, we are not restricted to small time steps size as is the case for the forward or explicit Euler method.

An important property of an integration method is that they are stable in the entire left-hand-side plane. Then stable left-hand-side poles lead to a stable solution as expected. Such methods are called A-stable [55], defined as follows:

Definition 10 ***(A-stability)*** *A numerical method is said to be A-stable or Absolute stable if its region of absolute stability contains the entire left-hand half-plane* $\mathrm{Re}(\Delta t\, p) < 0$ *where* $\Delta t > 0$.

It is clear that A-stability is the important property of an integration method for an EM time domain solver.

13.6.2 Impact of Numerical Damping of Integration Method

For a FW problem with delay, like a PEEC solution, it is desirable to use so-called *one-step* integration methods due to the discontinuities at the delay points. An interesting method that is related to the implicit Euler formula is the θ-method given in Table 2.5.

In Chapter 2, we considered another important issue that is the *n*umerical damping of a numerical integration method. Essentially, an oscillatory solution of an integration method may be damped by the integration method, especially at very high frequencies. Of course, the numerical damping could impact the accuracy of an oscillatory solution.

Exactly, the opposite issue may be important for the time domain stability and passivity of the (FW)PEEC solution, which includes the integration method. It is evident from the previous work that the frequencies that are larger than $10 f_{max}$ do not contain relevant information about the actual solution. Hence, the stability can be further improved if the time integration method does exhibit a low-pass behavior.

It is clear that we want the integration method to help to suppress very high frequencies if this is possible. This is represented by the stability function (13.44), which is given in the following definition [53]:

Definition 11 ***(L-stability)*** *A method is called* ***L-stable*** *if it is A-stable and in addition* $\lim_{(\Delta t\, p)\to\infty} R(\Delta t\, p) = 0$.

L-stability of an integration method is very desirable for time domain PEEC solvers since it helps to filter out high frequencies. A practical point of view is that any decrease in the high-frequency response sufficiently above f_{max} is helpful. This is accomplished by increasing the time step size Δt for the implicit integration method since the numerical damping increases with the increased time step size.

13.6.3 Digital Waveform Filtering

Numerical solutions are necessarily performed as discrete step systems in both domains. In the time domain, digital filters can be used to impact high-frequency responses at different levels of the solution. We chose a set of discrete equally spaced time points $p = 0, 1, 2, \ldots, P-1, P$ are specified where the *present* time $t_p = p\ \Delta t$ and $T_{max} = P\ \Delta t$. Here P is the total number of time points.

We observe that the representation of a waveform $x(t)$ represented in a discrete form by $x(t_p) = x_p$ is frequency limited. If we try to approximate a highest frequency f_M in a system that can be represented by a discrete sinusoidal waveform approximation using the above points, we find that

$$f_M = \frac{1}{2\ \Delta t}. \tag{13.45}$$

Hence, we need to choose Δt small enough such that the highest frequency limit f_M for the response can be transmitted. As an example, a time step of $\Delta t = 0.005$ ns corresponds to a largest frequency of $f_l = 1/(2\ \Delta t) = 100$ GHz.

Digital filtering techniques were employed in several works for stabilization. Early on, these techniques were not known as digital filters, for example, Refs [1–4]. At that time, these techniques were used in conjunction with explicit integration techniques.

In general, a digital filter can be represented as

$$a_1\, y_p + a_2\, y_{p-1} + a_3\, y_{p-3} + \cdots = b_1\, x_p + b_2\, x_{p-1} + b_3\, x_{p-3} + \cdots\,. \tag{13.46}$$

If $y_{p-1}, y_{p-2}, \ldots$ are all zero, then we say that they do not have feedback. Filters without the feedback are called *nonrecursive FIR* systems, whereas the digital filters with feedback are called *recursive IIR*.

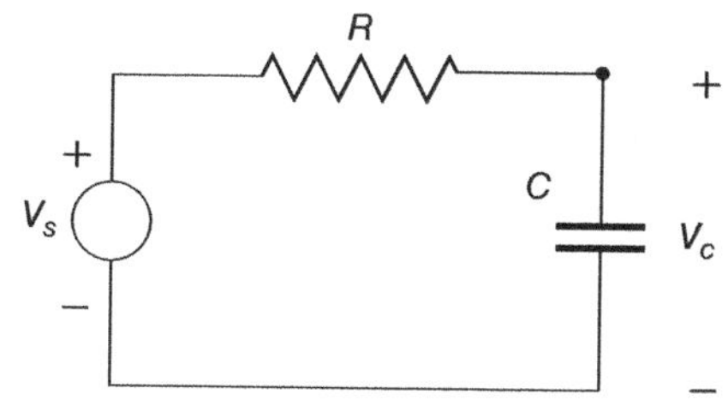

FIGURE 13.32 Low-pass circuit used for digital filter.

To enhance stability and passivity, the digital filter needs to have a low-pass behavior. We give a specific small example of a low-pass digital filter based on the circuit representation shown in Fig. 13.32 with the circuit equations

$$-V_s(t) + RC\frac{dV_c}{dt} + V_c = 0 \tag{13.47}$$

and by using the BE numerical integration formula, which is also called BD1 in Table 2.5, it can be transformed into a digital filter type equation of the form

$$\left[1 + \frac{RC}{\Delta t}\right] V_{c,\,p} = V_{s,\,p} + \frac{RC}{\Delta t} V_{c,\,p-1}, \tag{13.48}$$

where we can adjust the cutoff frequency by the choice of the value of $R\,C$. It is clear that we can insert such a filter in many different locations of the PEEC equivalent circuit for time domain stability and passivity enhancements.

The implementation of a filter requires convolution in the time domain since the multiplication in the frequency domain requires convolution in the time domain.

We note from (13.46) that the feedback *IIR* filter in (13.48) can be described as

$$a_1 y_p + a_2 y_{p-1} = b_1 x_p \tag{13.49}$$

results in a recursive transfer function where $a_1 = (1 + RC/T_s)$ and where the sampling time interval $T_s = \Delta t$. The other constants can be determined by comparing with (13.50).

Convolution is required to include the filter into the time solution. Specifically, the transfer for function y_p is used with the input x_p as

$$v_{c,\,p} = -a_2/a_1 v_{c,\,p-1} + b_1/a_1 v_{s,\,p}. \tag{13.50}$$

We should note that the digital filters can add delays to the waveform. Importantly, this delay has to be taken into account if it is applied for the modeling of delayed partial elements.

Published results are available on the use of digital filters in the early works [1–4]. Also, FIR digital filters have been used in Ref. [56].

ACKNOWLEDGMENT

A. Ruehli acknowledges the insightful discussions with M. Nakhla, Carleton University, on the subject of this chapter.

PROBLEMS

13.1 *Impedance of two-loop PEEC model*
Assume that the equivalent circuit model in Fig. 6.8 in Section 6.3.7 consists of two loops by adding another node and partial inductance Lp_{22} and a capacitive model with Pp_{33}. This configuration was studied in Ref. [5] for the zero thickness conductor in Fig. 13.33. The values of the partial elements are given in Table 13.1. Write the frequency domain MNA solution for the small PEEC circuit. Check the real part of the input impedance at the current source for the various combinations of the delays specified in Table 13.1. See if you can determine a negative real part in the impedance for some delays. The delays are obtained by moving the delay points on the respective cells. Use the normalized microelectronics units in Appendix A for all variables.

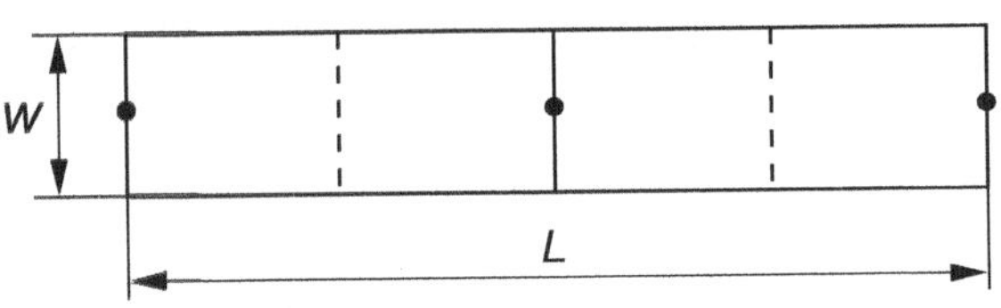

FIGURE 13.33 Example of two-cell zero thickness sheet.

TABLE 13.1 Values of partial elements for example problem.

Partial elements	Values	Minimum delay (ns)	Nominal delay	Maximum delay (ns)
Lp_{11}	0.022362 μH			
Lp_{12}	0.006314 μH	0.0	1/6	1/3
$Pp_{11} = P_{33}$	1.19143 1/pF			
Pp_{22}	0.80392 1/pF			
$Pp_{12} = P_{21}$	0.300756 1/pF	0.0	1/6	1/4
Pp_{13}	0.121378 1/pF	1/6	1/3	1/2

13.2 *Stability of the two-loop PEEC model*
Use the circuit from the first problem and perform a time domain solution rather than a frequency domain solution. Use triangular input current pulse with the same rise t_r and fall t_f times. Use the BD2 and the θ methods to solve the circuit from Table 2.5 check if the waveform is stable for different values of t_r. (Hint: Use smaller subnanosecond values for $t_r = t_f$.)

13.3 *Numerical damping due to integration methods*
To illustrate the impact of the numerical integration method on the stability and numerical damping, we solve the following small problem. We set up the circuit equations for the equivalent circuit in Fig. 13.34 where $L = 0.001$ μH and $C = 1$ pF. Choose the triangular pulse with an amplitude of 10 mA such that the duration is less than the period of oscillation. This circuit is clearly lossless and it should oscillate without damping. Choose the time step for the back Euler (BE) method in Table 2.5 such that you can observe the numerical damping of the integration method. Other integration methods result in less damping. The Theta method is a one-step method which is suitable for time integration for PEEC models with delay. The method is also listed in Table 2.5. Test the method with $\theta = 0.675$. We should note that unstable methods lead to amplification or instability rather than damping even if the PEEC model is stable.

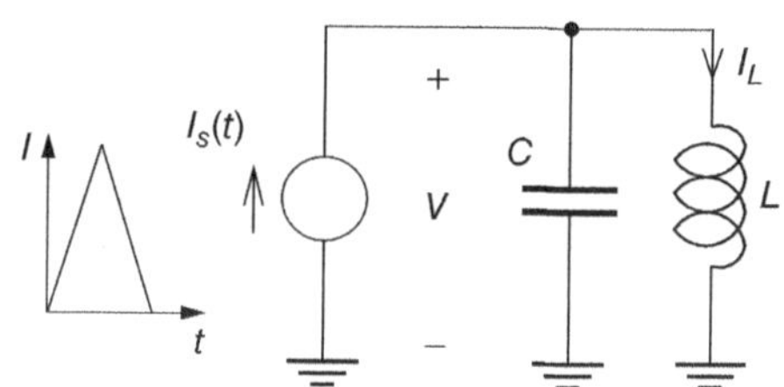

FIGURE 13.34 Test circuit for numerical damping of integration method.

13.4 *Digital filter*
We assume that the time step for a problem is x ps. Construct a digital filter as described in Section 13.6.3 with a cutoff frequency of 50 GHz. Test which step of size x ps is needed for a 10 ps ramp input waveform to get a good accuracy for the waveforms.

REFERENCES

1. B. P. Rynne and P. D. Smith. Stability of time marching algorithms for the electric field integral equation. *Journal of Electromagnetic Waves and Applications*, 4(12):1181–1205, January 1990.
2. P. D. Smith. Instabilities in time marching methods for scattering: cause and rectification. *Electromagnetics*, 10(4):439–451, October 1990.
3. B. P. Rynne. Time domain scattering from arbitrary surfaces using the electric field integral equation. *Journal of Electromagnetic Waves and Applications*, 5(1):93–112, 1991.
4. B. P. Rynne. Comments on a stable procedure in calculating the transient scattering by conducting surfaces of arbitrary shape. *IEEE Transactions on Antennas and Propagation*, APP-41(4):517–520, April 1993.
5. A. E. Ruehli, U. Miekkala, and H. Heeb. Stability of discretized partial element equivalent EFIE circuit models. *IEEE Transactions on Antennas and Propagation*, 43(6):553–559, June 1995.
6. A. Bellen, N. Guglielmi, and A. Ruehli. Methods for linear systems of circuit delay differential equations of neutral type. *IEEE Transactions on Circuits and Systems*, 46(1):212–216, January 1999.

7. J. Nitsch, F. Gronnwald, and G. Wollenberg. *Radiating Nonuniform Transmission-Line Systems and the Partial Element Equivalent Circuit Method*. John Wiley and Sons, Inc., New York, 2009.
8. G. Antonini and P. Pepe. Input-to-state stability analysis of partial element equivalent circuit models. *IEEE Transactions on Circuits and Systems I*, 56(3):673–684, March 2009.
9. B. H. Jung, T. K. Sarkar, S. W. Ting, Y. Zhang, Z. Mei, Z. Ji, M. Yuan, A. De, M. Salazar-Palma, and S. M. Rao. *Time and Frequency Domain Solutions of EM Problems Using Integral Equations and a Hybrid Methodology*. John Wiley and Sons, Inc., New York, 2010.
10. L. W. Nagel. SPICE: a computer program to simulate semiconductor circuits. Electronics Research Laboratory Report, ERL M520, University of California, Berkeley, CA, May 1975.
11. V. Vlasimirescu. *The SPICE Book*. Wiley, Piscataway, NJ, 1994.
12. P. Triverio, S. Grivet-Talocia, M. S. Nakhla, F. G. Canavero, and R. Achar. Stability, causality, and passivity in electrical interconnect models. *IEEE Transactions on Advanced Packaging*, 30(4):795–808, November 2007.
13. A. E. Ruehli and A. C. Cangellaris. An overview of computer modeling for large EMC chip and package problems. In *Proceedings International Zurich Symposium on EMC*, Volume 14, Zürich, Switzerland, pp. 253–258, February 2001.
14. C. Ho, A. E. Ruehli, and P. Brennan. Interactive circuit analysis and design using APL. In *Proceedings of the IEEE International Symposium on Circuits and Systems*, May 1975.
15. A. Semlyen and B. Gustavsen. A half-size singularity test matrix for fast and reliable passivity assessment of rational models. *IEEE Transactions on Power Delivery*, 24(1):345–451, January 2009.
16. A. E. Ruehli, G. Antonini, and L. Jiang. Passivization of EM PEEC solutions in the frequency and time domain. In *ICEAA International Conference on EM in Advanced Applications*, Volume 22, Torino, Italy, p. 266, September 2013.
17. G. Antonini, A. E. Ruehli, D. Romano, and L. Jiang. Tests for time domain EM solvers for stability and towards passivity. In *Digest of Electrical Performance of Electronic Packaging Systems*, Volume 22, San Jose, CA, pp. 263–266, October 2013.
18. B. Gustavsen and A. Semlyen. Rational approximation of frequency domain responses by vector fitting. *IEEE Transactions on Power Apparatus and Systems*, 14(3):1052–1061, July 1999.
19. B. Gustavsen. Enforcing passivity for admittance matrices approximated by rational functions. *IEEE Transactions on Power Delivery*, 16(1):97–104, February 2001.
20. S. Grivet-Talocia. An adaptive sampling technique for passivity characterization and enforcement of large interconnect macromodels. *IEEE Transactions on Advanced Packaging*, 30(2):226–236, May 2007.
21. E.-P. Li. *Electrical Modeling and Design for 3D System Integration*. John Wiley and Sons, Inc., New York, Hoboken, NJ, 2012.
22. M. Swaminathan and A. E. Engin. *Power Integrity Modeling and Design for Semiconductors and Systems*. Prentice Hall, Upper Saddle River, NJ, 2007.
23. J. Morsey and A. C. Cangellaris. Prime: passive reduction of interconnect models from measured data. In *Digest of Electrical Performance of Electronic Packaging*, Cambridge, MA, pp. 47–50, 2001.
24. L. Daniel and J. Phillips. Model order reduction for strictly passive and causal distributed systems. *Proceedings of the Design Automation Conference*, pp. 46–48, June 2002.
25. B. Gustavsen. Computer code for passivity enforcement of rational macromodels by residue perturbation. In *IEEE Transactions on Advanced Packaging*, Volume 30, pp. 209–215, May 2007.
26. P. Triverio and S. Grivet-Talocia. Robust causality characterization via generalized dispersion relation. *IEEE Transactions on Advanced Packaging*, 31(3):579–592, August 2008.
27. F. Ferranti, L. Knockaert, and T. Dhaene. Passivity-preserving parametric macromodeling by means of scaled and shifted state-space systems. *IEEE Transactions on Microwave Theory and Techniques*, 59(10):2394–2403, October 2011.

28. H. W. Bode. *Network Analysis and Feedback Amplifier Design*. Van Nostrand, Princeton, NJ, 1945.

29. R. W. Newcomb. *Linear Multiport Synthesis*. McGraw-Hill, New York, 1966.

30. K. S. Yee. Numerical solution of initial boundary value problems involving Maxwell's equations in isotropic media. *IEEE Transactions on Antennas and Propagation*, 14(5):302–307, May 1966.

31. T. Weiland. Eine Methode zur Loesung der Maxwellschen Gleichungen fuer sechskompoentige Felder auf diskreter Basis. *Archiv der Elektrischen Ubertragung*, 31:116–120, 1977.

32. J. M. Jin. *The Finite Element Method in Electromagnetics*. John Wiley and Sons, Inc., New York, 2nd edition, 2002.

33. Y. Zhu and A. Cangellaris. *Multigrid Finite Element Methods for Electromagnetic Field Modeling*. John Wiley and Sons, Inc., New York, 8th edition, 2006.

34. E. S. Kuh and R. A. Rohrer. *Theory of Linear Active Circuits*. Holden-Day, Inc., San Francisco, CA, 1968.

35. R. Rohrer and H. Nosrati. Passivity considerations in stability studies of numerical integration algorithms. *IEEE Transactions on Circuits and Systems*, 28(9):857–866, September 1981.

36. O. Brune. Synthesis of a finite two-terminal network whose driving-point impedance is a prescribed function of frequency. *Journal of Mathematical Physics*, 10(1–4):191–236, April 1931.

37. B. D. Anderson and S. Vongpanitlerd. *Network Analysis and Synthesis*. John Wiley and Sons, Inc., Englewood Cliffs, NJ, 1973.

38. A. Odabasioglu, M. Celik, and L. T. Pileggi. Prima: passive reduced-order interconnect macromodeling algorithm. *IEEE Transactions on Computer-Aided Design*, 17(8):645–654, August 1998.

39. R. Achar and M. S. Nakhla. Simulation of high-speed interconnects. *Proc IEEE*, 89(5):693–728, 2001.

40. M. Celik, L. Pileggi, and A. Odabasioglu. *IC Interconnect Analysis*. Kluwer Academic Publishers, Boston/Dordrecht/London, 2002.

41. U. M. Ascher and L. R. Petzold. *Computer Methods for Ordinary Differential Equations and Differential Algebraic Equations*. SIAM, Philadelphia, PA, 1998.

42. K. J. Kerns and A. T. Yang. Preservation of passivity during RLC network reduction via split congruence transformation. *IEEE Transactions on Computer-Aided Design*, 17(7):582–591, July 1998.

43. K. Ogata. *State Space Analysis of Control Systems*. Prentice-Hall, Inc: Englewood cliffs, NJ, 1967.

44. J. Garrett, A. Ruehli, and C. R. Paul. Accuracy and stability improvements of integral equation models using the partial element equivalent circuit PEEC approach. *IEEE Transactions on Antennas and Propagation*, 46(12):1824–1832, December 1998.

45. B. Gustavsen and A. Semlyen. Fast passivity assessment for S-parameter rational models via a half-size test matrix. *IEEE Transactions on Microwave Theory and Techniques*, 56(12):2701–2708, December 2008.

46. S. Chakraborty, V. Jandhyala, and J. Pingenot. Polar integration for exact space-time quadrature in time-domain integral equations. *IEEE Transactions on Antennas and Propagation*, 54(10):3037–3042, October 2006.

47. G. Antonini, D. Deschrijver, and T. Dhaene. Broadband rational macromodeling based on the adaptive frequency sampling algorithm and the partial element equivalent circuit method. *IEEE Transactions on Electromagnetic Compatibility*, 50(1):128–137, February 2008.

48. G. Wollenberg and A. Görisch. Analysis of 3-D interconnect structures with PEEC using SPICE. *IEEE Transactions on Electromagnetic Compatibility*, 41(2):412–417, November 1999.

49. W. Gander and W. Gautschi. Adaptive quadrature - revisited. *BIT Numerical Mathematics*, 40(1):84–101, March 2000.

50. A. E. Ruehli, G. Antonini, and L. Jiang. Skin-effect model for round wires in PEEC. In *IEEE EMC Europe, International Symposium on EMC*, Rome, Italy, September 2012.

51. A. Taflove Ed. *Computational Electrodynamics*. Artech House, Inc., Norwood, MA, 1998.

52. L. Lapidus and J. H. Seinfeld. *Real and Complex Analysis, Numerical Solution of Ordinary Differential Equations*. ACADPR, Ljubljana, 1971.

53. E. Hairer and G. Wanner. *Solving Ordinary Differential Equations II, Stiff and Differential Algebraic Problems*. Springer-Verlag, New York, 1991.

54. F. N. Najm. *Circuit Simulation*. John Wiley and Sons, Inc., New York, 2010.

55. W. Gander, M. Gander, and F. Kwok. *Scientific Computing: An Introduction Using Maple and MATLAB*. Springer-Verlag, Berlin and New York, 2014.

56. A. Sadigh and E. Arvas. Treating the instabilities in marching-on-in-time methods from a different perspective. *IEEE Transactions on Antennas and Propagation*, APP-41(12):1695–1702, December 1993.

57. S. Grivet-Talocia and B. Gustavsen. *Passive macromodeling: Theory and Applications*. John Wiley and Sons, Inc, Hoboken, New Jersey, 2016.

A

TABLE OF UNITS

A.1 COLLECTION OF VARIABLES AND CONSTANTS FOR DIFFERENT APPLICATIONS

We provide two tables, depending on the application. Fundamentally, we would like to use units that keep each element value close to 1 (unity). The purpose of this is to keep the element values in the matrices as close as possible. It is intuitively obvious that an element in a matrix of 10^{-20} is ignorable small compared to 1. Using the appropriate normalized units is a good way to keep the elements in the same orders of magnitude. Here, we give Tables A.1 and A.2, one for low-frequency power electronics and one for microelectronics and high- frequency circuits.

TABLE A.1 Important variable and constants in SI units.

Variable	Value	Units
Distance d, length ℓ	1	m
Voltage V	1	V
Current I	1	A
Volume current density J_{ev}	1	A/m^2
Surface current density J_{es}	1	A/m
Magnetic current I_m	1	V
Magnetic current density J_m	1	V/m^2
Surface current density J_{ms}	1	V/m
Power P	1	W
Frequency f	1	Hz
Time t	1	s
Resistance R	1	Ω

(*continued overleaf*)

Circuit Oriented Electromagnetic Modeling Using the PEEC Techniques, First Edition.
Albert E. Ruehli, Giulio Antonini, and Lijun Jiang.

TABLE A.1 (*Continued*)

Variable	Value	Units
Capacitance C	1	$F = s/\Omega$
Inductance L	1	$H = s\ \Omega$
Charge Q	1	A s
Magnetic field H	1	A/m
Magnetic flux density B	1	$V\ s/m^2$
Magnetic flux Ψ	1	V s
Electric scalar potential Φ	1	V
Magnetic scalar potential Φ_m	1	A
Magnetic vector potential A	1	V s/m
Electric vector potential F	1	A s/m
Vacuum permeability μ_0	$4\pi\ 10^{-7}$	$H/m = s\ \Omega/m$
Vacuum permittivity ε_0	$8.854\ 10^{-12}$	$F/m = s/(\Omega m)$
Speed of light in vacuum	$2.997925\ 10^{8}$	m/s
Conductivity of copper σ	$5.78\ 10^{7}$	$1/(\Omega\ m)$
Resistivity of copper ρ	$1.724\ 10^{-8}$	$\Omega\ m$
Skin depth for copper δ	$0.066/\sqrt{f}$	m
Conductivity of aluminium σ	$3.72\ 10^{7}$	$1/(\Omega\ m)$
Resistivity of aluminium ρ	$2.288\ 10^{-8}$	$\Omega\ m$
Skin depth for aluminium δ	$0.0826/\sqrt{f}$	m
Conductivity of brass σ	$1.57\ 10^{7}$	$1/(\Omega\ m)$
Resistivity of brass ρ	$6.37\ 10^{-8}$	$\Omega\ m$
Skin depth for brass δ	$0.127/\sqrt{f}$	m
Conductivity of lead-free solder σ	$6.7\ 10^{6}$	$1/(\Omega\ m)$
Resistivity for lead-free solder ρ	$0.15\ 10^{-6}$	$\Omega\ m$
Skin depth for lead free solder δ	$0.19/\sqrt{f}$	m

TABLE A.2 Important variable and constants in normalized microelectronics units.

Variable	Value	Units
Distance d, length ℓ	$1\ (10^{-3})$	mm
Voltage V	1	V
Current I	$1\ (10^{-3})$	mA
Power P	$1\ (10^{-3})$	mW = V mA
Frequency f	$1\ (10^{+9})$	GHz
Time t	$1\ (10^{-9})$	ns
Resistance R	$1\ (10^{+3})$	$k\Omega$
Capacitance C	$1\ (10^{-12})$	$pF = ns/k\Omega$
Inductance L	$1\ (10^{-6})$	$\mu H = ns\ k\Omega$
Charge Q	$1\ (10^{-12})$	pC = mA ns
Magnetic field H	1	mA/mm
Magnetic flux density B	$1\ (10^{-3})$	$V\ ns/mm^2 = ns\ \mu H/mm^2$
Magnetic flux Ψ	$1\ (10^{-9})$	V ns
Vacuum permeability μ_0	$4\pi\ 10^{-4}$	$\mu H/mm = ns\ k\Omega/mm$

(*continued overleaf*)

TABLE A.2 (*Continued*)

Variable	Value	Units
Vacuum permittivity ε_0	$8.854\ 10^{-3}$	pF/mm = ns/(kΩ mm)
Speed of light in vacuum	299.7925	mm/ns
Conductivity of copper σ	$5.78\ 10^{7}$	1/(kΩ mm)
Resistivity of copper ρ	$1.724\ 10^{-8}$	kΩ mm
Skin depth for copper δ	$0.0021/\sqrt{f}$	mm
Conductivity of aluminium σ	$3.72\ 10^{7}$	1/kΩ mm
Resistivity of aluminium ρ	$2.288\ 10^{-8}$	kΩ mm
Skin depth for aluminium δ	$0.0026/\sqrt{f}$	mm
Conductivity of brass σ	$1.57\ 10^{7}$	1/(kΩ mm)
Resistivity of brass ρ	$6.37\ 10^{-8}$	kΩ mm
Skin depth for brass δ	$0.004/\sqrt{f}$	mm
Conductivity of lead-free solder σ	$6.7\ 10^{6}$	1/(kΩ mm)
Resistivity for lead-free solder ρ	$0.15\ 10^{-6}$	kΩ mm
Skin depth for lead-free solder δ	$0.006/\sqrt{f}$	mm

B

MODIFIED NODAL ANALYSIS STAMPS

B.1 MODIFIED NODAL ANALYSIS MATRIX STAMPS

The purpose of this appendix is to give a summary of modified nodal analysis (MNA) stamps [1]. They show how to enter circuit elements into the MNA matrix. In this book, we concentrated on linear elements. For most circuit solvers, the linear part is an important subset of a program. Further, details and references on this subject can be found in Refs [1–5]. In general, we attempt to adhere to the basic SPICE language statements [6] as a command language for the overall solution approach. The fundamental MNA concepts are presented in Chapter 2. For the convenience of the reader, we duplicate the circuit element Table B.1. The stamps represent contribution of the circuit elements in the table to the MNA matrix.

To repeat the key issue, the element-by-element stamping concept is based on the incomplete Kirchhoff's circuit law in Section 2.3.4. Hence, for each node, we stamp a circuit element contribution at the time into the matrix. The Kirchhoff's current law (KCL) is satisfied at each node, as soon as all the elements that belong to the node are stamped into the circuit matrix. The stamps not only add the elements to the MNA matrix, but they also add the entries needed in to the connection matrices $\mathbf{A}$ and $\mathbf{A}^\mathbf{T}$. In general, if one of the nodes is connected to ground or reference node, then the contribution of this part in the stamp is eliminated. This usually results in the elimination of a row and a column in the element stamp.

B.1.1 Resistor

Two different stamps can be used for the resistor shown in Fig. B.1 depending on whether we would like to have the current as an output variable or not. An example of the SPICE statement for the resistor is

```
RName Node1Name Node2Name 5 kOhms
```

Circuit Oriented Electromagnetic Modeling Using the PEEC Techniques, First Edition.
Albert E. Ruehli, Giulio Antonini, and Lijun Jiang.

TABLE B.1 Small list of circuit elements for which we need MNA matrix stamps.

Circuit elements	Frequency domain	Time domain
R Resistance	$V(s) = R\,I(s)$	$v(t) = R\,i(t)$
C Capacitance	$I(s) = s\,CV(s)$	$i(t) = C\,dv(t)/dt$
L Inductance	$V(s) = sLI(s)$	$v(t) = L\,di(t)/dt$
Current controlled voltage source (CCVS)	$V(s) = h\,I(s)$	$v(t) = h\,i(t)$
Current controlled current source (CCCS)	$I(s) = f\,I(s)$	$v(t) = f\,i(t)$
Voltage controlled voltage source (VCVS)	$V(s) = e\,V(s)$	$v(t) = e\,v(t)$
Voltage controlled current source (VCCS)	$I(s) = g\,V(s)$	$i(t) = g\,v(t)$

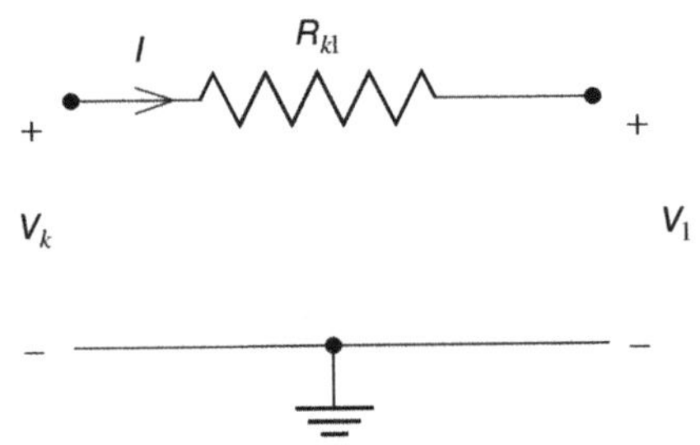

FIGURE B.1 Resistor connected to two nodes in circuit.

TABLE B.2 Matrix element stamp for resistor $R_{k\ell} = 1/G_{k\ell}$ for $R_{k\ell} \neq 0$.

Row/column	V_k	V_ℓ	RHS
k	$G_{k\ell}$	$-G_{k\ell}$	0
ℓ	$-G_{k\ell}$	$G_{k\ell}$	0

where the first character of the name needs to be R and $Node1Name = V_k$ and $Node2Name = V_\ell$ as an example for the notation used below. In our example, the resistance value is 5 kΩ. The first stamp is used if the current through the resistor is not needed as an unknown output. Also, we assume that the value of the resistor is never set to zero since the form of the equation is $I = G_{k\ell}(V_k - V_\ell)$ where $G_{k\ell} = 1/R_{k\ell}$. The resistor stamp is given in Table B.2.

The stamp where the current is specified as an unknown is given in Table B.3.

TABLE B.3 Matrix element stamp for resistor $R_{k\ell}$ with current.

Row/Column	V_k	V_ℓ	I	RHS
k			+1	0
ℓ			−1	0
Aux	1	−1	$-R_{k\ell}$	

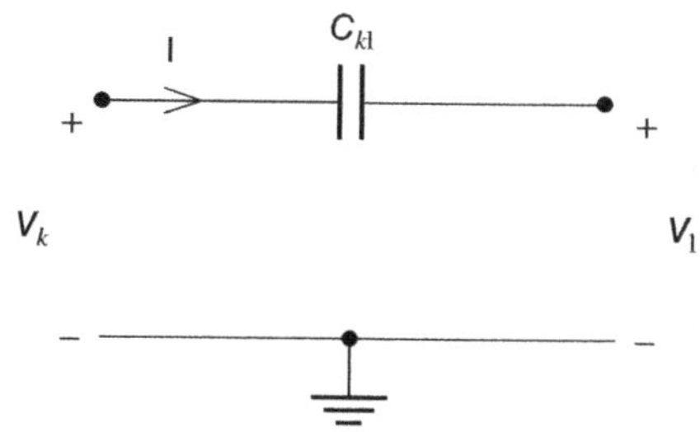

FIGURE B.2 Capacitance connected between two nodes.

TABLE B.4 Matrix stamp for capacitor in the frequency domain.

Row/Column	V_k	V_l	RHS
k	$s\,C_{kl}$	$-s\,C_{kl}$	0
l	$-s\,C_{kl}$	$s\,C_{kl}$	0

TABLE B.5 Matrix stamp for capacitor in the time domain.

Row/Column	V_k	V_l	RHS
k	$k_p\,C_{kl}$	$-k_p\,C_{kl}$	$C_{kl}\,k_{p-1}(-V_k^{p-1} + V_l^{p-1})$
l	$-k_p\,C_{kl}$	$k_p\,C_{kl}$	$C_{kl}\,k_{p-1}(-V_l^{p-1} + V_k^{p-1})$

Note that in the stamp that includes the unknown current in Table B.3, we use $R_{k\ell}$ rather than $G_{k\ell}$. Then, the entry does not lead to a singular circuit matrix for the case if $R_{k\ell} = 0$. The additional equation added to the stamp is called Aux.

B.1.2 Capacitor

Next, we consider a capacitor for which we need element stamps for the time as well as the frequency domain. An example for the use of a capacitance stamp is given in Section 2.7. The capacitor circuit element connected between two nodes k and l is shown in Fig. B.2. The frequency domain element stamp is closely related to the resistance stamp (Table B.4). To obtain the capacitance stamp we can replace the conductance $G = 1/R$ with the admittance $Y = sC$, which leads to the matrix stamp for capacitor as shown in Table B.4.

The details for the time domain formulation are given in Section 2.4. The time domain element stamp for a capacitor is shown in Table B.5.

B.1.3 Independent Voltage Source

Independent voltage sources are very often used as input sources in circuits. The general case for such a source is shown in Fig. B.3.

An interesting aspect is that the basic stamp is the same for the frequency and the time domains. Only the values of the source itself represent if it is a time or frequency domain source or both. We notice that the stamp automatically has the current of the source as an unknown (Table B.6).

B.1.4 Independent Voltage Source with Series Elements

In many cases, voltage sources also include a series resistance or an impedance that represents a typical source. For this case, the element stamp for the source is augmented with an impedance $Z(s)$ that can consist of several elements in the frequency domain. Hence, examples for Z are simple R or an $R - L$ series connection circuit models. Of course, some elements can also be included in the time domain. The interesting aspect is the fact that a single stamp can be used for the combination. The equivalent circuit is given in Fig. B.4.

The stamp for this is given in the frequency domain, which is similar to the zero impedance source (Table B.7).

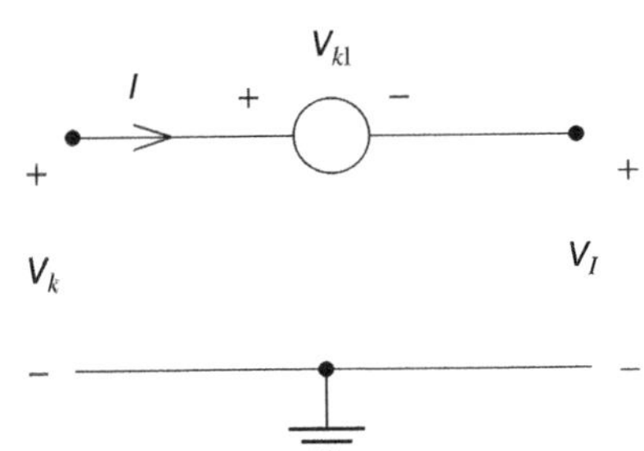

FIGURE B.3 Independent voltage source connected between two nodes.

TABLE B.6 Matrix element stamp for voltage source.

Row/column	V_k	V_l	I	RHS
k			1	
l			−1	
Aux	1	−1	0	V_{kl}

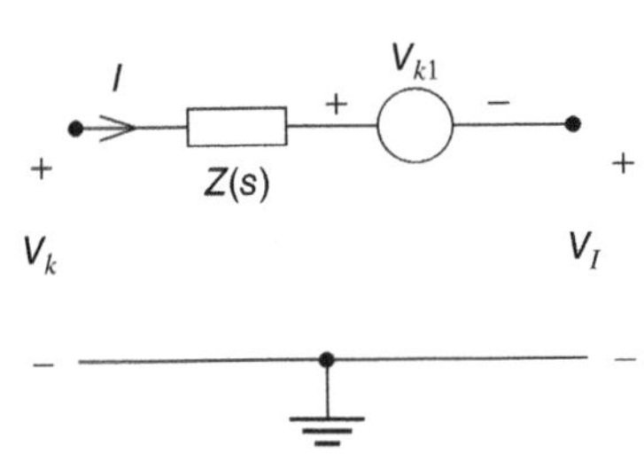

FIGURE B.4 Voltage source connected between one or two nodes.

TABLE B.7 Matrix element stamp for voltage source.

Row/Column	V_k	V_l	I	RHS
k			1	0
l			-1	0
Aux	1	-1	$-Z(s)$	V_{kl}

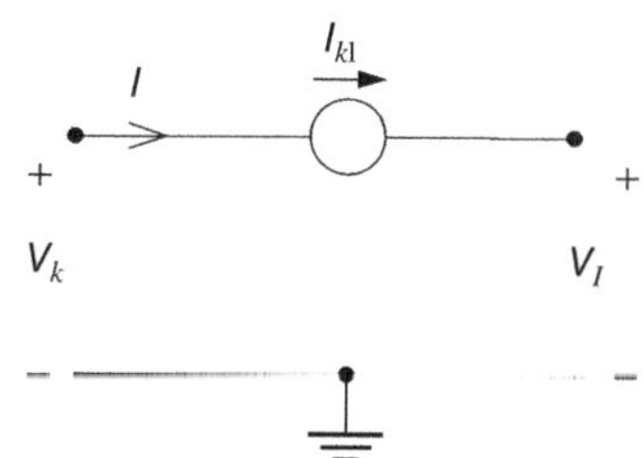

FIGURE B.5 Independent current source connected to two nodes.

TABLE B.8 Matrix stamp for current source.

Row/column	V_k	V_ℓ	RHS
k			$-I_{k\ell}$
ℓ			$I_{k\ell}$

B.1.5 Independent Current Source

Independent current sources also frequently serve as input sources as shown in Fig. B.5. A short circuit is also used as an element that does not alter the circuit's performance, but it is also frequently used as a way to obtain the output of an additional current that does not appear in the conventional output vector.

The entry into the right-hand side for the matrix system is given for the case when both nodes of the source are part of the circuit. One of the two entries is eliminated for the case when one node of the source is connected to ground. We note that this stamp is entered into the RHS. As we can see, there are no entries into the MNA matrix for a current source (Table B.8).

B.1.6 Short Circuit Connection

For some applications, short circuits are required. An interesting application is the connection of different subparts from different model parts. Figure B.6 shows the equivalent circuit for the short if we set the source voltage to zero. This shows that the voltage source model (Table B.3) can also be used for a short (Table B.9).

As we can see, the stamp for a short also introduces the current through the short. Hence, it also can be used to determine the current through a wire without introducing a series impedance. Hence, it will not change the circuit behavior.

It is clear that the stamp for the short is the same as a voltage source with zero volts.

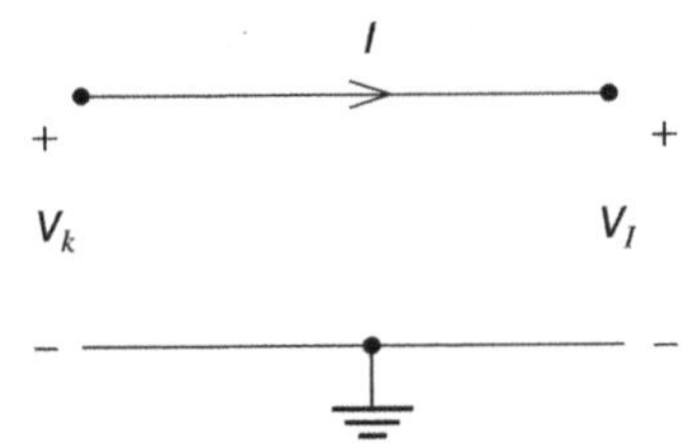

FIGURE B.6 Short circuit connected between two nodes.

TABLE B.9 Matrix stamp for short circuit.

Row/column	V_k	V_l	I	RHS
k			1	0
l			−1	0
Aux	1	−1	0	0

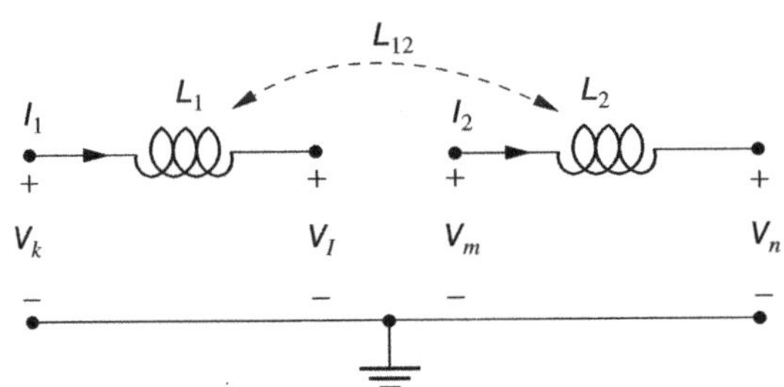

FIGURE B.7 Equivalent circuit for coupled inductances.

B.1.7 Coupled Inductances

For most PEEC circuits, inductors are of fundamental importance. Of equal importance is the inclusion of mutual inductances as is evident from Chapter 5. We consider two cases. One is conventional inductances. The second case, with the same type of stamp, includes unconventional mutual inductances that include delays such that (FW)PEEC models can be represented.

The SPICE statements for mutual inductances, unfortunately, involve a coupling factor as an inheritance from the transformer equations for two inductors shown in Fig. B.7. An example for the SPICE representation is given by:

```
Lp11 Node1Name Node2Name 0.001uH
Lp22 Node3Name Node4Name 0.001uH
KL12 Lp11 Lp22 0.76
```

where K represents the coupling factor, which is given by $K = Lp12/\sqrt{(}Lp1\ Lp2)$. The computation of the coupling factor is a costly operation for millions of inductances. This computation is unnecessary, especially for potentially very large inductance matrices.

A more appropriate statement for mutual inductance can be used for the mutual inductance input descriptions. Due to the symmetry $Lp_{12} = Lp_{21}$, a better statement for the mutual coupling is

```
KL12 Lp11 Lp22 0.0005uH TD=0.132ns
```

TABLE B.10 Matrix element stamps for inductances and mutual inductances.

Row/column	V_k	V_l	V_m	V_n	I_1	I_2	RHS
k					1		
l					−1		
m						1	
n						−1	
Aux	1	−1			$-s\,L_{11}$	$-s\,L_{12}$	
Aux			1	−1	$-s\,L_{21}$	$-s\,L_{22}$	

We also included a time delay example in this statement. This would be the delay for mutual inductances. The stamp for this case is given in Table B.10. The structure of the stamp is the same for the time domain.

For the time domain, we replace s with the appropriate numerical integration method in the auxiliary equations.

B.1.8 Ideal Transformer Model

The ideal transformer is an important element in circuit theory that finds use for circuit synthesis and other applications. Importantly, this circuit element does not contribute to the states of the MNA descriptor system and it is also lossless. We can write the conventional circuit equations in terms of the ratio of the primary to secondary voltage ratio β with the current ratio of $1/\beta$ such that power is conserved (Fig. B.8).

The stamp for the ideal two winding transformer is given in Table B.11.

The ideal transformer stamp for multiple windings is a straightforward extension of the two winding model given here.

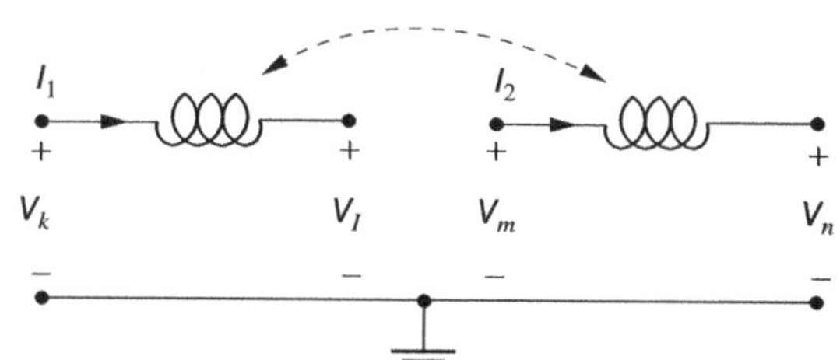

FIGURE B.8 Equivalent circuit for ideal transformer with two windings.

TABLE B.11 Matrix element stamps for two winding ideal transformer.

Row/column	V_k	V_l	V_m	V_n	I_1	I_2	RHS
k					1		0
l					−1		0
m						1	0
n						−1	0
Aux	0	0	0	0	1	$-\beta$	0
Aux	$-\beta$	β	1	−1	0	0	0

B.2 CONTROLLED SOURCE STAMPS

Controlled sources are required for a large class of circuit solutions. For PEEC models, current controlled current sources (CCCSs) are used as well as voltage controlled voltage sources (VCVSs). We consider the four combinations between current and voltage sources in the following section.

B.2.1 Current Controlled Voltage Source (CCVS)

The first controlled source stamp we consider is the current controlled voltage source (CCVS). The equation for the voltage is $V_m - V_n = hName\ I_{k\,l}$, where *Name* represents the individual name assigned to the *h* factor (Fig. B.9).

As can be observed, the stamp is given for the case where the current is measured between nodes *k* and *l* is for the case of a short current measuring connection. The stamp can be simplified if we use a branch current as the measuring element. This part is clearly borrowed from the current measuring in the short stamp in Table B.12.

B.2.2 Voltage Controlled Voltage Source (VCVS)

Next, we consider the matrix stamp for a VCVS. The auxiliary equation for this case is $V_m - V_n = kName\ (V_k - V_l)$. This leads to the matrix stamp of Table B.13 (Fig. B.10).

B.2.3 Current Controlled Current Source (CCCS)

Next, we consider the matrix stamp for a CCCS (Table B.14). The current $I_{mn} = fxxx\ I_{kl}$. Here, I_{mn} is the current of the source from node *m* to node *n* (Fig. B.11).

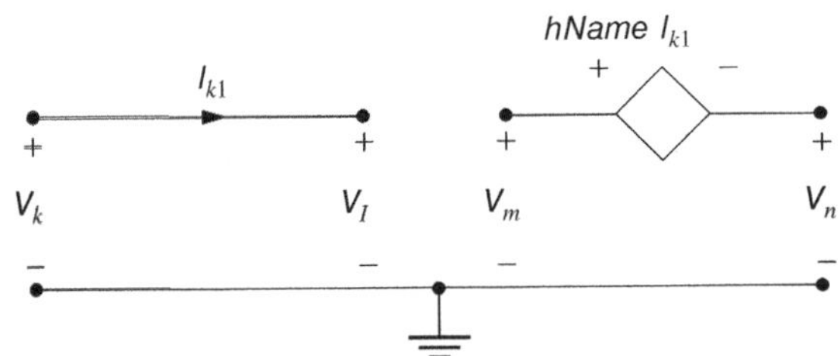

FIGURE B.9 Equivalent circuit for current controlled voltage source.

TABLE B.12 Matrix stamp for current controlled voltage source.

Row/column	V_k	V_l	V_m	V_n	I_{kl}	I_{mn}	RHS
k					1		0
l					−1		0
m						1	0
n						−1	0
Aux	1	−1					0
Aux			1	−1	*hxxx*		0

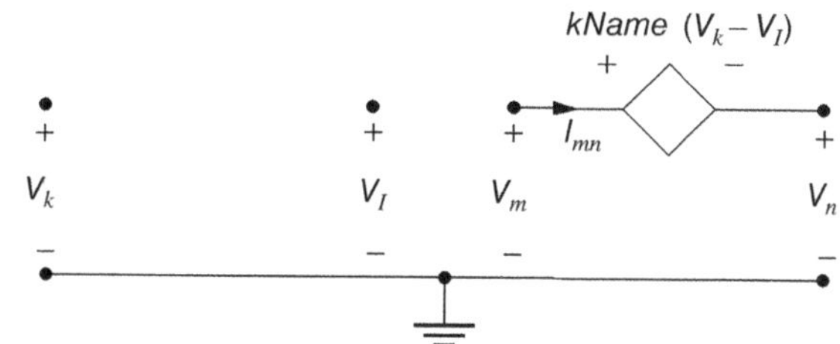

FIGURE B.10 Equivalent circuit for voltage controlled voltage source.

TABLE B.13 Matrix stamp for voltage controlled voltage source.

Row/column	V_k	V_l	V_m	V_n	I_{mn}	RHS
k						
l						
m					1	
n					−1	
Aux	−*kName*	*kName*	1	−1		

TABLE B.14 Matrix stamp for current controlled current source.

Row/column	V_k	V_l	V_m	V_n	I_{kl}	RHS
k					1	
ℓ					−1	
m					*fName*	
n					−*fName*	
Aux	1	−1				

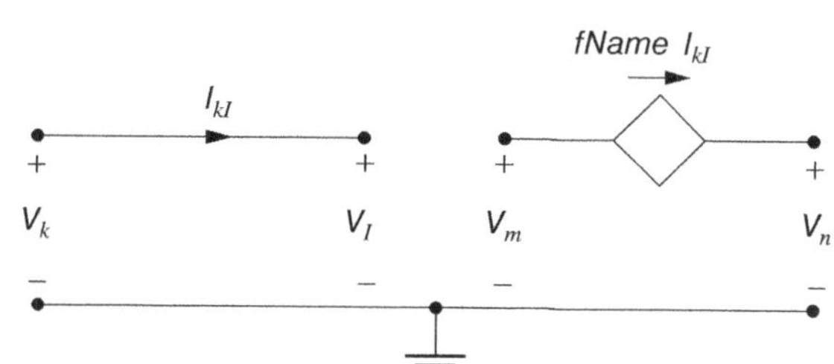

FIGURE B.11 Equivalent circuit for current controlled current source.

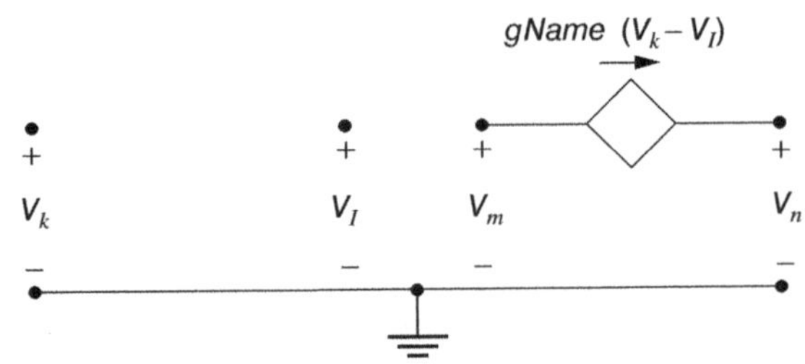

FIGURE B.12 Equivalent circuit for voltage controlled current sources.

TABLE B.15 Matrix stamp for VCCS.

Row/column	V_k	V_l	V_m	V_n	RHS
k					
l					
m	$gName$	$-gName$			
n	$-gName$	$gName$			

B.2.4 Voltage Controlled Current Source (VCCS)

The matrix stamp for a voltage controlled current source (VCCS) is used to represent some devices (Fig. B.12, Table B.15). The voltage controlled current is given by $V_{kl} = gxxx\ (V_k - V_\ell)$. These stamps, together with Chapter 2, should yield a good understanding on how to form the appropriate circuit equations.

REFERENCES

1. C. Ho, A. Ruehli, and P. Brennan. The modified nodal approach to network analysis. *IEEE Transactions on Circuits and Systems*, 22(6):504–509, June 1975.
2. J. Vlach and K. Singhal. *Computer Methods for Circuit Simulation*. Van Nostrand Reinhold Company, New York, 1983.
3. A. Ruehli Ed. *Circuit Analysis, Simulation and Design, Part 1*. Elsevier Science Publishers B.V., North-Holland, 1986.
4. L. Pillegi, R. Rohrer, and C. Visweswariah. *Electronic Circuits and System Simulation Methods*. McGraw-Hill Book Company, New York, 1995.
5. F. N. Najm. *Circuit Simulation*. John Wiley and Sons, Inc., New York, 2010.
6. L. W. Nagel. SPICE: a computer program to simulate semiconductor circuits. Electronics Research Laboratory Report, ERL M520, University of California, Berkeley, CA, May 1975.

C

COMPUTATION OF PARTIAL INDUCTANCES

The computation of partial inductances represents a key component for almost all partial element equivalent circuit (PEEC) models, perhaps even more than the partial coefficients of potential in the following section. Many lower-frequency high-current problems can be solved with inductance-only models. Some references on this subject are [1–3] and also [4–7].

The evaluation of partial inductances Lp for the most general case involves the evaluation of a sixfold integral of the form (C.1) which is (5.15), or

$$Lp_{12} = \frac{1}{\mathcal{A}_1}\frac{1}{\mathcal{A}_2}\frac{\mu}{4\pi}\int_{\mathcal{A}_1}\int_{\mathcal{A}_2}\int_{\ell_1}\int_{\ell_2}\frac{\hat{\imath}_1 \cdot \hat{\imath}_2}{R_{1,2}}\,d\ell_2\,d\ell_1\,d\mathcal{A}_2\,d\mathcal{A}_1, \tag{C.1}$$

where the details are given in Chapter 5.

The example formula (C.1) applies to the case where conductors 1 and 2 are bars or cells that are not necessarily parallel to each other. Obviously, for the partial self-inductances Lp_{11}, the conductors 1 and 2 are the same. This case leads to a singularity if the points coincide such that $R_{1,2} = 0$. An advantage of an analytical solution for Lp_{11} is that the singularity can be eliminated if two of the integrals for the same surface can be analytically evaluated.

In general, we need to consider a larger class of configurations for the conductors or cells to cover all important cases. The evaluation of partial inductances for nonorthogonal structures is even more challenging. In the frequency domain for very high frequencies and sufficiently large cells, we may need to include the retardation in the integral that makes the analytical integration even more complicated. This issue is considered in Section 5.8.

The most simple and useful configurations are rectangular bars in a rectangular Manhattan-type coordinate system. As is evident from Chapters 5 and 7, it is much more challenging to find efficient formulas for the nonorthogonal case. Still, analytical formulas are very desirable for the larger aspect ratio dimensions that are often used in PEEC

Circuit Oriented Electromagnetic Modeling Using the PEEC Techniques, First Edition.
Albert E. Ruehli, Giulio Antonini, and Lijun Jiang.

models. It is desirable to treat the largest physical dimensions of a cell analytically. This issue is considered in Appendix E. However, analytical solutions may also have limitations if the dimensions are too extreme as shown in Section 5.5.2. Fortunately, the use of double precision arithmetic used in most computers today is helpful for minimizing this problem.

In general, either analytical formulas are exclusively used or, alternatively, a combination of analytical formulas in conjunction with some numerical techniques are employed. This aspect is discussed in Appendix E. Key factors are sufficient accuracy as well as the reduction of compute time. Compute time is of importance since we may need to compute millions of Lp's for large problems.

Most analytical formulations for partial electrical element computations are for Manhattan-type rectangular geometries, which may be parallel to the x axis and/or being parallel to the xOy plane, where O is the origin. To be able to use the given analytical formula for many situations, it is necessary to rotate or shift the coordinate system so that the placement of the objects agrees with the orientation in the formula. For most situations, rotation and/or shifting is not necessary. We can reposition objects to the required location given by the formulas below in the x, y, and z coordinates.

We give the rotation matrices for all three cases, where a point x, y, z in space is rotated about an axis to a new location x', y', z'. Specifically, the rotation occurs about one axis by an angle the axis is left unchanged by the rotation. An example for the rotation about the x-axis is given in Fig. C.1.

The first case considered is the rotation about the x-axis, which is illustrated in Fig. C.1. If the point is rotated by ψ degrees, the point will be moved to (x, y', z') from (x, y, z) using the matrix

$$\begin{bmatrix} x' \\ y' \\ z' \end{bmatrix} = \begin{bmatrix} 1 & 0 & 0 \\ 0 & \cos\psi & -\sin\psi \\ 0 & \sin\psi & \cos\psi \end{bmatrix} \begin{bmatrix} x \\ y \\ z \end{bmatrix}, \tag{C.2}$$

where $x' = x$.

The coordinates are rotated around the y-axis by the angle ψ to the new location (x', y, z') using

$$\begin{bmatrix} x' \\ y' \\ z' \end{bmatrix} = \begin{bmatrix} \cos\psi & 0 & -\sin\psi \\ 0 & 1 & 0 \\ \sin\psi & 0 & \cos\psi \end{bmatrix} \begin{bmatrix} x \\ y \\ z \end{bmatrix}, \tag{C.3}$$

where $y' = y$.

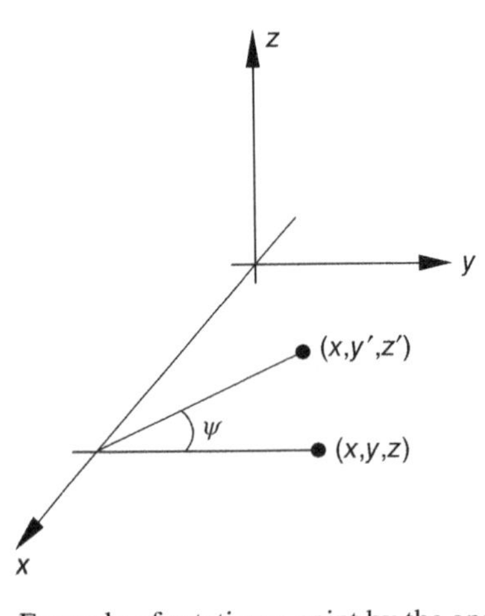

FIGURE C.1 Example of rotating a point by the angle ψ for x-axis.

Finally, if the coordinate is rotated around the z-axis for ψ degrees, the rotated coordinates (x', y', z) can be obtained from (x, y, z)

$$\begin{bmatrix} x' \\ y' \\ z' \end{bmatrix} = \begin{bmatrix} \cos\psi & -\sin\psi & 0 \\ \sin\psi & \cos\psi & 0 \\ 0 & 0 & 1 \end{bmatrix} \begin{bmatrix} x \\ y \\ z \end{bmatrix}, \tag{C.4}$$

where $z' = z$. In general, by combining these three steps, we can rotate the coordinates to align with any point in the coordinate system.

The situation is much simpler for the case where we have to shift a point in space in the coordinate direction. We simply add a constant Δ in the appropriate direction. For example, we shift a point by Δx, Δy, Δz to new point by simply adding the changes, or

$$\begin{bmatrix} x' \\ y' \\ z' \end{bmatrix} = \begin{bmatrix} \Delta x \\ \Delta y \\ \Delta z \end{bmatrix} + \begin{bmatrix} x \\ y \\ z \end{bmatrix}. \tag{C.5}$$

The above rotation and the shifting operations together allow changes to be made for the general case. Hence, it can be used for any one of the formulas given below.

To reorient the rectangular coordinate system, we simply exchange the coordinates in a cyclic way. An example is $x \to y$, $y \to z$, and $z \to x$. The issue is more complicated for the nonorthogonal formulas since we also may require to perform translations.

C.1 PARTIAL INDUCTANCE FORMULAS FOR ORTHOGONAL GEOMETRIES

This appendix lists important formulas for the computation of the orthogonal partial inductance values. First, we want to repeat that partial inductance with orthogonal currents do not couple. This results in $Lp_{12} = 0$.

For clarity, we present all the formulas in the same coordinate system. Some of the formulas may need numerical solutions while others are analytically exact and are complete. Approximate formulas may not be accurate enough for some applications such as skin-effect computations like the volume filament (VFI) model in Chapter 9, where the cells are closely located. For these applications, at least 4–5 digits of accuracy is required. We should be aware that some of the approximate formulas in Ref. [1] have limited accuracy.

For several analytical formulas in this appendices we include very small numbers in the order of 10^{-20} to 10^{-40} to avoid possible artificial singularities for some combinations of input parameters. This is a conventional numerical technique to avoid singularity problems.

C.1.1 Lp_{12} for Two Parallel Filaments

A simple important geometry is the wire-to-wire or filament-to-filament problem shown in Fig. C.2. Of course, the result is singular if the wires coincide. Filaments are very important building blocks, which we use for the combined analytical and numerical integration. The current direction in the wires is shown to be along the length.

The partial inductance formula for Lp_{12} for two current filaments shown in Fig. C.2 reduces to

$$Lp_{12} = \frac{\mu_0}{4\pi} \int_{x_{s1}}^{x_{e1}} \int_{x_{s2}}^{x_{e2}} \frac{1}{R_{1,2}}\, dx_2\, dx_1, \tag{C.6}$$

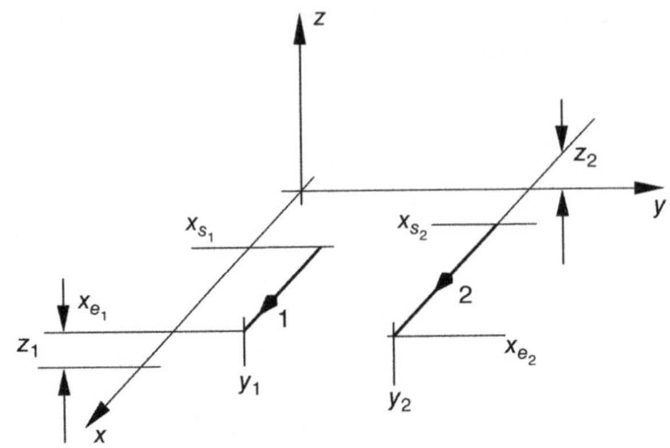

FIGURE C.2 Partial mutual inductance for two filament wires.

where

$$R_{1,2} = \sqrt{(x_1 - x_2)^2 + (y_1 - y_2)^2 + (z_1 - z_2)^2}. \tag{C.7}$$

The analytical form of the partial inductance is given for this case by a simple integration step as

$$Lp_{12} = \frac{\mu_0}{4\pi} \sum_{k=1}^{4} (-1)^{k+1} \left[a_k \log \left(a_k + \sqrt{a_k^2 + r_{12}^2} \right) - \sqrt{a_k^2 + r_{12}^2} \right] \tag{C.8}$$

with

$$r_{12} = \sqrt{(y_2 - y_1)^2 + (z_2 - z_1)^2} \tag{C.9}$$

$$a_1 = x_{s2} - x_{e1} \tag{C.10}$$

$$a_2 = x_{e2} - x_{e1} \tag{C.11}$$

$$a_3 = x_{e2} - x_{s1} \tag{C.12}$$

$$a_4 = x_{s2} - x_{s1}. \tag{C.13}$$

Note that e represents the *end* of the object and s is used for the *start*.

Equation (C.8) can also be used as an approximation to the round wire partial inductance especially if the wire diameter is small compared to the spacing. Of course, other cross section can be approximated efficiently with several filaments and numerical integration in the cross-section direction.

C.1.2 Lp_{11} for Round Wire

We give two formulas for the partial self-inductance for round shapes. The popular equation for the partial self-inductance for the section of round wire shown in Fig. C.3 is

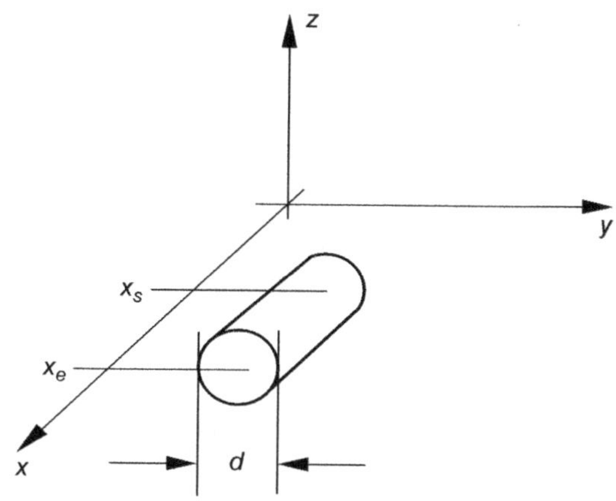

FIGURE C.3 Partial self-inductance for round wire section.

$$Lp_{11} = \frac{1}{\mathcal{A}_1^2}\frac{\mu_0}{4\pi}\int_{x_{s1}}^{x_{e1}}\int_{\mathcal{A}_1}\int_{x'_{s1}}^{x'_{e1}}\int_{\mathcal{A}'_1}\frac{1}{R_{1,1'}}\,d\mathcal{A}'_1\,dx'_1\,d\mathcal{A}_1\,dx_1, \tag{C.14}$$

where $R_{1,1'}$ is the usual distance between the source and observation points. $\mathcal{A}_1$ and $\mathcal{A}'_1$ are the cross sections.

The first formula is a simple approximation of a wire partial self- inductance used in Refs [1, 2], which is based on the assumption that the above filament formula can be used to represent a round wire by placing one of the filaments at the center of the wire, and the second one on the surface of the wire at distance $a = d/2$ away. The approximate result for this case is

$$Lp_{11} \simeq \frac{\mu_0}{2\pi}\ell\left[\ln\left(\frac{\ell}{a}+\sqrt{\left(\frac{\ell}{a}\right)^2+1)}\right)-\sqrt{1+\left(\frac{a}{\ell}\right)^2}+\frac{a}{\ell}\right], \tag{C.15}$$

where $\ell = (x_e - x_s)$ is the length of the segment. We could view this approximation technique as the most simple case of numerical integration based on the analytical result for the longitudinal direction.

A second analytical formula is for the partial inductance of a zero thickness cylindrical tube with a radius $a = d/2$ shown in Fig. C.3. The integrals to evaluate are

$$Lp_{11} = \frac{\mu_0}{4\pi^2}\int_{\phi=0}^{\pi}\int_{x=0}^{\ell}\int_{x'=0}^{\ell}\frac{dx'\,dx\,d\phi}{\sqrt{4a^2\sin^2(\phi/2)+(x-x')^2}}. \tag{C.16}$$

by recognizing that symmetry can be used to reduce the fourfold integral to the threefold integral. We were able to analytically solve the integrals for the case of interest where the length ℓ of the wire is longer than the diameter $d = 2\,a$ [8]. The result for the tube conductor is given by

$$Lp_{11} = \frac{\mu_0 \ell}{4} \left[\left(\frac{k^2}{480} + \frac{k^4}{1280} + \frac{1}{3600} \right) \pi^3 + \left(\frac{1}{18} - \frac{k^2}{24} \right) \pi \right.$$
$$\left. + \left(2 - 2\log(\ell) + 6\log(2) + 2\log(a) - 4\log(k\pi) \right) \frac{1}{\pi} + \frac{8a}{\ell \pi^2} \right], \tag{C.17}$$

where $k = 2a/\ell$. We should note that the equation is also used in equation (9.29) for skin-effect models [9].

C.1.3 Lp_{12} for Filament and Current Sheet

Another important partial mutual inductance represents the case where one conductor is approximated with a filament and the other with a zero thickness sheet as shown in Fig. C.4. Of course, the current in the sheet is in the same x-direction as in the wire as is shown in Fig. C.4 and the integral is given by

$$Lp_{12} = \frac{1}{(y_{e1} - y_{s1})} \frac{\mu}{4\pi} \int_{y_{s1}}^{y_{e1}} \int_{x_{s1}}^{x_{e1}} \int_{x_{s2}}^{x_{e2}} \frac{1}{R_{1,2}} \, dx_2 \, dx_1 \, dy_1. \tag{C.18}$$

The analytical solution for the above integrals is

$$Lp_{12} = \frac{\mu_0}{4\pi} \frac{1}{w_1} \sum_{k=1}^{4} \sum_{m=1}^{2} (-1)^{k+m} \left[\frac{a_k^2 - Z^2}{2} \log(b_m + r_{\mathrm{km}} + \epsilon) \right.$$
$$\left. + a_k b_m \log(a_k + r_{\mathrm{km}} + \epsilon) - a_k Z \tan^{-1}\left(\frac{a_k b_m}{Z r_{\mathrm{km}}} \right) - \frac{b_m}{2} r_{\mathrm{km}} \right] \tag{C.19}$$

with

$$a_1 = x_{s2} - x_{e1}, \quad a_2 = x_{e2} - x_{e1} \tag{C.20}$$

$$a_3 = x_{e2} - x_{s1}, \quad a_4 = x_{s2} - x_{s1} \tag{C.21}$$

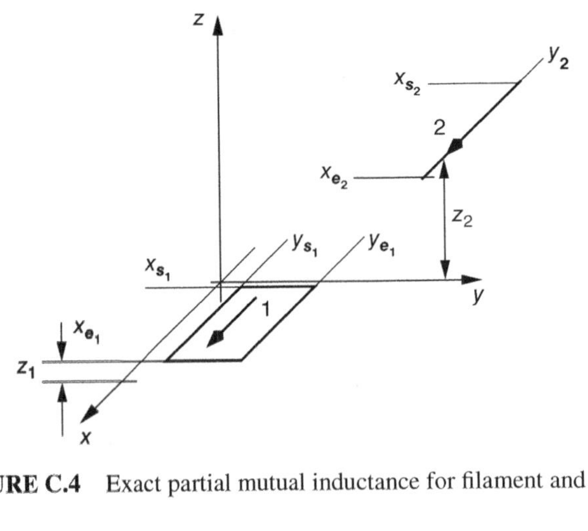

FIGURE C.4 Exact partial mutual inductance for filament and sheet.

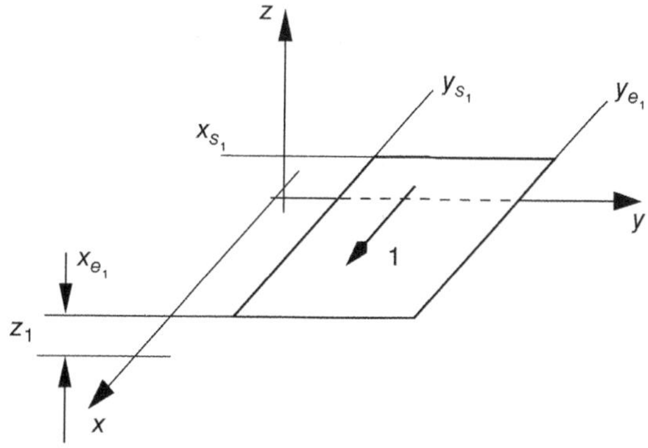

FIGURE C.5 Single zero thickness conductor sheets.

$$b_1 = y_2 - y_{s1}, \qquad b_2 = y_2 - y_{e1} \tag{C.22}$$

$$Z = z_2 - z_1, \qquad r_{\rm km} = \sqrt{a_k^2 + b_m^2 + Z^2}. \tag{C.23}$$

This formula is useful for the case where the two conductors are of a different shape. Other conductor shapes can be taken into account by combining this formula with numerical integration for the cross-section dimension of one or both inductive cells.

C.1.4 Lp_{11} for Rectangular Zero Thickness Current Sheet

Fortunately, the analytical Lp for a zero thickness rectangular sheet shown in Fig. C.5 leads to a relatively simple formula.

The result is

$$Lp_{11} = \frac{\mu_0 \ell}{6\,\pi}\left[3\,\log\left[u + (u^2+1)^{1/2}\right] + u^2 + \frac{1}{u} + 3\,u\,\log\left[\frac{1}{u} + \left(\frac{1}{u^2}+1\right)^{1/2}\right] - \left[u^{4/3} + \left(\frac{1}{u}\right)^{2/3}\right]^{3/2}\right], \tag{C.24}$$

where $\ell = x_{e1} - x_{s1}$ and $u = \ell/(y_{e1} - y_{s1})$.

We note that this relatively simple formula is very useful for semianalytical solutions in layered models. Importantly, it eliminates the singular behavior problem for the partial self-inductance.

C.1.5 Lp_{12} for Two Parallel Zero Thickness Current Sheets

Two zero thickness parallel current sheets are shown in Fig. C.6. Fortunately, several analytical formulas are available, for example, Ref. [2]. The integral to be solved is

$$Lp_{12} = \frac{1}{(y_{e1} - y_{s1})(y_{e2} - y_{s2})}\,\frac{\mu}{4\pi}\int_{y_{s1}}^{y_{e_1}}\int_{x_{s1}}^{x_{e1}}\int_{y_{s2}}^{y_{e2}}\int_{x_{s2}}^{x_{e2}} \frac{1}{R_{1,2}}\,dx_2\,dy_2\,dx_1\,dy_1. \tag{C.25}$$

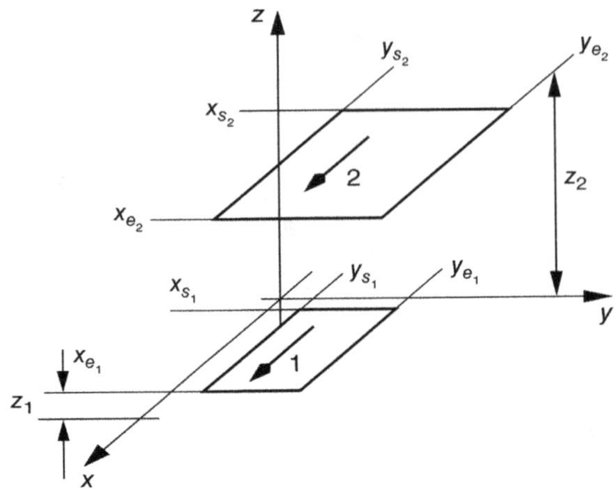

FIGURE C.6 Two parallel zero thickness conductors example.

The closed solution of integral (C.25) is

$$Lp_{12} = \frac{\mu_0}{4\pi} \frac{1}{(y_{e1} - y_{s1})(y_{e2} - y_{s2})} \sum_{k=1}^{4} \sum_{m=1}^{4} (-1)^{m+k}$$

$$\left[\frac{b_m^2 - Z^2}{2} a_k \log(a_k + r_{\text{km}} + \epsilon) + \frac{a_k^2 - Z^2}{2} b_m \log(b_m + r_{\text{km}} + \epsilon) \right.$$

$$\left. - \frac{1}{6}(b_m^2 - 2Z^2 + a_k^2) r_{\text{km}} - b_m Z a_k \tan^{-1}\left(\frac{a_k b_m}{r_{\text{km}} Z}\right) \right], \tag{C.26}$$

where

$$a_1 = x_{s2} - x_{e1}, \qquad a_2 = x_{e2} - x_{e1} \tag{C.27}$$

$$a_3 = x_{e2} - x_{s1}, \qquad a_4 = x_{s2} - x_{s1} \tag{C.28}$$

$$b_1 = y_{s2} - y_{e1}, \qquad b_2 = y_{e2} - y_{e1} \tag{C.29}$$

$$b_3 = y_{e2} - y_{s1}, \qquad b_4 = y_{s2} - y_{s1} \tag{C.30}$$

$$Z = z_2 - z_1 + \epsilon, \qquad r_{\text{km}} = \sqrt{a_k^2 + b_m^2 + Z^2}, \tag{C.31}$$

where we added $\epsilon = 10^{-37}$ to the equation as a very small number.

C.1.6 Lp_{12} for Two Orthogonal Rectangular Current Sheets

As is shown in Fig. C.7, two zero thickness sheets are at an angle of 90° while the current flow is in the x-direction for both sheets. This formula is again very useful. Of course, the

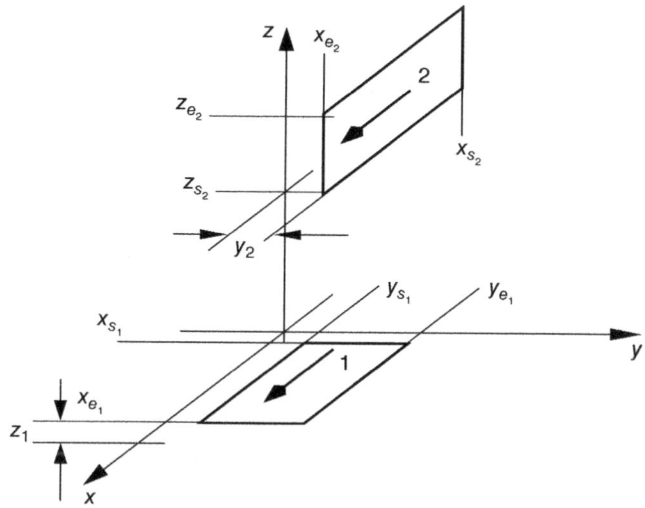

FIGURE C.7 Two zero thickness conductors at a 90° angle.

currents in the sheets as indicated in Fig. C.7 need to be parallel to each other for a nonzero partial inductance.

Its original definition is

$$Lp_{12} = \frac{1}{(y_{e1} - y_{s1})(z_{e2} - z_{s2})} \frac{\mu_0}{4\pi} \int_{y_{s1}}^{y_{e1}} \int_{x_{s1}}^{x_{e1}} \int_{z_{s2}}^{z_{e2}} \int_{x_{s2}}^{x_{e2}} \frac{1}{R_{1,2}} \, dx_2 \, dz_2 \, dx_1 \, dy_1, \tag{C.32}$$

where the final result for Lp_{12} for the two orthogonal current sheets is

$$\begin{aligned} Lp_{12} = \frac{\mu_0}{4\pi} \frac{1}{(y_{e1} - y_{s1})(z_{e2} - z_{s2})} \sum_{k=1}^{4} \sum_{m=1}^{2} \sum_{\ell=1}^{2} (-1)^{\ell+m+k+1} \\ \left[\left(\frac{a_k^2}{2} - \frac{c_\ell^2}{6} \right) c_\ell \log(b_m + r_{\text{km}\ell} + \epsilon) + \left(\frac{a_k^2}{2} - \frac{b_m^2}{6} \right) b_m \log(c_\ell + r_{\text{km}\ell} + \epsilon) \right. \\ + a_k b_m c_\ell \log(a_k + r_{\text{km}\ell} + \epsilon) - \frac{b_m c_\ell}{3} r_{\text{km}\ell} \\ \left. - \frac{a_k^3}{6} \tan^{-1} \left(\frac{b_m c_\ell}{a_k \, r_{\text{km}\ell}} \right) - \frac{b_m^2 a_k}{2} \tan^{-1} \left(\frac{a_k c_\ell}{b_m \, r_{\text{km}\ell}} \right) - \frac{a_k c_\ell^2}{2} \tan^{-1} \left(\frac{a_k b_m}{c_\ell \, r_{\text{km}\ell}} \right) \right] \end{aligned} \tag{C.33}$$

with

$$a_1 = x_{s2} - x_{e1}, \quad a_2 = x_{e2} - x_{e1} \tag{C.34a}$$

$$a_3 = x_{e2} - x_{s1}, \quad a_4 = x_{s2} - x_{s1} \tag{C.34b}$$

$$b_1 = y_2 - y_{s1}, \qquad b_2 = y_2 - y_{e1} \tag{C.34c}$$

$$c_1 = z_{e2} - z_1, \qquad c_2 = z_{s2} - z_1 \tag{C.34d}$$

and

$$r_{\text{km}\ell} = \sqrt{a_k^2 + b_m^2 + c_\ell^2}. \tag{C.34e}$$

C.1.7 Lp_{11} for Rectangular Finite Thickness Bar

The partial self-inductance of a rectangular bar shown in Fig. C.8 represents an important building block for many problems. The equation for this case is given by

$$Lp_{11} = \frac{\mu_0}{4\pi} \frac{1}{\mathcal{A}^2} \int_{\mathcal{A}} \int_{\mathcal{A}'} \int_{x_{s1}}^{x_{e1}} \int_{x'_{s1}}^{x'_{e1}} \frac{1}{R_{1,1'}}\, dx'_1\, dx_1\, d\mathcal{A}'\, d\mathcal{A}, \tag{C.35}$$

where the cross-section area is given by $\mathcal{A} = (y_{e1} - y_{s1})(z_{e1} - z_{s1})$ and $x'_{e1} = x_{e1}$ and $x'_{s1} = x_{s1}$. The solution of this sixfold integral is in general obtained by introducing new variable differences for x, y, z. The closed form expression [4] is

$$\begin{aligned}
\frac{Lp_{11}\pi}{2\mu_0\ell} &= \frac{w^2}{24\,u}\left[\log\left(\frac{1+a_2}{w}\right) - a_5\right] + \frac{1}{24\,u\,w}\left[\log(w + a_2) - a_6\right] \\
&+ \frac{w^2}{60\,u}[a_4 - a_3] + \frac{w^2}{24}\left[\log\left(\frac{u+a_3}{w}\right) - a_7\right] \\
&+ \frac{w^2}{60\,u}(w - a_2) + \frac{1}{20\,u}(a_2 - a_4) + \frac{u}{4}a_5 - \frac{u^2}{6\,w}\tan^{-1}\left(\frac{w}{u\,a_4}\right) \\
&+ \frac{u}{4\,w}a_6 - \frac{w}{6}\tan^{-1}\left(\frac{u}{w\,a_4}\right) + \frac{a_7}{4} - \frac{1}{6\,w}\tan^{-1}\left(\frac{u\,w}{a_4}\right) \\
&+ \frac{1}{24\,w^2}[\log(u + a_1) - a_7] + \frac{u}{20\,w^2}(a_1 - a_4) \\
&+ \frac{1}{60\,w^2 u}(1 - a_2) + \frac{1}{60\,u\,w^2}(a_4 - a_1) + \frac{u}{20}(a_3 - a_4) \\
&+ \frac{u^3}{24\,w^2}\left[\log\left(\frac{1+a_1}{u}\right) - a_5\right] + \frac{u^3}{24\,w}\left[\log\left(\frac{w+a_3}{u}\right) - a_6\right] \\
&+ \frac{u^3}{60\,w^2}[(a_4 - a_1) + (u - a_3)]
\end{aligned} \tag{C.36}$$

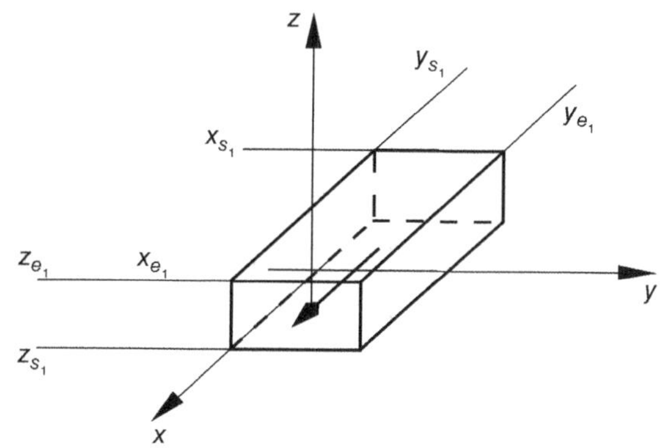

FIGURE C.8 Lp_{11} for rectangular bar with thickness.

with

$$u = (x_{e1} - x_{s1})/(y_{e1} - y_{s1}), \quad w = (z_{e1} - z_{s1})/(y_{e1} - y_{s1}) \tag{C.37a}$$

$$a_1 = \sqrt{1 + u^2}, \qquad a_2 = \sqrt{1 + w^2} \tag{C.37b}$$

$$a_3 = \sqrt{w^2 + u^2}, \qquad a_4 = \sqrt{1 + w^2 + u^2} \tag{C.37c}$$

$$a_5 = \log\left(\frac{1 + a_4}{a_3}\right), \qquad a_6 = \log\left(\frac{w + a_4}{a_1}\right) \tag{C.37d}$$

and

$$a_7 = \log\left(\frac{u + a_4}{a_2}\right). \tag{C.37e}$$

The evaluation of (C.36) should for accuracy reasons be performed by summing from top to bottom, where the new terms are added to the sum of the previous terms. Test results show that the errors become large for very large values of u and for small values of w as considered in Section 5.5.2.

A possible solution for this is the computation of Lp_{11} by breaking the length into several segments as is illustrated in Section 5.6.1. This is required only for very long conductors.

The accuracy problem for this case has also been observed in Ref. [10]. It is also pointed out that neighboring partial mutual inductances can be computed by using only partial self-inductance computations. Two neighboring block cells are shown in Fig. C.9. The partial self-inductance of the two neighbors cells shown can be computed using formula (C.36).

If we also compute the partial self-Lp_{33} for a combined block 1,2, then the result is

$$Lp_{33} = Lp_{11} + Lp_{22} + 2\, Lp_{12}. \tag{C.38}$$

Hence, we can compute Lp_{12} totally based on partial self- inductance only, or

$$Lp_{12} = 0.5\, (Lp_{33} - Lp_{11} - Lp_{22}). \tag{C.39}$$

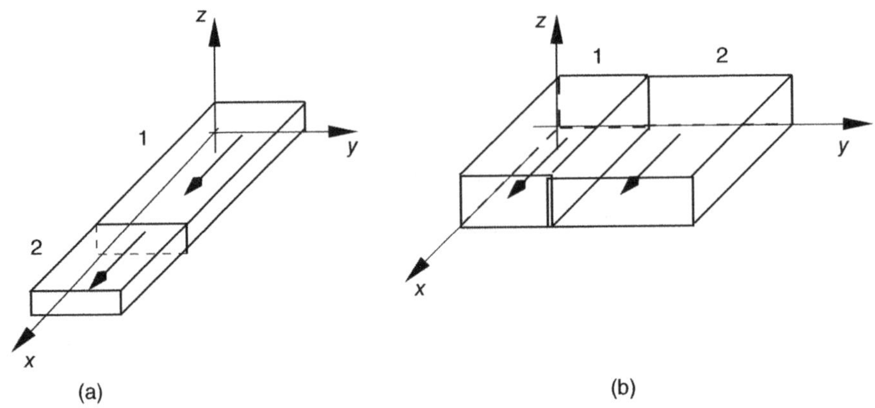

FIGURE C.9 Partial inductance for two rectangular conductors.

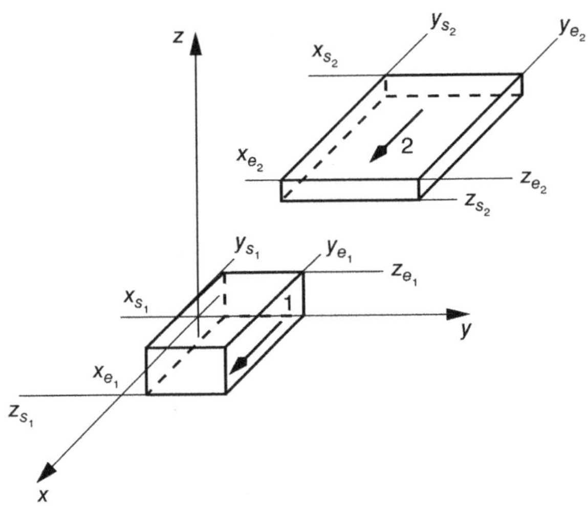

FIGURE C.10 Partial inductance for two rectangular conductors.

This can be very helpful for skin-effect models where the accuracy is an issue due to the closely located neighbors. Hence, three partial self-inductance computations are required. This helps, since we can more accurately compute a partial mutual inductance Lp_{12} for some geometries. We want to point out that this technique can also be applied to conductors in parallel as shown in Fig. C.9 (b). We leave the computation of Lp_{12} as an exercise to the reader.

C.1.8 Lp_{12} for Two Rectangular Parallel Bars

An important structure for rectangular PEEC models is the case of two parallel rectangular bars as is shown in Fig. C.10. The analytical formula for the mutual inductance between two rectangular bars is given in Ref. [3]. In our coordinate system setup, the formula is

$$Lp_{12} = \frac{\mu_0}{4\pi} \frac{1}{\mathcal{A}_1 \mathcal{A}_2} \sum_{k=1}^{4} \sum_{\ell=1}^{4} \sum_{m=1}^{4} (-1)^{k+\ell+m+1} \left[-\frac{b_\ell\, c_m\, a_k^3}{6} \tan^{-1}\left(\frac{b_\ell\, c_m}{a_k\, R}\right) \right.$$

$$
-\frac{b_\ell\, c_m^3\, a_k}{6}\tan^{-1}\left(\frac{b_\ell\, a_k}{c_m\, R}\right)-\frac{b_\ell^3\, c_m\, a_k}{6}\tan^{-1}\left(\frac{c_m\, a_k}{b_\ell\, R}\right)
$$

$$
+a_k\left(\frac{b_\ell^2 c_m^2}{4}-\frac{b_\ell^4}{24}-\frac{c_m^4}{24}\right)\log\left(\frac{a_k+R}{\sqrt{b_\ell^2+c_m^2}}\right)
$$

$$
+b_\ell\left(\frac{c_m^2\, a_k^2}{4}-\frac{c_m^4}{24}-\frac{a_k^4}{24}\right)\log\left(\frac{b_\ell+R}{\sqrt{c_m^2+a_k^2}}\right)
$$

$$
+c_m\left(\frac{b_\ell^2 a_k^2}{4}-\frac{b_\ell^4}{24}-\frac{a_k^4}{24}\right)\log\left(\frac{c_m+R}{\sqrt{a_k^2+b_\ell^2}}\right)
$$

$$
+\frac{1}{60}\,(b_\ell^4+c_m^4+a_k^4-3b_\ell^2c_m^2-3c_m^2a_k^2-3a_k^2b_\ell^2)\,R\Bigg] \tag{C.40}
$$

with

$$
a_1 = x_{s2}-x_{e1}, \qquad a_2 = x_{e2}-x_{e1} \tag{C.41a}
$$

$$
a_3 = x_{e2}-x_{s1}, \qquad a_4 = x_{s2}-x_{s1} \tag{C.41b}
$$

$$
b_1 = y_{s2}-y_{e1}, \qquad b_2 = y_{e2}-y_{e1} \tag{C.41c}
$$

$$
b_3 = y_{e2}-y_{s1}, \qquad b_4 = y_{s2}-y_{s1} \tag{C.41d}
$$

$$
c_1 = z_{s2}-z_{e1}, \qquad c_2 = z_{e2}-z_{e1} \tag{C.41e}
$$

$$
c_3 = z_{e2}-z_{s1}, \qquad c_4 = z_{s2}-z_{s1} \tag{C.41f}
$$

$$
\mathcal{A}_1 = (y_{e1}-y_{s1})(z_{e1}-z_{s1}), \quad \mathcal{A}_2 = (y_{e2}-y_{s2})(z_{e2}-z_{s2}) \tag{C.41g}
$$

and

$$
R(a_k, b_\ell, c_m) = \sqrt{a_k^2+b_\ell^2+c_m^2}. \tag{C.41h}
$$

We added this formula for completeness. Unfortunately, even using double precision arithmetic leads only to limited accuracy for reasonably large parameters in this formulation. For this reason, using thin layer approximations in Section C.1.5 with numerical integration along the thickness leads to more stable results.

C.1.9 $1/R^3$ Kernel Integral for Parallel Rectangular Sheets

Partial elements are of importance for other formulation such as the ones in Chapter 11 so that other material properties can be taken into account. In some of these formulations, integrals with a higher order Green's function need to be solved. Specifically, the integral for a $1/R^3$ kernel is important. Here, we consider the geometry to be two rectangles shown in Fig. C.11.

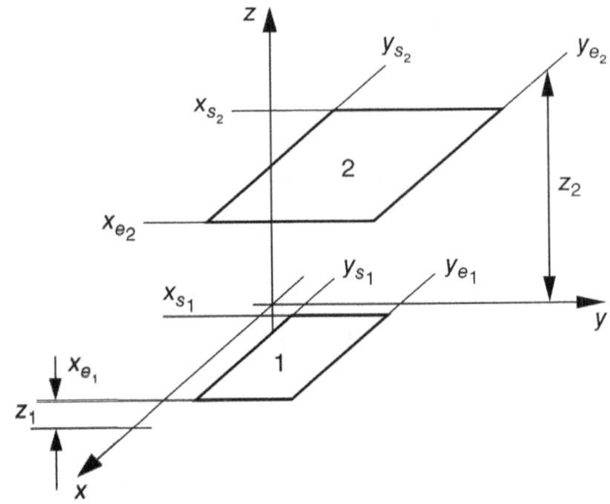

FIGURE C.11 Zero thickness conductors example.

$$Ip_{12} = \int_{x_{s1}}^{x_{e1}} \int_{y_{s1}}^{y_{e1}} \int_{x_{s2}}^{x_{e2}} \int_{y_{s2}}^{y_{e2}} \frac{(z_1 - z_2)}{R^3}\, dy_2\, dx_2\, dy_1\, dx_1, \tag{C.42}$$

where, $R^3 = [(x_1 - x_2)^2 + (y_1 - y_2)^2 + (z_1 - z_2)^2]^{3/2}$.

Then the final formulation of Ip_{12} for two orthogonal current sheets is

$$\begin{aligned} Ip_{12} = \sum_{m=1}^{2}\sum_{n=1}^{2}\sum_{i=1}^{2}\sum_{j=1}^{2} (-1)^{m+n+i+j}(z_1 - z_2) \Bigg\{ &- R_{mnij} \\ &+ (x_{vm} - x_{vn}) \ln[(x_{vm} - x_{vn}) + R_{mnij}] \\ &+ (x_{vm} - x_{vn})(y_{vi} - y_{vj}) \tan^{-1}\left[\frac{(x_{vm} - x_{vn})(z_1 - z_2)}{(y_{vi} - y_{vj})^2 + (z_1 - z_2)^2 + (y_{vi} - y_{vj})R_{mnij}}\right] \\ &+ (x_{vm} - x_{vn})(y_{vi} - y_{vj}) \tan^{-1}\left[\frac{z_1 - z_2}{x_m - x_n}\right] + (z_1 - z_2)^2(y_{vi} - y_{vj}) \\ &\cdot \ln\left[\frac{\sqrt{((y_{vi} - y_{vj})^2 + (z_1 - z_2)^2 + R_{mnij}(y_{vi} - y_{vj}))^2 + (x_{vm} - x_{vn})^2(z_1 - z_2)^2}}{\sqrt{(x_{vm} - x_{vn})^2 + (z_1 - z_2)^2}}\right]\Bigg\} \end{aligned} \tag{C.43}$$

$$R_{mnij} = \sqrt{(x_{vm} - x_{vn})^2 + (y_{vi} - y_{vj})^2 + (z_1 - z_2)^2},$$

where the subscripts are for

$$m = 1: \quad vm = s1; \text{ and for } m = 2: \quad vm = e1; \tag{C.44}$$

$$n = 1: \quad vn = s2; \text{ and for } n = 2: \quad vn = e2; \tag{C.45}$$

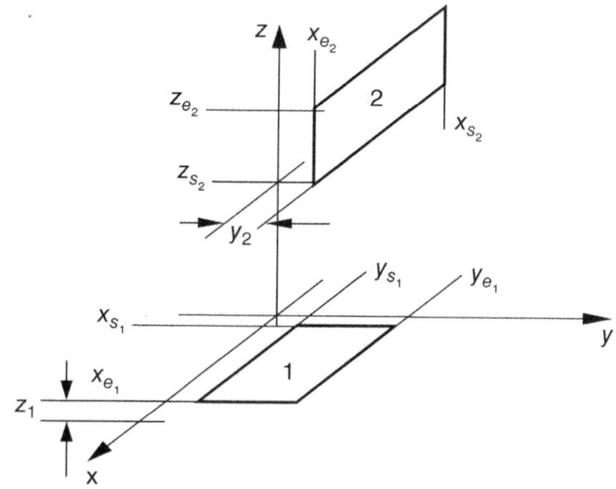

FIGURE C.12 Two parallel zero thickness conductors.

$$i = 1 : \quad vi = s_1; \text{ and for } \quad i = 2 : \quad vn = e1; \tag{C.46}$$

$$j = 1 : \quad vj = s_2; \text{ and for } \quad j = 2 : \quad vn = e2. \tag{C.47}$$

C.1.10 $1/R^3$ Kernel Integral for Orthogonal Rectangular Sheets

A second important integral with a $1/R^2$ Green's function is for two orthogonal rectangles shown in Fig. C.12. For this case, we want to give a solution for the following integral

$$Ip_{12} = \int_{x_{s_1}}^{x_{e_1}} \int_{y_{s_1}}^{y_{e_1}} \int_{x_{s_2}}^{x_{e_2}} \int_{z_{s_2}}^{z_{e_2}} \frac{z_1 - z_2}{R^3} \, dx_1 \, dy_1 \, dx_2 \, dz_2, \tag{C.48}$$

where, $R^3 = [(x_1 - x_2)^2 + (y_1 - y_2)^2 + (z_1 - z_2)^2]^{3/2}$.

Then the final mutual high order coupling formulation Ip_{12} for two orthogonal current sheets is

$$\begin{aligned} Ip_{12} = \sum_{m=1}^{2} \sum_{n=1}^{2} \sum_{i=1}^{2} \sum_{j=1}^{2} (-1)^{m+n+i+j} \Bigg\{ & -0.5(y_2 - y_{vi}) R_{mnij} \\ & + (y_2 - y_{vi})(x_{vn} - x_{vm}) \tanh^{-1} \left(\frac{R_{mnij}}{x_{vn} - x_{vm}} \right) \\ & + 0.5[(x_{vn} - x_{vm})^2 - (z_{vj} - z_1)^2] \ln \, [(y_2 - y_{vi}) + R_{mnij}] \\ & + (x_{vn} - x_{vm})(z_{vj} - z_1) \tan^{-1} \frac{(z_{vj} - z_1)}{y_2 - y_{vi}} \\ & - (x_{vn} - x_{vm})(z_{vj} - z_1) \tan^{-1} \frac{(z_{vj} - z_1)^2 + (x_{vn} - x_{vm})^2 + R_{mnij}(x_{vn} - x_{vm})}{(y_2 - y_{vi})(z_{vj} - z_1)} \Bigg\} \end{aligned} \tag{C.49}$$

where the subscripts are for

$$m = 1 : \quad vm = s_1; \text{ and for } \quad m = 2 : \quad vm = e_1; \tag{C.50}$$

$$n = 1 : \quad vn = s_2; \text{ and for } \quad n = 2 : \quad vn = e_2; \tag{C.51}$$

$$i = 1 : \quad vi = s_1; \text{ and for } \quad i = 2 : \quad vi = e_1; \tag{C.52}$$

$$j = 1 : \quad vj = s_2; \text{ and for } \quad j = 2 : \quad vj = e_2. \tag{C.53}$$

And

$$R_{mnij} = \sqrt{(x_{vn} - x_{vm})^2 + (y_2 - y_{vi})^2 + (z_{vj} - z_1)^2}.$$

C.2 PARTIAL INDUCTANCE FORMULAS FOR NONORTHOGONAL GEOMETRIES

This section is dedicated to the computation of nonorthogonal partial inductance values.

The general integrals for the computation of partial inductance for nonorthogonal such as nonorthogonal hexahedral elements are derived in Chapter 7, equation (7.18) as

$$Lp_{aa'} = \mu_0 \int_c \int_b \int_a \int_{c'} \int_{b'} \int_{a'} (\hat{\mathbf{a}} \cdot \hat{\mathbf{a}}') \, h_{a'} \;\; g(\mathbf{r}_g; \mathbf{r}'_g) \, da' \, db' \, dc' h_a \, da \, db \, dc. \tag{C.54}$$

It is clear that analytical solutions are even more challenging for nonorthogonal geometries. Still, analytical integration needs to be used as much as possible, especially for the longer dimensions of the cells. Since many problems are in this class, we want to efficiently solve problems with relatively long cells such as a thin wire-type structure. To accomplish this, we pursue a mixed solution using analytical results wherever possible. Few formulas will provide a simple, complete answer but rather are important building blocks that lead to a combined analytical/numerical solution as is shown in Appendix E.

C.2.1 Rotation for Different Nonorthogonal Conductor Orientations

The formulas for the translation and rotation of the coordinate systems are given at the beginning of this chapter in (C.2)–(C.5).

Many models given in this section are constructed using non-orthogonal wire models. In these cases we need to rotate the orientation of the structures. The problem at hand has to be rotated into the orientation of the solution given in this text. To achieve this, we have to use the following steps:

- The coordinate system is shifted such that one wire is placed at the right distance from the origin. Next, find the angle ψ of the second wire's projection on the xOy plane, where O is the origin.
- Rotate the original coordinate system around the $\hat{z}$ axis by the angle $-\psi$ such that the second wire is placed on the yOz plane of the new coordinate system.
- Shift the coordinate system along the $\hat{z}$ axis so that one end of the first wire is moved the right distance from the xOy plane.

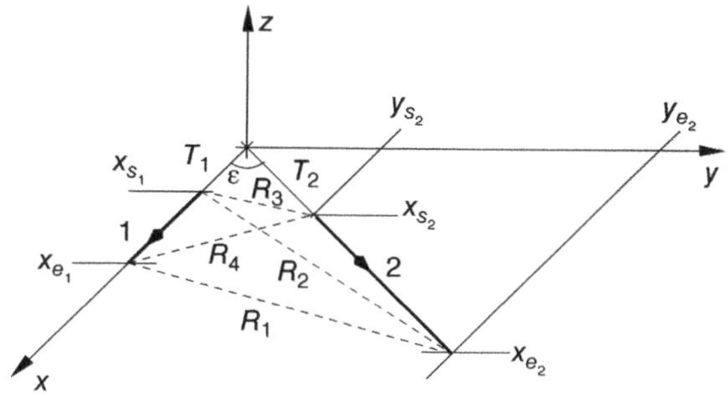

FIGURE C.13 Partial mutual inductance for two non-orthogonal filament wires in plane.

C.2.2 *Lp* for Arbitrary Oriented Wires in the Same Plane $z = 0$

We provide several solutions for filaments since wire filaments are useful basic elements for non-orthogonal Lp's. These filaments are used for non orthogonal cells in conjunction with numerical integrations in the cross section.

The first situation consists of two wires in the $z = 0$ plane as shown in Fig. C.13.

The integrals for two filaments are

$$Lp_{12} = \frac{\mu_0}{4\pi} \int_{\ell_1} \int_{\ell_2} \frac{\hat{t}_1 \cdot \hat{t}_2}{R_{1,2}} \, d\ell_2 \, d\ell_1 \tag{C.55}$$

where ℓ_1 and ℓ_2 are the two wire segments shown in Fig. C.13 and the dot product between the tangential vectors is $\hat{t}_1 \cdot \hat{t}_2$. This first result is derived from [1, 2]. The partial inductance between the filaments 1 and 2 is given by

$$\begin{aligned} Lp_{12} = \frac{\mu_0}{4\pi} C_E \Bigg[& 2\,(T_2 + \ell_2) \tanh^{-1}\left(\frac{\ell_1}{R_1 + R_2}\right) \\ & + 2\,(T_1 + \ell_1) \tanh^{-1}\left(\frac{\ell_2}{R_1 + R_4}\right) \\ & - 2\,T_2 \tanh^{-1}\left(\frac{\ell_1}{R_3 + R_4}\right) \\ & -2\,T_1 \tanh^{-1}\left(\frac{\ell_2}{R_2 + R_3}\right) - \frac{W\,d}{S_E}\Bigg]. \end{aligned} \tag{C.56}$$

where

$$\begin{aligned} W = \Bigg[& tan^{-1}\left(\frac{C_{DQE} + (T_2 + \ell_2)(T_1 + \ell_1)\, S_{EQ}}{R_1\, S_{ED}}\right) \\ & - tan^{-1}\left(\frac{C_{DQE} + (T_2 + \ell_2)\, T_1\, S_{EQ}}{R_2\, S_{ED}}\right) \end{aligned}$$

$$+ \tan^{-1}\left(\frac{C_{DQE} + T_1\, T_2\, S_{EQ}}{R_3\, S_{ED}}\right)$$
$$\left. - \tan^{-1}\left(\frac{C_{DQE} + (T_1 + \ell_1)\, T_2\, S_{EQ}}{R_4\, S_{ED}}\right)\right] \tag{C.57}$$

The following definitions are used for the two filaments:

$$a_1 = x_{e2} - x_{e1} \quad , \quad a_2 = x_{e2} - x_{s1} \tag{C.58}$$
$$a_3 = x_{s2} - x_{s1} \quad , \quad a_4 = x_{e1} - x_{s2} \tag{C.59}$$
$$a_5 = x_{e1} - x_{s1} \quad , \quad a_6 = x_{e2} - x_{s2} \tag{C.60}$$
$$b_2 = (y_{e2} - y_{s2})\,(1 - 10^{-8}) \quad , \tag{C.61}$$
$$\ell_1 = a_5 \quad , \quad \ell_2^2 = a_6^2 + b_2^2 \tag{C.62}$$
$$R_1^2 = a_1^2 + y_{e2}^2 \quad , \quad R_2^2 = a_2^2 + y_{e2}^2 \tag{C.63}$$
$$R_3^2 = a_3^2 + y_{s2}^2 \quad , \quad R_4^2 = a_4^2 + y_{s2}^2 \tag{C.64}$$
$$S_1^2 = R_4^2 - R_3^2 + R_2^2 - R_1^2 \quad , \quad S_2^2 = R_2^2 - R_3^2 - \ell_2^2 \tag{C.65}$$
$$S_3^2 = R_4^2 - R_3^2 - \ell_1^2 \quad , \quad S_4^2 = 4\,\ell_1^2\,\ell_2^2 - (S_1^2)^2 \tag{C.66}$$

The translations for the start of the wires are given by

$$T_1 = \ell_1 \frac{2\,\ell_2^2\,S_3^2 + S_1^2\,S_2^2}{S_4^2} \quad , \quad T_2 = \ell_2 \frac{2\,\ell_1^2\,S_2^2 + S_1^2\,S_3^2}{S_4^2}. \tag{C.67}$$

were we use the notation for the angle ε:

$$C_E = \cos(\varepsilon) = \frac{S_1^2}{2\,\ell_1\,\ell_2}, \quad S_{EQ} = \sin^2(\varepsilon) = 1 - C_E^2 \tag{C.68a}$$
$$S_E = \sqrt{S_{EQ}}, \quad C_{DQE} = d^2\, C_E, \quad S_{ED} = d\, S_E \tag{C.68b}$$

where $d^2 = R_3^2 - T_2^2 - T_1^2 + 2\,T_1\,T_2\,C_E$ (like in equation C. 79).

Again, this filament-to-filament partial mutual inductance is used in combination with numerical integration for different cross-sections, for example, [16].

This equation should be used if the wires are located in the same plane since this formula is simpler than the ones in the next sections. Please note that we did add a small number of the orientation of the wire such that parallel wires can be treated by this formula.

C.2.3 *Lp* for Wire Filaments with an Arbitrary Direction

In this section, the filaments are oriented in any mutual orientation as shown in Fig. C.14. This allows an arbitrary relative orientation of the filaments. Using the rotation and translation operations in (C.2) to (C.5), we can place the two wires in any location in the global rectangular coordinate system. Hence, this result will be ideally suited for the filament representation of non-orthogonal quadrilateral and hexahedral shapes.

The integral for this case can be set up as

$$Lp_{12} = \frac{\mu_0 \cos\,\varepsilon}{4\pi} \int_{x_{s1}}^{x_{e1}} \int_{\alpha=0}^{\alpha=1} \frac{d\alpha\; dx}{|\boldsymbol{r}_1 - \boldsymbol{r}_2|} \tag{C.69}$$

where the angle ε is defined as in [1] for the dot product and the vector is

$$\boldsymbol{r}_2 = (1-\alpha)\,\boldsymbol{r}_{s2} + \alpha\,\boldsymbol{r}_{e2} \tag{C.70}$$

where $\boldsymbol{r}_{s2}$ and $\boldsymbol{r}_{e2}$ are the start and end point of filament 2, respectively, in the global coordinate system.

It is important to minimize the number of divisions and multiplications. For this reason, we present the formulation in a more compute friendly form

$$\begin{aligned} Lp_{12} = \frac{\mu_0}{4\pi}\, C_E \bigg[& 2\,(T_2 + \ell_2)\tanh^{-1}\left(\frac{\ell_1}{R_1+R_2}\right) \\ & + 2\,(T_1+\ell_1)\tanh^{-1}\left(\frac{\ell_2}{R_1+R_4}\right) \\ & - 2\,T_2\tanh^{-1}\left(\frac{\ell_1}{R_3+R_4}\right) \\ & - 2\,T_1\tanh^{-1}\left(\frac{\ell_2}{R_2+R_3}\right) - \frac{W\,d}{S_E}\bigg] \end{aligned} \tag{C.71}$$

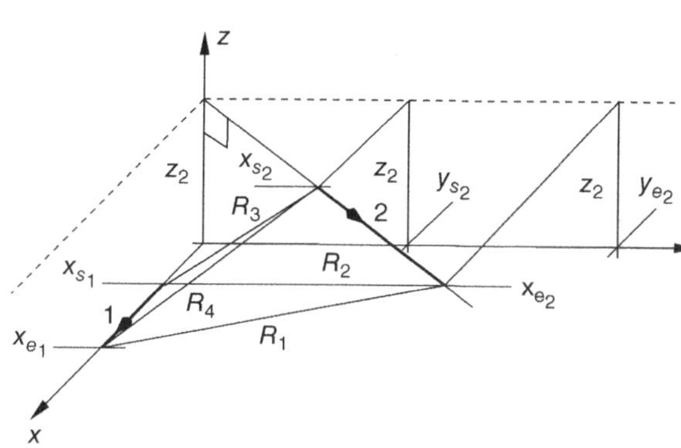

FIGURE C.14 Partial mutual inductance for two wires in arbitrary relative directions.

where

$$W = \left[tan^{-1} \left(\frac{C_{DQE} + (T_2 + \ell_2)(T_1 + \ell_1)\, S_{EQ}}{R_1\, S_{ED}} \right) \right.$$
$$- tan^{-1} \left(\frac{C_{DQE} + (T_2 + \ell_2)\, T_1\, S_{EQ}}{R_2\, S_{ED}} \right)$$
$$+ tan^{-1} \left(\frac{C_{DQE} + T_1\, T_2\, S_{EQ}}{R_3\, S_{ED}} \right)$$
$$\left. -tan^{-1} \left(\frac{C_{DQE} + (T_1 + \ell_1)\, T_2\, S_{EQ}}{R_4\, S_{ED}} \right) \right] \tag{C.72}$$

The following definitions are used for the arbitrary filaments:

$$a_1 = x_{e2} - x_{e1} \quad , \quad a_2 = x_{e2} - x_{s1}$$
$$a_3 = x_{s2} - x_{s1} \quad , \quad a_4 = x_{s2} - x_{e1} \tag{C.73}$$
$$a_5 = x_{e1} - x_{s1} \quad , \quad a_6 = x_{e2} - x_{s2}$$

$$\ell_1 = a_5 \quad , \quad \ell_2^2 = a_6^2 + b_2^2 + c_2^2 \tag{C.74}$$

$$b_2 = y_{e2} - y_{s2} \quad , \quad c_2 = z_{e2} - z_{s2}$$
$$R_1^2 = a_1^2 + y_{e2}^2 + z_{e2}^2 \quad , \quad R_2^2 = a_2^2 + y_{e2}^2 + z_{e2}^2$$
$$R_3^2 = a_3^2 + y_{s2}^2 + z_{s2}^2 \quad , \quad R_4^2 = a_4^2 + y_{s2}^2 + z_{s2}^2 \tag{C.75}$$

$$S_1^2 = R_4^2 - R_3^2 + R_2^2 - R_1^2 \quad , \quad S_2^2 = R_2^2 - R_3^2 - \ell_2^2 \tag{C.76}$$

$$S_3^2 = R_4^2 - R_3^2 - \ell_1^2 \quad , \quad S_4^2 = 4\ell_1^2\ell_2^2 - (S_1^2)^2 \tag{C.77}$$

The translations for the start of the wires are given by

$$T_1 = \ell_1 \frac{2\,\ell_2^2\, S_3^2 + S_1^2\, S_2^2}{S_4^2} \quad , \quad T_2 = \ell_2 \frac{2\,\ell_1^2\, S_2^2 + S_1^2\, S_3^2}{S_4^2} \tag{C.78}$$

Another definition from Grover we use is

$$d^2 = R_3^2 - T_2^2 - T_1^2 + 2\,T_1\,T_2\,C_E \tag{C.79}$$

were we use the notation for the angle ε:

$$C_E = cos\,(\varepsilon) = \frac{S_1^2}{2\,\ell_1\,\ell_2}, \; S_{EQ} = sin^2(\varepsilon) = 1 - C_E^2 \tag{C.80a}$$

$$S_E = \sqrt{S_{EQ}}, \quad C_{DQE} = d^2\, C_E, \quad S_{ED} = d\, S_E \tag{C.80b}$$

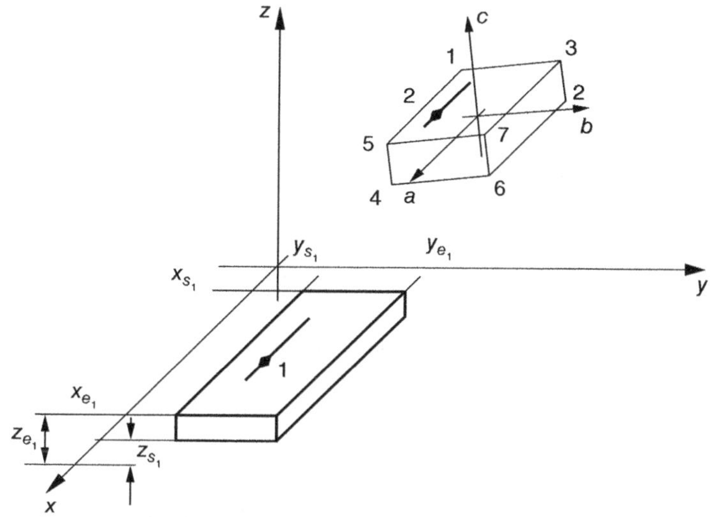

FIGURE C.15 Arbitrary oriented rectangular bars.

Again, this filament-to-filament partial mutual inductance is used in combination with numerical integration for different cross-sections, for example, [16]. Please note that a small number could be added to the coordinates for parallel wires as we did in the formula for the in plane case in the previous section.

C.2.4 *Lp* for Two Cells or Bars with Same Current Direction

In this example, we assume that all the current filaments are in the same direction, even if conductor cross sections are not parallel to the y–z coordinates as shown in Figure C.15. In this case, the unit vectors are the same or, $\hat{\boldsymbol{x}} = \hat{\boldsymbol{a}}$. Hence, the wire-to-wire formula (C.8) or the sheet-to-wire inductance in Section C.1.3 can be utilized.

Then, we can represent the partial inductance in the form

$$Lp_{12} = \frac{\mu_0}{4\pi} \int_{x_{s1}}^{x_{e1}} \int_{y_{s1}}^{y_{e1}} \int_{z_{s1}}^{z_{e1}} \int_{x_2} \int_{b_2} \int_{c_2} \frac{1}{R_{1,2}}\, dx_1\, dy_1\, dz_1\, dx_2\, db_2\, dc_2, \qquad \text{(C.81)}$$

where $R_{1,2} = R[r_1(x_1, y_1, z_2), r_2(x_2, b_2, c_2)]$. In this specific example, conductor 1 is in orthogonal coordinates and conductor 2 is in local coordinates.

Of course, if we apply the partial inductance for a sheet and a wire in (C.69), then the numerical integration for the sheet results in the numerical integration in the z-direction for conductor 1 and in the b' and c' direction for conductor 2.

C.2.5 *Lp* for Arbitrary Hexahedral Partial Self-Inductance

Progress in this area is also based on the Gauss law (3.33). Specifically, it was shown in Ref. [11] that the result is similar to quadrilateral surfaces (Section D.2.2). The resultant surface integral is obtained by starting with

$$\frac{1}{R(\boldsymbol{r},\boldsymbol{r}')} = \frac{-1}{2}\, \nabla' \times \nabla\, R(\boldsymbol{r},\boldsymbol{r}'). \qquad \text{(C.82)}$$

With this, the integration over the surface of the hexahedral shape replaces the integration over the volume as in Refs. [12] and [13] similar to (7.34) for surfaces

$$Ip_{11} = -\frac{1}{2}\sum_{i=1}^{6}\sum_{j=1}^{6}\int_{S_i'}\int_{S_j} R\,(\hat{\boldsymbol{u}}_j \cdot \hat{\boldsymbol{u}}_i')\, dS_j\, dS_i', \tag{C.83}$$

where in this case, $\hat{\boldsymbol{u}}_i$ is the normal to the surface i. We should note that the dot product between the current directions for the partial self-terms is given by currents in the same direction.

The Gauss law was applied in the local coordinate domain in Ref. [14] and good results have been obtained using the Gauss law as well as other results using approximate shapes for some of the integrals yielded very good results in Ref. [13].

C.2.6 *Lp* for Arbitrary Hexahedral Partial Mutual Inductance

The partial self- and mutual inductances for arbitrary hexahedral shapes are more challenging. We can use local coordinates to compute the partial mutual inductance for two arbitrarily placed hexahedral elements. An example for such elements is given in Refs [15, 16].

Figure C.16 shows two volume conductor cells. We assume that the local coordinate system for the first conductor is (a, b, c) and (a', b', c') for the second one. A vector in the global coordinates located on the first conductor is $\boldsymbol{r} = (x, y, z)$ for the first hexahedral element, and another one is $\boldsymbol{r}' = (x', y', z')$. The current directions are indicated by dashed line are pointing in the $\hat{\boldsymbol{a}}$ and $\hat{\boldsymbol{a}}'$ directions as is shown in Fig. C.16 and

$$Lp_{12} = \mu_0 \int_a\int_b\int_c\int_{a'}\int_{b'}\int_{c'} \hat{\boldsymbol{a}} \cdot \hat{\boldsymbol{a}}' \left|\frac{\partial \boldsymbol{r}}{\partial a}\right| \left|\frac{\partial \boldsymbol{r}'}{\partial a'}\right| g(\boldsymbol{r}, \boldsymbol{r}')\, dv\, dv', \tag{C.84}$$

where g is the scalar Green's function

$$g(\boldsymbol{r}, \boldsymbol{r}') = \frac{e^{-j\beta|\boldsymbol{r}-\boldsymbol{r}'|}}{4\pi|\boldsymbol{r}-\boldsymbol{r}'|} \quad \text{or, for a quasi-static case } g(\boldsymbol{r}, \boldsymbol{r}') = \frac{1}{4\pi|\boldsymbol{r}-\boldsymbol{r}'|}. \tag{C.85}$$

From the local coordinate system, we can determine a set of filaments in the global coordinate using the transform equations in section (8.2.6) to find the end points of each wire filament. Section D.2.2 can be used together with the numerical Gaussian integration. In Appendix E, the numerical integration is discussed which can be applied to the cross-sections to yield the partial inductance of interest.

To give more details, the local coordinate system presented in Chapter 8 is used to represent the hexahedral element shown in Fig. C.16. The purpose of the local coordinates is to identify the location of the points for the filaments in terms of the variables a, b, c where $a \in [-1, +1]$ and where a can represent $a = a, b, c$. The purpose is to uniquely map a point a, b, c into a point in the global coordinates $\boldsymbol{r}$. The local coordinates for the filament are used to map them to the global x, y, z coordinates. Mapping a point in the above hexahedron

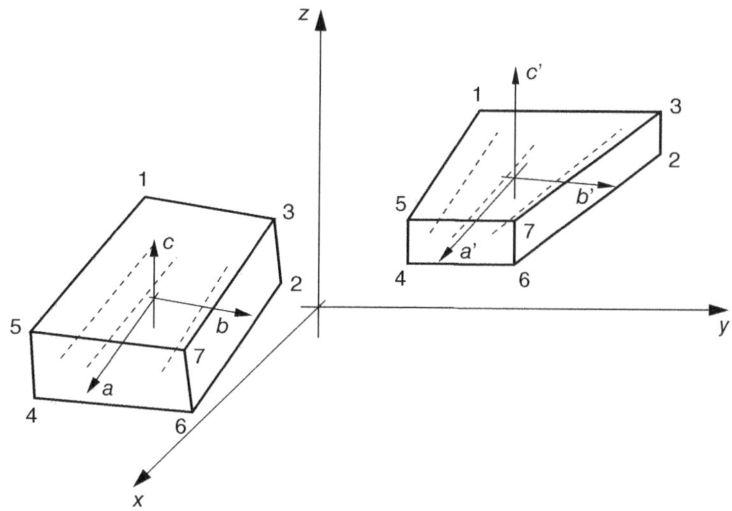

FIGURE C.16 General nonorthogonal hexahedral elements.

from a local coordinate point a, b, c to a global coordinate point x, y, z is described by

$$x = \sum_{k=0}^{7} N_k(a,b,c)\, x_k, \tag{C.86}$$

which is applied for $x = x, y, z$. The coefficients in (C.86) are repeated here as

$$N_0 = 1/8(1-a)(1-b)(1-c) \tag{C.87a}$$

$$N_1 = 1/8(1-a)(1-b)(1+c) \tag{C.87b}$$

$$N_2 = 1/8(1-a)(1+b)(1-c) \tag{C.87c}$$

$$N_3 = 1/8(1-a)(1+b)(1+c) \tag{C.87d}$$

$$N_4 = 1/8(1+a)(1-b)(1-c) \tag{C.87e}$$

$$N_5 = 1/8(1+a)(1-b)(1+c) \tag{C.87f}$$

$$N_6 = 1/8(1+a)(1+b)(1-c) \tag{C.87g}$$

$$N_7 = 1/8(1+a)(1+b)(1+c), \tag{C.87h}$$

where $a \in [-1, +1]$ and again $a = a, b, c$. The close relation to the binary variables has been given before.

With this, we are in a position to also express the tangential vectors with respect to the local coordinates as

$$\frac{\partial \boldsymbol{r}}{\partial a} = \frac{\partial x}{\partial a}\hat{\boldsymbol{x}} + \frac{\partial y}{\partial a}\hat{\boldsymbol{y}} + \frac{\partial z}{\partial a}\hat{\boldsymbol{z}}, \tag{C.88}$$

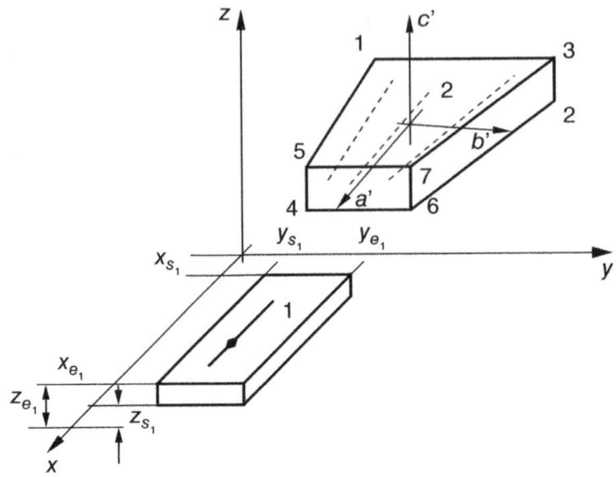

FIGURE C.17 Combined rectangular and quadrilateral cell example.

where the derivatives are found from (C.86). Finally, the magnitude of the tangential vector $h_a = |\partial \mathbf{r}_g/\partial a|$ where the position–dependent unit vectors can be determined from $\hat{\boldsymbol{a}} = (\partial \mathbf{r}_g/\partial a)/h_a$ where again $a = a, b, c$.

With this transformation, we can simplify the sixfold integration into a two-area integration over the current directions $\hat{\boldsymbol{a}}$ and $\hat{\boldsymbol{a}}'$, respectively

$$Lp_{aa'} = \mu_0 \int_a \int_b \int_c \int_{a'} \int_{b'} \int_{c'} \delta(b - b_0, b' - b_0', c - c_0, c' - c_0')$$

$$\hat{\boldsymbol{a}} \cdot \hat{\boldsymbol{a}}' \left| \frac{\partial \mathbf{r}}{\partial a} \right| \left| \frac{\partial \mathbf{r}'}{\partial a'} \right| g(\mathbf{r},\mathbf{r}')\, dv\, dv'$$

$$= Lp_{ff'}(b_0, b_0', c_0, c_0'), \tag{C.89}$$

where f and f' represent the two filaments. This is the partial mutual inductance between the two filaments pointing in $\hat{\boldsymbol{a}}$ and $\hat{\boldsymbol{a}}'$ directions. Hence, for the volumes of the two conductors, the total partial mutual inductance between two hexahedrons is

$$Lp_{aa'} = \int_b \int_c \int_{b'} \int_{c'} Lp_{ff'}(b, b', c, c')\, db\, db'\, dc\, dc', \tag{C.90}$$

where the integration over the cross sections is performed with Gaussian numerical integration. This section is added to point out the difference between the local and global coordinates. Compute time can be saved if the numerical integration is mostly used for the smaller dimensions of the cell sides. Local coordinates are efficient for the mapping of the dimensions to the normalized units used for numerical integration methods.

Local coordinates are only required for nonorthogonal conductor cell dimensions such as quadrilateral or hexahedral shapes. For the example in Fig. C.17, we can use mixed coordinates. Global coordinates are used for conductor cell 1 while local coordinates are used for conductor 2. The integrals are

$$Lp_{12} = \frac{\mu_0}{4\pi} \int_{x_{s1}}^{x_{e1}} \int_{y_{s1}}^{y_{e1}} \int_{z_{s1}}^{z_{e1}} \int_{a_2} \int_{b_2} \int_{c_2} \frac{1}{R_{1,2}} \, dx_1 \, dy_1 \, dz_1 \, da_2 \, db_2 \, dc_2. \tag{C.91}$$

Besides the change in the coordinate systems, the filament representation is the same as for the case in previous section. Representation is the same as for the case in the previous section. More details on nonorthogonal systems is given in Chapter 7.

REFERENCES

1. F. W. Grover. *Inductance Calculations: Working Formulas and Tables*. Dover Publications, New York, 1962.
2. C. Paul. *Inductance, Loop and Partial*. John Wiley and Sons, Inc., New York, 2010.
3. C. Hoer and C. Love. Exact inductance equations for rectangular conductors with applications to more complicated geometries. *Journal of Research National Bureau of Standards, Section C: Engineering and Instrumentation*, 69(C):127–137, April 1965.
4. A. E. Ruehli. Inductance calculations in a complex integrated circuit environment. *IBM Journal of Research and Development*, 16(5):470–481, September 1972.
5. P. K. Wolff and A. E. Ruehli. Inductance computations for complex three dimensional geometries. In *Proceedings of the IEEE International Symposium on Circuits and Systems*, pp. 16–19, 1981.
6. A. Djordjevic and T. K. Sakkar. Computation of inductance of simple vias between two striplines above a ground plane. *IEEE Transactions on Microwave Theory and Techniques*, MTT-33(3):265–269, March 1985.
7. R.-B. Wu, C.-N. Kuo, and K. K. Chang. Inductance and resistance computations for three-dimensional multiconductor interconnection structures. *IEEE Transactions on Microwave Theory and Techniques*, MTT-40(2):263–270, February 1992.
8. A. E. Ruehli, G. Antonini, and L. Jiang. Skin-effect model for round wires in PEEC. In *IEEE EMC Europe, Interantional Symposium on EMC*, Rome, Italy, September 2012.
9. A. E. Ruehli, G. Antonini, and L. Jiang. Skin effect loss models for time and frequency domain PEEC solver. *Proceedings of the IEEE*, 101(2):451–472, February 2013.
10. G. Zhong and C.-K. Koh. Exact closed-form formula for partial mutual inductances rectangular conductors. *IEEE Transactions on Circuits and Systems*, 50(10):1349–1353, October 2003.
11. L. Knockaert. A general Gauss theorem for evaluating singular integrals over polyhedral domains. *Electromagnetics*, 11(2):269–280, April 1991.
12. R. Y. Zhang, J. K. White, and J. G. Kassakian. Fast simulation of complicated 3-D structures above lossy magnetic media. *IEEE Transactions on Magnetics*, 50(10):2377–3384, October 2014.
13. Y. Hackl, P. Scholz, W. Ackermann, and T. Weiland. Multifunction approach and specialized numerical integration algorithms for fast inductance evaluations in nonorthogonal PEEC systems. *IEEE Transactions on Electromagnetic Compatibility*, 57(5):1155–1163, October 2015.

14. M. A. Cracraft. Mobile array design with ANSERLIN antennas and efficient, wide-band PEEC models for interconnect and power distribution network analysis. Ph.D. Dissertation [Online]. Available: *http://hdl.handle.net/10355/29582*, Missouri University Science and Technology, USA, 2007.
15. A. Muesing and J. W. Kolar. Efficient partial element calculation and the extension to cylindrical elements for the PEEC method. in *2008 11th Workshop on Control and Modeling for Power Electronics, COMPEL*, Zurich, August 2008.
16. A. Muesing, J. Ekman, and J. W. Kolar. Efficient calculation of non-orthogonal partial elements for the PEEC method. *IEEE Transactions on Magnetics*, 45(3):1140–1142, March 2009.

D

COMPUTATION OF PARTIAL COEFFICIENTS OF POTENTIAL

The second set of partial elements needed for PEEC models are partial coefficients of potential. The computation of these coefficients is fundamental for the capacitance calculations in Chapter 4 and also PEEC equivalent circuits in Chapter 6. This appendix is dedicated to the evaluation of the potential coefficients for surface charge cells. For clarity and ease of use, we again transformed all formulas to the same coordinate system and orientation. The basic formula for a quasistatic partial potential coefficient is given by (4.26) as

$$\begin{aligned} Pp_{12} &= \frac{1}{\epsilon\, S_1\, S_2} \int_{S_2} \int_{S_1} g(\boldsymbol{r}, \boldsymbol{r}')\, dS_1\, dS_2 \\ &= \frac{1}{4\pi\epsilon\, S_1\, S_2} \int_{S_2} \int_{S_1} \frac{1}{|\boldsymbol{r}_1 - \boldsymbol{r}_2|}\, dS_1\, dS_2, \end{aligned} \tag{D.1}$$

where the details on the surface integrals are given in Section 6.3.3.

Here, $g(\boldsymbol{r}, \boldsymbol{r}')$ is a scalar Green's function, which is quasi-static unless the retardation in it is included as done in Section 5.8. The frequency domain retardation is not considered inside of the integrals in this appendix, while the time domain fundamentally includes delays externally. Methods for including the frequency domain delay are considered in Chapters 5 and 13. The partial coefficients of potential consist of normalized symmetric integrations over two cell surfaces S_1 and S_2 in (D.1).

We want to point out that the normalization by the surface areas S_1, S_2 is such that the related quantities are total charge on the cells rather charge densities. This leads to a circuit matrix that is better conditioned [1].

Some formulas are derived based on specific approximations, while others are general analytical results. Where analytical results could not be found or are too complex,

Circuit Oriented Electromagnetic Modeling Using the PEEC Techniques, First Edition.
Albert E. Ruehli, Giulio Antonini, and Lijun Jiang.

appropriate numerical integration methods are used, which are given in Appendix E. We found that analytical results are usually faster to evaluate.

As considered for partial inductance formulas, in Section 5.5.2 inaccuracies can be introduced by very complex analytical results for large parameter aspect ratios. However, the potential coefficients are in general more accurate since they involve only surface integrals rather than volume integrals as is the case for partial inductances. Fortunately, double-precision arithmetic is used for most computers, which helps with the accuracy of the formula evaluation.

It is clear from Chapter 8 that it is desirable to allow much larger aspect ratio mesh cells for PEEC than those generally used in other approaches since it reduces compute time due to the smaller number of unknowns. For this reason, we also attempt to compute the partial elements with sufficient accuracy for large aspect ratio cells. A mixed, analytic, and numerical approach we call *multifunction* is presented in Appendix E.

D.1 PARTIAL POTENTIAL COEFFICIENTS FOR ORTHOGONAL GEOMETRIES

D.1.1 Pp_{12} for Two Parallel Wires

One of the simplest coefficients is the wire-to-wire (filament-to-filament) partial potential, which we call Pp_{12}. Of course, this result is singular if the wires touch.

The integrals for the wire-to-wire partial potential Pp_{12} for the geometry shown in Fig. D.1 is

$$Pp_{12} = \frac{1}{4\pi\epsilon}\frac{1}{\ell_1\ell_2}\int_{x_{s1}}^{x_{e1}}\int_{x_{s2}}^{x_{e2}}\frac{1}{R_{1,2}}\,dx_2\,dx_1, \tag{D.2}$$

where

$$R_{1,2} = \sqrt{(x_1 - x_2)^2 + (y_1 - y_2)^2 + (z_1 - z_2)^2} \tag{D.3a}$$

and the wire lengths are given by

$$\ell_1 = x_{e1} - x_{s1}, \quad \ell_2 = x_{e2} - x_{s2}. \tag{D.3b}$$

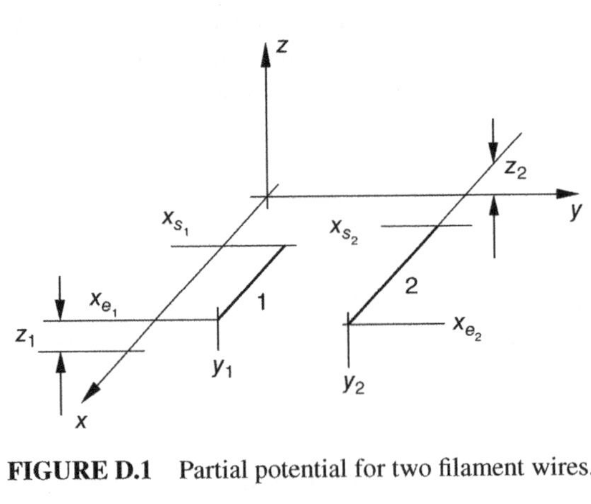

FIGURE D.1 Partial potential for two filament wires.

The closed-form solution for (D.2) is easy to obtain.

$$Pp_{12} = \frac{1}{4\pi\epsilon}\frac{1}{\ell_1\ell_2}\sum_{k=1}^{4}(-1)^{k+1}[a_k \log(a_k + r_k) - r_k] \tag{D.4}$$

with

$$a_1 = x_{s2} - x_{e1}, \quad a_2 = x_{e2} - x_{e1} \tag{D.5a}$$

$$a_3 = x_{e2} - x_{s1}, \quad a_4 = x_{s2} - x_{s1} \tag{D.5b}$$

$$r_k = \sqrt{a_k^2 + (y_2 - y_1)^2 + (z_2 - z_1)^2}. \tag{D.5c}$$

This exact equation is used extensively for wire-type models and also for numerical integration over cross-sectional-type geometries.

D.1.2 $1/R^3$ Integral Ip_{12} for Two Parallel Wires

For some applications, such as the surface formulation in Section 11.4.2, we also need integrals with an $1/R^3$ kernel. The geometry is the same as the one in the previous section in Fig. D.1. Hence, the integral to be solved is

$$Ip_{12} = \frac{1}{\ell_1\ell_2}\int_{x_{s1}}^{x_{e1}}\int_{x_{s2}}^{x_{e2}}\frac{1}{R_{1,2}^3}\,dx_2\,dx_1, \tag{D.6}$$

where

$$R_{1,2} = \sqrt{(x_1 - x_2)^2 + (y_1 - y_2)^2 + (z_1 - z_2)^2} \tag{D.7a}$$

and

$$\ell_1 = x_{e1} - x_{s1}, \quad \ell_2 = x_{e2} - x_{s2}. \tag{D.7b}$$

Then, the closed form of this partial integral is

$$Ip_{12} = \frac{1}{\ell_1\ell_2\, r_q}\sum_{k=1}^{4}(-1)^{k+1}\sqrt{a_k^2 + (y_2 - y_1)^2 + (z_2 - z_1)^2} \tag{D.8}$$

with

$$a_1 = x_{e2} - x_{s1}, \quad a_2 = x_{s2} - x_{s1} \tag{D.9a}$$

$$a_3 = x_{s2} - x_{e1}, \quad a_4 = x_{e2} - x_{e1}. \tag{D.9b}$$

We treat these integrals in the same symmetrical way we do the other partial elements.

D.1.3 $1/R^3$ Integral Ip_{12} for Two Orthogonal Filaments

Unlike partial inductances, the potentials are coupled for the orthogonal conductors shown in Fig. D.2. The same is true for other coupling elements.

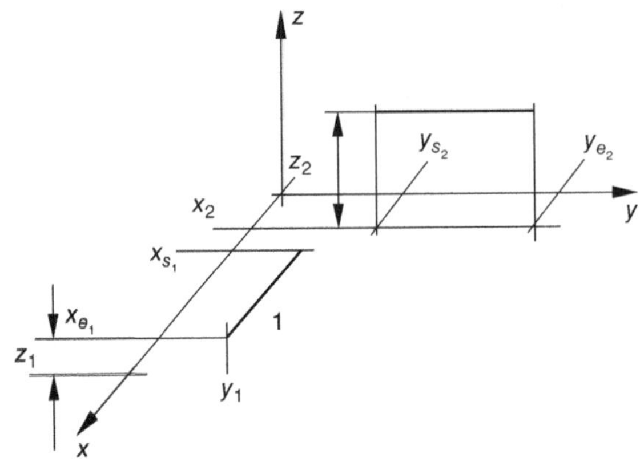

FIGURE D.2 Two orthogonal wires.

The kernel for this case is of the $1/R^3$ form and the integral to be solved is

$$Ip_{12} = \frac{1}{\ell_1 \ell_2} \int_{x_{s1}}^{x_{e1}} \int_{y_{s2}}^{y_{e2}} \frac{1}{R_{1,2}^3} \, dx_2 \, dy_1, \tag{D.10}$$

where

$$R_{1,2} = \sqrt{(x_1 - x_2)^2 + (y_1 - y_2)^2 + (z_1 - z_2)^2} \tag{D.11a}$$

and

$$\ell_1 = x_{e1} - x_{s1}, \quad \ell_2 = y_{e2} - y_{s2}. \tag{D.11b}$$

Then, the closed form of this partial integral is

$$Ip_{12} = \frac{1}{4\pi\epsilon} \frac{1}{\ell_1 \ell_2} (I_e - I_s) \tag{D.12}$$

with

$$I_i = \frac{|x_{e1} - x_2|}{2(x_{e1} - x_2)\, Z} \left[\sin^{-1}\left(\frac{(a_i^2 - Z^2)\, b_e - 2\, a_i^2\, Z^2}{b_e |a_i + Z|} \right) - \sin^{-1}\left(\frac{(a_i^2 - Z^2)\, b_s - 2\, a_i^2\, Z^2}{b_s\, |a_i + Z|} \right) \right], \tag{D.13}$$

where we replace $i = e, s$, for the two parts, and use the constants

$$a_e = x_{e1} - x_2, \quad b_e = (y_{e2} - y_{s1})^2 + Z^2 \tag{D.14a}$$

$$a_s = x_{s1} - x_2, \quad b_s = (y_{s2} - y_1)^2 + Z^2 \tag{D.14b}$$

$$Z = z_2 - z_1. \tag{D.14c}$$

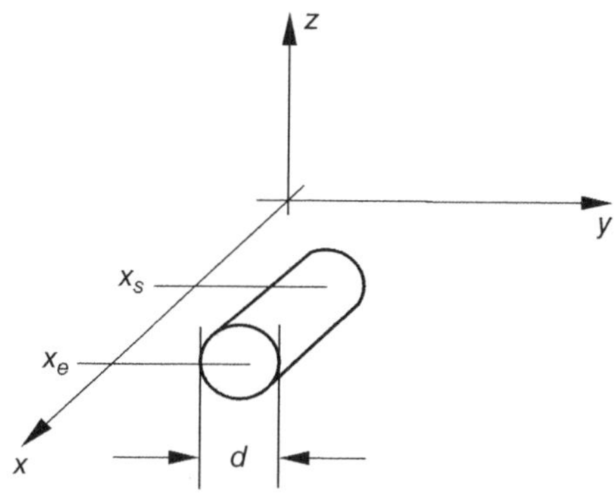

FIGURE D.3 Partial potential coefficient for round tube.

D.1.4 Pp_{11} for Round Tube Cell Shape

We give a formula for the partial self-potential coefficient for a round tube cell, which is of importance for several models such as thin wire-type structures. Here, a new formula is used to evaluate the coefficient. However, the approach is similar to the one in Ref. [2]. The partial coefficient of potential for a zero thickness cylindrical tube with a radius $a = d/2$ is similar to the partial inductance in Section C.1.2 for Lp (Fig. D.3).

The integral that is evaluated in this case is

$$Pp_{11} = \frac{1}{4\pi\epsilon\ S^2}\int_S\int_{S'}\frac{1}{R_{1,1'}}\,dS'\,dS, \tag{D.15}$$

where S represents the surface of the tube and where

$$R_{1,1'} = \sqrt{(x_1 - x_1')^2 + (y_1 - y_1')^2 + (z_1 - z_1')^2}. \tag{D.16}$$

Cylindrical coordinates and symmetry are used to reduce the fourfold integral to a threefold integral as was done for the partial inductance. The solution of the integrals is based on the assumption that the length ℓ of the wire is longer than the diameter $d = 2a$. The result for the tube conductor is

$$Pp_{11} = \frac{1}{4\,\epsilon_0\ell}\Bigg[\left(\frac{k^2}{480} + \frac{k^4}{1280} + \frac{1}{3600}\right)\pi^3 + \left(\frac{1}{18} - \frac{k^2}{24}\right)\pi + \Big(2 - 2\log(\ell) + 6\log(2) + 2\log(a) - 4\log(k\pi)\Big)\frac{1}{\pi} + \frac{8a}{\ell\pi^2}\Bigg], \tag{D.17}$$

where $k = d/\ell$ and $\ell = x_e - x_s$. Hence, its evaluation is not difficult.

D.1.5 Pp_{12} for a Sheet and a Filament

Another important building block for the partial potential coefficient conductor is a filament and a uniformly charged conducting sheet or cell as shown in Fig. D.4. Of course, the wire

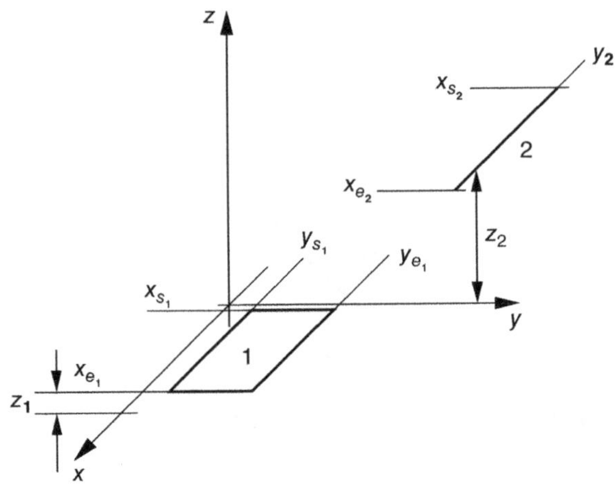

FIGURE D.4 Partial potential coefficient for filament and sheet.

may only be part of another surface. The integration for the partial potential coefficients is given by

$$Pp_{12} = \frac{1}{S_1 \ell_2} \frac{1}{4\pi\epsilon} \int_{y_{s1}}^{y_{e1}} \int_{x_{s1}}^{x_{e1}} \int_{x_{s2}}^{x_{e2}} \frac{1}{R_{1,2}} \, dx_2 \, dx_1 \, dy_1. \tag{D.18}$$

The analytical solution for the above formula is

$$Pp_{12} = \frac{1}{4\pi\epsilon} \frac{1}{S_1 \, (x_{e2} - x_{s2})} \sum_{k=1}^{4} \sum_{m=1}^{2} (-1)^{m+k} \left[\frac{a_k^2 - Z^2}{2} \log(b_m + r_{km} + \widehat{\epsilon}\,) \right.$$
$$\left. + \; a_k \, b_m \log(a_k + r_{km} + \widehat{\epsilon}\,) - a_k \, Z \tan^{-1}\left(\frac{a_k \, b_m}{Z \, r_{km}}\right) - \frac{b_m}{2} \, r_{km} \right] \tag{D.19}$$

with

$$a_1 = x_{s2} - x_{e1}, \quad a_2 = x_{e2} - x_{e1} \tag{D.20a}$$

$$a_3 = x_{e2} - x_{s1}, \quad a_4 = x_{s2} - x_{s1} \tag{D.20b}$$

$$b_1 = y_2 - y_{s1}, \quad b_2 = y_2 - y_{e1} \tag{D.20c}$$

$$Z = z_2 - z_1, \quad r_{km} = \sqrt{a_k^2 + b_m^2 + Z^2}. \tag{D.20d}$$

This formula is useful directly if one conductor can be approximated by a thin wire and also for the case where the second conductor is of a different shape. Please note that the $\widehat{\epsilon}$ inside the brackets in equation (D.19) represents a very small number to prevent the singular behavior of the numerical evaluation of the equation. It is also useful for the case of a wide and and a narrow capacitive cell.

D.1.6 Pp_{11} for Rectangular Sheet Cell

The analytical Pp for a zero thickness rectangular cell shown in Fig. D.5 leads to a relatively simple formula.

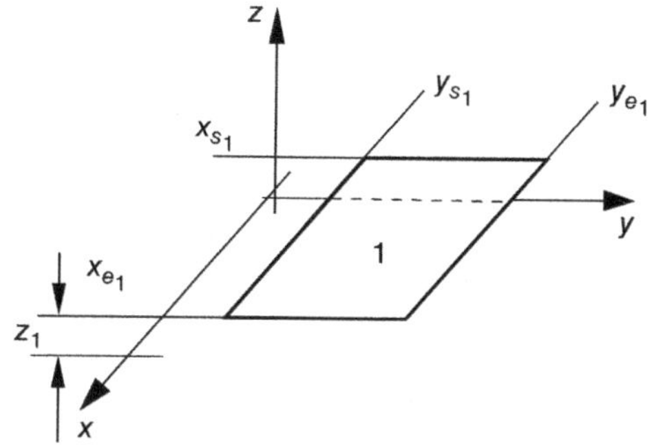

FIGURE D.5 Single zero thickness conductor sheets.

The result is

$$Pp_{11} = \frac{1}{6\,\pi\ell\epsilon}\left[\,3\log\left(u+\sqrt{u^2+1}\right)+u^2+\frac{1}{u}\right.$$
$$\left.+\;3\,u\log\left(\frac{1}{u}+\sqrt{\frac{1}{u^2}+1}\right)-\left(u^{4/3}+\frac{1}{u^{2/3}}\right)^{3/2}\right], \qquad \text{(D.21)}$$

where $u = \ell/(y_{e1} - y_{s1})$ and $\ell = x_{e1} - x_{s1}$.

Importantly, this result is computationally useful since it is simple. It also eliminates the singularity problem for the partial self-potential term.

D.1.7 Pp_{12} for Two Parallel Rectangular Sheet Cells

The geometry for two parallel conducting sheets is shown in Fig. D.6, which was used in Ref. [3]. Here, we transformed it into our standard coordinate system.

The solution for this case is given by

$$Pp_{12} = \frac{1}{4\pi\epsilon}\,\frac{1}{S_1S_2}\int_{y_{s1}}^{y_{e1}}\int_{x_{s1}}^{x_{e1}}\int_{y_{s2}}^{y_{e2}}\int_{x_{s2}}^{x_{e2}}\frac{1}{R_{1,2}}dx_2dy_2dx_1dy_1, \qquad \text{(D.22)}$$

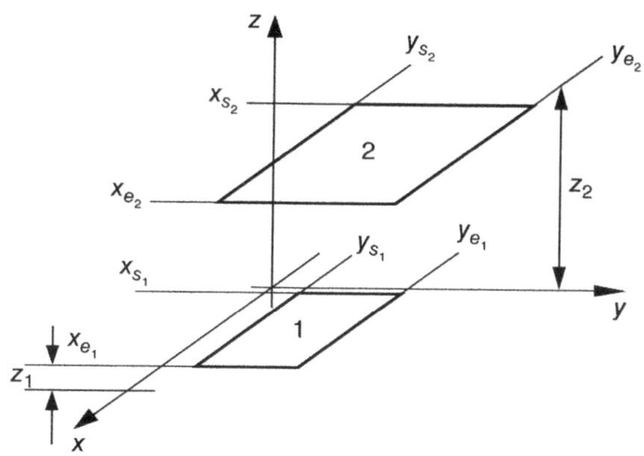

FIGURE D.6 Two parallel rectangular conducting sheets.

where

$$S_1 = (x_{e1} - x_{s1})(y_{e1} - y_{s1}) \tag{D.23a}$$

$$S_2 = (x_{e2} - x_{s2})(y_{e2} - y_{s2}) \tag{D.23b}$$

$$R_{1,2} = \sqrt{(x_1 - x_2)^2 + (y_1 - y_2)^2 + (z_1 - z_2)^2}. \tag{D.23c}$$

Then, the closed form of this potential coefficient is

$$\begin{aligned} Pp_{12} = \frac{1}{4\pi\epsilon} \frac{1}{S_1 S_2} \sum_{k=1}^{4} \sum_{m=1}^{4} (-1)^{m+k} & \left[\frac{b_m^2 - Z^2}{2} a_k \log(a_k + r_{km} + \hat{\epsilon}) \right. \\ & + \frac{a_k^2 - Z^2}{2} b_m \log(b_m + r_{km} + \hat{\epsilon}) - \frac{1}{6}(b_m^2 - 2Z^2 + a_k^2) r_{km} \\ & \left. - b_m Z a_k \tan^{-1}\left(\frac{a_k b_m}{r_{km} Z}\right) \right], \end{aligned} \tag{D.24}$$

where

$$a_1 = x_{s2} - x_{e1}, \qquad a_2 = x_{e2} - x_{e1} \tag{D.25a}$$

$$a_3 = x_{e2} - x_{s1}, \qquad a_4 = x_{s2} - x_{s1} \tag{D.25b}$$

$$b_1 = y_{s2} - y_{e1}, \qquad b_2 = y_{e2} - y_{e1} \tag{D.25c}$$

$$b_3 = y_{e2} - y_{s1}, \qquad b_4 = y_{s2} - y_{s1} \tag{D.25d}$$

$$Z = z_2 - z_1 + \hat{\epsilon}, \qquad r_{km} = \sqrt{a_k^2 + b_m^2 + Z^2}. \tag{D.25e}$$

It is noted that $\hat{\epsilon} = 10^{-37}$ is a very small number. This expression can be applied to both the self- and mutual potential coefficients. However, it is computationally more complex for the self-term in the previous section.

D.1.8 Pp_{12} for Two Orthogonal Rectangular Sheet Cells

Two rectangular cell sheets at an angle of 90° are shown in Fig. D.7. The partial potential is similar to (D.6) given by

$$Pp_{12} = \frac{1}{4\pi\epsilon} \frac{1}{S_1 S_2} \int_{y_{s1}}^{y_{e1}} \int_{x_{s1}}^{x_{e1}} \int_{z_{s2}}^{z_{e2}} \int_{x_{s2}}^{x_{e2}} \frac{1}{R_{1,2}} \, dx_2 \, dz_2 \, dx_1 \, dy_1, \tag{D.26}$$

where

$$S_1 = (x_{e1} - x_{s1})(y_{e1} - y_{s1}) \tag{D.27a}$$

$$S_2 = (x_{e2} - x_{s2})(z_{e2} - z_{s2}) \tag{D.27b}$$

$$R_{1,2} = \sqrt{(x_1 - x_2)^2 + (y_1 - y_2)^2 + (z_1 - z_2)^2}. \tag{D.27c}$$

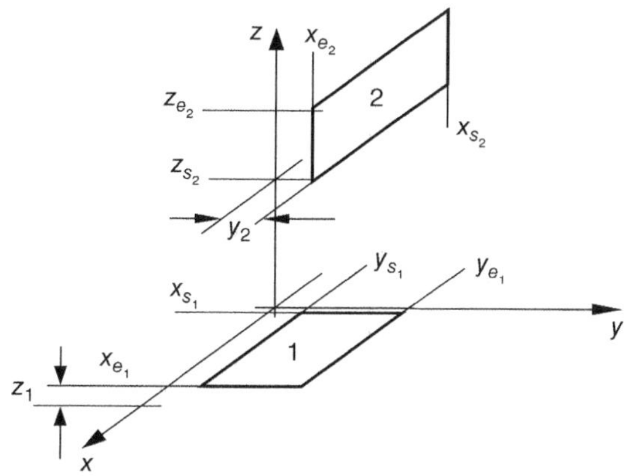

FIGURE D.7 Two orthogonal rectangular conducting sheets.

Then, the final formulation of Pp_{12} for two orthogonal charge sheets is

$$Pp_{12} = \frac{1}{4\pi\epsilon}\frac{1}{S_1S_2}\sum_{k=1}^{4}\sum_{m=1}^{2}\sum_{\ell=1}^{2}(-1)^{\ell+m+k+1}\left[a_k b_m c_\ell \log(a_k + r_{km\ell} + \hat{\epsilon}) + \left(\frac{a_k^2}{2} - \frac{c_\ell^2}{6}\right) c_\ell \log(b_m + r_{km\ell} + \hat{\epsilon}) - \frac{b_m c_\ell}{3} r_{km\ell} + b_m\left(\frac{a_k^2}{2} - \frac{b_m^2}{6}\right)\log(c_\ell + r_{km\ell} + \hat{\epsilon}) - \frac{a_k c_\ell^2}{2}\tan^{-1}\left(\frac{a_k b_m}{c_\ell\, r_{km\ell}}\right) - \frac{a_k^3}{6}\tan^{-1}\left(\frac{b_m c_\ell}{a_k\, r_{km\ell}}\right) - \frac{b_m^2 a_k}{2}\tan^{-1}\left(\frac{a_k c_\ell}{b_m\, r_{km\ell}}\right)\right], \tag{D.28}$$

where

$$a_1 = x_{s2} - x_{e1}, \quad a_2 = x_{e2} - x_{e1} \tag{D.29a}$$

$$a_3 = x_{e2} - x_{e1}, \quad a_4 = x_{s2} - x_{s1} \tag{D.29b}$$

$$b_1 = y_2 - y_{s1}, \quad b_2 = y_2 - y_{e1} \tag{D.29c}$$

$$c_1 = z_{e2} - z_1, \quad c_2 = z_{s2} - z_1 \tag{D.29d}$$

$$r_{km\ell} = \sqrt{a_k^2 + b_m^2 + c_\ell^2}. \tag{D.29e}$$

Again, $\hat{\epsilon} = 10^{-37}$ is a very small number to prevent singularities.

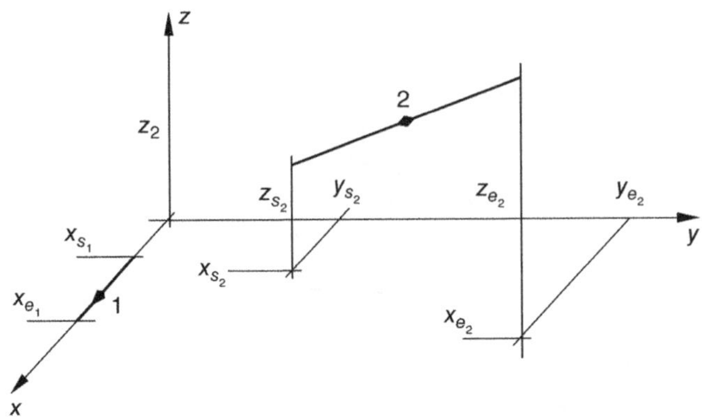

FIGURE D.8 Partial mutual inductance for two wires in arbitrary relative directions.

D.2 PARTIAL POTENTIAL COEFFICIENT FORMULAS FOR NONORTHOGONAL GEOMETRIES

D.2.1 Pp for Wire Filaments with an Arbitrary Direction

In this section, the filaments are oriented in any mutual orientation as shown in Fig. D.8. This allows an arbitrary relative orientation of the filaments. Using the rotation and translation operations in (C.2) to (C.5), we can place the two wires in any location in the global rectangular coordinate system. Hence, this result will be ideally suited for the filament representation of non-orthogonal quadrilateral and hexahedral shapes. It is important to minimize the number of divisions and multiplications. For this reason, we present the formulation in a more compute friendly form

$$
\begin{aligned}
Pp_{12} = \frac{1}{4\pi\varepsilon\ell_1\ell_2} & \left[2(T_2+\ell_2)\tanh^{-1}\left(\frac{\ell_1}{R_1+R_2}\right) \right. \\
& + 2(T_1+\ell_1)\tanh^{-1}\left(\frac{\ell_2}{R_1+R_4}\right) \\
& - 2T_2\tanh^{-1}\left(\frac{\ell_1}{R_3+R_4}\right) \\
& \left. -2T_1\tanh^{-1}\left(\frac{\ell_2}{R_2+R_3}\right) - \frac{Wd}{S_E} \right]
\end{aligned}
\tag{D.30}
$$

where

$$
\begin{aligned}
W = & \left[tan^{-1}\left(\frac{C_{DQE}+(T_2+\ell_2)(T_1+\ell_1)S_{EQ}}{R_1 S_{ED}}\right) \right. \\
& - tan^{-1}\left(\frac{C_{DQE}+(T_2+\ell_2)T_1 S_{EQ}}{R_2 S_{ED}}\right)
\end{aligned}
$$

$$+\tan^{-1}\left(\frac{C_{DQE}+T_1T_2S_{EQ}}{R_3S_{ED}}\right)$$

$$\left.-\tan^{-1}\left(\frac{C_{DQE}+(T_1+\ell_1)T_2S_{EQ}}{R_4S_{ED}}\right)\right] \tag{D.31}$$

The following definitions are used for the arbitrary filaments:

$$\begin{aligned} a_1 &= x_{e2}-x_{e1} \quad , \quad a_2 = x_{e2}-x_{s1} \\ a_3 &= x_{s2}-x_{s1} \quad , \quad a_4 = x_{s2}-x_{e1} \\ a_5 &= x_{e1}-x_{s1} \quad , \quad a_6 = x_{e2}-x_{s2} \end{aligned} \tag{D.32}$$

$$\ell_1 = a_5 \quad , \quad \ell_2^2 = a_6^2+b_2^2+c_1^2 \tag{D.33}$$

$$b_2 = y_{e2}-y_{s2} \quad , \quad c_1 = z_{e2}-z_{s2}$$

$$\begin{aligned} R_1^2 &= a_1^2+y_{e2}^2+z_{e2}^2 \quad , \quad R_2^2 = a_2^2+y_{e2}^2+z_{e2}^2 \\ R_3^2 &= a_3^2+y_{s1}^2+z_{s2}^2 \quad , \quad R_4^2 = a_4^2+y_{s1}^2+z_{s2}^2 \end{aligned} \tag{D.34}$$

$$S_1^2 = R_4^2-R_3^2+R_2^2-R_1^2 \quad , \quad S_2^2 = R_2^2-R_3^2-\ell_2^2 \tag{D.35}$$

$$S_3^2 = R_4^2-R_3^2-\ell_1^2 \quad , \quad S_4^2 = 4\ell_1^2\ell_2^2-(S_1^2)^2 \tag{D.36}$$

The translations for the start of the wires are given by

$$T_1 = \ell_1\frac{2\ell_2^2S_3^2+S_1^2S_2^2}{S_4^2} \quad , \quad T_2 = \ell_2\frac{2\ell_1^2S_2^2+S_1^2S_3^2}{S_4^2} \tag{D.37}$$

Another definition from Grover we use is

$$d^2 = R_3^2-T_2^2-T_1^2+2T_1T_2C_E \tag{D.38}$$

were we use the notation for the angle ε:

$$C_E = \cos\varepsilon = \frac{S_1^2}{2\ell_1\ell_2}, S_{EQ} = \sin^2(\varepsilon) = 1-C_E^2 \tag{D.39a}$$

$$S_E = \sqrt{S_{EQ}}, \quad C_{DQE} = d^2C_E, \quad S_{ED} = dS_E \tag{D.39b}$$

Again, this filament-to-filament partial coefficients of potential is used in combination with numerical integration to represent capacitive cell surfaces.

D.2.2 Pp_{12} for a Pair of General Quadrilaterals on Same Plane

It is clear that nonorthogonal partial potential coefficients are difficult to compute analytically. As always, the most challenging parts are the self-coefficients and the immediate neighbor conductors.

The analytic solution for two quadrilateral shapes is treated in Section 7.3.1. The quadrilaterals are located on a plane xOy, which we assume to be at $z = 0$ as shown in Fig. D.9. The two nonorthogonal quadrilaterals 1 and 2 with surfaces S_1 and S_2 represent cells in local coordinates a, b. However, the solution is obtained in the global orthogonal coordinate system, x, y, z. The integral to be evaluated is

$$Pp_{12} = \frac{1}{4\pi\epsilon S_1 S_2} \int_{S_1} \int_{S_2} \frac{1}{|\mathbf{r} - \mathbf{r}'|} dS_1 dS_2 = \frac{1}{4\pi\epsilon S_1 S_2} \int_{S_1} \int_{S_2} \frac{1}{R} dS_1\, dS_2 \qquad \text{(D.40)}$$

with

$$R = \sqrt{(x - x')^2 + (y - y')^2 + (z - z')^2}, \qquad \text{(D.41)}$$

where $\boldsymbol{r}$ and $\boldsymbol{r}'$ denote the source and observation points on S_1 and S_2, respectively. By using the Gauss and Stokes theorems, the double surface integral of (D.40) is transformed into the sum of 16 line integrals

$$Pp_{12} = -\frac{1}{4\pi\epsilon\, SS'} \sum_{i=1}^{4} \sum_{j=1}^{4} \int_{l_i'} \int_{l_j} R\, \hat{\mathbf{u}}_j \hat{\mathbf{u}}_i'\, dl_j\, dl_i' = -\frac{1}{4\pi\epsilon\, SS'} \sum_{i=1}^{4} \sum_{j=1}^{4} P_{ij}, \qquad \text{(D.42)}$$

where $\hat{\boldsymbol{u}}$ and $\hat{\boldsymbol{u}}'$ vectors are orthogonal to the circumference of the quadrilateral surfaces as shown in Fig. D.9. The results for the evaluation of the partial potential coefficients is given in Section 7.3.1. The integrals are performed over the lines shown in Fig. D.10. This section illustrates how analytic integration techniques can be helpful in the evaluation of some of the nonorthogonal partial coefficients of potential. It is surprising that the integrals for the nonsingular kernel R are also very complex. These integrals with the kernel R also have applications for Taylor expansion of a retardation term. This is considered in Section 5.8 for nonorthogonal filaments. Integrals with an R-kernel will result, which are considered in Section 7.3.1 such as the two filaments shown in Fig. D.10. Result for the integrals is given in Section 7.3.2.

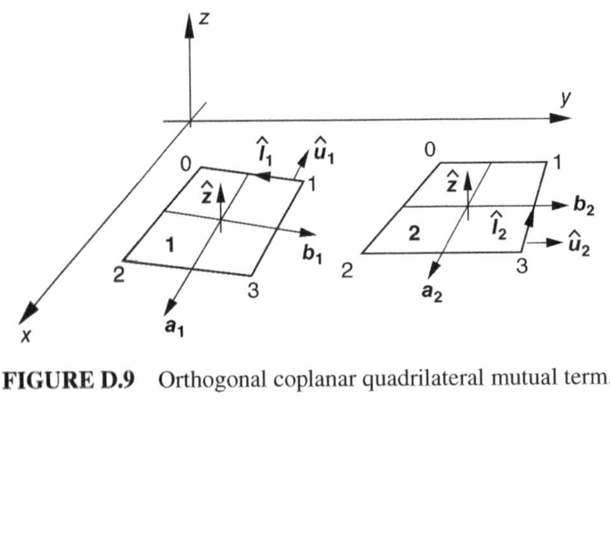

FIGURE D.9 Orthogonal coplanar quadrilateral mutual term.

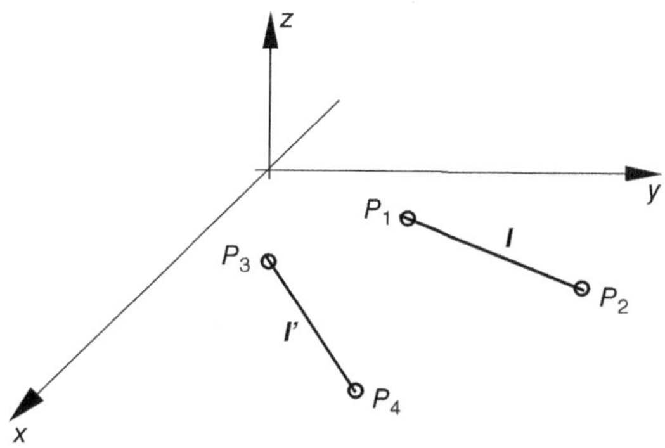

FIGURE D.10 Relevant geometry for the double-line integrals.

REFERENCES

1. A. E. Ruehli and P. A. Brennan. Accurate metallization capacitances for integrated circuits and packages. *IEEE Journal Solid-State Circuits*, 4(4):289–290, August 1973.
2. A. Muesing and J. W. Kolar. Efficient partial element calculation and the extension to cylindrical elements for the PEEC method. In *2008 11th Workshop on Control and Modeling for Power Electronics, COMPEL*, Zurich, August 2008.
3. A. E. Ruehli and P. A. Brennan. Efficient capacitance calculations for three-dimensional multiconductor systems. *IEEE Transactions on Microwave Theory and Techniques*, 21(2):76–82, February 1973.
4. F. W. Grover. *Inductance Calculations: Working Formulas and Tables*. Dover Publications, New York, 1962.

E

AUXILIARY TECHNIQUES FOR PARTIAL ELEMENT COMPUTATIONS

E.1 MULTI-FUNCTION PARTIAL ELEMENT INTEGRATION

An important part of the overall solution is a mixture of techniques for the efficient partial element evaluation. We call the overall approach multifunction method (MFM) [1, 2] since it involves different algorithms or functions with the main goal of keeping the compute time for the partial elements low.

The partial Lp and Pp coefficients formulations given in Appendices C and D represent an important collection of integrals for analytic solutions. The analytic solutions usually lead to the efficient evaluation of the coefficients. However, it is obvious from Appendices C and D that the efficient analytic evaluation of nonorthogonal coefficients is much more challenging.

The overall approach also needs to include numerical integration methods especially for cases where the analytic integration may not exist or is too complicated. Fortunately, more work has been conducted that contributes to the multifunction approach [2] and to the solution of total or partially numerically evaluated coefficients, for example, Ref. [3].

The efficient numerical computation of nonorthogonal partial inductances is very challenging especially if the length-to-width ratio between the sides as large as 1000:1. Numerical integration is more efficient for the smallest dimensions of conductor of the cell. For most nonorthogonal cells, a mixed numerical and analytical solution works best. Therefore, we need to consider techniques suitable for the shorter lengths of the cells.

An important aspect – which is often overlooked – is physically *dense* parts of the PEEC models. For example, skin-effect models in Chapter 9 require much higher accuracy for the partial inductance evaluation. In general, numerical compute cost can be larger for the evaluation of skin-effect models due to the higher accuracy required.

Another issue of importance is the evaluation of the singular partial self-coefficient Lp_{11} and Pp_{11}. The singularity can be eliminated with two analytic integrations over the same surface. Interesting techniques have been worked out for the solution of the singularity

Circuit Oriented Electromagnetic Modeling Using the PEEC Techniques, First Edition.
Albert E. Ruehli, Giulio Antonini, and Lijun Jiang.

problem. For example in Ref. [2], a combination of rectangular and hexahedral shapes is used and nonsingular rectangular subcells are embedded to take care of the singular parts.

The singularity is weaker – if not eliminated – by some of the integrations which are performed analytically. One of the most used approaches is the use of analytical filaments or sheet results given in Appendices C and D for complex orthogonal cases [4]. Quite complex cases can be evaluated for rectangular uniform shapes since the partial self-inductances Lp_{11} are available for zero thickness sheets and rectangular cross sections. Recently, approximate formulas for partial self-inductances with nonrectangular cross-sectional shapes have been considered [5].

For nonorthogonal shapes, the filament approach has been successfully applied in Refs [6] and [3] where for the partial self-inductances Lp_{11} an adaptive Simpson numerical integration method is used. In Ref. [2], an adaptive Gauss–Legendre numerical approach is used.

The accuracy issue for close conductors is considered in Section 5.5.2. The observation is that we need four to five digits of accuracy for closely located partial inductances. Compute time is a key issue for these very closely located cells. It has been shown, for example, Ref. [7], that many of the approximate partial inductance formulas [8] are not accurate enough for the near coefficients in physically dense meshing regions.

The partial coefficients of potential in the circuit matrix solution are of the form pp_{ij}/pp_{ii}. The number of digits of accuracy in the pp_{ij} coefficients can be reduced due to the declining ratio for distant elements. We need to make sure that the coefficient ratio is sufficiently small. Hence, we reduce the accuracy or the number of significant digits in the accuracy of off-diagonal elements to speed up the evaluation.

In the multifunction approach, a mixture is used to calculate the integrals for the different geometrical shapes. The challenge is to choose number of subdivisions such that the best compromise is made between accuracy and compute time. For far-coefficients we can reduce the relative error from five digits to four or even three digits. Since the majority of the coupling coefficient is of the physical distant type, this results in a large reduction in the overall compute time.

E.1.1 Appropriate Numerical Integration Methods

Rather than presenting a general overview over integration techniques, we restrict ourselves to suitable approach for the numerical integration for partial element integrals. Unfortunately, compute time for the numerical evaluation of the 3D integrals increases rapidly with the number of subdivisions. Hence, using as few integration points as possible is an important goal. Also, wherever possible, we want to use an analytical evaluation for the overlapping self-integrals that lead to the singularities.

Numerical integration methods are required for some of the integrals of functions without explicit antiderivatives or whose antiderivatives are too complex. Hence, numerical techniques play an important role for integrals that are difficult to solve.

We use an appropriate class of numerical quadrature techniques for the solution. Specifically, each single, general integral numerical quadrature can be represented as

$$\int_a^b f(x)dx = \sum_{i=0}^{N} c_i f(x_i) + \epsilon(f), \tag{E.1}$$

where x_i are the integration points, c_i is the weight, and $\epsilon(f)$ is the residual error of the numerical integration. All integration points x_i are defined within the integration region

$[a, b]$. For some integration methods such as the Simpson and the trapezoidal rule, the end points are included in x_i. However, in another class of methods such as the midpoint rule and Gaussian integration, the end points are not part of the discretization points. This is important for nearest neighboring cells, but not necessarily for partial self-element, which will be considered in Section E.1.2.

A category of numerical quadrature methods is called the *Newton–Cotes formulas*. They are derived by approximating the integrand with interpolating polynomials. The error term in the interpolating polynomial of degree n involves the $(n+1)$th derivatives of the function to be approximated. This type of quadrature method is exact if the approximating polynomial is of a lesser degree than or equal to n.

Newton–Cotes formulas [9] use values of the function at equally spaced points. For a polynomial of order n, there are $(n+1)$ coefficients. Polynomials with $(n+1)$ coefficients result in, for $(n+1)$ equally spaced, fixed quadrature points with an accuracy up to n.

Alternatively, if both the quadrature points x_i and the weights c_i are adjustable, higher accuracy can be achieved due to the $2n$-freedoms for such a method. A polynomial of order $(2n-1)$ has $2n$ coefficients, and it is possible to achieve an accurate integration of a polynomial with an order of $2n-1$ by using computed quadrature points as well as weights. For this reason, we employ Gaussian integration methods that are of this type.

The Gaussian quadrature methods [9, 10] are based on optimally spaced quadrature points, rather than equally spaced ones. The nodes are $x_1, x_2, \ldots, x_n$ and the weights are $c_1, c_2, \ldots, c_n$ in

$$\int_a^b f(x)\,dx = \sum_{i=1}^{n} c_i f(x_i) + O(f^{2n}) \tag{E.2}$$

are precomputed and where the error term indicates that Gaussian quadrature can integrate polynomials accurately up to the order of $(2n-1)$.

The following theorem provides the theoretical support of the Gaussian quadrature method.

Theorem 1 *Suppose that $x_1, x_2, \ldots, x_n$ are the roots of the nth Legendre polynomial $P_n(x)$ and that for each $i = 1, 2, \ldots, n$, the numbers c_i are defined by*

$$c_i = \int_{-1}^{1} \prod_{j=1, j\neq i}^{n} \frac{x - x_j}{x_i - x_j} dx. \tag{E.3}$$

If $P(x)$ is any polynomial of degree less than $2n$, then

$$\int_{-1}^{1} P(x)dx = \sum_{i=1}^{n} c_i P(x_i). \tag{E.4}$$

The Gaussian quadrature points are the roots of the order n Legendre polynomial $P_n(x)$. Also, the weight c_i is the corresponding integration results of Lagrange polynomials. Fortunately, the coefficients are precalculated. We provide a table of Gaussian quadrature coefficient in Table E.1. More elaborate Tables are available, such as [11].

A fundamental issue is the mapping of the integration interval from $[a, b]$ to the interval between $[-1, 1]$. The following transformations are used for the mapping

$$\int_a^b f(x)dx = \int_{-1}^{1} f\left[\frac{(b-a)\,v + (b+a)}{2}\right] \frac{(b-a)}{2}\,dv. \tag{E.5}$$

If the Gaussian quadrature points between $[-1, 1]$ are mapped back to the region $[a, b]$, the weights can be scaled so that a customized set of Gaussian quadrature points and weights can be used directly as follows:

$$\int_a^b f(x)dx = \sum_{i=1}^{n} c_i f(x_i). \tag{E.6}$$

We should note that the mappings for the solution of nonorthogonal geometries in Chapter 8 also result in the interval $[-1, 1]$ (Table E.1).

In PEEC, the potential coefficients and partial inductances need the evaluation of multiple integrals. The integrands are frequently related to antiderivative. Hence, Gaussian quadratures can conveniently be applied to get accurate numerical approximate solutions. One example for the multiple integral is a double integral defined over a rectangular region, which is a surface integral defined for a rectangular cell surface

$$\int_{cell} f(x, y)\, dS = \int_a^b \int_c^d f(x, y)\, dx\, dy = \sum_{i=1}^{n} c_i \sum_{j=1}^{m} c_j f(x_i, y_j), \tag{E.7}$$

where the weights are given in Table E.1 correspond to (E.7).

TABLE E.1 Table for Gaussian quadrature coefficients.

N	x_i	c_i
2	∓0.577350	1.00000
3	0	0.888889
	∓0.7745967	0.5555556
4	∓0.3399810	0.6521452
	∓0.8611363	0.3478548
5	0	0.5688889
	∓0.5384693	0.47862867
	∓0.9061798	0.23692689
6	∓0.2386192	0.4679139
	∓0.6612094	0.3607616
	∓0.9324695	0.1713244
10	∓0.14887434	0.29552422
	∓0.433395390	0.26926672
	∓0.679409570	0.21908636
	∓0.865063370	0.14945134
	∓0.97390653	0.06667134

E.1.2 Numerical Solution for Singular Self-Coefficients Lp_{11} or Pp_{11}

The singular situation clearly occurs when the Green's function in the self-coefficients becomes infinite since both integrations are over the same cell. In some cases, the singular behavior can be avoided by using a change of variables given in Ref. [12] or Appendix D in Ref. [13]. Another approach is the numerical integration over the singularity such as an averaging technique [14].

Most of the Green's functions in this text are singular of the form $g(x) = \beta|x - x'|^{-1}$ if $x = x'$. However, if one of the integrations is accomplished analytically, and the singularity is reduced to the basic form $f(x) = \alpha \log |x - x'|$. If we compare the behavior between the two singularities by using the L'Hôpital rule [15] for this case is

$$\lim_{x \to x'} \frac{f'(x - x')}{g'(x - x')} \approx \lim_{x \to x'} \alpha/\beta(x - x') = 0, \tag{E.8}$$

where we assume that $\alpha(x - x')$ and $\beta(x - x')$ are constant. Due to the additional analytical integrations, a milder logarithmic singularity results. This is a more suitable singular behavior for numerical integration.

The numerical solution of the still singular integrals is a separate issue. It is clear that such a logarithmic singularity is more suitable for a numarical integration solution. This can be accomplished by so-called *adaptive* methods. In these methods, the iterative improvement helps the solution of mildly singular problems.

The iterative improvement is difficult for the conventional Gauss integration method presented above due to choice of the steps. For this reason, the adaptive refinement is applied to other methods such as the adaptive Simpson and the adaptive Lobatto methods [10, p. 565]. For example, the Simpson method uses the end point a and b of the range where the first approximation of the integral I_1 for $f(x)$ is given by

$$I_1 = \frac{h}{1.5}\left[f(a) + 4f\left(\frac{a+b}{2}\right) + f(b)\right], \tag{E.9}$$

where $h = (b - a)/4$.

In the next iteration, two additional points are added, $fml = f(a + h)$ and the right point $fmr = f(b - h)$. With this, the next improved solution can be added:

$$I_2 = \frac{h}{3}\left[f(a) + 4(fml + fmr) + 2f\left(\frac{a+b}{2}\right) + f(b)\right]. \tag{E.10}$$

The two Simpson results I_1 and I_2 can be combined to find an improved I_1 result using the Romberg scheme [10, p537]

$$I_1 = (16\, I_2 - I_1)/15. \tag{E.11}$$

More steps are taken by further reducing the subdivision h. In each step, we apply the same scheme for part of the range only. Hence, the same steps are applied to two half regions $[a, m]$ and $[m, b]$, and so on.

More details and examples are given in Ref. [10] for the solution of moderately singular integrals. The numerical integration of the milder singularity is acceptable if it yields a solution with about five digit accuracy.

E.1.3 Analytical and Numerical Integral Solutions with Variable Subdivisions for Nonself-Partial Elements

If a partial coefficient matrix is of size N, then the efficient evaluation of the $N(N-1)$ partial off-diagonal coupling elements can be very compute time intensive. Of course, the symmetry of the partial elements can be used to reduce the compute time by about a factor of two. In this section, we consider the reduction of the steps h needed for the numerical integrations.

This part of the MFM algorithm is based on the above-outlined concepts [1]. Again, the majority of applications are for nonorthogonal cells. We assume that the appropriate meshing or space discretization into cells has been applied. For the mid- and high-frequency (FW)PEEC cases, the size of the largest inductive and capacitive cell is less than $\lambda_{min}/20$, where λ_{min} is the minimum wavelength corresponding to the maximum frequency in the excitation spectrum.

For the prediction algorithm, geometrical quantities are required such as the distance between the cells of interest and the cell size. The first quantity of interest is the distance between the cells in comparison to the size of the larger cell. This is used to determine the number of subdivision for the numerical integration. This part is given as follows:

1. Find the maximum size, *maxSize*, of the two cells.
2. Find the center-to-center distance, R_{c2c}, between the two cells.
3. Find the ratio called *farRatio*=R_{c2c}/*maxSize*.
4. *If*(*farRatio* > 30), center-to-center point approximation can be used for the distance-oriented part of the integral. Of course retardation must still be considered for full wave solutions.

For smaller distances, a more elaborate algorithm is required, which are considered next.

We should note that this part of the algorithm can be improved to include the retarded element. Fortunately, for sufficiently large problems, the majority of the coefficients fall into the above large distance class. Also, we do not need to store the computed coupling coefficients for very small compute times. The purpose of this is to save storage space.

Algorithms for *close* spaced cells where the *farRatio* ≤ 30 are more challenging. We give an outline of a more elaborate example for such a case. Unless the distance between the cells is very small, the Gauss–Legendre quadrature approach is used, where the subdivisions are determined by the algorithm.

In all cases, the gap between the edges of two close cells, we call $EdgeDist_i$, is evaluated in each direction $i = x, y, z$. Then, a parameter is evaluated for each direction i as $distRat_i = EdgeDist_i/Size_i$ for $i = x, y, z$. The following decision steps can be applied to determine the integration order used for the evaluation of the near distance coefficients:

1. $if(distRat_i > 20)$ then $integrationOrder_i = 2$
2. $elseif(distRat_i > 10)$ then $integrationOrder_i = 3$
3. $elseif(distRat_i > 3)$ then $integrationOrder_i = 4$

4. *elseif*$(distRat_i > 1)$ then $integrationOrder_i = 5$
5. else $integrationOrder_i = 6$

for $i = x, y, z$, for each cell where required. The $integrationOrder_i$ determines the number n of the subdivisions used in the numerical integration.

The information obtained is utilized in the MFM algorithm to speed up partial element calculations. For very far units, other criteria can be used based on information such as the distance between entire blocks of conductor cells if the distance to obtain further speedup in the partial element evaluations since less checks have to be performed.

More can be accomplished with additional ways to evaluate partial elements as is discussed throughout the book. For example, cells can be divided into several groups for faster evaluation, such as orthogonal or nonorthogonal cells. Evaluation done for groups rather than a cell-by-cell level can lead to further reductions in compute time. We should note that in this section we assumed that numerical integration techniques are used. However, an analytical solutions in cooperation with numerical solutions where necessary yield the best results. An example is a long cell where the length is integrated analytically using formulas like the one in section C.1.1. Then, the arbitrary cross-section may be added using the variable analytic numerical integration outlined in this section where the number of integration steps may depend on the distance to its neighbors.

REFERENCES

1. G. Antonini and A. E. Ruehli. Fast multipole and multi-function PEEC methods. *IEEE Transactions on Mobile Computing*, 2(4):288–298, October 2003.
2. Y. Hackl, P. Scholz, W. Ackermann, and T. Weiland. Multifunction approach and specialized numerical integration algorithms for fast inductance evaluations in nonorthogonal PEEC systems. *IEEE Transactions on Electromagnetic Compatibility*, 57(5):1155–1163, October 2015.
3. A. Muesing, J. Ekman, and J. W. Kolar. Efficient calculation of non-orthogonal partial elements for the PEEC method. *IEEE Transactions on Magnetics*, 45(3):1140–1142, March 2009.
4. A. E. Ruehli. Inductance calculations in a complex integrated circuit environment. *IBM Journal of Research and Development*, 16(5):470–481, September 1972.
5. R. De Smedt. Difference between the partial self-inductance at DC and HF. *IEEE Transactions on Electromagnetic Compatibility*, 57(4):702–708, April 2015.
6. A. Muesing and J. W. Kolar. Efficient partial element calculation and the extension to cylindrical elements for the PEEC method. In *2008 11th Workshop on Control and Modeling for Power Electronics, COMPEL*, August 2008.
7. A. E. Ruehli and G. Antonini. On modeling accuracy of EMI problems using PEEC. In *Proceedings of the IEEE International Symposium on Electromagnetic Compatibility*, Boston, MA, August 2003.
8. F. W. Grover. *Inductance Calculations: Working Formulas and Tables*. Dover Publications, New York, 1962.
9. R. W. Hamming. *Numerical Methods for Scientists and Engineers*. Dover Publications, New York, 1986.
10. W. Gander, M. Gander, and F. Kwok. *Scientific Computing, An introduction using Maple and MATLAB*. Springer-Verlag, Berlin and New York, 2014.
11. M. Abramowitz and I. A. Stegun. *Handbook of Mathematical Functions*. U.S. Government Printing Office, Washington, DC, 1964.

12. M. G. Duffy. Quadrature over a pyramid or cube of integrands with a singularity at a vertex. *SIAM Journal on Numerical Analysis*, 19(6):1260–1262, December 1982.
13. J. M. Jin. *The Finite Element Method in Electromagnetics*. John Wiley and Sons, Inc., New York, 2nd edition, 2002.
14. M. A. Cracraft. Mobile array design with ANSERLIN antennas and efficient, wide-band PEEC models for interconnect and power distribution network analysis. Ph.D. Dissertation [Online]. Available: *http://hdl.handle.net/10355/29582*, Missouri University Science and Technology, USA, 2007.
15. D. V. Widder. *Advanced Calculus*. Prentice Hall, Englewood Cliffs, NJ, 1961.

INDEX

Circuit Oriented Electromagnetic Modeling Using the PEEC Techniques, First Edition.
Albert E. Ruehli, Giulio Antonini, and Lijun Jiang.

Printed and bound by CPI Group (UK) Ltd, Croydon, CR0 4YY

17/09/2026

14964289-0004